NEW ADVANCES IN HYDROGENATION PROCESSES - FUNDAMENTALS AND APPLICATIONS

Edited by **Maryam Takht Ravanchi**

New Advances in Hydrogenation Processes - Fundamentals and Applications
http://dx.doi.org/10.5772/62820
Edited by Maryam Takht Ravanchi

Contributors
Sunil Babaji Shinde, Raj Deshpande, Akito Hara, Kuninori Kitahara, Zuzeng Qin, Yuwen Zhou, Zili Liu, Hongbing Ji, Alejandra López-Suárez, Sandrine Bouquillon, Nadège Ferlin, Safa Hayouni, Bawadi Abdullah, Syed Anuar Faua Ad Syed Muhammad, Nik Azmi Nik Mahmood, Roxana Rusu-Both, Mattia Bartoli, Marco Frediani, Werner Oberhauser, Luca Rosi, Elisa Passaglia, Monica Quiroga, Nicolás Carrara, Juan Manuel Badano, Carolina Betti, Cecilia R. Lederhos, Mariana Busto, Carlos Vera, Domingo Liprandi, Edgardo Cagnola, Adam McCluskey, Miroslav Stanković, Euzebiusz Jan Dziwinski, Jolanta Iłowska, Bartłomiej Bereska, Józef Lach, Agnieszka Bereska, Michał Szmatoła, Petr Kačer, Martin Kindl, Ondřej Matuška, Devinder Singh, R.S. Tiwari, O.N. Srivastava

Published by InTech
Janeza Trdine 9, 51000 Rijeka, Croatia

Publishing Process Manager Maja Bozicevic
Technical Editor SPi Global
Cover InTech Design team

First published January, 2017
Printed in Croatia

Additional hard copies can be obtained from orders@intechopen.com

New Advances in Hydrogenation Processes - Fundamentals and Applications, Edited by Maryam Takht Ravanchi
p. cm.
Print ISBN 978-953-51-2869-4
Online ISBN 978-953-51-2870-0

For EU product safety concerns:
IN TECH d.o.o., Prolaz Marije Krucifikse Kozulić 3, 51000 Rijeka, Croatia,
info@intechopen.com or visit our website at intechopen.com.

Meet the editor

Dr. Maryam Takht Ravanchi is a senior researcher at Petrochemical Research and Technology Co. (NPC-RT) with about 15 years of experience. She received her PhD degree in Chemical Engineering from Amirkabir University of Technology, Tehran, Iran. Her principal areas of interest and expertise include natural gas conversion and environmental researches; specifically she focused on hydrogenation processes, carbon dioxide utilization and paraffin dehydrogenation. To date, she has published about 50 scientific papers in national and international journals and presented more than 50 papers in national and international conferences. Moreover, she has published four national patents and seven books.

Contents

Preface

Hydrogenation is a chemical reaction between hydrogen and another element usually occurred in the presence of a catalyst such as nickel (Ni), palladium (Pd) or platinum (Pt). Hydrogenation process is commonly used to saturate or reduce organic compounds and has three components: an unsaturated substrate, hydrogen (source) and a catalyst. Catalytic hydrogenation has versatile industrial applications, such as food industry, petrochemical industry, organic chemistry, hydrogenation of coal, etc. In the present book, some of the recent advances of this process are provided.

In Chapter 1, the application of end-functionalized isotactic poly (lactic acid) as support for Pd-based catalysts was evaluated for double and triple bond hydrogenation reactions. In Chapter 2, selective hydrogenation of 1-heptyne using supported mono and bimetallic catalysts was studied. In Chapter 3, asymmetric transfer hydrogenation as an attractive tool for synthesis of enantio-enriched compounds was presented. In Chapter 4, catalytic hydrogenation of CO_2 to methane by the use of Ni-based and Co-based catalyst was reviewed. As CO_2 is a greenhouse gas, its utilization is of high environmental importance. In Chapter 5, the behaviour of hydrogen atom in polycrystalline silicon thin film was investigated. In Chapter 6, hydrogenation of benzoic acid using mono and bimetallic Ru, Pd, Co and Re catalyst was investigated. In Chapter 7, hydrogenation behaviour of Zr-based quasi-crystalline alloys was evaluated. The effect of Ti addition on H_2 storage characteristics of quasi-crystalline alloys was studied. In Chapter 8, synthesis of Ni-based catalysts for hydrogenation of edible oils was presented. In Chapter 9, new strategies for the synthesis of egg-shell Pd/composite catalyst were presented. These catalysts were applicable in selective hydrogenation of styrene, 1-heptyne, 3-hexyne and 2,3-butanone. In Chapter 10, hydrogen storage capacity of pure and alloyed Ti was provided. In Chapter 11, a new plasticizer was introduced which was obtained by catalytic hydrogenation in the presence of Ni catalyst. In Chapter 12, the application of organometallics as a tool for surface phenomena was presented. The structure of organometallics resembled complexes adsorbed on the surface of heterogeneous catalysts. In Chapter 13, recent applications of flow chemistry in the field of medicinal chemistry via hydrogen reactions were provided. In Chapter 14, hydrogenation of lignocellulosic biomass for bio-fuel production was provided. In Chapter 15, the concept of bio-based ionic liquid was introduced, and catalytic hydrogenation of (poly)-alkenes or unsaturated ketones in different bio-based and non-bio-based ionic liquids was investigated In Chapter 16, a modern control algorithm for complex and high-risk petrochemical processes was presented, and production of 2-ethyl-hexanol through liquid phase hydrogenation of 2-ethyl-hexenal was studied as well.

I wish to express my sincere appreciation to InTech for giving us the opportunity to publish a book on the topic of hydrogenation. I would like to thank all the authors for their significant contribution and providing high-quality research to share worldwide. I also want to express my sincere gratitude to Ms. Maja Bozicevic for her help during the entire publication process.

Maryam Takht Ravanchi, PhD
Petrochemical Research and Technology Co.
Tehran, Iran

Chapter 1

Palladium-Based Catalysts-Supported onto End-Functionalized Poly(lactide) for C–C Double and Triple Bond Hydrogenation Reactions

Marco Frediani, Werner Oberhauser, Luca Rosi,
Elisa Passaglia and Mattia Bartoli

Additional information is available at the end of the chapter

http://dx.doi.org/10.5772/65447

Abstract

Poly(lactide) (PLA), which is the polymer that will be discussed in this chapter, was functionalized with nitrogen containing aromatic groups by means ring opening polymerization (ROP) reactions. The obtained macroligands were successfully employed to coordinate Pd(II), which chemoselectively hydrogenated $\alpha\beta$–unsaturated carbonyl compounds to give the saturated counterparts as main product. The catalyst could be easily recycled upon a simple filtration process.

End-functionalized isotactic PLA chains were used to generate stereocomplexes, featured by a higher thermal and hydrolytic resistance. The chemical nature of the nitrogen ligand introduced at the chain end of PLA was exploited to stabilize Pd-nanoparticles (Pd-NP_s) by interactions of the aromatic or nitrogen functionality with the NP_s– surface. The polymer-anchored Pd-NP_s were generated by a classical approach which comprised the coordination of Pd(II) by the functional group followed by reduction with hydrogen or by the metal vapor synthesis technique. The polymer-supported Pd-NP_s were successfully applied to catalyze the selective hydrogenation of alkynes to the corresponding alkene and cinnamylalcohol to the corresponding saturated aldehyde. In this book chapter the synthesis and structure-performance relationship of the new hydrogenation catalysts will be discussed.

Keywords: poly(lactide), palladium, nanoparticles, alkynes, hydrogenation

1. Introduction

In the realm of hydrogenation reactions, the selective hydrogenation of unsaturated carbon-carbon bonds is very attractive [1]. Asymmetric and partial hydrogenation reactions are pivotal for the access of important compounds employed as pharmaceuticals, food additives, pesticides, flavors, and fragrances [2]. In this respect, α,β-unsaturated carbonyl compounds present an interesting group of chemicals whose enantio- [3], chemo [4], and regio-selectivities [5] are highly challenging. For instance, Lindlar's [6] and Lindlar types' catalysts [7] are widely used to chemoselectively hydrogenate alkynes to the corresponding alkene. In parallel, a variety of different selective hydrogenation methods have been developed [8]. In particular, bimetallic catalysts [9], noble metal-based catalysts supported onto carbon-based materials [10] or inorganic matrixes [11], in combination with supercritical solvents [12] and "*flow*" reactors [13], have been studied. Also for the partial alkyne hydrogenation, a great deal of catalytic systems were proposed based on the use of noble metals and different approaches like reactions conducted in gas [14] and liquid phase [15], combined with ultrasound techniques [16] and biphasic catalytic systems [17].

The interest on polymer-based catalysts applied for hydroformylation reactions [18], hydrotreatment [19], and strong acid/base [20, 21] catalyzed reactions is steadily increasing. Since functional polymer-based catalytic systems generally exhibit notable advantages over traditional catalysts such as the tunable solubility and improved catalytic selectivity [22], the biodegradable polyester, poly(lactide) (PLA) was used as a polymer matrix [23], which is rather straightforwardly end-capped as reported by Helle et al. [24]. The possibility to produce carboxylic end-capped PLA chains functionalized with nitrogen containing moieties, called macroligands by Giachi et al. [25], allowed the synthesis of well-defined macrocomplexes containing metal cations, such as Pd(II). The latter Pd-based macrocomplexes were efficiently used to catalyze hydrogenation reactions conducted in the homogeneous phase followed by an easy separation step, as reported by Giachi et al. [26] and Bartoli et al. [27]. Isotactic end-functionalized PLA with opposite tacticity can be combined to form a supramolecular structure named stereocomplex [28]. Functionalized PLA stereocomplexes show high resistance against hydrolysis and thermal degradation and in addition, a poor solubility in all organic solvents. This latter property was exploited by Petrucci et al. [29] and Oberhauser et al. [30] for catalytic purposes to easily separate PLA-stereocomplex-based catalysts from catalytic solution.

In the following sections, synthesis, characterization, and catalytic application of homogeneous and heterogeneous Pd-catalysts supported on end-functionalized PLA were reported.

2. Pd-Nps supported onto PLA stereocomplexes: synthesis and characterization

The synthesis of PLA can be achieved by different synthetic pathways such as polycondensation of lactic acid [31] or through ring opening polymerization (ROP) of lactide (i.e., the cyclic dimer of lactic acid) [32]. ROP is the adopted synthetic methodology, which allows a carboxylic-end functionalization along with a strict control of the resulting molar mass. The ROP can be either performed by cationic, anionic, or carbene-type mechanism or by a coordination insertion mechanism [33], which shows an advancement over the former mechanisms regarding the efficiency of the polymer chain growth and racemization of the stereogenic center of the lactic acid units. In addition, functionalized PLA at the carboxylic chain-end are straightforwardly obtained by using a suitable initiator molecule for the polymerization. ROP can be performed in solution or in the bulk of the monomer except for some glycolides that decompose at the melting temperature [34]. Typically, PLA is polymerized in bulk due to the low melting temperature of the corresponding cyclic monomers. The mechanism of ROP, through a coordination-insertion mechanism, is mainly based on organic salts of Sn(II). In particular, $Sn(Oct)_2$ (Oct = octanoate) is one of the most employed catalysts for PLA synthesis due to its relatively low cost, solubility in the monomer, which is important for bulk reactions, high catalytic activity (yields over 90%) and a low racemization tendency of the starting glycolide. The mechanism regarding the $Sn(Oct)_2$-catalyzed ROP is shown in **Figure 1**.

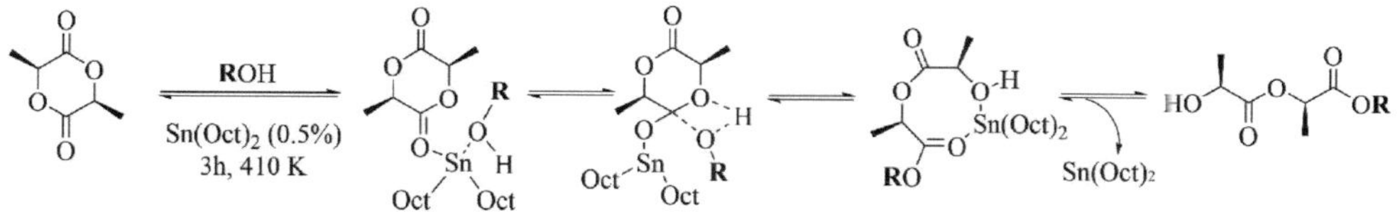

Figure 1. $Sn(Oct)_2$-catalyzed ROP of lactide in the presence of a hydroxylic initiator molecule (ROH).

As shown in **Figure 1**, the ROP of lactide starts with a simultaneous coordination of lactide and the initiator molecule to the metal center of the catalyst. This *cis*-coordination of the two reagents is a prerequisite for the nucleophilic attack of the coordinated hydroxyl oxygen atom on the lactide carboxylic carbon atom. Afterwards, the metal center inserts into the six-membered ring giving an eight member metallacycle. The obtained end-functionalized linear form of the lactide acts as a new initiator in the following polymerization cycle. The ROH, where R contains the new functional group, is incorporated at the carboxylic acid PLA chain end. The presence of water traces in the lactide melt notably influences the final molecular weight due to the competition role of water as an initiator molecule leading to a carboxylic acid end group which coordinates as carboxylate to the metal center. Nitrogen containing initiators, such as pyridine and 2-2′-bipyridine derivatives, were used in order to obtain carboxylic end-functionalized PLA chains. Synthesis conditions and obtained molar masses are compiled in **Table 1**.

The end-functionalized PLA-based macroligands were mainly analyzed by gel-permeation–chromatography (GPC) equipped with UV-Vis and refraction index (RI) detectors and by ^1H NMR spectroscopy. The UV-Vis detector confirmed the presence of the corresponding chromophore in the polymer chain. The ^1H NMR spectra of the macroligands show, apart the typical multiplets stemming from the polymer chain (i.e., doublet for CH_3 and quartet for CH of the repeating lactide unit), signals that are typically in the range of aromatic hydrogen atoms.

The ratio of the ^1H NMR integrals of these aromatic hydrogen atoms with that of the methyl group assigned to the terminal lactic acid gives the average number molar weight (M_n) of the polymer obtained, which is in good agreement with the experimental results of GPC analyses, in those cases where the polydispersity of the obtained material is low.

Initiator	Polymer chain	ID	MW (g/mol)
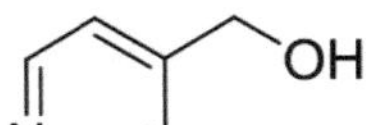 Py	**Py**-P(L)LA-OH	**L**[1]	5000
	Py-P(D)LA-OH	**L**[2]	5000
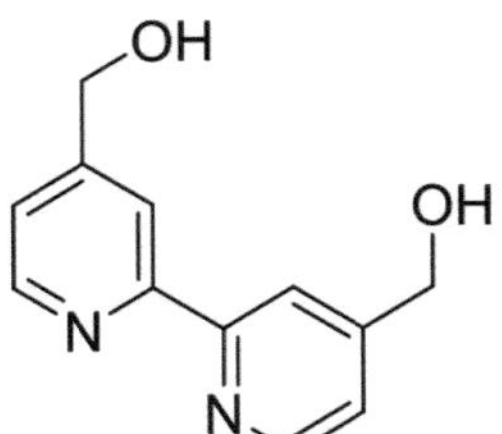 BipyOH$_2$	**BipyOH$_2$**-P(L)LA-OH	**L**[3]	10,000
	Bipy$_2$-P(D)LA-OH	**L**[4]	10,000
	Bipy-(PLA-P(L)LA-OH)$_2$	**L**[5]	10,000
	Bipy-(PLA-P(D)LA-OH)$_2$	**L**[6]	10,000
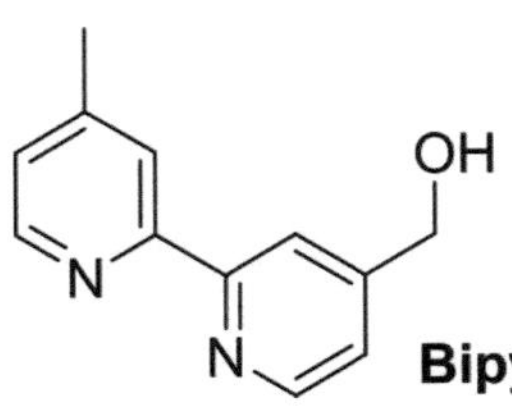 BipyOH	**BipyOH**-P(L)LA-OH	**L**[7]	10,000
	BipyOH -P(D)LA-OH	**L**[8]	10,000
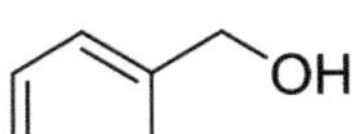 Bn	**Bn**-P(L)LA-OH	**L**[9]	10,000
	Bn-P(D)LA-OH	**L**[10]	10,000
H_2O	HOOC-P(L)LA-OH	**L**[11]	10,000
	HOOC-P(D)LA-OH	**L**[12]	10,000

[a]Theoretical of the molar mass.

[b]Bulk synthesis, 410 K, 3 h.

[c]R-PLA (i.e., atactic PLA) bulk synthesis, 410 K, 3 h, R-PLA-P(L or D)LA toluene reflux, 24 h.

[d]Toluene reflux, 24 h.

Table 1. Overview of the synthesized macroligands.

The obtained PLA-based macroligands were used to provide the corresponding stereocomplexes upon mixing equimolar CH_2Cl_2 solutions of functionalized PLA of opposite stereochemistry, followed by evaporation of the solvent. Differential Scanning Calorimetry (DSC) analyses conducted on the obtained material showed clearly the absence of the melting peak of characteristic PLA (ca. 435 K); new endothermic transition at higher values, typically around 495 K, attributable to the melting of PLA-stereocomplex was observed. The melting enthalpy changed in the same sense from around 38.0 (PLA) to 61.0 J/g (PLA-stereocomplex). These data support the formation of a supramolecular structure more thermally stable. Importantly, the type of end-group does not influence the powder X-ray diffraction (PXRD) pattern of functionalized PLA-stereocomplexes, while PLA and the corresponding stereocomplex are featured by a completely different PXRD pattern as shown in **Figure 2**.

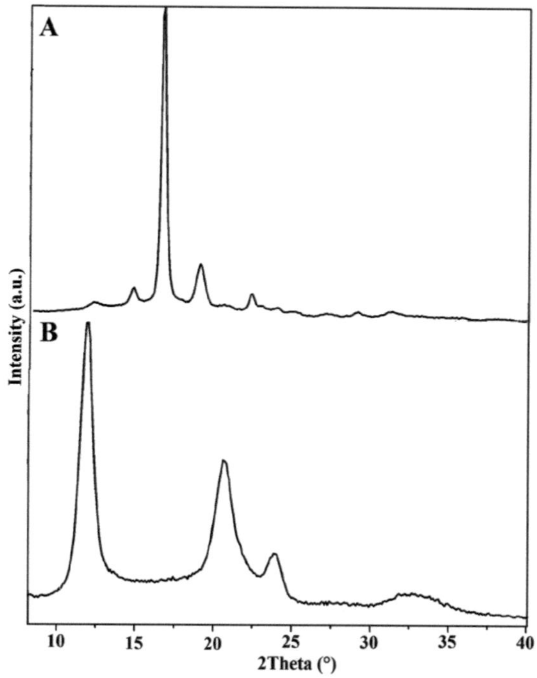

Figure 2. PXRD patterns of isotactic PLA (A) and of the corresponding stereocomplex (B).

The 2,2′-bipyridine or pyridine end-group of the macroligands easily react with $Pd(OAc)_2$ (OAc = acetate) in a 1:1 and 2:1 molar ratio, respectively, giving the expected *cis*-coordination of the nitrogen atoms to Pd(II) in case of 2,2′-bipyridine or *trans*-coordination in case of pyridine. $Pd(OAc)_2$ is a highly suitable Pd(II) precursor since it is soluble in noncoordinating solvents, for instance CH_2Cl_2, which is the solvent of choice for suspending PLA-stereocomplexes. In addition, coordinated $Pd(OAc)_2$ is easily reduced in the presence of hydrogen, giving the acetic acid as a removable side product from reaction medium.

The coordination of either the end-functionalized macroligands or stereocomplexes to Pd(II) was proved by ^1H NMR spectroscopy conducted in CD_2Cl_2. The use of $CDCl_3$ to characterize Pd-acetate macrocomplexes was avoided due to the easy protonation of acetate by trace amounts of acidity present in $CDCl_3$. The occurred coordination of the macroligands to Pd(II) was proved by ^1H NMR, UV-Vis spectroscopy and by comparison of this latter spectroscopic data with that obtained from an analogous model compound (i.e., molecular species without the polymer part), which is exemplified in **Figure 3**.

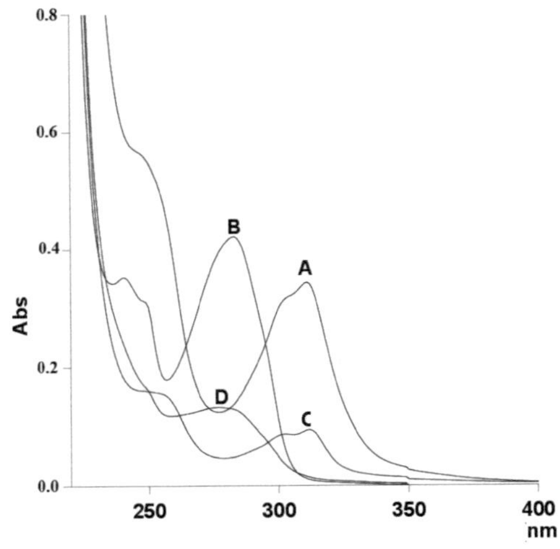

Figure 3. UV-Vis spectra of (A) $Pd(OAc)_2$(BipyOH), (B) BipyOH, (C) $Pd(OAc)_2$ $\mathbf{L}^7$, and (D) $\mathbf{L}^7$.

For the synthesis of well-defined Pd-nanoparticles (Pd-NPs), the size and shape of which strongly influence their catalytic activity and chemoselectivity [35], different synthetic approaches were applied:

- Synthesis of Pd-NPs by coordination of $Pd(OAc)_2$ to the functional groups of PLA-stereocomplexes, followed by metal reduction in the presence of hydrogen pressure (p = 1.5 MPa) at 305 K.
- Synthesis of Pd-NPs by the MVS technique [36], which has the great advantages over the former synthesis approach that the final metal content can be adjusted upon choosing the desired metal concentration of the solvated metal particles, the size of which are defined by the type of solvent chosen, which acts as a ligand.

Transmission electron microscopic (TEM) analyses carried out on Pd-NPs supported onto PLA-stereocomplexes clearly confirmed the influence of the synthetic procedure on the size of NPs and their distribution and the effect of the nitrogen-containing functional groups on the dispersion of Pd-NPs on the polymer-based support as shown in **Figure 4**.

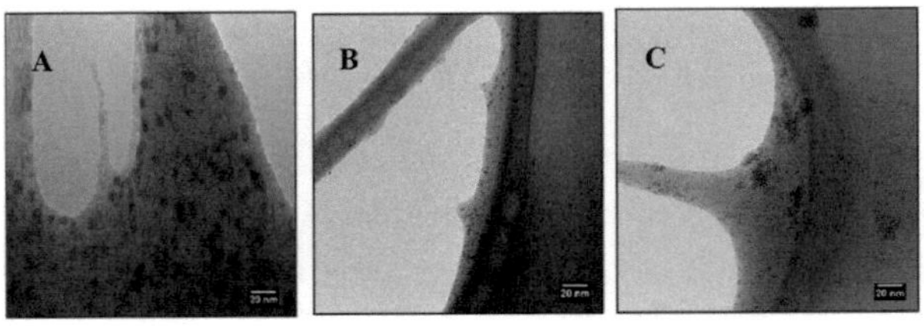

Figure 4. Comparison of TEM analysis of Pd@$\mathbf{L}^{7/8}$ obtained through MSV technique (A), average diameter of 2.0 nm, reduction of Pd precursors (B), average diameter of 3.8 nm, and Pd@$\mathbf{L}^{11/12}$ obtained through MSV technique (C), which showed a strong aggregation.

The thermal gravimetric analysis (TGA) of $\mathbf{L}^{11/12}$ and Pd@$\mathbf{L}^{11/12}$ showed that the presence of Pd-Np strongly affected the thermal stability of stereocomplex structures [30] as shown in **Figure 5**.

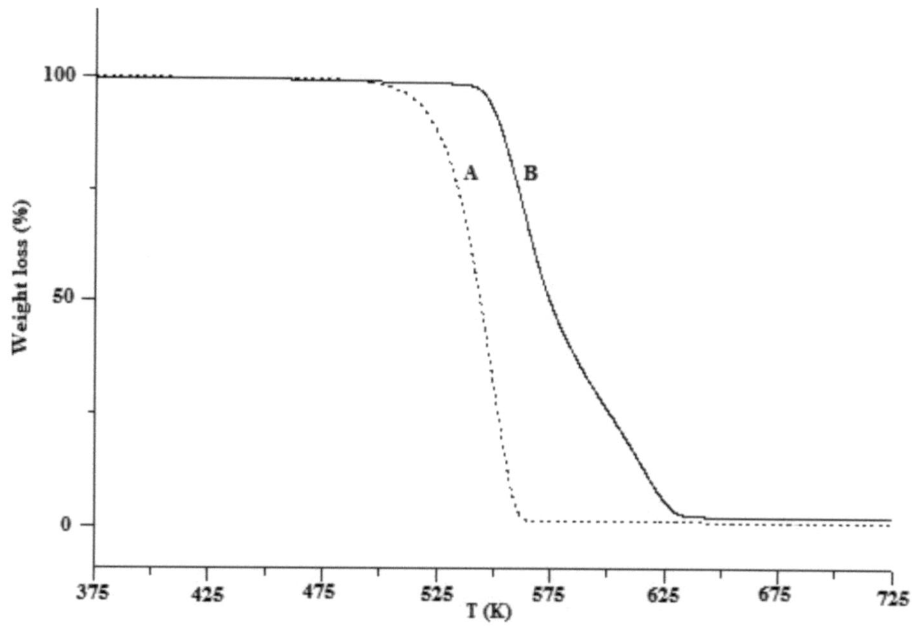

Figure 5. Comparison of TGA of $\mathbf{L}^{11/12}$ (A) and Pd@$\mathbf{L}^{11/12}$ (B).

The decomposition process of $\mathbf{L}^{11/12}$ started at 552 K instead of Pd@$\mathbf{L}^{11/12}$ that did not show a meaningful loss of weight up to 575 K. So, not only PLA-stereocomplexes prevent the aggregation of Pd-NPs but the latter increase the thermal stability of the stereocomplex structure.

3. Hydrogenation of C–C double and triple bonds

The chemoselective hydrogenation of selected α,β-unsaturated carbonyl compounds was carried out in toluene in the presence of the homogeneous precatalyst *trans*-$Pd(OAc)_2(\mathbf{L}^1)_2$ at 303 K. In all cases the corresponding saturated carbonyl compound was observed in a range of chemoselectivity between 46 and 100%, as shown in **Table 2** [27].

The obtained chemoselectivity was found to be strongly structure dependent and with the employed catalyst in some cases even higher chemoselectivities compared to other palladium-based catalysts were obtained [35]. The low chemoselectivity of substrate (IV) is due to the formation of the corresponding saturated alcohol, which is mainly explained by steric effects exerted by the α-methyl group of compound IV. Importantly, in all hydrogenation reactions carried out with the selected substrates (**Table 2**), the corresponding allyl alcohol was not formed.

Substrate	Conversion (%)	Selectivity (%)
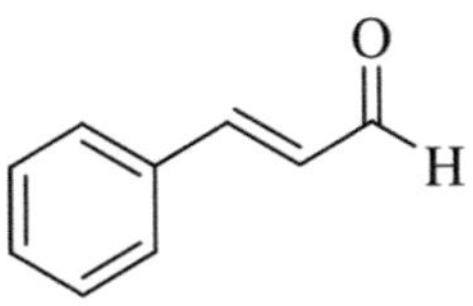 (I)	96	94
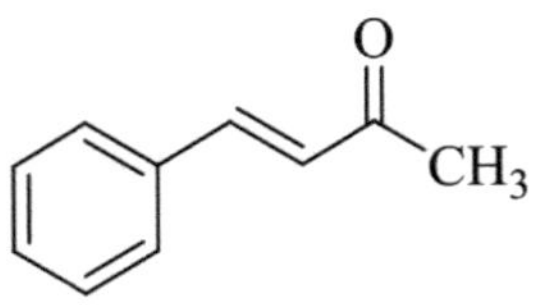 (II)	>99	>99
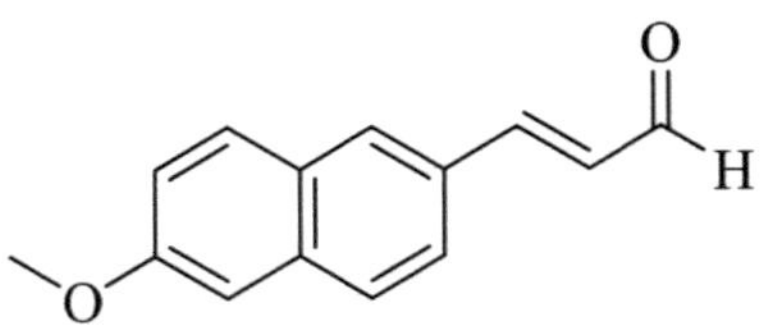 (III)	24	100
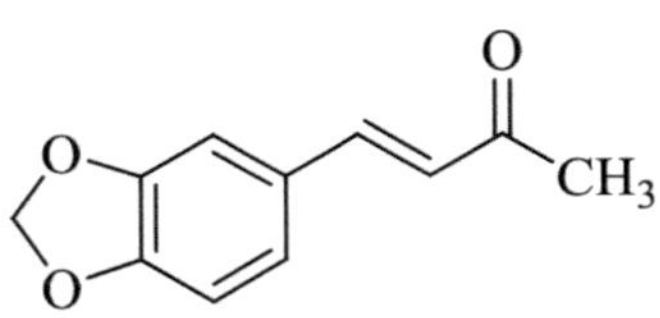 (IV)	13	46

Reaction conditions adopted: catalyst (5×10^{-2} mmol Pd), cinnamaldehyde = 10.5 mmol; toluene = 10 mL; p(H_2) = 0.1 MPa; and $T = 305$ K.

[a]After 24 h.

[b]After 2 h.

[c]After 2 h at 0.2 MPa.

Table 2. Catalytic performances of *trans*-Pd$(OAc)_2(L^1)_2$.

The main advantage of the catalytic system used is the complete solubility in the reaction medium (toluene) and the easy separation of the catalyst as solid upon addition of either MeOH or *n*-hexane to the catalytic solution, exploiting the fact that PLA is not soluble in both solvents. The occurrence of the catalytic reaction in the homogeneous phase mediated by a polymer anchored molecular Pd-based catalyst and not by Pd-NPs, was tested by performing catalytic reactions in the presence of Hg(0), which poisons heterogenous catalyst (colloidal and nanoparticle-based catalysts). As a result, the catalytic activity remained unchanged. The stability of the Pd-catalyst as Pd(II) in the presence of hydrogen was proved by X-ray photo-electron spectroscopy (XPS) of Pd3d, showing no occurred reduction of Pd(II) to Pd (0).

Pd-NPs synthesized by the MVS-technique and stabilized by differently end-functionalized PLA-stereocomplexes (i.e., $\mathbf{L}^{1/2}$, $\mathbf{L}^{7/8}$, $\mathbf{L}^{9/10}$, and $\mathbf{L}^{11/12}$) with pyridine, BipyOH, Bn, and OH-end groups, respectively (**Table 1**) were tested in the chemoselective C═C double bond cinnamaldehyde in Tetrahydrofuran (THF) [30]. As a result, the following order of decreasing catalytic activity as well as chemoselectivity was found: $\mathbf{L}^{7/8} > \mathbf{L}^{1/2} > \mathbf{L}^{9/10} > \mathbf{L}^{11/12}$ (**Table 3**), which is in line with the decreasing Pd-NPs' dispersion on the polymer surface.

Catalyst	Conversion (%)	TOF (h^{-1})	Selectivity (%)
Pd@$L^{1/2}$	54	485	88
Pd@$L^{1/2}$	30	269	68
Pd@$L^{7/8}$	74	664	97
Pd@$L^{7/8}$	72	646	96
Pd@$L^{7/8}$	88	n.d	90
Pd@$L^{9/10}$	30	n.a.	65
Pd@$L^{11/12}$	14	n.a.	60

Catalytic conditions: Pd (1.2×10^{-3} mmol), cinnamaldehyde (141.2 μmol), THF (10.0 mL), p(H_2) = 1 MPa, t = 2 h, and T = 335K.

[a]Turn-over frequency (TOF) referred to substrate accessible Pd-NPs' sites.

[b]Chemoselective for 3-phenylpropanal.

[c]Fourth recycling experiment.

[d]Solventless catalytic reaction (Pd = 1.4×10^{-2} mmol, cinnamaldehyde = 24.0 mmol, and t = 6 h).

Table 3. Catalytic performances of different Pd@$L^{x/y}$ in the selective hydrogenation of cinnamaldehyde.

Pd@$\mathbf{L}^{7/8}$ gave a 95% chemoselectivity at 100% conversion which is notably higher compared to other heterogeneous polymer based catalysts. The role of the polymer-anchored 2,2′-bipyridine functionality in $\mathbf{L}^{7/8}$ is that of an efficient small Pd-NP size (i.e., 2 nm) stabilizer, which leads to the high chemoselectivity found for 3-phenylpropanale, due to the efficient C═C double bond coordination to small Pd-NPs [30]. High chemoselectivity for Pd@ $\mathbf{L}^{7/8}$ was also found under solventless catalytic conditions (90% chemoselectivity for 3-phenylpropanal at 88% substrate conversion) conducted with a catalyst to a substrate ratio of 1:1714.

Pd-Nps generated by a stepwise synthesis which comprised the coordination of $Pd(OAc)_2$ to end-functionalized $\mathbf{L}^{3/4}$ and $\mathbf{L}^{5/6}$ followed by a successive reduction with hydrogen in either THF or CH_2Cl_2 [29]. As a result, CH_2Cl_2 reveled the solvent of choice, since the PLA-stereocomplex is much better suspended compared to THF and as a consequence the size of the obtained Pd-Nps in CH_2Cl_2 were of 3.11 nm in either cases. Pd@ $\mathbf{L}^{3/4}$ and Pd@$\mathbf{L}^{5/6}$ were employed to partially hydrogenate phenylacetylene and diphenylacetylene employing THF as the reaction medium. From a comparison of the catalytic data, shown in **Table 4**, emerges that Pd@$\mathbf{L}^{5/6}$, which is build up by the crystalline PLA-stereocomplex unit and amorphous (i.e., syndiotactic PLA) part, where the functional groups are located.

Catalyst	Substrate	Time (h)	Conversion (%)	TOF (h^{-1})	Chemoselectivity (%)	Regioselective (%)
Pd@$L^{3/4}$	Phenylacetylene	1	14	2340	88	n.a.
		2	28	2340	89	n.a.
Pd@$L^{5/6}$		1	44	6560	95	n.a.
		2	62	4623	95	n.a.
Pd@ $L^{3/4}$	Diphenylacetylene	1	28	9195	90	91.
		2	39	6404	92	92
Pd@ $L^{5/6}$		1	76	11335	90	94
		2	100	n.a.	85	95

Catalytic conditions: Pd (8.34×10^{-4} mmol), alkyne (4.0 mmol), THF (10.0 mL), and $p = 0.3$ MPa.

[a]TOF referred to substrate accessible Pd sites.

[b]Chemoselectivity referred to the corresponding alkene.

[c]$T = 298$ K.

[d]$T = 333$ K.

Table 4. Catalytic performances of different Pd@$L^{x/y}$ in the partial hydrogenation alkynes.

Hence the higher catalytic activity and chemoselectivity of Pd@$L^{5/6}$ compared to Pd@$L^{3/4}$ were explained in terms of a much easier access of Pd-Nps situated in an amorphous polymer phase. In addition, the easier access of the substrate to the Pd-NPs in Pd@$L^{5/6}$ removes more efficiently styrene (i.e., partial hydrogenation product of phenylacetylene) thus avoiding the styrene coordination to the Pd-Nps and hence the hydrogenation to ethylbenzene [37]. The partial hydrogenation of diphenylacetylene to *cis*-stilbene occurred at 333K. A comparison of the performance of Pd@$L^{5/6}$ with other heterogeneous Pd catalysts, such as Pd-Nps supported onto carbon without additional organic modifiers led to much lower chemoselectivities (i.e., styrene <80% and *cis*-stilbene 85%) [38]. The heterogeneous PLA-stereocomplex-based Pd catalysts were easily recovered from the catalytic solution simple by centrifugation of the catalytic THF solution followed by decantation of the solution in air atmosphere. ICP-OES analysis of the THF solution confirmed a very low level of Pd leaching (i.e., >0.5 ppm). HRTEM images acquired of the different recovered heterogeneous catalyst, showed that Pd@$L^{7/8}$ is the most promising support for Pd-NPs, avoiding Pd-Nps' stabilization property of $L^{7/8}$ is due to interactions between the Pd-Nps' surface and the bipyridine functionality of $L^{7/8}$ as shown by IR-spectroscopy [35].

4. Conclusions

PLA based biopolymers have found to be easily functionalized at the carboxylic acid end. The introduced functional group which is located at the chain end of the polymer chain can be exploited to coordinate metal ions, which are successively reduced to form metal nanoparticles, the growth and stabilization of which is controlled by the nature of the introduced functional group. The stereocomplexation of these latter macroligands leads to a polymer-based support which is featured by a much more thermal and hydrolytic stability compared to PLA, due to many hydrogen bond-based interactions between PLA-chains of opposite stereochemistry. In addition, the stereocomplex is insoluble in most organic solvents, so that recycling of the support is easily done upon a simple filtration process. We found that the end-functionalized of the stereocomplex with 2,2′-bipyridine (**L**$^{7/8}$) leads to an ideal support for Nps' surface. The stabilization of small Pd-Nps (2.0 nm) generated by the MSV technique, with **L**$^{7/8}$ in the course of the catalytic hydrogenation of α,β-unsaturated carbonyl compounds brings about the notable catalytic activity and chemoselectivity found for the corresponding saturated carbonyl compounds.

Author details

Marco Frediani[1*], Werner Oberhauser[2], Luca Rosi[1], Elisa Passaglia[3] and Mattia Bartoli[1]

*Address all correspondence to: marco.frediani@unifi.it

1 Department of Chemistry "Ugo Schiff," University of Florence, Sesto Fiorentino, Italy

2 Institute of Chemistry of Organometallic Compounds (CNR-ICCOM), Sesto Fiorentino, Italy

3 Institute of Chemistry of Organometallic Compounds (CNR-ICCOM), UOS Pisa, Area della Ricerca, Pisa, Italy

References

[1] De Vries, J. Elsevier, editors. The Handbook of Homogeneous Hydrogenation. 1st ed. Weinheim, Germany: WILEY-VCH Verlag GmbH & Co. KGaA; 2013. 1568 p.

[2] Blaser, HU. Heterogeneous catalysis for fine chemicals production. *Catal Today* 2000, 60(3), 161–165.

[3] Teichert J, den Hartog T, Hanstein M, Smit C, ter Horst B, Hernandez-Olmos V, Feringa B, Adriaan Minnaard. Organocatalytic reduction of carbon-carbon double bonds in racemization-sensitive compounds. *ACS Catalysis* 2011, 1(4), 309–315.

[4] Li X, Li L, Tang Y, Zhong L, Cun L, Zhu J, Liao J, Deng J. Chemoselective conjugate reduction of α,β-unsaturated ketones catalyzed by rhodium amido complexes in aqueous media. *J. Org. Chem.* 2010, 75(9), 2981–2988.

[5] Mahoney W, Stryker J. Hydride-mediated homogeneous catalysis. Catalytic reduction of α,β-unsaturated ketones using [$(Ph_3P)CuH$]6 and H_2. *J. Am. Chem. Soc.* 1989, 11(24), 8818–8823.

[6] Lindlar H. A new catalyst for selective hydrogenation. *Helv. Chim. Acta* 1952, 35(2), 446–450.

[7] Cramer D, Allinger N. Macro rings. XIII. Synthesis and properties of 1,7-cyclododecadiyne and related compounds. *J. Am. Chem. Soc.* 1956, 78(11), 2518–2524.

[8] Busca G. Chapter 9 - Metal Catalysts for Hydrogenations and Dehydrogenations. In: Heterogeneous Catalytic Materials. 2nd ed. Amsterdam: Elsevier; 2014. pp. 297–343.

[9] Prakash MG, Mahalakshmy R, Krishnamurthy KR, Viswanathan B. Studies on Ni–M (M = Cu, Ag, Au) bimetallic catalysts for selective hydrogenation of cinnamaldehyde. *Catal. Today* 2016, 263, 105–111.

[10] Matsubara EY, Rosolen JM, Donate PM, Gunnella R. Palladium decoration of hybrid carbon nanotubes/charcoal composite and its catalytic behavior in the hydrogenation of trans-cinnamaldehyde. *J. Mol. Catal. A: Chem.* 2015, 410, 34–40.

[11] Wu Q, Zhanga C, Zhanga B, Lia X, Yinga Z, Liua T, Lina W, Yua Y, Chenga H. Highly selective Pt/ordered mesoporous TiO_2–SiO_2 catalysts for hydrogenation of cinnamaldehyde: The promoting role of Ti^{2+}. *J. Colloid Interface Sci.* 2016, 463, 75–82.

[12] Piquerasa M, Puccia V, Vegab DA, Volpe MA. Selective hydrogenation of cinnamaldehyde in supercritical CO_2 over Me–CeO2 (Me = Cu, Pt, Au): Insight of the role of Me–Ce interaction. *Appl. Catal. B.* 2016, 185, 265–271.

[13] Müller A, Ludwig M, Arlit M, Lange R. Evaluation of reactor concepts for the continuous production of fine chemicals using the selective hydrogenation of cinnamaldehyde over palladium catalysts. *Catal. Today* 2015, 241, 214–220.

[14] McCuea AJ, Guerrero-Ruizb A, Rodríguez-Ramosc I, Andersona JA. Palladium sulphide—A highly selective catalyst for the gas phase hydrogenation of alkynes to alkenes. *J. Catal.* 2016, 340, 10–16.

[15] Sasaki T, Bhanage BM. Silica supported palladium phosphine as a robust and recyclable catalyst for semi-hydrogenation of alkynes using syngas. *J. Mol. Catal. A: Chem.* 2016, 414, 78–86.

[16] Wu Z, Cravotto G, Gaudino EC, Giacomino A, Medlock J. Ultrasonically improved semi-hydrogenation of alkynes to (Z-)alkenes over novel lead-free Pd/Boehmite catalysts. *Ultrason. Sonochem.* 2016, 16, 30161–30164.

[17] Ogweno AO, Ojwach SO, Akerman MP. Cationic pyridyl(benzoazole) ruthenium(II) complexes: Efficient and recyclable catalysts in biphasic hydrogenation of alkenes and alkynes. *Appl. Catal. A.* 2014, 486, 250–258.

[18] Kontkanen M, Vlasova L, Suvanto S, Haukka M. Microencapsulated rhodium/cross-linked PVP catalysts in the hydroformylation of 1-hexene. *Appl. Catal. A.* 2011, 401(1–2), 141–146.

[19] Klimov OV, Nadeina KA, Dik PP, Koryakina GI, Pereyma VY, Kazakov MO, Budukva SV, Gerasimov E, Prosvirin IP, Kochubey DI, Noskov AS. CoNiMo/Al_2O_3 catalysts for deep hydrotreatment of vacuum gasoil. *Catal. Today* 2016, 271, 56–63

[20] Reddya MV, Reddyb GCS, Jeong YT. Polystyrene-supported p-toluenesulfonic acid (PS/PTSA): As a highly active and reusable heterogeneous bronsted acid catalyst for the synthesis of novel 1H-indol-3-yl-4H-chromene-3-carbonitriles under neat conditions. *Tetrahedron Lett.* 2016, 57(11), 1289–1292.

[21] Shan R, Chena G, Yana B, Shia J, Liu C. Porous CaO-based catalyst derived from PSS-induced mineralization for biodiesel production enhancement. *Energ. Convers. Manage.* 2015, 106, 405–413.

[22] Leadbeater NE. Chapter 12.14—Polymer-supported Organometallic Catalysts. In: Reference Module in Chemistry, Molecular Sciences and Chemical Engineering. 2nd ed. Oxford: Elsevier; 2007. pp. 663–753.

[23] Meng X, Shi G, Wu C, Chen W, Xin Z, Shi Y, Sheng Y. Chain extension and oxidation stabilization of Triphenyl Phosphite (TPP) in PLA. *Polym. Degrad. Stab.* 2015, 124, 112–118.

[24] Heller M, Schubert US. Optically active supramolecular poly(L-lactide)s end-capped with terpyridine. *Macromol. Rapid Commun.* 2001, 22 (16), 1358–1363.

[25] Giachi G, Oberhauser W, Frediani M, Passaglia E. Pd(II)-pyridine macrocomplexes based on poly(lactide). *J. Polym. Sci., Part A: Polym. Chem.* 2011, 49(21), 4708–4713.

[26] Giachi G, Oberhauser W, Frediani M, Passaglia E. Aerobic alcohol oxidation catalyzed by polyester-based Pd(II) macrocomplexes. *J. Polym. Sci., Part A: Polym. Chem.* 2012, 50 (13), 2725–2731.

[27] Bartoli M, Rosi L, Petrucci G, Armelao L, Oberhauser W, Frediani M, Piccolo O, Rathod VD, Paganelli S. An easily recoverable and recyclable homogeneous polyester-based Pd catalytic system for the hydrogenation of α,β-unsaturated carbonyl compounds. *Catal. Comm.* 2015, 65, 228–233.

[28] Tsuji H. Poly(lactide) stereocomplexes: Formation, structure, properties, degradation, and applications. *Macromol. Biosci.* 2005, 5(7), 569–597.

[29] Petrucci G, Oberhauser W, Bartoli M, Giachi G, Frediani M, Passaglia E, Capozzoli L, Rosi L. Pd-nanoparticles supported onto functionalized poly(lactic acid)-based stereocomplexes for partial alkyne hydrogenation. *Appl. Catal. A.* 2014, 469,132–138.

[30] Oberhauser W, Evangelisti C, Jumde RP, Petrucci G, Bartoli M, Frediani M, Mannini M, Capozzoli L, Passaglia E, Rosi L. Palladium-nanoparticles on end-functionalized

poly(lactic acid)-based stereocomplexes for the chemoselective cinnamaldehyde hydrogenation: Effect of the end-group. *J. Catal.* 2015, 330, 187–196.

[31] Turner SR, Liu Y. 5.14 - Chemistry and Technology of Step-Growth Polyesters- Matyjaszewski, Krzysztof. In: Möller, Martin, editors. Polymer Science: A Comprehensive Reference. 1st ed. Amsterdam: Elsevier; 2012. pp. 311–331.

[32] Duda A. 4.11 - ROP of Cyclic Esters. Mechanisms of Ionic and Coordination Processes. In: Möller, Martin, editors. Polymer Science: A Comprehensive Reference. 1st ed. Amsterdam: Elsevier; 2012. pp. 213–246.

[33] Löfgren A, Albertsson A, Dubois P, Jérôme R. Recent advances in ring-opening polymerization of lactones and related compounds. *Macromol. Sci. Rev. Macromol. Chem. Phys.* 1995, 35(3), 379–418.

[34] Liu T, Simmons TL, Bohnsack DA, Mackay ME, Smith III MR, Baker GL. Synthesis of polymandelide: A degradable polylactide derivative with polystyrene-like properties. *Macromolecules.* 2007, 40(17), 6040–6047.

[35] Evangelisti C, Panziera N, D'Alessio A, Bertinetti L, Botavina M, Vitulli G. New monodispersed palladium nanoparticles stabilized by poly-(N-vinyl-2-pyrrolidone): Preparation, structural study and catalytic properties. *J. Catal.* 2010, 212(12), 246–252.

[36] Dyson PJ. Arene hydrogenation by homogeneous catalysts: Fact or fiction? *J. Chem. Soc. Dalton Trans.* 2003, 15, 2964–2974.

[37] Domínguez S, Berenguer-Murcia A, Pradhan BK, Linares-Solano A, Cazorla-Amorós D. Semihydrogenation of phenylacetylene catalyzed by palladium nanoparticles supported on carbon materials. *J. Phys. Chem. C* 2008, 112(10), 3027–3834. DOI: 10.1021/jp710693u

[38] Starodubtseva EV, Vinogradova MG, Turova OV, Bumagin NA, Rakov EG, Sokolov VI. Palladium(0) supported on carbon nanotubes as an efficient catalyst of the C triple bond; length of m dash C bond hydrogenation. *Catal. Comm.* 2009, 10(10), 1441–1442.

Chapter 2

Alkyne Selective Hydrogenation with Mono- and Bimetallic-Anchored Catalysts

Cecilia Lederhos, Carolina Betti, Domingo Liprandi, Edgardo Cagnola and Mónica Quiroga

Additional information is available at the end of the chapter

http://dx.doi.org/10.5772/64866

Abstract

Partial hydrogenation of alkynes has industrial and academic relevance on a large scale; industries such as petrochemical, pharmacological and agrochemical use these compounds as raw material. Finding an economic, active and selective catalyst for the production of alkenes through partial hydrogenation of alkynes is thus an important challenge. Mono- and bimetallic catalysts (palladium, ruthenium and nickel) were synthetized by the incipient wetness technique using gamma alumina and an activated carbon as supports. The catalysts were characterized by inductively coupled plasma, hydrogen chemisorption, temperature-programmed reduction and X-ray photoelectronic spectroscopy (XPS). The objective of this work is to study 1-heptyne-selective hydrogenation using supported catalysts influenced by different factors: (a) pretreatment reduction temperature, (b) reaction temperature, (c) type of support, (d) metal loading, (e) precursor salt and (f) addition of a second metal to monometallic palladium catalyst. The Lindlar commercial catalyst, commonly used in these types of reactions, was used for comparative purposes. XPS technique allowed verifying that the presence of electron-deficient species on the catalyst surface with high metal loading affects the conversion and selectivity to the desired product. Nevertheless, the influence of geometrical effects and/or mixed active sites in the catalysts, as well as metal-metal and metal-support interactions, cannot be neglected.

Keywords: selective hydrogenation, alkyne, alkene, heterogeneous catalysts, mono- and bimetallic catalysts

1. Introduction

Petroleum cuts contain mixtures of unsaturated and aromatic compounds. Among them, acetylenic hydrocarbons are very unstable; so they must be transformed to olefins. Alkenes have industrial and academic relevance on a large scale; industries such as petrochemical, pharmaceutical and agrochemical use these compounds as raw materials. Partial hydrogenation reactions using catalytic materials allow a reduction in operational costs and also enable high selectivity to alkenes. Specifically, the catalytic selective hydrogenation of alkynes using either homogeneous or heterogeneous catalysts has been widely studied in the past several years [1]. The hydrogenation of any alkyne conduces naturally to the alkene formation as the former trends to bind more strongly than the latter on the supported metal catalyst, thus blocking the possibility of the alkene readsorption or displacing it from the support surface. Many natural products, such as biologically active compounds [2], are synthesized through this kind of reactions.

Classical heterogeneous catalysts used to hydrogenate multiple carbon-carbon bonds contain noble metals such as Pd, Pt, Ru and Rh, which are highly active and selective [3, 4]. Many authors have found that supported palladium catalysts present the highest catalytic activity for the partial hydrogenation of alkynes to alkenes [5, 6]. One of the most used catalytic systems for these kinds of reactions is the Lindlar catalyst ($Pd/CaCO_3$ modified with $Pb(OAc)_2$), developed in 1953 [7]. During decades, much research has been carried out modifying this type of catalyst in order to increase the activity and selectivity to alkenes of low molar weight. Several materials have been used as supports, and they are usually classified as organic (macroreticular/macroporous polymers) or inorganic (silica, alumina, zeolites and clays). Besides, modified palladium [8] or nanoparticles of Pd have also been investigated [9]. Another kind of material, not clearly included in any of these groups, is carbonaceous species, whose outstanding properties as a catalyst support are well recognized [10], among them are the possibility of modifying the specific surface area, the porosity and the surface chemistry; moreover, carbon supports present the advantage of being inert in liquid reaction media [11]. Many catalysts, mono- or bimetallic as well as complexes of several transition metals, have also been proposed for these kinds of reactions [2, 12–14]. A major part of research efforts have been devoted to the partial hydrogenation of short-chain alkynes such as ethyne [15], with few works related to longer chain alkynes. As Pd has increased its cost, it is a challenge to synthesize cheaper catalysts. In this context, using nickel catalysts during the selective hydrogenation of alkyne is less studied and has been recently researched [12].

Based on the above considerations, the objectives of this chapter are to evaluate the effects of different factors on the activity and selectivity during the selective hydrogenation of 1-heptyne, a long-chain terminal alkyne. The factors studied are (a) pretreatment reduction temperature, (b) reaction temperature, (c) type of support, (d) metal loading, (e) precursor salt and (f) addition of a second metal (such as nickel) to monometallic palladium catalyst. Last but not least, all the catalytic performances are compared against those obtained with the commercial Lindlar catalyst.

2. Experimental details

2.1. Catalysts preparation

All the monometallic catalysts used in this work were prepared by the incipient wetness technique, and for bimetallic catalyst the successive impregnation technique was used following a procedure previously indicated [12]. γ-Al_2O_3 (Ketjen CK 300, cylinders of 1.5 mm diameter) and a pelletized activated carbon (AC: GF-45 provided by NORIT) were used as supports.

All of the supports were impregnated with acid aqueous solutions of $PdCl_2$ or $Pd(NO_3)_2$ (Fluka, purity of >99.98%), and $Ni(NO_3)_2.6H_2O$ (Fluka, purity of >98.5%) was used as the precursor salt for the bimetallic catalyst.

The monometallic alumina-based catalysts were calcined at 823 K for 3 h. Before its catalytic evaluation, all the catalysts were reduced for 1 h under a hydrogen stream at 573 K. Only the High loaded Pd catalyst prepared with chlorine precursor salt was reduced at 373 and 573 K to study the effect of pretreatment temperature on the catalytic behaviour. Besides, the bimetallic catalyst was reduced for 1 h at 673 K.

2.2. Catalysts characterization

Physical adsorption of gases (N_2 at 77 K and CO_2 at 273 K) and mercury porosimetry were used to determine the pore volume distribution and the specific surface area for both supports and the specific surface area [16]. Both adsorptives (N_2 and CO_2) allow estimating the pore volume distribution for pores with a diameter up to 7.5 nm approximately. From Dubinin-Radushkevich equation to the CO_2 isotherm at 273 K, it is possible to obtain the micropore volume (V_{micro}), while the supermicropore volume (V_{sm}) is attained by the subtraction of V_{micro} to the volume calculated by the same equation but applied to the N_2 adsorption isotherm at 77 K. By mercury porosimetry using a Carlo Erba 2000 equipment, macropore volume (V_{macro}) and part of the mesopore volume (V_{meso}) were determined. The rest of V_{meso} was obtained from N_2 adsorption isotherm and Hg porosimetry as described previously [17].

Using ICP Perkin Elmer equipment, metal loadings of the catalysts were measured.

Micromeritics AutoChem II 2920 was used to determine the H_2 chemisorption of Pd at 303 K using 0.2 g of catalyst. *In situ* sample reduction was performed using a H_2/Ar stream (5% v/v), ensuring the absence of beta-phase palladium hydride on the catalyst surface. The samples were degassed *in situ* for 2 h in an argon flow (AGA purity of 99.99%) and cooled up to 303 K. Then, the hydrogen uptake was measured by sending calibrated pulses. The metal dispersion was calculated assuming a H:Pd = 1 stoichiometry [18].

Temperature programmed reduction, TPR, of the samples was performed using a Micromeritics AutoChem II 2920. Gamma-alumina-supported catalysts were pretreated for 30 min at 673 K under an O_2 stream, and cooled with an Ar stream (AGA purity of 99.99%). The palladium anchored on GF-45 was pretreated with an argon flow at the same operational conditions used for alumina-based catalysts. The TPR profiles were obtained increasing the temperature up to 1223 K at 10 K min^{-1} in a 5% (v/v) hydrogen/argon stream.

Electronic states of Pd, Cl or N were measured using a VG Microtech Multilab equipment following a procedure elsewhere published [19]. The reference-binding energies (BEs) were Al 2p at 74.4 eV and C 1s peak at 284.5 eV for alumina and carbonaceous catalysts, respectively.

2.3. Catalytic evaluations

The performed reaction test was the selective hydrogenation of 1-heptyne. **Figure 1** shows the scheme of 1-heptyne hydrogenation reaction; the possible products obtained are 1-heptene (desired product) and n-heptane.

A stainless steel reactor was used during the catalytic evaluations. Catalyst of 0.75 g, a hydrogen pressure of 150 kPa and temperatures of 298 and 303 K were used. Seventy-five millilitres of 5% (v/v) 1-heptyne (Fluka, purity of >98%) in toluene (Merck, purity of >99%) solution was used as feed. The possibility of diffusional limitations during the catalytic tests was assessed following a procedure previously described [20]. Experiments were carried out at different stirring velocities in the 180–1400 rpm range. Stirring velocities higher than 500 rpm allowed obtaining identical values of activity and selectivity for all of the catalysts. On the other hand, the catalyst particle size was varied verifying that the conversion and selectivity were the same (within experimental error) than those for the non-crushed catalysts. Therefore, chemical control ensures neither external nor internal diffusional limitations during the catalytic tests.

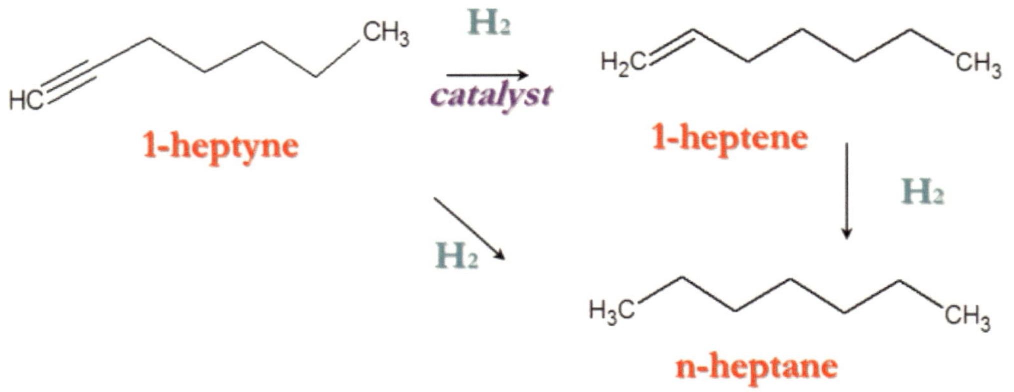

Figure 1. Scheme for the 1-heptyne hydrogenation.

Gas chromatography (GC) using a flame ionization detector (FID) with a capillary column (Chrompack CP WAX 52 CB) was used to analyse 1-heptyne, 1-heptene and n-heptane.

3. Results and discussion

3.1. Catalysts characterization

Table 1 presents the Brunauer-Emmett-Teller (BET) surface area and micro-, supermicro-, meso- and macropore volumes (S_{BET}, V_{micro}, V_{sm}, V_{meso} and V_{macro}, respectively) of the γ-$Al_2O_3$and AC supports. It can be observed that the activated carbon (AC) includes almost the same amounts of the four types of pores, with a large proportion of pore volume in the range of supermicro-, micro-, meso- and macropores, the so-called transport pores while γ-Al_2O_3 is a mesoporous solid having a poor contribution of supermicro-, micro- and macropores.

Support	S_{BET} (m^2 g^{-1})	V_{micro} (mL g^{-1})	V_{sm} (mL g^{-1})	V_{meso} (mL g^{-1})	V_{macro} (mL g^{-1})
Al_2O_3	180	0.048	0.030	0.487	0.094
AC	1718	0.345	0.498	0.449	0.400

Table 1. BET surface area and pore volume distribution of the supports [16].

Table 2 contains the catalyst notation used according to the precursor salt, metal loading or reduction temperature used during the catalysts preparation. Besides for all of the studied catalysts, dispersions (D), XPS results and activity values (turnover frequency (TOF)) calculated considering a kinetic of zero order are reported in **Table 2**.

Catalysts	Precursor salt	Metal loading (wt%)	Reduction Temp (K)	D (%)	XPS		TOF (s^{-1})
					Pd $3d_{5/2}$ BE (eV)	Cl/Pd or N/Pd (at/at)	
$Pd_{(5\%)}$/Al_373	$PdCl_2$	4.69 Pd	373	42	337.0	0.6	9.1×10^{-3}
$Pd_{(5\%)}$/Al	$PdCl_2$	4.69 Pd	573	28	336.5	0.2	1.8×10^{-2}
$Pd_{(5\%)}$/AC	$PdCl_2$	4.82 Pd	573	32	336.8	0.3	4.9×10^{-3}
$Pd_{(0.4\%)}$/Al	$PdCl_2$	0.44 Pd	573	58	334.9	–	5.1×10^{-2}
$PdN_{(0.4\%)}$/Al	$Pd(NO_3)_2$	0.37 Pd	573	32	334.9	–	1.1×10^{-1}
Pd-Ni/Al	$Pd(NO_3)_2$ $Ni(NO_3)_2$	0.37 Pd 0.89 Ni	673	31	$334.2^{(54\%)}$ $335.3^{(46\%)}$	–	1.6×10^{-1}

Table 2. Catalysts naming convention, precursor salt, metal loading, reduction temperature, dispersion values, XPS BE of Pd $3d_{5/2}$, Cl/Pd or NPd atomic ratios and TOF values for all of the catalysts.

As it can be seen in **Table 2**, the dispersion values for $Pd_{(5\%)}$/Al catalyst decrease as the reduction temperature increases owing to the agglomeration of Pd particles. Also, at the same reduction temperature (573 K), palladium supported on AC presents slightly higher dispersion than when alumina is used as support possibly because of the high surface area of AC [21] or to the different porosity of the supports. Besides, a decrease in the metal loading, using the same precursor salt, produces a significant increase in the dispersion value as the palladium active sites are most exposed on the catalyst surface. On the other hand, at the same reduction temperature, the change of precursor salt produces a higher dispersion when $PdCl_2$ is used because of complex oxychlorinated species formation during the calcination process. These species, $Pd^{\delta+}O_xCl_y$, present a stronger interaction with the support than that showed by PdO, thus improving the metal dispersion [22].

Additionally, as nickel monometallic catalysts do not consume hydrogen during the chemisorption analysis, the dispersion value of bimetallic Pd-Ni/Al catalyst was calculated taking into account only the Palladium active sites. It can be noted that the dispersions of $PdN_{(0.4\%)}$/Al and Pd-Ni/Al catalysts are very similar, confirming that nickel in the bimetallic catalyst does not chemisorb H_2 during the analysis. Last but not least, the dispersion values in **Table 2** are in total accordance with those reported by other authors [23].

In **Table 2**, Pd $3d_{5/2}$ BE and the Cl/Pd or N/Pd superficial atomic ratios obtained by XPS technique are also listed. Low-loaded monometallic catalysts, $Pd_{(0.4\%)}$/Al and $PdN_{(0.4\%)}$/Al, present the Pd $3d_{5/2}$ peak with a binding energy equal to 334.9 eV that corresponds to Pd° [24], whereas high-loaded monometallic catalysts, $Pd_{(5\%)}/Al_{\_373}$, $Pd_{(5\%)}$/Al and $Pd_{(5\%)}$/AC, displayed peak values of BE at 337.0, 336.5 and 336.8 eV, respectively. These higher binding energy values indicate that the metal is electron-deficient ($Pd^{\delta+}$) possibly because of the presence of non-reduced Pd oxychloride species formed during the calcination process [22] or to non-reduced Pd species stabilized by neighbouring Cl atoms [23, 25]. The XPS BE results indicate that the reduction temperature influences the electronic state of palladium in $Pd_{(5\%)}$/Al catalysts: the one reduced at 373K has a peak at 337.0 eV, whereas that reduced at 573 K presents a peak shifted to 336.5 eV. Therefore, the low reduction temperature used generates a more electron-deficient Pd (337.0 eV) with a high Cl superficial content. On the other hand, high-loaded monometallic catalysts present Pd with a slightly higher electron deficiency ($Pd^{\delta+}$ species) and a higher concentration of superficial Cl when Pd is anchored on GF-45.

In the case of the Pd-Ni/Al bimetallic catalyst after the deconvolution of the Pd $3d_{5/2}$ BE, two signals can be seen at 334.2 eV (54% of all of the metal species, atomic basis) and 335.3 eV (46 at/at%) palladium species. These values suggest the presence of two type of Pd species in simultaneous, represented by $Pd^{\delta-}$ (electron-rich species) and slightly electron-deficient palladium species ($Pd^{\delta+}$, with δ close to 0), respectively. The former could be attributed to the formation of metallic bonds or alloy, occurring at low temperatures [26, 27]. Additionally, for Pd-Ni/Al catalyst, the BE of Ni $2p_{3/2}$ peak appears at 856.4 eV, which is attributed to electron-deficient species (Ni^{n+}, with n close to 2) probably corresponding to different interactions between nickel and aluminium (from the support) [28], or to the formation of intermediate Pd-Ni-Al_2O_3 surface species [12].

The XPS spectra of high-loaded monometallic catalysts prepared from chlorine precursors show a peak at ca. 198.5 eV that corresponds to C1 $2p_{3/2}$ The peak was associated to surface chloride species [24] that were not completely eliminated after reduction. Besides, neither Cl nor N was detected by XPS on the surface for the low-loaded monometallic catalysts.

The TPR profiles of palladium mono- and bimetallic catalysts are shown in **Figure 2**. In this figure, it can be seen that all the prepared catalysts present a main reduction peak at low temperatures, between 259 and 358 K, that can be attributed to the reduction of palladium oxidized species (PdO_x) to Pd° [29]. Besides, the low-loaded catalysts present the reduction peak shifted to lower temperatures, indicating that Pd species are more easily reduced than those present on high-loaded catalysts. These shifts to higher temperatures are due to different types of the metal-support interactions.

For the low-loaded catalysts, $Pd_{(0.4\%)}$/Al, $PdN_{(0.4\%)}$/Al and Pd-Ni/Al, the profiles in **Figure 2** show an inverted peak between 335 and 339 K, which could be assigned to the decomposition of the β-PdH phase that is formed during the reduction of the PdO_x particles at low temperatures [22, 29, 30].

The TPR profiles for the high-loaded catalysts prepared from $PdCl_2$ salt ($Pd_{(5\%)}$/Al and $Pd_{(5\%)}$/AC) show a second broad peak between 400 and 600 K, which is attributed to the reduction of $Pd^{\delta+}O_xCl_y$ species [19, 22].

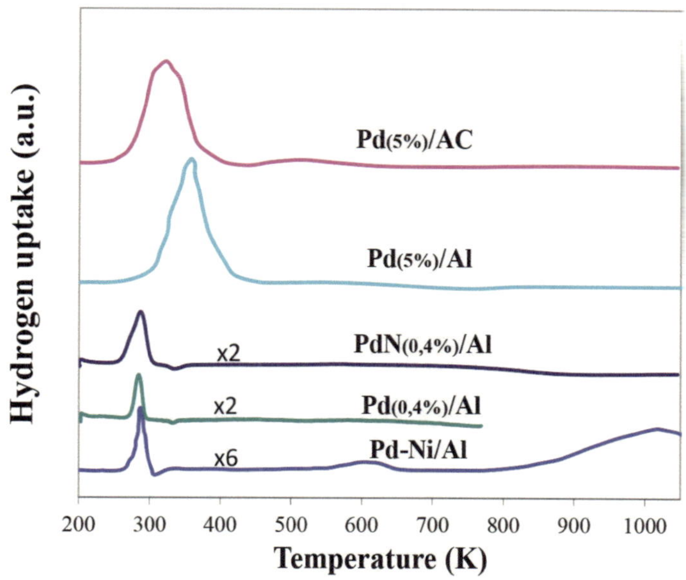

Figure 2. TPR profiles for $Pd_{(5\%)}$/Al, $Pd_{(5\%)}$/AC, $Pd_{(0.4\%)}$/Al, $PdN_{(0.4\%)}$/Al and Pd-Ni/Al.

The TPR profile of the bimetallic catalyst Pd-Ni/Al is also presented in **Figure 2**. Up to 500 K, the reduction profile is very similar to the monometallic $PdN_{(0.4\%)}$/Al catalyst, having its main reduction peak at 286 K corresponding to PdO_x reduction. The decomposition of the β-PdH phase is also present at lower temperature, 307 K; the shift of this signal suggests that the decomposition of the β-PdH phase is more easily accomplished in the bimetallic catalyst. Furthermore, as shown in **Figure 2**, the bimetallic catalyst has a second peak at 621 K, which is attributed to the reduction of NiO species to Ni° [31–33]. It is well known that nickel monometallic catalysts prepared from nitrate salts are reduced at temperatures between 600 and 1000 K when the contact between NiO and alumina is intimate [33–35]. The patterns of reduction depend on the nature of the metal-support interactions, which can be modified by the calcination temperature employed during the preparation of the monometallic nickel catalysts [34]. Besides, a broad peak is also present in this profile with a maximum at 1000 K, which is attributed to the reduction of nickel aluminates, $NiAl_2O_4$, showing a strong metal-support interaction [35, 36]. According to the calculated degree of reduction, determined by TPR, the bimetallic catalyst has a low percentage of reduced Ni (7%) and Pd (74%). This suggests the presence of strong Pd-Ni intermetallic interaction in the catalyst; however, the interaction of Pd and Ni with the support cannot be neglected.

When palladium-supported catalysts are used during the alkyne hydrogenations, the β-phase hydride acts as a hydrogen source that promote over hydrogenation to obtain mainly the alkane, decreasing the selectivity to the alkene. The disappearance of the β-PdH phase is very important because it could impact directly on the activity and selectivity [37]. These authors state that the disappearance of the β-PdH phase considerably decreases alkynes hydrogenation rate to alkanes, thus increasing the selectivity to alkenes formation. For the prepared mono- and bimetallic catalysts, the palladium β-phase Pd hydride is not present as it is proved by the TPR profiles at the pretreatment reduction temperature adopted.

According to XPS and TPR characterizations, it can be concluded that after pretreatment Pd° is present in the low-loaded monometallic catalysts, while $Pd^{\delta+}O_xCl_y$ species are formed in the high-loaded Pd catalysts. On the other hand, on the bimetallic catalyst, two kinds of palladium species ($Pd^{\delta+}$, with δ close to 0, and $Pd^{\delta-}$) and Ni^{n+} (with n close to 2) are present on the surface.

3.2. Catalytic evaluations

3.2.1. Effect of the reduction temperature

Figure 3 presents 1-heptyne total conversion and selectivity to 1-heptene as a function of time for $Pd_{(5\%)}/Al$ catalyst pretreated for 1 h in a hydrogen flow at 373 and 573 K, the reaction temperature was 303 K. In the figure, it can be seen that the total conversion increases as the reduction temperature is increased, while the selectivity is slightly lower at the higher reduction temperature (≥90%). From the TOF values displayed in **Table 2**, it may be concluded that the catalyst reduced at higher temperature is nearly twice more active than when it is reduced at 373 K. The activity results can be explained taking into account the electronic state of Pd in each catalyst: the more electron-deficient palladium species the less active is the catalyst for the hydrogenation of 1-heptyne. It is probably that the $Pd^{\delta+}$ species inhibit the interaction between the metal and 1-heptene due to an electronic effect, decreasing its electron-donor character. Therefore, it can say that the presence of electron-deficient $Pd^{\delta+}O_xCl_y$ species improves the selectivity to the desired product. Given that the chlorine is not completely removed from the $Pd_{(5\%)}/Al$ catalyst after the heat treatments, it seems likely that the role of the remaining chlorine on the catalyst surface could be to stabilize the positively charged palladium structures, resulting in less active but more selective catalysts (electronic and steric factors). Therefore, our results suggest a correlation between the reduction temperature, electron deficiency of palladium species and chloride content, with total conversion and

http://dx.doi.org/10.5772/64866

selectivity. In total accordance with our results, an increase in selectivity was found by some authors [38] when smaller and more electron-deficient Pd clusters were used. Other authors [39] found that 1-heptene is more weakly adsorbed than 1-heptyne on electron-deficient Pd species and, once formed, the 1-heptene molecules are more easily desorbed than 1-heptyne. This effect was previously found for other partial hydrogenation reactions using monometallic Ru-supported catalysts [40].

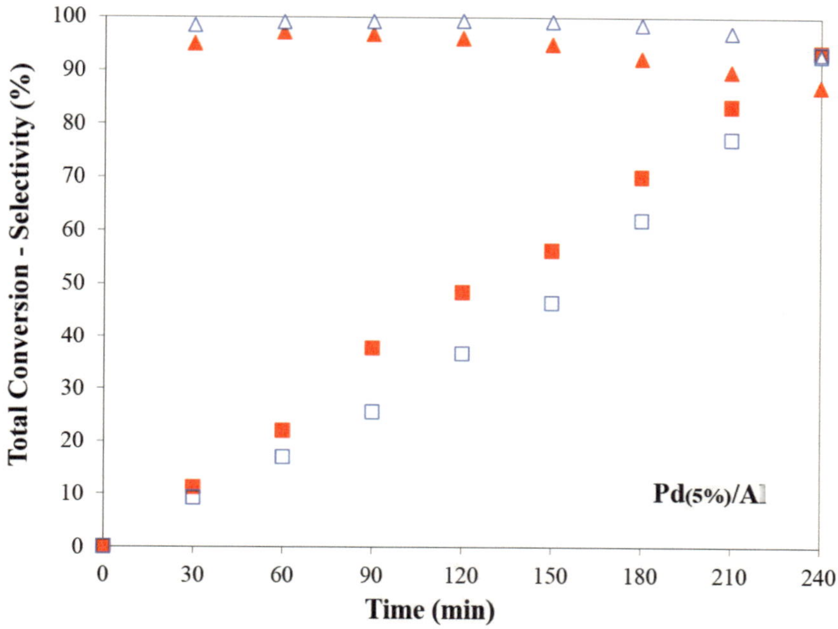

Figure 3. *Effect of reduction temperature*: Total conversion of 1-heptyne (■, □) and selectivity to 1-heptene (▲,△) as a function of time for $Pd_{(5\%)}/Al$ catalyst: reduced at 573 K (filled symbols) and reduced at 373 K (opened symbols). Reaction temperature: 303 K.

According to the obtained results, 573 K was selected as the reduction temperature of the monometallic catalysts because higher values of total conversions were obtained.

3.2.2. *Effect of reaction temperature*

Another important factor to evaluate is the reaction temperature. Alkyne hydrogenation reactions must be carefully controlled, especially at large scale when large amount of catalyst are used as in these kinds of exothermic reactions. It is well known that above 313 K, a complete hydrogenation of the alkyne compounds occurs [5]. In order to evaluate this effect, the reaction tests were performed at 280 and 303 K using the $Pd_{(5\%)}/Al$ catalyst reduced for 1 h at 573 K (optimal reduction temperature). In **Figure 4**, the total conversion and selectivity to 1-heptene as a function of time at the mentioned temperatures are presented.

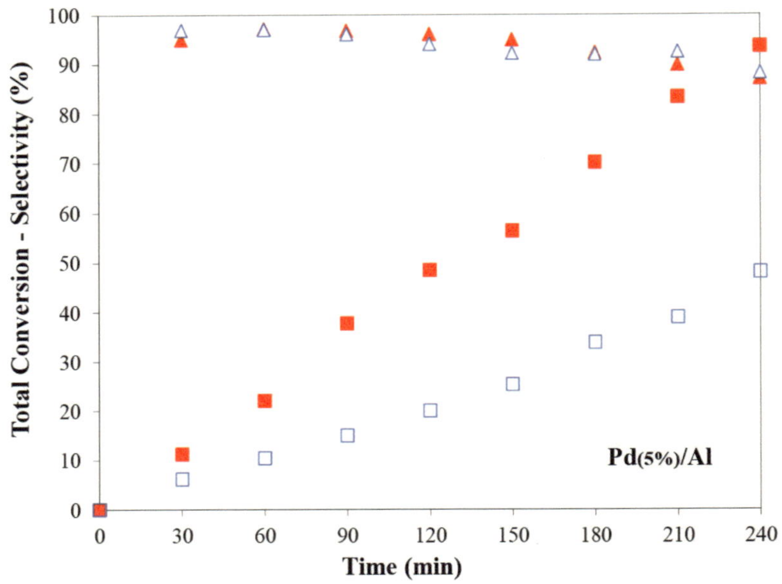

Figure 4. *Effect of reaction temperature*: Total conversion of 1-heptyne (■, □) and selectivity to 1-heptene (▲,Δ) as a function of time for $Pd_{(5\%)}$/Al catalyst: 303 K (filled symbols) and 280 K (opened symbols). Reduction temperature: 573 K.

The obtained results shown in **Figure 4** indicate a marked difference in the total conversion of 1-heptyne when the reaction temperature is increased. The total conversion is markedly higher when the reaction is carried out at 303 K, while similar selectivity to 1-heptene values is obtained at both temperatures. Other authors have found similar results while studying the partial hydrogenation of several alkynes [41, 42].

As hydrogenation reactions are extremely exothermic, an increase in the reaction temperature does not favour the reaction thermodynamically, but it improves the kinetic of the reaction. Experimental results show that the optimum reaction temperature is 303 K because higher total conversions of 1-heptyne are obtained with high selectivity to 1-heptene (≥90%).

3.2.3. *Effect of support*

It is important to compare different materials employed as supports of the active phases, so two kinds of supports were selected: alumina (inorganic) and activated carbon (a carbonaceous material, called AC). The evaluated catalysts were $Pd_{(5\%)}$/Al and $Pd_{(5\%)}$/AC, reduced for 1 h at 573 K and run at 303 K. The total conversion and selectivity to 1-heptene versus time are shown in **Figure 5**.

Analysing the data presented in **Figure 5** and the TOF values shown in **Table 2**, it can be noted that better performance is achieved when Al_2O_3 is used as support. Initially, the obtained selectivity values with both supports are higher than 90%, but for total conversion higher than 60% the selectivity to 1-heptene slightly decays.

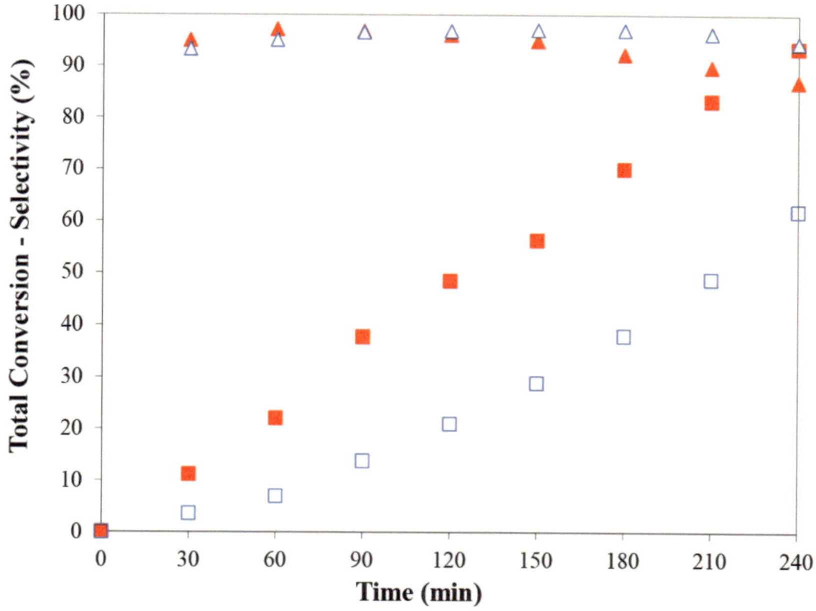

Figure 5. *Effect of support*: Total conversion of 1-heptyne (■, □) and selectivity to 1-heptene (▲,△) as a function of time for $Pd_{(5\%)}$/Al (filled symbols) and $Pd_{(5\%)}$/AC (opened symbols). Reduction and reaction temperatures: 573 and 303 K, respectively.

As both catalysts present quite similar dispersion values as well as $Pd^{\delta+}$ electron-deficient species and Cl/Pd atomic ratios, the observed differences in activity must be assigned to the characteristic of the support. The influence of the support on the physicochemical properties and, therefore, on the catalytic behaviour of metals is well established in the literature [43].

During the preparation step of the catalysts, the specific support properties of the carbons (such as chemical nature, texture, pore structure, surface state, etc.) can modify the morphology and/or localization of the metal particles, electronic structure of the surface metal atoms, adsorption-desorption equilibrium of reactants, and so on. This can generate differences in the conversion and selectivity values. Thus, as our results suggest, the activity and selectivity of palladium-supported catalysts is a complex property of the whole catalyst and cannot be related to a single parameter. Considering the above, the slightly higher selectivity to 1-heptene at the highest conversion values found for $Pd_{(5\%)}$/AC catalyst could be associated to shape selectivity induced by the porous support. In this way, this might be due to the 1-heptene molecule has a planar end, unlike the more voluminous end of the fully saturated n-heptane. The increase of the selectivity to the desired product may be associated with the localization of the Pd species in narrow pores (micro- and supermicropores) in the carbon support. If this is the case, it could also be suggested that the lower total conversion of $Pd_{(5\%)}$/AC catalyst is due to the narrower porosity of the activated carbon, as it is probable that fewer 1-heptyne molecules could reach the active sites located in the supermicropores. If a significant fraction of the active species are located in pores of a particular size (larger supermicropores, practically absent in $Pd_{(5\%)}$/Al catalyst, and mesopores), the concentration of 1-heptene in the neighbourhood of the Pd species could be enhanced, thus favouring the consecutive hydrogenation of 1-heptene to heptane. Although the surface chemistry of GF-45 support is quite unlike that of alumina, the similar dispersions and electronic states of palladium on $Pd_{(5\%)}/Al$ and $Pd_{(5\%)}/AC$ reinforce the idea that their different catalytic behaviours are related to the differences in the support porosities.

3.2.4. *Effect of the metal loading*

The price of a metal, its toxicity, easy handling and safety are a set of properties to take into account during the preparation of the catalysts. Industrially profitable processes with highly active, selective and cheaper catalysts are continually researched. As the cost of the catalyst is important, a reduction of the metal loading on the final catalyst was considered and its effect on activity and selectivity during the hydrogenation reaction was assessed.

In **Figure 6**, the total conversion and selectivity to 1-heptene for the palladium catalysts supported on alumina with a metal loading of 5 and 0.4 wt% of Pd are presented. The used precursor salt was $PdCl_2$, the catalysts were reduced at 573 K and the catalytic tests were carried out at 303 K.

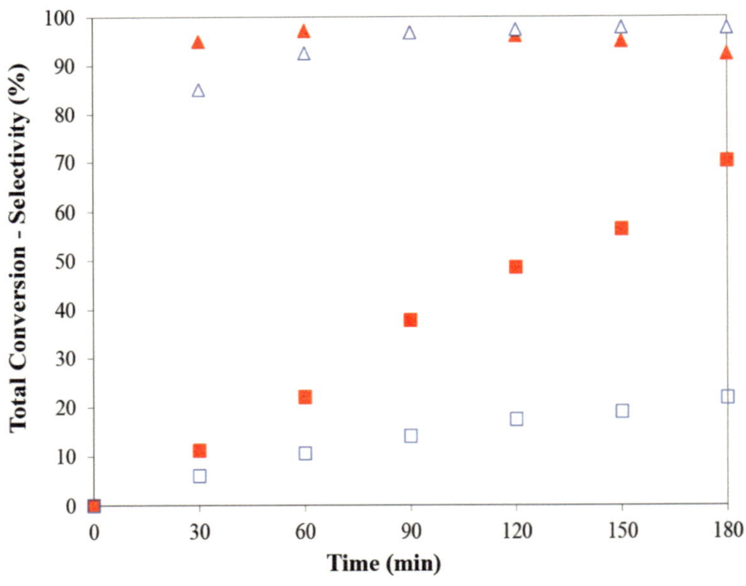

Figure 6. *Effect of metal loading*: Total 1-heptyne conversion (■, □) and selectivity to 1-heptene (▲,△) as a function of time for $Pd_{(5\%)}/Al$ (filled symbols) and $Pd_{(0.4\%)}/Al$ (opened symbols). Reduction and reaction temperatures: 573 and 303 K, respectively.

In the figure, it can be seen that decreasing the metal concentration of the catalyst, at identical operational conditions, decreases the total conversion (70–22% at 180 min). On the other hand, considering TOF values of both catalyst shown in **Table 2**, it can be concluded that the low-loaded $Pd_{(0.4\%)}/Al$ catalyst is 2.8 times more active than the $Pd_{(5\%)}/Al$. The selectivities to the

http://dx.doi.org/10.5772/64866

desired product (1-heptene) were 95% average without large changes when the metal loading is decreased. This fact is very important from an economic and industrial point of view.

It is well known that during hydrogenation reactions, metallic centres rich in electrons can cleave the bond in H_2 by means of the interaction of a filled ***d*** metal orbital with the empty sigma antibonding molecular orbital of H_2 [44]. The rupture of the hydrogen bond is more easily done on metals with a high amount of available electrons in the external ***d*** orbital, as it is the case of $Pd_{(0.4\%)}$/Al (with Pd°). This rupture should be less likely on metals with fewer ***d*** electrons, as in the case of $Pd_{(5\%)}$/Al (with $Pd^{\delta+}$ species). Therefore, the differences in activity between the reported catalysts could be partly attributed to differences in the electronic density of the external ***d*** orbital of each metal (electronic factor). So, the high activity of $Pd_{(0.4\%)}$/Al could be attributed to different factors: (a) palladium totally reduced Pd°, which favours the dissociative adsorption of hydrogen (electronic factor), (b) high dispersion of the low-loaded catalyst indicating that a high amount of small active sites are present on the surface (geometric factor) and (c) the absence of chlorine, a bulky and electronegativity element, which prevent the adsorption of the alkyne (steric factor).

3.2.5. *Effect of precursor salt*

The effect of precursor salt (palladium chloride and nitrate) on the total conversion and selectivity to 1-heptene during the 1-heptyne partial hydrogenation was studied using $Pd_{(0.4\%)}$/Al and $PdN_{(0.4\%)}$/Al catalysts reduced in a hydrogen flow at 573 K and tested at 303 K. The results of total conversion and selectivity to 1-heptene are shown in **Figure 7**.

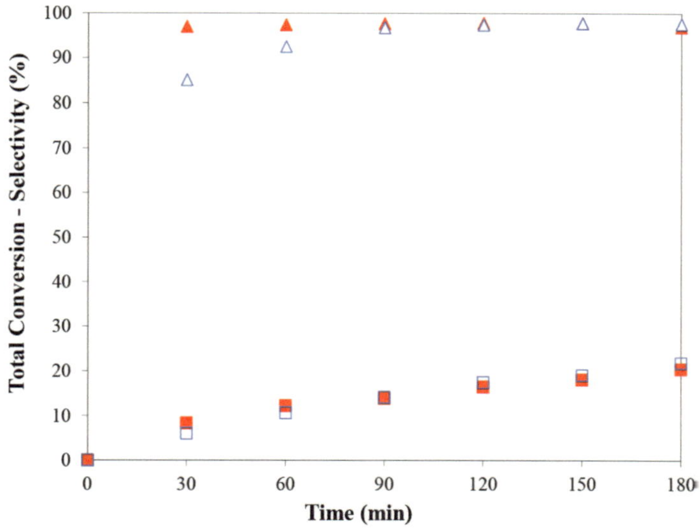

Figure 7. *Effect of precursor salt.* Total conversion of 1-heptyne (■, □) and selectivity to 1-heptene (▲,△) as a function of time for $PdN_{(0.4\%)}$/Al (filled symbols) and $Pd_{(0.4\%)}$/Al (opened symbols). Reduction and reaction temperatures: 573 and 303 K, respectively.

In **Figure 7**, it can be observed that both catalysts present very similar behaviour on 1-heptyne total conversions and on selectivity to 1-heptene. As shown in Section 3.1, the characterization techniques indicated the presence of reduced palladium after the pretreatment with hydrogen flow at 573 K, and also there is neither chlorine nor nitrogen species on the surface of both catalysts (absence of electronic and steric factor). As different dispersions were obtained with chemisorption analysis, differences in TOF (shown in **Table 2**) indicate that $PdN_{(0.4\%)}/Al$ is twice active than $Pd_{(0.4\%)}/Al$. So, geometric factors are responsible of the higher activity of $PdN_{(0.4\%)}/Al$.

3.2.6. Effect of the addition of a second metal: bimetallic catalyst

As the low-loaded palladium catalyst prepared with the nitrate precursor salt was more active than that prepared with chloride salt, the addition of Ni to the monometallic $PdN_{(0.4\%)}/Al$ catalyst was evaluated in order to improve 1-heptyne total conversion. In **Figure 8**, 1-heptyne total conversion and selectivity to 1-heptene as a function of time obtained during the hydrogenation test for the monometallic $Pd_{(0.4\%)}/Al$ and bimetallic Pd-Ni/Al catalysts are plotted. The bimetallic catalyst was reduced for 1 h at 673 K, and the temperature of the catalytic evaluations was 303 K.

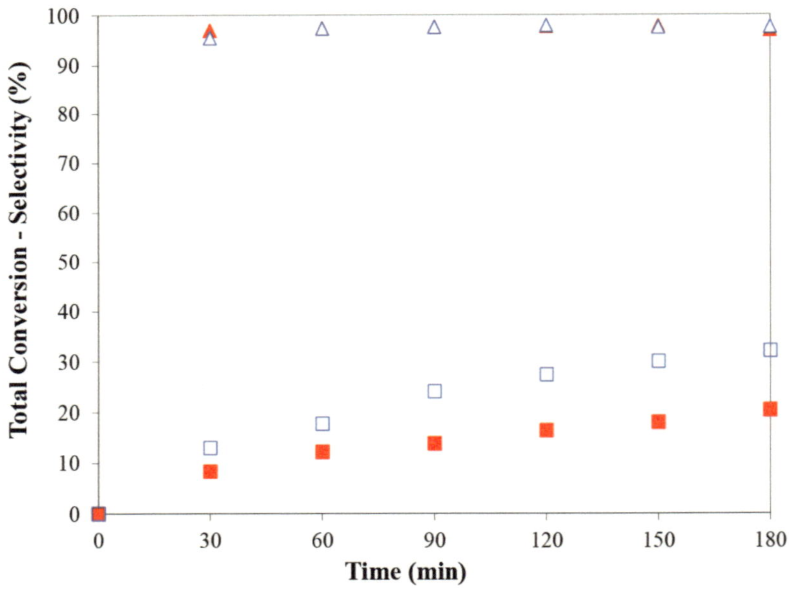

Figure 8. *Effect of the addition of Ni*: Total conversion of 1-heptyne (■, □) and selectivity to 1-heptene (▲,Δ) as a function of time for $PdN_{(0.4\%)}/Al$ (filled symbols) and Pd-Ni/Al (opened symbols). Reduction and reaction temperatures: 573 and 303 K, respectively.

According to **Figure 8**, the bimetallic and monometallic catalysts present similar and very high selectivities to the desired product (97% c.a.). Therefore, it can be said that the addition of Ni as a second metal to the alumina-supported palladium catalysts has no influence on the selectivity to 1-heptene. Therefore, the disappearance of the beta-phase palladium hydride could be the cause, at least in part, of the high selectivities found for mono- and bimetallic palladium catalysts. Besides, the nickel addition to the palladium monometallic catalyst improves 1-heptyne total conversion. TOF values indicate that Pd-Ni/Al is 1.5 times more active than $Pd_{(0.4\%)}$/Al. There is not a simple interpretation to explain the influence of the second metal on the Pd performance. Alkyne partial hydrogenation reactions are more or less sensitive to geometrical and electronic effects, the latter being the most important ones.

As observed by XPS, the nickel addition to the palladium catalyst promotes electron deficiency of Ni (Ni^{n+}) and Pd ($Pd^{\delta+}$), and generates electron-rich palladium species ($Pd^{\delta-}$). Since the bimetallic catalyst has a high amount of available electrons in the ***4d*** Pd orbital than monometallic $PdN_{(0.4\%)}$/Al, the hydrogen cleavage would be favoured on this catalyst. This would explain the higher activity of the bimetallic Pd-Ni catalyst. The modification of the electronic state of Pd could be responsible for the better catalytic behaviour, moreover considering the high electronic density of the triple bond. However, the influence of geometrical effects and/or mixed sites on the activity of the bimetallic catalyst cannot be discarded.

3.2.7. *Commercial Lindlar catalyst*

The Lindlar catalyst is very often used as a reference for the selective hydrogenation of alkynes, so in this work the performance of the commercial catalyst at 303 K was obtained. **Figure 9** presents the total conversion of 1-heptyne and selectivity to 1-heptene versus reaction time for this catalyst.

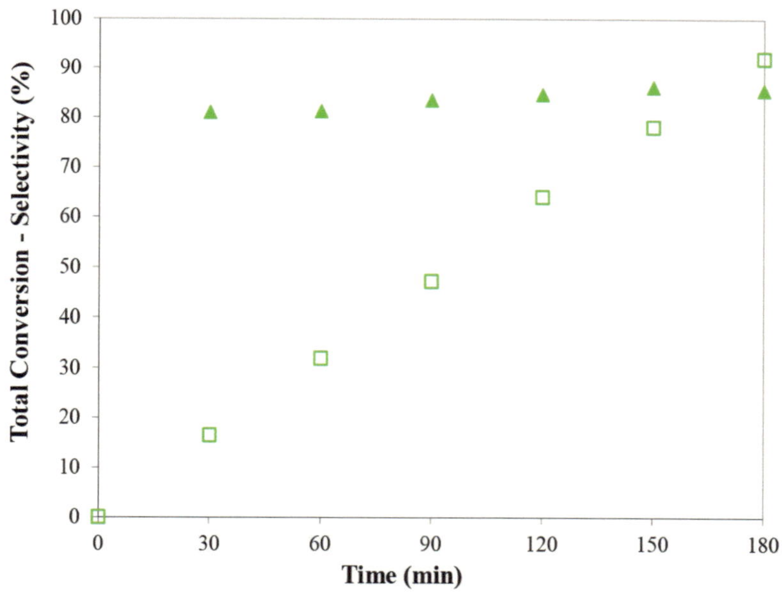

Figure 9. *Commercial Lindlar catalyst*: Total conversion to 1-heptyne (□) and selectivity to 1-heptene (▲) as a function of time for Lindlar catalyst. Reaction temperature: 303 K.

In **Figure 9**, it can be observed that the Lindlar catalyst allows obtaining high values of 1-heptyne total conversion with high selectivity to 1-heptene, between 82% and 86%. Comparing the results obtained with the prepared high-loaded catalysts (5 wt% of Pd, similar to that of Lindlar), a slightly better yield to 1-heptene is obtained with the $Pd_{(5\%)}/Al$ catalyst than that obtained with Lindlar catalyst. Besides, it must be remarked that the prepared catalyst has the advantage of being a pelletized material. On the other hand, the lowest yield to 1-heptene is obtained with $Pd_{(5\%)}/AC$. Differences in total conversion can be associated to electronic effects, as the surface palladium species on the Lindlar catalyst are more electron-deficient than those present on $Pd_{(5\%)}/Al$ or $Pd_{(5\%)}/AC$.

3.3. Final considerations

In **Figure 10**, the initial 1-heptyne hydrogenation rates (calculated per gram of palladium) and the selectivities to the desired product (1-heptene) at 180 min for all of the synthesized catalysts against Lindlar catalyst results are compared.

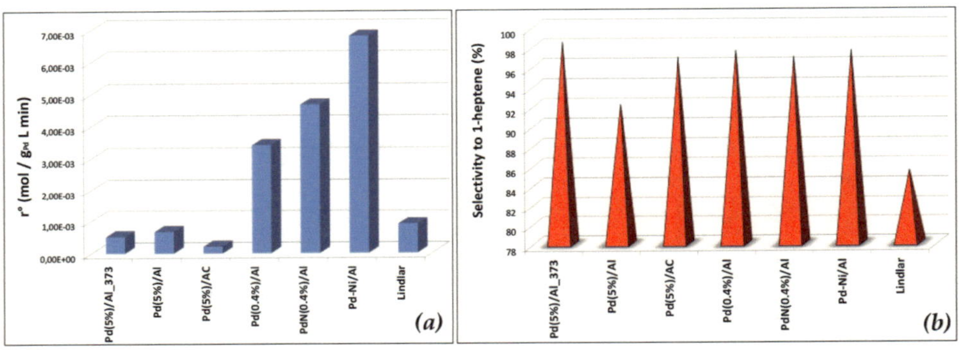

Figure 10. Comparison of initial 1-heptyne hydrogenation rates (a) and selectivities to 1-heptene at 180 min (b) for the catalytic systems evaluated in the present chapter.

According to the results obtained and shown in **Figure 10**, the tested catalysts showed the following order of activity:

$$Pd\text{-}Ni/Al > PdN_{(0.4\%)}/Al > Pd_{(0.4\%)}/Al >> Lindlar > Pd_{(5\%)}/Al > Pd_{(5\%)}/Al\_{373} > Pd_{(5\%)}/AC$$

With the following order of selectivities:

$$Pd_{(5\%)}/Al\_{373} \approx Pd\text{-}Ni/Al = Pd_{(0.4\%)}/Al \approx PdN_{(0.4\%)}/Al = Pd_{(5\%)}/AC > Pd_{(5\%)}/Al >> Lindlar$$

From **Figure 10**, it can be concluded that the prepared pelletized catalysts are more active and selective than the Lindlar catalyst.

Finally, pelletized catalysts have additional advantages against the Lindlar catalyst: the easy and cheap way of recovering the catalytic systems from the remaining solution at the end of the industrial process.

4. Conclusions

A series of palladium mono- and bimetallic catalysts were synthetized, characterized and tested during the partial hydrogenation of a terminal alkyne. All the prepared catalysts are active and highly selective for the partial hydrogenation of 1-heptyne under mild reaction conditions. Various factors were studied, and the behaviour of the catalysts on activity and selectivity during the hydrogenation of 1-heptyne was evaluated. The studied factors and the principal conclusions obtained are condensed below:

1. *Variation of reaction and reduction temperatures*: $Pd_{(5\%)}/Al$ catalyst was reduced at two temperatures (373 and 573 K) before its catalytic evaluation and was tested at 280 and 303 K. The optimum reaction temperature to obtain high activity with high yield of 1-heptene (desired product) is 303 K. On the other hand, high activity values are accessed when the reduction temperature is 573 K.

2. *Different support*: Alumina and activated carbon were used as supports of the palladium catalysts. The results show that higher activities are obtained when alumina is used as support. The differences in the support porosities could be the cause of the differences between $Pd_{(5\%)}/Al$ and $Pd_{(5\%)}/AC$ catalytic behaviour. Nevertheless, more experimental work is necessary to reach a better understanding about the effect of the support on the catalytic behaviour of the catalysts studied.

3. *Metal loading*: The decrease of the metal loading leads to cheaper catalysts, particles more dispersed on the catalyst surface, metal species easier to reduce, with better activities, maintaining high selectivity values.

4. *Precursor salt*: In this particular case of 1-heptyne partial hydrogenation, between nitrate and chlorine precursors, a better activity is obtained with $PdN_{(0.4\%)}/Al$. It can be remarked that this catalyst is the most active system. However, it should be noted that the chlorine species are necessary in order to increase the selectivity (steric and electronic factors due to chloride species).

5. *Bimetallic catalyst*: The addition of a second metal, as nickel to palladium monometallic catalyst, modifies the electronic state of palladium leaving it with electron-deficient and electron-rich species, even though the influence of geometrical effects and/or mixed sites cannot be neglected. Nickel proved to be a good promoter for the palladium monometallic catalyst as it caused an increment in the activity without modifying the high selectivity to 1-heptene characteristic of Pd monometallic catalysts. Besides, this bimetallic catalyst presents the advantages of low Pd loading and low cost of the nickel salt which leads to a cheaper and highly active and selective catalyst.

6. *Comparison with commercial Lindlar catalyst*: All catalysts synthesized during the present study show better selectivity values to the desired product (1-heptene) than Lindlar catalyst. This makes them valuable catalytic materials for the alkyne hydrogenation reactions and as an alternative option to the expensive commercial catalyst. In addition, Lindlar presents the disadvantage of being a non-pelletized material and must be operated

under slurry conditions; therefore, the reagent solution must be purified after the reaction through an uneconomic method to recover the catalyst.

Finally, considering XPS results, this chapter allows concluding that the presence of electron-deficient species on the catalyst surface of the catalysts with high metal loading is an important factor and influences on the activity and selectivity to the desired product. Nevertheless, the influence of geometrical effects and/or mixed sites in the catalysts as well as metal-metal and metal-support interactions cannot be neglected.

Acknowledgements

The experimental assistance of F. Coloma-Pascual is greatly acknowledged. The financial assistance of CAI+D (UNL), CONICET (PIP 410) and ANPCyT are also acknowledged.

Author details

Cecilia Lederhos[1*], Carolina Betti[1], Domingo Liprandi[2], Edgardo Cagnola[2] and Mónica Quiroga[1,2]

*Address all correspondence to: clederhos@fiq.unl.edu.ar

1 Institute of Catalysis and Petrochemistry Research INCAPE (UNL-CONICET), Santa Fe, Argentina

2 Inorganic Chemistry, Department of Chemistry, Faculty of Chemical Engineering, National University of Littoral (UNL), Santa Fe, Argentina

References

[1] L'Argentière PC, Cagnola EA, Liprandi DA, Román- Martínez MC, Salinas-Martínez de Lecea C. Carbon-supported Pd complex as catalyst for cyclohexene hydrogenation. Appl. Catal. A: Gen. 1998; 172: 41–48.

[2] Ulan JG, Wilhelm FM. Mechanism of 2-hexyne hydrogenation on heterogeneous palladium. J. Mol. Catal. 1989; 54 (2): 243–261.

[3] Teschner D, Vass E, Hävecker M, Zafeiratos S, Schnörch P, Sauer H, Knop-Gericke A, Schlögl R, Chamam M, Wootsch A, Canning AS, Gamman JJ, Jackson D, McGregor J, Gladden LF. Alkyne hydrogenation over Pd catalysts: A new paradigm. J. Mol. Catal. 2006; 242: 26–37. DOI:10.1016/j.jcat.2006.05.030.

[4] Kaĉer P, Kuzma M, Ćervenỳ L. The molecular structure effects in hydrogenation of cycloalkylsubstituted alkynes and alkenes on Pt and Pd catalysts. App Catal. A: Gen. 2004; 259: 179–183. DOI:10.1016/j.apcata.2003.09.025.

[5] Nishimura S. Handbook of Heterogeneous Catalytic Hydrogenation for Organic Synthesis. Canada: John Wiley & Sons, Inc.; 2001, pp. 148–169. ISBN 0-47139698-2.

[6] Chen B, Dingerdissen U, Krauter JGE, Lansink Rotgerink HGJ, Möbus K, Ostgard DJ, Panster P, Riermeir TH, Seebald S, Tacke T, Trauthwein H. New developments in hydrogenation catalysis particularly in synthesis of fine and intermediate chemicals. App. Catal. A: Gen. 2005; 280: 17–46. DOI: 10.1016/j.apcata.2004.08.025.

[7] Lindlar H, Dubuis R, Jones FN, McKusick BC. Palladium catalyst for partial reduction of acetylenes. Org. Synth. 1966; 46: 89–92.

[8] Anderson JA, Mellor JL, Wells RPK. Pd catalysed hexyne hydrogenation modified by Bi and by Pb. J. Catal. 2009; 261: 208–216. DOI: 10.1016/j.jcat.2008.11.023.

[9] Evangelisti C, Panziera N, D'Alessio A, Bertinetti L, Botavina M, Vitulli G. New monodispersed palladium nanoparticles stabilized poly-(N-vinyl-2-pyrrolidone): Preparation, structural study and catalytic properties. J. Catal. 2010; 272: 246–252. DOI: 10.1016/j.jcat.2010.04.006.

[10] Radovic LR, Rodríguez-Reinoso F. Carbon Materials in Catalysis. In Chemistry and Physics of Carbon. Thrower PA, 14 Ed., Marcel Dekker: New York, NY; 1996, Vol. 25, p. 243. ISBN 9780824793685

[11] Salinas-Martínez de Lecea C, Linares-Solano A, Díaz-Auñón JA, L'Argentière PC. Improvement of activity and sulphur resistance of Pd complex catalysts using carbon-coated γ-Al_2O_3 and activated carbon supports. Carbon 2000; 38: 157–160. DOI: 10.1016/S0008-6223(99)00230-4.

[12] Lederhos CR, Badano JM, Quiroga ME, Coloma-Pascual F, L'Argentière PC. Influence of Ni addition to a low-loaded palladium catalyst on the selective hydrogenation of 1-heptyne. Quim. Nova. 2010; 33(4): 816–820.

[13] Quiroga M, Liprandi D, Cagnola E, L'Argentiere P. 1-heptyne semihydrogenation catalyzed by palladium or rhodium complexes. Influence of: metal atom, ligands and the homo/heterogeneous condition. Appl. Catal. A: Gen. 2007; 326: 121–129. DOI: 10.1016/j.apcata.2007.03.035.

[14] Lennon D, Marshall R, Webb G, Jackson SD. The effect of hydrogen concentration on propyne hydrogenation over a carbon supported palladium catalyst studied under continuous flow conditions. Stud. Surf. Sci. Catal. 2000; 130: 245–250.

[15] Hamilton CA, Jackson SD, Kelly GJ, Spence R, de Bruin D. Competitive reactions in alkyne hydrogenation. App. Catal. A: Gen. 2002; 237: 201–209. DOI: 10.1016/S0926-860X(02)00332-0.

[16] Díaz-Auñón JA, Román-Martínez MC, L'Argentière PC, Salinas-Martínez de Lecea C. Catalytic lifetime of amine metal complexes supported on carbons in cyclohexene hydrogenation. Stud. Surf. Sci. Catal. 2000; 130C: 2075–2080. DOI: 10.1016/S1381-1169 (99)00355-6.

[17] Díaz-Auñón JA, Román-Martínez MC, Salinas-Martínez de Lecea C, L'Argentière PC, Cargnola EA, Liprandi D, Quiroga ME. [$PdCl_2(NH_2(CH_2)_{12})CH_3)_2$] supported on an active carbón: effect of the carbon properties on the catalytic activity of cyclohexene hydrogenation. J. Mol. Catal. A: Chem. 2000; 153: 243–256. DOI: 10.1016/S1381-1169(99) 00355-6.

[18] Wang C-B, Lin H-K, Ho C-M. Effects of the addition of titania on the thermal characterization of alumina-supported palladium. J. Mol. Catal. A: Chem. 2002; 180: 285–291. DOI: 10.1016/S1381-1169(01)00450-2.

[19] Lederhos C, L'Argentière PC, Fígoli NS. 1-Heptyne selective hydrogenation over Pd supported catalysts. Ind. Eng. Chem. Res. 2005; 44: 1752–1756.

[20] Liprandi DA, Quiroga ME, Cagnola EA, L'Argentière PC. A new more sulfur-resistant rhodium complex as an alternative to the traditional Wilkinson's catalyst. Ind. Eng. Chem. Res. 2002; 41: 4906–4910.

[21] Li L, Zhu ZH, Yan ZF, Lu GQ, Rintoul L. Catalytic ammonia decomposition over Ru/ carbon catalysts: The importance of the structure of carbon support. Appl. Catal. A: Gen. 2007; 320: 166–172. DOI: 10.1016/j.apcata.2007.01.029.

[22] Gaspar AB, Dieguez LC. Dispersion stability and methylcyclopentane hydrogenolysis in Pd/Al_2O_3 catalysts. Appl. Catal. A: Gen. 2001; 201: 241–251. DOI: 10.1016/S0926-860X (00)00442-7.

[23] L'Argentière PC, Cañón MM, Fígoli NS, Ferrón. AES and XPS studies of the influence of Ni addition on Pd/Al_2O_3 catalytic activity and sulfur resistance. J. Appl. Surf. Sci. 1993; 68: 41–47. DOI: 10.1016/0169-4332(93)90214-V.

[24] NIST X-ray Photoelectron Spectroscopy Database NIST Standard Reference Database 20, Version 3.5 (Web Version), National Institute of Standards and Technology, USA; 2007.

[25] Pârvulescu VI, Filoti G, Pârvulescu V, Grecu N, Angelescu E, Nicolescu IV. Styrene hydrogenation on supported Pd, Fe and Pd-Fe/γ-Al_2O_3 catalysts. J. Mol. Catal. 1994; 89: 267–282. DOI: 10.1016/0304-5102(93)E0312-5.

[26] Abu Bakar NHH, Bettahar MM, Abu Bakar M, Monteverdi S, Ismail J. Low temperature activation of Pt/Ni supported MCM-41 catalysts for hydrogenation of benzene. J. Mol. Catal. A: Chem. 2010; 333: 11–19. DOI: 10.1016/j.molcata.2010.10.007.

[27] Álvarez-Rodríguez J, Rodríguez-Ramos I, Guerrero-Ruiz A, Arcoya. A Selective hydrogenation of citral over Pt/KL type catalysts doped with Sr, La, Nd and Sm. Appl. Catal. A: Gen. 2011; 401: 56–64. DOI: 10.1016/j.apcata.2011.04.045.

[28] Hillebrecht FU, Fuggle JC, Bennett PA, Zolnierek Z. Electronic structure of Ni and Pd alloys. II. X-ray photoelectron core-level spectra. Phys. Rev. B 1983; 27: 2179–2193.

[29] Kennedy DR, Webb G, Jackson SD, Lennon D. Propyne hydrogenation over alumina-supported palladium and platinum catalysts. Appl. Catal. A: Gen. 2004; 259: 109–120. DOI: 10.1016/j.apcata.2003.09.018.

[30] Wang C-B, Lin H-K, Ho C-M. Effects of the addition of titania on the thermal characterization of alumina – supported palladium. J. Molec. Catal. A: Chem. 2002; 180: 285–291. DOI: 10.1016/S1381-1169(01)00450-2.

[31] Mukainakano Y, Yoshida K, Okumura K, Kunimori K, Tomishige K. Catalytic performance and QXAFS analysis of Ni catalysts modified with Pd for oxidative steam reforming of methane. Catal. Today 2008; 132: 101–108. DOI: 10.1016/j.cattod.2007.12.031.

[32] Li G, Hu L, Hill JM. Comparison of reducibility and stability of alumina-supported Ni catalysts prepared by impregnation and co-precipitation. Appl. Catal. A: Gen. 2006; 301: 16–24. DOI: 10.1016/j.apcata.2005.11.013.

[33] Hou Z, Yokota O, Tanaka T, Yashima T. Characterization of Ca-promoted Ni/α-Al_2O_3 catalyst for CH_4 reforming with CO_2. Appl. Catal. A: Gen. 2003; 253: 381–387. DOI: 10.1016/S0926-860X(03)00543-X.

[34] Abelló S, Verboekend D, Bridier B, Pérez-Ramírez J. Activated takovite catalysts for partial hydrogenation of ethyne, propyne, and propadiene. J. Catal. 2008; 259: 85–95. DOI: 10.1016/j.jcat.2008.07.012.

[35] Hoffer BW, van Langeveld AD, Janssens JP, Bonné RLC, Lok CM, Moulijn JA. Stability of highly dispersed Ni/Al_2O_3 catalysts: effects of pretreatment. J. Catal. 2000; 192: 432–440. DOI: 10.1006/jcat.2000.2867.

[36] Dias JAC, Assaf JM. Autothermal reforming of methane over Ni/γ-Al_2O_3 promoted with Pd. The effect of the Pd source in activity, temperature profile of reactor and in ignition. Appl. Catal. A 2008; 334: 243–250. DOI: 10.1016/j.apcata.2007.10.012.

[37] Coq B, Figueras F. Bimetallic palladium catalysts: influence of the co-metal on the catalyst performance. J. Mol. Catal A: Chem. 2001; 173: 117–134. DOI: 10.1016/S1381-1169(01)00148-0.

[38] Mallat T, Baiker A. Selectivity enhancement in heterogeneous catalysis induced by reaction modifiers. Appl. Catal. A: Gen. 2000; 200: 3–22. DOI: 10.1016/S0926-860X(00)00645-1.

[39] Nijhuis TA, van Koten G, Moulijn JA. Optimized palladium catalyst systems for the selective liquid-phase hydrogenation of functionalized alkynes. Appl. Catal. A: Gen. 2003; 238: 259–271. DOI: 10.1016/S0926-860X(02)00372-1.

[40] Lederhos C, L'Argentière PC, Coloma-Pascual F, Fígoli NS. A study about the effect of the temperature of hydrogen treatment on the properties of Ru/Al_2O_3 and Ru/C and

their catalytic behavior during 1-heptyne semi-hydrogenation. Catal. Lett. 2006; 110: 23–28. DOI: 10.1007/s10562-006-0085-z.

[41] L'Argentiere PC, Cagnola EA, Quiroga ME, Liprandi DA. A palladium tetra-coordinated complex as catalyst in the selective hydrogenation of 1-heptyne. Appl. Catal. A: Gen. 2002; 226: 253–263. DOI: 10.1016/S0926-860X(01)00911-5.

[42] Liprandi DA, Cagnola EA, Quiroga ME, L'Argentière PC. Influence of the reaction temperature on the 3-hexyne semi-hydrogenation catalyzed by a palladium(II) complex. Catal. Lett. 2009; 128: 423–433. DOI: 10.1007/s10562-008-9768-y.

[43] Mouaddib M, Perrichon V. Influence of the support in the alcohols synthesis. In Proceedings of the 9th International Congress on Catalysis. Phillips MJ, Ternant M, Eds., The Chemical Institute of Canada: Ottawa, Canada; 1988, Vol. 2, p. 521.

[44] Shriver DF, Atkins PW, Langford CH. Inorganic Chemistry, 3rd edn. WH Freeman and Co., New York, NY; 1994, p. 258.

Chapter 3

Asymmetric Transfer Hydrogenation of C=O and C=N Bonds Catalyzed by [Ru(η^6 arene)(diamine)] Complexes: A Multilateral Study

Ondřej Matuška, Martin Kindl and Petr Kačer

Additional information is available at the end of the chapter

http://dx.doi.org/10.5772/110623

Abstract

In these days, asymmetric transfer hydrogenation (ATH) is a very attractive method for synthesis of enantioenriched chiral compounds, especially fine chemicals such as drugs or agrochemicals. In this review, several topics related to the asymmetric transfer hydrogenation of ketones and cyclic or acyclic imines are discussed. Initially, the reaction mechanism of the ATH of ketones and imines, mainly 3,4-dihydroisoquinoline derivates, is examined. Next, typical reaction conditions, structural effects of the catalyst and a substrate, and analytical methods used for ATH monitoring and practical applications of the ATH in the chemical industry are described.

Keywords: ruthenium, asymmetric hydrogenation, imines, dihydroisoquinolines, catalysis

1. Introduction

Synthesis of biologically active chiral compounds such as drugs or pesticides goes hand in hand with the necessity of a high optical purity. Chiral molecules play an important role in many basic functions of living organisms, e.g., molecular recognition in biological systems is achieved by chiral environment (e.g., receptors), and thus the response induced by one enantiomer can be completely different from the other enantiomer of the same compound.

Commonly used synthetic approaches are not stereoselective and lead to a racemic mixture of the products from which the desired enantiomer is to be separated using specific separation methods, such as racemic cleavage. This method is relatively broadly applied in the chemical industry but it can hardly be considered as ideal. Frequently, the yield of this process is limited only to 50% since the second (undesired) enantiomer is not recyclable and ends up as a waste. This fact represents a significant limitation, especially, during the synthesis of fine chemicals, e.g., drugs, where every loss can have a significant impact on profitability of the production process. These economical aspects are one of the most important reasons why the methods of enantioselective (asymmetric) synthesis are still in the forefront of modern synthetic chemistry.

From a historical point of view, the first described enantioselective reaction was asymmetric hydrogenation (AH). Mainly homogenous catalysts, represented by coordinated compounds with optically pure ligands, carrying asymmetric information, have been used in this type of reaction. These catalysts can be divided into several subgroups, e.g., according to the ligand structure, function groups, central atom, or mechanistic aspects. Nevertheless, the hydrogen source plays most prominent role. Meanwhile the classical asymmetric hydrogenation used gaseous hydrogen, while asymmetric transfer hydrogenation (ATH) focused on utilizing substances contained in the reaction mixture, such as propane-2-ol or azeotropic mixture of formic acid and triethylamine. The absence of gaseous hydrogen in the case of ATH enabled to skip the requirement of pressure reactors, which lowered the overall cost of the process and minimized the explosion hazard.

First, homogeneous catalysts to be applied in ATH of prochiral ketone and imine compounds were introduced by the group of professor Noyori between 1995 and 1996 [1, 2] (**Figure 1**). The catalysts contained ruthenium(II) as the central atom, enantioenriched chiral diamine ligand, such as *N* (2-amino-1,2-diphenylethyl)-4-toluenesulfonylamide, in short TsDPEN and η^6 aromatic ligand (e.g., benzene, *p*-cymene or mesitylene). The main advantage of these complexes lied in their high modularity. The structure of these complexes can be relatively easily modified in order to enhance their catalytic properties to better fit the hydrogenated substrate.

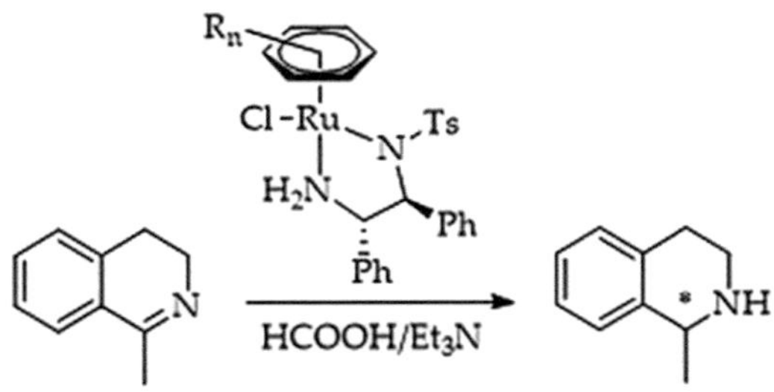

Figure 1. A scheme of asymmetric transfer hydrogenation of 1-methyl-3,4-dihydroisoquinoline.

This work is predominantly focused on the ATH of compounds with C=N and C=O double bonds in its structure using Noyori's ruthenium catalytic complexes. Additionally, several related topics are discussed such as the mechanism of asymmetric transfer hydrogenation of imines and ketones, the modification of the catalyst structure, the influence of the reaction conditions, and its application to the chemical industry and the synthesis of pharmaceutical substances.

2. Mechanism of ATH catalyzed by [RuCl(η^6-arene)TsDPEN]

The first study regarding mechanistic aspects of the reaction was performed by Noyori et al. in 2001 [3]. This work was focused on the ATH of ketones. By means of molecular modeling methods, it was demonstrated that the reaction proceeds *via* a six-membered cyclic transition state. For the purpose of these calculations, formaldehyde as the substrate and simplified structure of the catalytic complex, i.e., Ru(H)(η^6-benzene)(ethylenediamine) were used (**Figure 2**). According to their findings, the substrate primarily formed C=O···H-N intermediate with the ruthenium complex and then evolved into six-membered cyclic transition state. Simultaneously, a proton transfer from the NH group took place to the carbonyl oxygen and hydride transfer onto the C=O carbon atom.

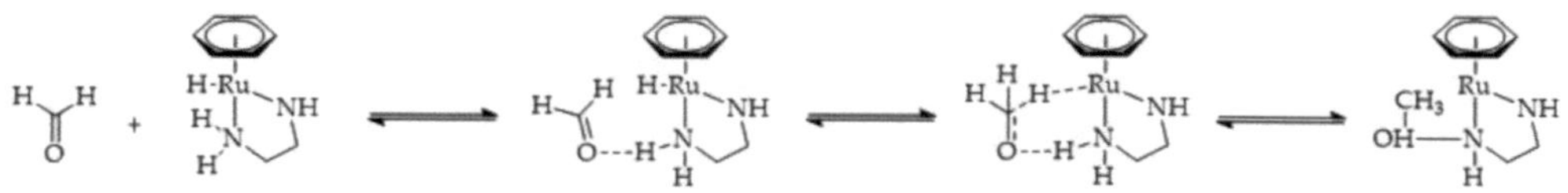

Figure 2. The original mechanistic concept published by Noyori et al. [3].

However, in 2013 Dub and Ikariya extended the previous Noyori's study and reported the detailed density functional theory (DFT) study [4], which showed that the hydrogenation of ketones occurred *via* a two-step pathway including the solvent as an important part of the hydrogen transfer mechanism.

From this perspective, the catalyst containing a ligand with (*S*, *S*) configuration provided (*S*) product. Nevertheless, Wills et al. [5] pointed out that this concept was not applicable on the ATH of imines, since the reaction was catalyzed by the catalyst with (*S*, *S*) configuration of the ligand providing (*R*)-product. An interesting fact was reported by Åberg et al. [6], who, in their study, proved that ATH of imines could be performed only in the presence of acids (for instance, an azeotropic mixture of formic acid and triethylamine), whereas ketones were possible to be hydrogenated also in alkaline medium. Imine, contrary to the ketone, enters the reaction in a protonated form and thus to perform the reaction itself, an acid addition is necessary. Consequently, Wills et al. [5] proposed several transition state structures, the so-called ionic mechanism, and presented their own experimental data.

2.1. Asymmetric transfer hydrogenation of cyclic imines (dihydroisoquinolines)

Theoretical work accompanied with the computational study with the application of DFT to investigate the ionic mechanism concept in ATH of 1-methyl-3,4-dihydroisoquinoline resulted in a series of interesting proposals. All calculations, having employed [Ru(Cl)(η^6 *p* cymene)(*S*, *S*)-TsDPEN] as the catalyst, differentiated from the previous computational studies by the application of asymmetric η^6-*p*-cymene ligand instead of typically used symmetric η^6 benzene or η^6 1,2,3,4,5,6-hexamethylbenzene. The main outcome lied in the calculated structures of transition states, their energy minima followed by the construction of two energy diagrams where the most probable structures have been studied. Furthermore, the resulting high enantioselectivity of the ATH of cyclic imines was explained.

It was assumed that a protonated substrate is attached to the molecule of the catalyst by the hydrogen bond between the hydrogen of the protonated substrate and the oxygen of the sulfonyl amide group of the catalyst and by that transition state is stabilized. Nonbinding interactions between π electrons of aromatic ring of the substrate and hydrogen atoms of *p*-cymene (i.e., CH/π interactions [7]), which also stabilize the favorable transition state, play a crucial role in the determination of enantiomeric excess (*ee*). Depending on energetic differences between "favorable" and "not favorable" transition states, geometry of the transition state determines *ee* of the product at the end of the reaction.

Besides, corresponding calculations allowed to suggest the pathway toward ATH of 1 methyl-3,4-dihydroisoquinoline using formic acid/triethylamine azeotrope as the source of hydrogen (**Figure 3**).

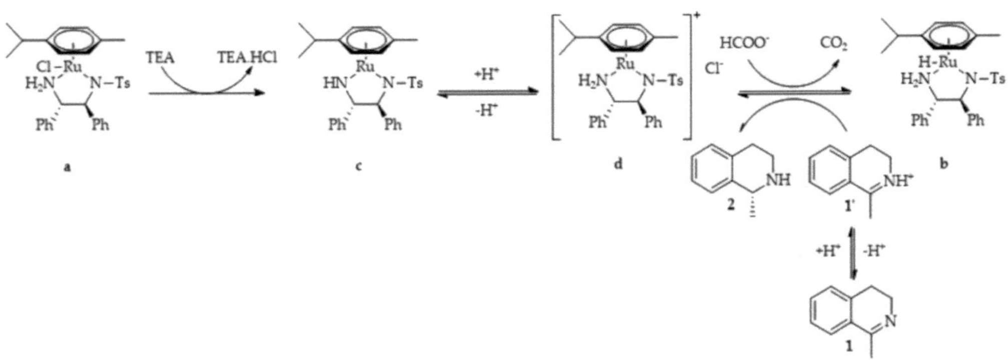

Figure 3. Suggested scheme of the catalytic cycle in ATH of 1-methyl-3,4 dihydroisoquinoline.

The proposed cycle starts with the transformation of the [Ru(Cl)(η^6 *p* cymene)(*S,S)* TsDPEN] complex (**a**) onto ruthenium-hydride (**b**), where this action is accompanied by the release of HCl, further neutralized by the present base (triethylamine), leading to the 16 ecomplex (**c**). In the next step, the complex (**c**) receives a proton from formic acid to form the complex (**d**), followed by the complex (**d**) turning in the presence of formate ion into ruthenium-hydride (**b**) and releasing CO_2. Finally, ruthenium-hydride participates in the asymmetric transfer hydrogenation process of the protonated substrate (**1'**). With the transfer of hydrogen from the complex (**b**) to the substrate (**1'**) the catalytic complex turns back to the intermediate (**d**). By this, the whole cycle is closed.

2.2. Asymmetric transfer hydrogenation of acyclic imines (*N*-benzyl-1-phenylethan-1-imin)

Asymmetric transfer hydrogenation of acyclic imines proceeds rather differently than in the case of cyclic imines such as substituted 3,4-dihydroisoquinolines. ATH of *N* benzyl-1-phenylethan-1-imin using Noyori's based catalyst with (*S, S*) ligand configuration lead to the excess of the product with (*S*)-configuration, which is in conflict with the results obtained in ATH of cyclic imines. The application of computational methods can help to find out [8] specific differences between both reaction types. It has been pointed out that in opposite to the rigid structure of cyclic imines, acyclic imines are very flexible in conformation and can undergo isomerization between *E* and *Z* isomer. Both of these isomers are capable of an interaction with the catalysts in a different way, ergo this fact has an extended impact on the enantioselectivity of the whole reaction, since both of the isomers are hydrogenated over different transition states. Therefore, in the case of the ATH of cyclic amines as well as in this case the transition state is stabilized by the hydrogen bond between the oxygen group of the catalyst and hydrogen of the protonated acyclic substrate. However, ATH of acyclic imines proceeds *via* the transition state with a different geometry producing the products with opposite geometry than in the case of cyclic imines such as 3,4-dihydroisoquinolines. Among other issues, this work stated that major *E* isomer is hydrogenated with a high selectivity to the corresponding (*S*)-isomer. Meanwhile, *Z* isomer is hydrogenated with much lower selectivity that causes decreasing of an overall enantioselectivity of the reaction.

2.3. Influence of the base on the asymmetric transfer hydrogenation of imines

To clarify the role of the base in asymmetric transfer hydrogenation of cyclic imines using Noyori's Ru(Cl)(η^6-*p*-cymene)(*S,S*)-TsDPEN], catalyst is a rather complex problem [9]. The series of aliphatic (secondary, tertiary) or aromatic amines were employed into the formic acid/base system used for ATH. This work is divided into two parts, kinetic studies using *in situ* monitoring by nuclear magnetic resonance (NMR) spectroscopy and spectroscopic studies, including NMR spectroscopy, Fourier transform ion cyclotron resonance mass spectroscopy (FT-ICR MS), and vibrational circular dichroism (VCD) together with infrared (IR) spectroscopy. At first, during kinetic study of ATH of (*R*) 1,4 dimethyl-3,4-dihydroisoquinoline, substantially different reaction rates with various bases were observed. The highest reaction rates were measured after using tertiary bases, such as triethylamine (TEA), whose rate was over four times higher than *N, N* diisopropyl(ethyl)amine (DIPEA). Nevertheless, the secondary bases such as morpholine, piperidine, or pyrrolidine performed with even a lower reaction rate. Surprisingly, the reaction rate was rather high for aromatic amine, pyrrole and certainly low for pyridine. Both the reaction rate and asymmetric tendencies were significantly affected. In the case of diastereomeric excess (*de*), the lowest value was observed for DIPEA, followed by TEA. The secondary amines performed with a higher *de* over 60% and the highest value was obtained for pyrrolidine as one of the smallest molecules in the mixture. Also, *de* values for aromatic amines were in the range of 65–75%. These results suggested that differences in diastereoselectivity were triggered by the steric demands of the bases applied. A very similar procedure was performed using 1-methyl-3,4-dihydroisoquinoline as the substrate. As in the previous case, significant differences between additions of tertiary, secondary and aromatic bases were observed in terms of the reaction rate. Interestingly enough, a slightly higher reaction rate was observed using DIPEA, than TEA. However, the enantioselectivity was not significantly changed with the usage of various bases.

The spectroscopic methods were applied to understand the influence of the base in depth. For the NMR examinations, the mixture of [Ru(Cl)(η^6-*p*-cymene)(*S,S*)-TsDPEN] catalyst together with pure formic acid or the mixture of formic acid and TEA with different molar ratio were used. Employing pure formic acid was accompanied by the formation of the ruthenium-hydride species, which solely decomposed under acidic conditions. Furthermore, TEA addition has positive effect regarding the formation of catalytic hydride (**Figure 3**, complex **b**), positive effect was observed to the ratio of TEA/formic acid 5:1, and above this value no significant improvement in the case of hydride concentration was observed. Also, using ^{1}H-NMR spectrum, three intermediates were found in the mixture: (1) ruthenium-hydride species, (2) ruthenium formate complex [Ru($HCOO^-$)(η^6 *p* cymene)(*S,S*)-TsDPEN and (3) a species, which is assumed to be the second diastereomer of the ruthenium-hydride. After NMR experiments, ESI$^+$ FT-ICR MS to observe active ruthenium-hydride species using TEA/formic acid in molar ratio 5:2 was applied. The MS spectrum contained three clusters, the first one, with *m/z* 595.1500 which belonged to the 16 e^- complex (**Figure 3**, complex **c**), the second one, was determined as the cluster of active ruthenium-hydride associated with TEA. The last one, was assigned to the associate of a precatalyst (**Figure 3**, complex **a**) with TEA. From the results obtained, authors suggested that the binding of the base could be realized by three different ways: (1) the base is coordinated to the central Ru atom, (2) the base forms N···H-N bond with chiral ligand of the catalyst, or (3) the protonated base forms a N^+-H···O hydrogen bond with a sulfonyl group of the chiral ligand (**Figure 4**). To exactly determine which is the right way, the application of the VDC analytical method combined with IR spectroscopy concluded that the protonated base is connected with the catalytic complex *via* N^+-H···O hydrogen bond with a sulfonyl group of TsDPEN ligand. The consequence of these findings can be that the base selection has a huge impact on the reaction rate and enantioselectivity of the ATH of cyclic imines.

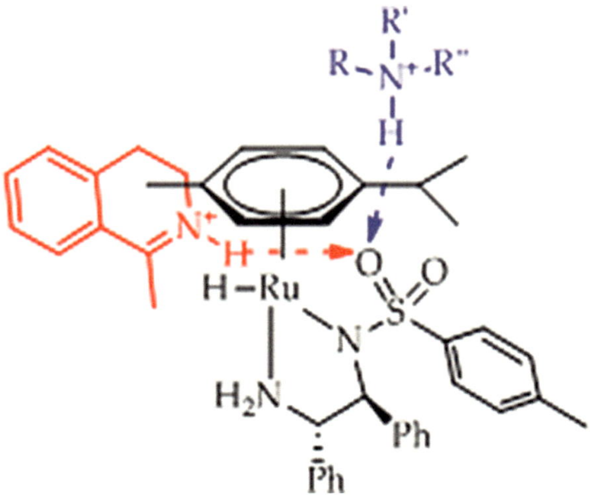

Figure 4. A scheme of binding of both substrate (red) and base (blue) to the active site of ruthenium-hydride species.

http://dx.doi.org/10.5772/110623

3. Influence of the reaction conditions

Similar to every chemical reaction, it is important to carry out asymmetric transfer hydrogenation under specific reaction conditions, particularly ratio between the reaction rate and the enantioselectivity optimized as much as possible. The first comprehensive parametric study [10] focused on the determination of the best possible reaction conditions for ATH of cyclic imines, especially 3,4-dihydroisoquinoline derivates. This study thus involved clarifying the best attainable value of temperature, reaction mixture concentration or substrate to catalyst molar ratio (S/C).

Primarily, the attention was paid to the concentration in the reaction mixture as one of the important parameters. Two experiments were performed with different S/C molar ratios, S/C = 100 and S/C = 200, and other crucial molar ratios set to: formic acid/TEA = 2.5, hydrogenation mixture/substrate = 8.8, and the temperature set to 30°C. The major difference was observed in reaction ratio. For S/C = 200 the reaction ratio was slightly lower and also the difference grew with an increasing concentration. Probably, this fact can be explained by a certain amount of the catalyst being blocked by the protonated base, resulting in a lower reaction rate than in the case of the experiment with S/C = 100. The final outcome of this first set of experiments is as follows: reaction rate increases with an increasing of the reaction mixture concentration. This can be probably explained that at higher concentrations, the reaction rate is no longer limited by the frequency of effective collisions between active ruthenium-hydride species and the protonated molecule of the substrate but by the total amount of ruthenium-hydride intermediate present in the reaction mixture. Furthermore, the differences of the reaction mixture, related to the mass of the catalyst with different S/C ratios, were examined afterward as the S/C ratio is parameter that generally affects the course of catalytic reactions. As a result, it was confirmed that modifying S/C ratio leads to different reaction rates, regardless of the influence of the catalyst amount.

The second basic important parameter for every chemical reaction is the temperature. For this reaction, the measurement of its effect on both, the reaction rate and the enantioselectivity was conducted in the range of 10–50°C (**Figure 5**). Increasing the reaction rate with an increasing temperature was expected and could be considered common. However, the decrease of enantioselectivity was observed with an increasing temperature. This fact can be explained by Yamakawa's supported theory that two transition states exist. One would lead to the preferred configuration of the product; the other would lead to the other configuration. The so-called unfavorable transition state will prevail at a higher temperature and the nonpreferred product would become more abundant.

The amount of the hydrogenation mixture is also important for the process of ATH. The mixture of formic acid and TEA provides hydrogen for the hydrogenation itself. The most commonly used molar ratio for these two is 2:5. The variation of this amount was expected to have a major influence on the course of the reaction and thus several hydrogenation mixture/substrate ratios were tested, where the concentrations were set to 7%, as well as S/C = 100, formic acid/TEA = 2.5, and the temperature to 30°C. Contrary to the original expectation, increasing the amount of the hydrogenation mixture leads to a decrease in the initial reaction rate (**Figure 5**). Therefore, two working hypothesis were considered to explain such behavior. The first, under a strong acidic condition, the catalyst's ligand became protonated, and subsequently deco-ordinated from the Ru atom, followed by the loss of catalytic activity and the second one, in a large excess of hydrogenation mixture, the protonated triethylamine is also in large excess over the protonated substrate and sterically hinders active site of the catalyst for substrate.

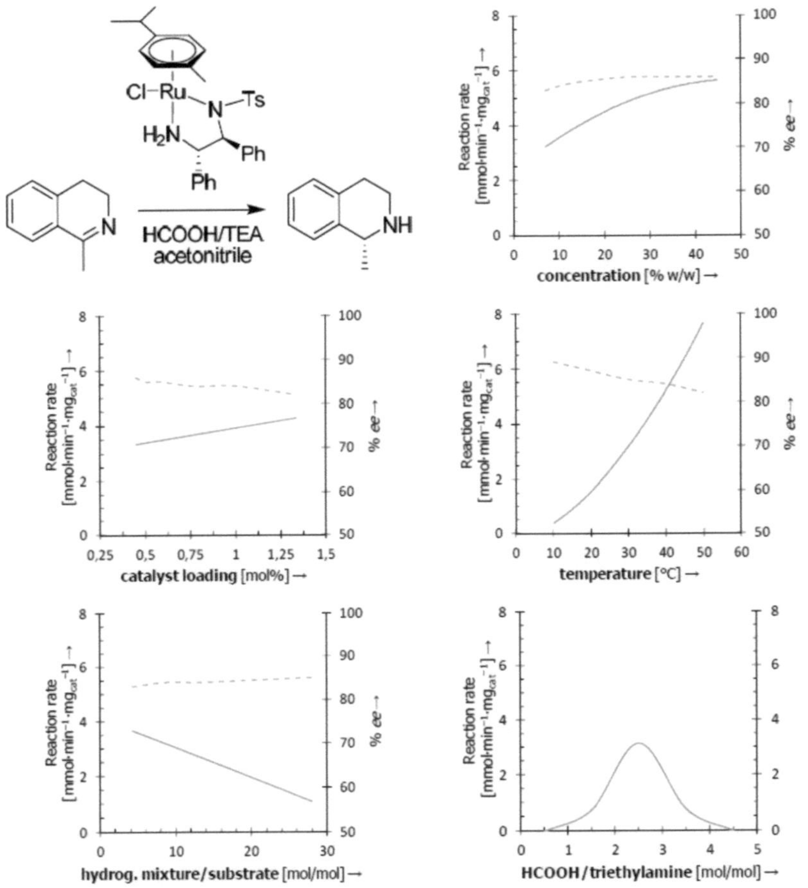

Figure 5. Graphical summary of the results obtained in parametric study of ATH.

The variation of the ratio between formic acid and triethylamine, the two components of the hydrogenation mixture, could provide an insight into several subtle aspects of the reaction mechanism. Also, several visual differences between reaction mixtures containing different molar ratios of TEA and formic acid were observed, yellow color for mixtures containing higher molar ratio between formic acid and triethylamine and orange color for the mixture with higher amount of base. This fact indicates that the catalytic complex undergo some significant changes in excess of acid followed by loss of activity of the catalyst. Although, using higher amount of the TEA also showed that the reaction perform much more slowly than usual.

Explanation for this phenomenon could be really simple; the excess of TEA probably neutralizes all of the formic acid and by this disable the reaction itself. However, according to the results obtained during the study, azeotropic mixture of formic acid/TEA (molar ratio 5:2) seems to be an optimal as the source of hydrogen for the purpose of ATH (**Figure 5**).

4. Analytical methods tailored to asymmetric transfer hydrogenation

For the purpose of monitoring of the reaction and determination of *ee* of the reaction, several effective methods were developed. Kinetic study of hydrogenation can be realized in NMR spectrometer [11] and for the determination of *ee* method using derivatization of the sample followed by the determination of diastereoisomers using gas chromatography [12] can be applied.

4.1. *In situ* kinetic study using NMR spectroscopy

Implementation of the kinetic measurements of ATH of imines in flask brings some drawbacks. One of them is relatively complicated preparation of the sample for the following gas chromatography analysis. This preparation includes alkalinization of the reaction mixture for the release of the basic product and salt from formic acid, extraction of organic compounds into the diethyl ether, evaporating of the ether and dissolving of the sample with acetonitrile and analysis itself. For the practical reasons it is not possible to perform efficient enough kinetic measurements. To remove these drawbacks practical *in situ* NMR method of monitoring can be implemented. The core of this method lies in mixing of the catalyst together with the hydrogenation mixture (formic acid/TEA) dissolved in deuterated solvents like acetonitrile or dimethyl sulfoxide in NMR tube. The reaction is started by the addition of solution of substrate and in timely manner H^1-NMR spectra are acquired by automated software. By this, high-quality kinetic profile of the reaction is obtained without an intervention from the ambient environment.

4.2. Determination of the enantiomeric excess (*ee*) of ATH

Typical instrumental method for determination of optical purity at products of reaction is gas or liquid chromatography, using columns with chiral stationary phase. However, these columns are relatively expensive. Another disadvantage is that these columns is higher sensitivity, which manifest itself during GC analysis by lower temperature limit and at LC analysis by impossible analysis of any chemicals or are limited by certain pH Hence, alternative methods using derivatization or chiral solvatation of the product are expedient.

4.2.1. Determination of ee by derivatization with (R)-menthyl-chloroformate

This method is based on a quantitative reaction of the product with (*R*)-menthyl-chloroformate. The result of the reaction is a mixture of two diastereomeric carbamates, which can be separated using gas chromatography without the necessity of using any chiral column. This method was successfully validated and nowadays is used for the analysis of many tetrahydroisoquinoline derivatives with a sufficiently low boiling point to GC analysis.

4.2.2. Determination of ee by chiral solvation by Pirkle's alcohol

Chiral solvation by Pirkle's alcohol ((*R*)-(-)-1-(anthracen-9-yl)-2,2,2-trifluoroethanol) was used to determine *ee* at tetrahydroisoquinoline derivatives with a high boiling point. The method is based on the fact that the products of ATH provide diastereomeric solvates by the reaction with Pirkle's alcohol and each product can be distinguished by NMR analysis.

5. Structural effects of the catalyst and the cyclic imine substrate

The main result, regarding ATH of dihydroisoquinolines, depends either on individual structural fragments of the catalyst itself or substitution on the molecule of the substrate. Even a small change either in the structure of catalyst or substrate can have a significant impact on the enantioselectivity, reaction rate or even feasibility of the process.

5.1. Modification of the structure of Noyori's catalysts

As previously mentioned, one of the main advantages of Noyori's [Ru(Cl)(η^6 arene)(*S, S*) TsDPEN] catalysts is their structural flexibility. The structure of these complexes can be divided into three components, while each of them can be synthetically modified to increase the catalytic activity and enantioselectivity of the reaction in ATH of specific substrates (**Figure 6**).

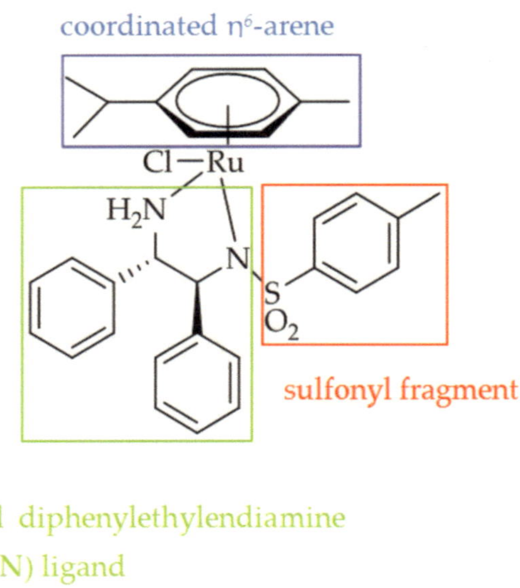

Figure 6. Scheme showing individual parts of the catalyst.

http://dx.doi org/10.5772/110623

5.1.1. *Modification of the coordinated η⁶-arene*

The core of enantioselectivity of asymmetric transfer hydrogenation lies in the weak bond between the hydrogen atom of η⁶-arene of the catalyst and π electrons of the aromatic ring of the substituted 3,4-dihydroisoquinoline substrate, e. g., CH/π interactions [7]. This interaction can assure the desired stabilization, providing the substrate molecule adopts a specific orientation permitting its formation. This interaction has also been one of the key reasons that gradually led to a certain abandoning of ketone-analogous mechanisms and formulations of the so-called ionic mechanism [5, 6, 13].

The work focused on the field of experimental testing of the Noyori based catalysts differentiated in their η⁶-arene in ATH of various substituted 3,4-dihydroisoquinolines, accompanied by a computational study [14]. For the evaluation of the effect of η⁶-arene ligand, a set of kinetic experiments was performed, followed by the determination of the enantiomeric purity of the products. For the purpose of this study, six cyclic imines (6, 7-dimethoxy-1-methyl-3,4-dihydroisoquinoline, 1-methyl-3,4-dihydroisoquinoline, 6 methoxy-1-methyl-3,4-dihydroisoquinoline, 7-methoxy-1-methyl-3,4-dihydroisoquinoline, 1-phenyl-3,4-dihydroisoquinoline, and 1-(4-trifluormethylphenyl)-3,4-dihydroisoquinoline) were tested in asymmetric transfer hydrogenation using four catalysts differing in their η⁶ arene ligand (*i.e.*, benzene, mesitylene, *p*-cymene and hexamethylbenzene). According to the results, catalysts bearing mesitylene and *p* cymene (**Figure 7**), could deliver higher *ee* values than the catalyst containing benzene as η⁶ arene ligand. The explanation of this fact was found out by authors during the molecular modeling. The calculation of transition states of the ATH with catalyst containing *p*-cymene and mesitylene suggested that for those two, it was possible to form a double CH/π interaction, whereas for the catalysts with a benzene ligand, only a single CH/π interaction could be formed that resulted in lower *ee* values.

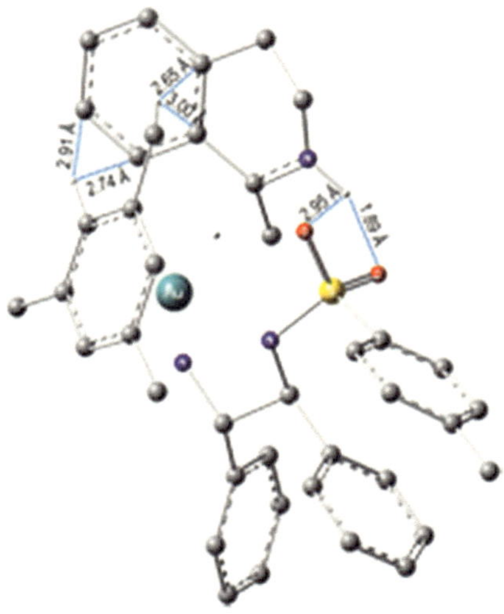

Figure 7. The so-called favorable transition state that occurred during the ATH of 1-methyl-3,4-dihydroisoquinoline.

Apart from the enantioselectivity, the modification of η^6-arene ligand affects also other reaction parameters. These include the turnover frequency (TOF), e.g., for the catalysts with hexamethylbenzene ligand, the TOF value was the smallest of all tested catalysts. Since homogeneous catalysis is involved, the solubility of the catalyst also plays a very important part of the synthesis. The modification of η^6-arene ligand significantly changes the solubility of the complex. The catalysts bearing mesitylene, *p* cymene and hexamethylbenzene were soluble in almost all polar solvents, whereas the catalyst with the benzene ligand was more or less soluble only in highly boiling solvents as dimethyl sulfoxide or *N, N* dimethylformamide.

5.1.2. Modification of the sulfonyl moiety

As mentioned in one of the previous sections, sulfonyl moiety of the catalyst, especially its oxygen atoms are important during the anticipated reaction mechanism where they serve as the active sites of the catalysts by interacting with both the protonated base and the protonated substrate with the use of hydrogen bonds. However, over a certain period of time, *N*-R-sulfonyl fragment was a target to a series of changes. Originally published catalysts [1, 2] contained *N*-aryl- or *N* alkylsulfonyl moieties of the TsDPEN ligand, typically mesityl (Mes), tosyl (Ts) or 1 naphthyl (Np). Later, the catalyst bearing methanesulfonyl fragment was found to be highly active in asymmetric hydrogenation of ketones and imines using gaseous hydrogen [15]. Among others, Wills and co-workers reported series of novel *N*-alkylated derivatives and tethered complexes, linking the arene and diamine ligand of the catalyst together [5, 16].

More profound modifications of *N*-R-sulfonyl fragment have been showed, aiming at substantial changes in catalyst's properties, e. g., for immobilization, or for conducting ATH in different media (water, ionic liquid, etc.) [17]. It is evident that the literature contains many interesting examples of structural variations. Nevertheless, the scope of available ligands is difficult to compare due to inconsistent reaction conditions.

5.1.2.1. Asymmetric transfer hydrogenation of 1-phenyl dihydroisoquinolines

Asymmetric transfer hydrogenation of 1-phenyl dihydroisoquinolines [18], represents a considerable challenge since the catalyst bearing the ligand of *N*-*p*-toluenesulfonyl-1, 2-diphenylethylenediamine (TsDPEN) failed to catalyze this reaction under standard reaction conditions. Although the original Noyori-type ATH catalysts always bore *N* arylsulfonyl as the substituent, several studies showed that Ir and Ru catalysts, containing methanesulfonyl-DPEN (MsDPEN), hydrogenated various aryl-*N*-benzyl imines, *N* sulfonylimines and aryl ketones with molecular hydrogen [5, 15, 19–21]. The ligand of methanesulfonyl DPEN is a part of a rather small group of ligands finding their use as a component of catalysts applicable for asymmetric (transfer) hydrogenation. Rather recently, another ligand bearing alkyl group, *N*-(camphor-10-sulfonyl)-DPEN (CsDPEN), was reported for its high efficiency in the ATH of a series of carbonyl compounds [22, 23]. However, no alkylsulfonyl diamines, such as 3,4 dihydroisoquinoline derivatives, have been applied in ATH of imines.

In addition, several Noyori-based Ru catalysts bearing *N*-naphtalene-1-sulfonyl-DPEN (NpsDPEN) and *N*-((*1S, 2S*-borneol-10-sulfonyl)-DPEN (CsDPEN) were tested in ATH of series of 1-phenyl-3,4-dihydroisoquinolines. The change of the original *N*-arylsulfonyl-DPEN from *p*-toluenesulfonyl-DPEN to naphthalene-1-sulfonyl in the catalysts structure, thus, enabled the catalytic activity in the ATH of aryl substituted 3,4-dihydroisoquinolines. The camphor fragment itself was indeed specific. The reaction which provided the desired catalytic complex was also an enantioselective reaction by itself. The preparation of the catalytic complex was performed in propane-2-ol, which also served as the hydrogen donor for ATH of ketones and thus the prepared complex also underwent transfer auto hydrogenation of the carbonyl group. The isolated crystalline material contained only one isomer. This isolated complex was tested in ATH of several 1-phenyl-3,4-dihydroisoquinolines derivates. Interestingly enough, in the reduction of aryl substrates containing electron-donating groups, the catalyst's performance was comparable with that measured with catalyst bearing naphthalene-1-sulfonyl-DPEN fragment. On the contrary, aryl substrates lacking electron-donating groups or even containing an electron withdrawing group displayed a very low catalytic activity. In the ATH of alkyl substrates, 1-methyl-3,4-dihydroisoquinoline and 6,7-dimethoxy-1-methyl-3,4-dihydroisoquinoline, the reactivity was much higher than with naphthalene-1-sulfonyl-DPEN, and slightly exceeding the activity of original Noyori's catalyst with toluenesulfonyl-DPEN ligand.

5.1.2.2. *Comparison of the different sulfonyl moieties in the ATH of alkyl-3,4-dihydroisoquinolines*

The role of sulfoonamide moiety of Noyori-based catalysts [Ru(II)Cl(η^6 *p* cymene)(*S,S*)-(*N*-arylsulfonyl-DPEN) in the asymmetric transfer hydrogenation of two cyclic imine substrates (1-methyl-3,4-dihydroisoquinoline and 6,7 dimethoxy-1-methyl-3,4-dihydroisoquinoline) was also investigated [24]. All together, nine complexes, differing in substitution of the aromatic ring of the ligand, were synthesized and characterized, most of which have not been previously reported and the majority of the corresponding ligands have not been described in imine ATH.

As the standard, Noyori's original catalysts bearing N-*p*-toluenesulfonyl-DPEN ligand was selected, which served as the benchmark for the kinetic study. The substitution of the sulfonyl part of the catalyst was divided into several parts. The first were the ligands with bulky groups (*p*-(*tert*-butyl)phenyl, mesityl and 1-naphtyl), the second aromatic ring bearing electron donating methoxy groups differing in their position on the aromatic ring. Others were ligands with (3,4-dichloro)phenyl substituent as the representative of halogenated aromatic ring. The last two complexes contained: first, a very bulky and electron-poor 3,5-bis(trifluoromethyl)phenyl group, and the second, a heteroatom-containing ring, thiophene, chosen as an alternative to the common aromatic substituents.

The results obtained in this study showed that the change of aryl substituent on the sulfonyl part of the catalyst had a great influence on the reaction rate in ATH of 3,4 dihydroisoquinolines. Especially, the halogenated and hetero-aromatic substituents delivered reasonable reactivity only for one of the two substrates. The sterically demanding naphthyl containing ligand was the least preferred one.

5.2. Structural effects of the substrate

Isoquinoline-based molecules belong to the most important naturally-occurring alkaloids encompassing a significant group of biologically active species. These compounds have various pharmacological effects, which are enabled by their structural similarity with endogenous neurotransmitters. Therefore, a kinetic NMR study of ATH of five different 3,4 dihydroisoquinolines (**Figure 8**) was performed [25]. Four of them differed in various methoxy-substitutions of the dihydroisoquinoline skeleton, while the fifth examined substrate was a precursor for the production of mivacurium, a muscle relaxant. With 7 methoxy derivates (**3, 4, 5**), the reaction rate was significantly higher than in the case where no methoxy group was present in the position 7. The ATH of 6-methoxy-1-methyl-3,4-dihydroisoquinoline (**2**), bearing the 6-methoxy group, proceeded with a considerably lower reaction rate. However, the positive influence of the 7-methoxy group seemed to have overcome the detrimental effect of the 6-methoxy substituent when these two were present together, as it was for instance in the case of 6,7-dimethoxy-1-methyl-3,4-dihyroisoquinoline. Similarly, the bulky 1 (3′,4′,5′ trimethoxybenzyl) moiety of the last substrate (**5**) further enhanced the reaction performance.

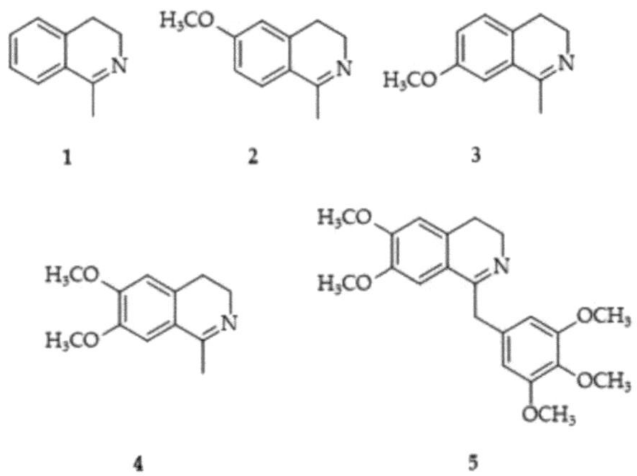

Figure 8. List of model substrates used to determine the effect of different substitution at the aromatic ring.

The substitution of the dihydroisoquinoline substrate with methoxy groups is followed by a much higher reaction rate and enantioselectivity than in the case of 1-methyl-3,4-dihydroisoquinoline (**1**). The presence of one methoxy group in the substrate molecule (**2** and **3**) led to an *ee* increase of 4–5 percentage points (89% (**2**) and 88% (**3**) versus 84% with the substrate **1**). With more methoxy groups affording *ees* higher by 9–10 percentage points (93% (**4**) and 94% (**5**) versus 84% with the substrate **1**). This observation could be attributed to substituent effects on the aromatic ring of the substrate, which affected the CH/π attraction between the catalyst and the substrate.

These experimental observations were supported by examining all involved substrates using molecular modeling, especially, the charge distribution (calculated Mulliken charges, NPA charges and grid-based Bader analysis) on the C = N double bond (**Figure 9**). However, no expected correlations were found.

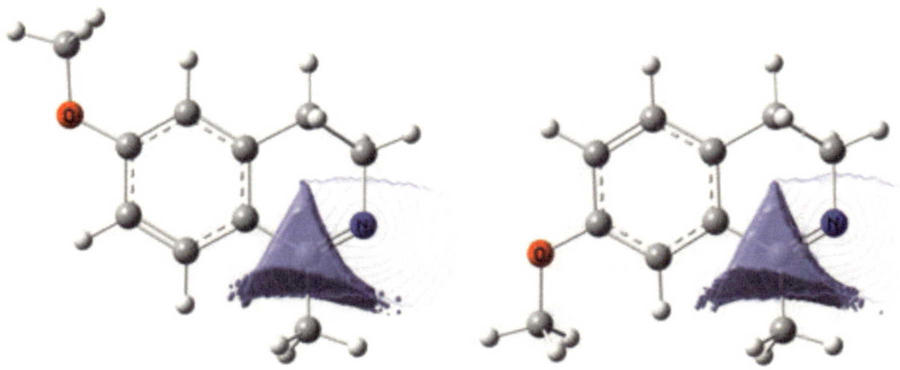

2 3

Figure 9. Partitions of total electron density on C and N atoms of imine bond of substrates containing one methoxy group.

To sum up, changing the position of single methoxy group resulted in drastic differences on the reaction performance in terms of both, the reaction rate and enantioselectivity. The presence of methoxy groups remarkably increased the reaction's enantioselectivity. Finally, an interesting conclusion could be drawn that less basic substrates were hydrogenated with higher reaction rates.

6. Practical industrial applications of asymmetric transfer hydrogenation

Even though asymmetric transfer hydrogenation was originally merely a subject of academic interest, nowadays this reaction finds its use also in several sectors of chemical industry. There are a large number of optically active amines and alcohols used as active substances and a suitable method for ATH preparation. As any method or technology, ATH inclusively has its pros and cons. The indisputable advantages, compared to asymmetric hydrogenation by gaseous hydrogen, are primarily the catalyst stability under air conditions (AH catalyst typically contains phosphine ligands, which easily undergo oxidation by atmospheric oxygen) and second, avoiding the use of gaseous hydrogen. These two facts actually permit testing many structurally different catalysts in a relatively short period of time. Eliminating the use of pressure hydrogen and thus the demanding apparatus for reactions at a high pressure noticeably simplifies the production facilities, which is an important aspect from the economical point of view.

Relatively small values of turnover frequencies (TOF) present one of the major drawbacks compared to AH catalysts. Extending the reaction time or using a higher amount of the catalyst represent potential solutions (since the catalysts for ATH are inexpensive compared to AH, this solution is economically acceptable). Increasing the reaction temperature is, however, not the optimal solution, since enantioselectivity is decreasing with a higher temperature. Eliminating metal residues from the product is not an issue in these days as applying commercially available methods allows reducing the number of residues to units of *ppm*.

In 1997, Avencia Company patented rhodium catalysts bearing diamine, aminoalcohol ligands, respectively, for asymmetric transfer hydrogenation of imines and ketones. These Rh complexes are analogues to original Noyori's ruthenium catalysts. However, except for the central atom, these catalysts also differ in their aromatic ligand, which is in the most cases η^5-pentamethylcyclopentadienyl. These catalysts are used for the production of several different types of chiral alcohols and amines. To name some examples, one of the running processes is ATH of tetralone to (*R*)-1-tetralol with the capacity of 200 dm^3 and the corresponding yield of 95 and 97% *ee*, than the process of preparation of (*S*)-1-(4-flurophenyl)ethanol with its yield of 85 and 98.4% *ee* or the production of (*R*)-N-diphenylphosphinyl-1-methylamine with 95% yield and 99% *ee* [26].

Asymmetric transfer hydrogenation has the potential to find use also in the production of fine chemicals such as drugs, where a high optical purity of the final products is demanded. To provide an example, the preparation of the precursor for the synthesis of muscle relaxant, mivacurium-chloride, can be mentioned [27]. In this case, the application of ATH for the preparation of mivacurium-chloride seems to be a more favorable, since the cleavage used in the classical preparation, capitalizing on using *L*-dibenzoyltartaric acid to separate both enantiomers, is rather ineffective and produce a lot of nonrecyclable waste.

7. Conclusion

The main purpose of this study was to describe selected parts of the asymmetric transfer hydrogenation of ketones and particularly imines. Our attention was predominantly aimed to the importance of the mechanistic aspects of the ATH or structural effects influencing the reaction course. All parts of this work show that the asymmetric transfer hydrogenation is the reaction which can find its use across all branches of the chemical industry.

Acknowledgements

This work was realized within the Operational Program Prague – Competitiveness (CZ. 2.16/3.1.00/22197) and "National Program of Sustainability" (NPU I LO1215) MSMT - 34870/2013); The Operational Program Prague–Competitiveness (CZ.2.16/3.1.00/21537) and "National Program of Sustainability" (NPU I LO1601) MSMT-43760/2015); The Operational Program Prague–Competitiveness (CZ.2.16/3.1.00/24501) and "National Program of Sustainability" (NPU I LO1613) MSMT-43760/2015) and GACR (GA15-08992S).

Author details

Ondřej Matuška, Martin Kindl and Petr Kačer*

*Address all correspondence to: kacerp@vscht.cz

University of Chemistry and Technology (UCT Prague), Prague, Czech Republic

References

[1] Fujii, A., et al., *Ruthenium(II)-catalyzed asymmetric transfer hydrogenation of ketones using a formic acid-triethylamine mixture*. Journal of the American Chemical Society, 1996. 118(10): 2521–2522. DOI: 10.1021/ja954126l

[2] Uematsu, N., et al., *Asymmetric transfer hydrogenation of imines*. Journal of American Chemical Society, 1996. 118(20): 4916–4917. DOI: 10.1021/ja960364k

[3] Noyori, R., M. Yamakawa, and S. Hashiguchi, *Metal-ligand bifunctional catalysis: A nonclassical mechanism for asymmetric hydrogen transfer between alcohols and carbonyl compounds*. The Journal of Organic Chemistry, 2001. 66(24): 7931–7944. DOI: 10.1021/jo010721w

[4] Dub, P.A. and T. Ikariya, *Quantum chemical calculations with the inclusion* of nonspecific *and specific solvation: Asymmetric transfer hydrogenation with bifunctional ruthenium catalysts*. Journal of the American Chemical Society, 2013. 135(7): 2604–2619. DOI: 10.1021/ja3097674

[5] Martins, J.E.D., G.J. Clarkson, and M. Wills, *Ru(II) complexes of N-alkylated TsDPEN ligands in asymmetric transfer hydrogenation of ketones and imines*. Organic Letters, 2009. 11(4): 847–850. DOI: 10.1021/ol802801p

[6] Åberg, J.B., J.S.M. Samec, and J.-E. Backvall, *Mechanistic investigation on the hydrogenation of imines by [p (Me$_2$CH)C$_6$H$_4$Me]RuH(NH$_2$CHPhCHPhNSO$_2$C$_6$H$_4$-p-CH$_3$). Experimental support for an ionic pathway*. Chemical Communications, 2006, 1(26): 2771–2773. DOI: 10.1039/b605838h

[7] Yamakawa, M., I. Yamada, and R. Noyori, *CH/π attraction: The origin of enantioselectivity in transfer hydrogenation of aromatic carbonyl compounds catalyzed by chiral η^6-arene-ruthenium(II) complexes*. Angewandte Chemie International Edition, 2001. 40(15): 2818–2821. DOI: 10.1002/1521-3773(20010803)40:15<2818::AID-ANIE2818>3.0.CO;2-Y

[8] Šot, P., et al., *Asymmetric transfer hydrogenation of acetophenone N-benzylimine using [Ru^{II}Cl((S,S)-TsDPEN)(η^6-p-cymene)]: A DFT study*. Organometallics, 2012. 31(17): 6496–6499. DOI: 10.1021/om300413n

[9] Kuzma, M., et al., *New insight into the role of a base in the mechanism of imine transfer hydrogenation on a Ru(II) half-sandwich complex*. Dalton Transactions, 2013. 42(14): 5174–5182. DOI: 10.1039/c3dT32733g

[10] Pecháček, J., et al., *Asymmetric transfer hydrogenation of imines catalyzed by a Noyori-type Ru(II) complex—a parametric study*. Tetrahedron: Asymmetry, 2013. 24(4): p. 233–239. DOI: 10.1016/j.tetasy.2013.01.010

[11] Václavík, J.; Pecháček, J.; Přech, J.; Kuzma, M.; Kačer, P.; červený, L., In situ monitoring of asymmetric transfer hydrogenation of Imines using NMR Spectrometry. Chem. Listy 2012, 106 (3), 206–210.

[12] Přech, J., et al., *Determination of enantiomeric composition of substituted tetrahydroisoquinolines based on derivatization with menthyl chloroformate*. American Journal of Analytical Chemistry, 2013. 4(3): 9. DOI: 10.1016/j.catcom.2013.03.004.

[13] Václavík, J., et al., *Asymmetric transfer hydrogenation of imines and ketones using Chiral RuIICl(η^6-p-cymene)[(S,S)-N-TsDPEN] as a catalyst: A computational study*. Organometallics, 2011. 30(18): 4822–4829. DOI: 10.1021/om200263d

[14] Šot, P., et al., *The role of the aromatic ligand in the asymmetric transfer hydrogenation of the CN bond on Noyori's chiral Ru catalysts*. Tetrahedron: Asymmetry, 2014. 25(18–19): 1346–1351. DOI: 10.1016/j.tetasy.2014.08.011

[15] Ohkuma, T., et al., *Asymmetric hydrogenation of α-hydroxy ketones catalyzed by MsDPEN –Cp*Ir(III) complex*. Organic Letters, 2007. 9(13): p. 2565-2567. DOI: 10.1021/ol070964w

[16] Morris, D.J., A.M. Hayes, and M. Wills, *The "reverse-tethered" ruthenium (II) catalyst for asymmetric transfer hydrogenation: Further applications*. The Journal of Organic Chemistry, 2006. 71(18): 7035–7044. DOI: 10.1021/jo061154l

[17] Václavík, J., et al., *Practical aspects and mechanism of asymmetric hydrogenation with chiral half-sandwich complexes*. Molecules, 2013. 18(6): 6804. DOI: 10.3390/molecules18066804

[18] Přech, J., et al., *Asymmetric transfer hydrogenation of 1-phenyl dihydroisoquinolines using Ru(II) diamine catalysts*. Catalysis Communications, 2013. 36: 67–70. DOI: 10.1016/j.catcom.2013.03.004

[19] Shirai, S.-y., et al., *Remarkable positive effect of silver salts on asymmetric hydrogenation of acyclic imines with Cp*Ir complexes bearing chiral N-sulfonylated diamine ligands*. Organometallics, 2009. 28(3): 802–809. DOI: 10.1021/om800926p

[20] Chen, F., et al., *Asymmetric hydrogenation of cyclic N-sulfonylimines with phosphine-free chiral cationic Ru-MsDPEN catalysts*. Chinese Journal of Chemistry, 2010. 28(9): 1529–1532. DOI: 10.1002/cjoc.201090260

[21] Chen, F., et al., *Asymmetric hydrogenation of N-alkyl ketimines with phosphine-free, chiral, cationic Ru-MsDPEN catalysts*. Chemistry—A European Journal, 2011. 17(4): 1109–1113. DOI: 10.1002/chem.201002846

[22] Li, X., et al., *An efficient Ir(III) catalyst for the asymmetric transfer hydrogenation of ketones in neat water*. Synlett, 2006. 2006(08): 1155–1160. DOI: 10.1055/s-2006-932490

[23] Yin, L., et al., *Efficient and promising asymmetric preparation of enantiopure tolvaptan via transfer hydrogenation with robust catalysts*. Tetrahedron: Asymmetry, 2010. 21(19): 2390–2393. DOI: 10.1016/j.tetasy.2010.08.016

[24] Matuška, O., et al., *Role of the sulfonamide moiety of Ru(II) half-sandwich complexes in the asymmetric transfer hydrogenation of 3, 4-dihydroisoquinolines*. Reaction Kinetics, Mechanisms and Catalysis, 2016. 118(1): 215–222. DOI: 10.1007/s11144-016-0991-z

[25] Václavík, J., et al., *Molecular structure effects in the asymmetric transfer hydrogenation of functionalized dihydroisoquinolines on (S,S)-[RuCl(η^6-p-cymene)TsDPEN]*. Catalysis Letters, 2013. 143(6): 555–562. DOI: 10.1007/s10562-013-0996-4

[26] Blacker, J.; Martin, J., *Scale-Up Studies in Asymmetric Transfer Hydrogenation*. In *Asymmetric Catalysis on Industrial Scale*, Wiley-VCH Verlag GmbH & Co. KGaA: Weinheim, Germany, 2004; pp 201–216. DOI: 10.1002/3527602151.ch11

[27] Vilhanová, B., et al., *Two optimized synthetic pathways toward a chiral precursor of Mivacurium chloride and other skeletal muscle relaxants*. Tetrahedron: Asymmetry, 2013. 24(1): 50–55. DOI: 10.1016/j.tetasy.2012.11.012

Chapter 4

Recent Advances in Heterogeneous Catalytic Hydrogenation of CO_2 to Methane

Zuzeng Qin, Yuwen Zhou, Yuexiu Jiang, Zili Liu and Hongbing Ji

Additional information is available at the end of the chapter

http://dx.doi.org/10.5772/65407

Abstract

With the accelerating industrialization, urbanization process, and continuously upgrading of consumption structures, the CO_2 from combustion of coal, oil, natural gas, and other hydrocarbon fuels is unbelievably increased over the past decade. As an important carbon resource, CO_2 gained more and more attention because of its converting properties to lower hydrocarbon, such as methane, methanol, and formic acid. Among them, CO_2 methanation is considered to be an extremely efficient method due to its high CO_2 conversion and CH_4 selectivity. However, the CO_2 methanation process requires high reaction temperatures (300–400°C), which limits the theoretical yield of methane. Thus, it is desirable to find a new strategy for the efficient conversion of CO_2 to methane at relatively low reaction temperature, and the key issue is using the catalysts in the process. The advances in the noble metal catalysts, Ni-based catalysts, and Co-based catalysts, for catalytic hydrogenation CO_2 to methane are reviewed in this paper, and the effects of the supports and the addition of second metal on CO_2 methanation as well as the reaction mechanisms are focused.

Keywords: catalytic hydrogenation, carbon dioxide, methanation, heterogeneous catalysis, noble metal catalyst, Ni-based catalyst, Co-based catalyst

1. Introduction

Over the past centuries, CO_2 has become the main carbon resource due to the decreases of limited resources such as coal, oil, and natural gas [1]. However, the CO_2 concentration in the atmosphere has consequently increased from ~280 ppm (preindustrial) to ~390 ppm in 2010 at a rate of ca. 1% per year [2], which arguably contributes to the "greenhouse effect," and increases the global temperatures and climate change. CO_2 emissions are still existing threat to humans; it is high time that effective measures should be taken to decrease the emission of CO_2.

Hence, the carbon capture and sequestration (CCS) system is considered to be an efficient method for CO_2 utilization [3, 4]. Nevertheless, the hydrogenation reaction is the most important chemical conversions of CO_2; moreover, which offers a good opportunity for sustainable development in the energy and environmental sectors. Indeed, the reaction process not only reduces the CO_2 amount in the atmosphere but also produces fuels and valuable chemicals [5].

As a promising fuel energy, methane, a simple hydrocarbon, has a wide range of applications in the industry and civil use, which also used to produce some downstream products, such as ethyne, hydrogen, and ammonia [6, 7]; therefore, the strategy of CO_2 methanation is significantly meaningful. Undeniably, the resources of fossil fuels are diminishing and fuel prices have undergone strong fluctuation in recent years. Therefore, developing alternative fuels from nonfossil fuel sources and processes are highly desirable. The products from CO_2 hydrogenation, such as methane, hydrocarbons, methanol, and dimethyl ether, are excellent fuels in internal combustion engines, and are easily stored and transported, but the literature studies showed that the CO_2 conversion to methanol and dimethyl ether is still very low (~20%) and it is difficult to achieve higher conversion of CO_2 [8, 9]. CO_2 methanation is a simple reaction, generating methane under atmospheric pressure with several advantages over other chemicals. Although the conversion was still very low, the CH_4 formation from CO_2 at low temperature has become an important breakthrough in the utilization of CO_2 [10].

CO_2 methanation is a significant catalytic hydrogenation process, as is shown in Eq. (1).

$$CO_2 + 4H_2 \rightarrow CH_4 + 2H_2O, \qquad \Delta H_{298K} = -252.9 KJ{\cdot}mol^{-1} \tag{1}$$

The methanation of CO_2 has a wide range of applications including the production of syngas and the formation of compressed natural gas [1]. A prototype CO_2 recycling plant to supply clean energy preventing global warming has been built in 1996 using these key materials and has been operating successfully [11]. Without doubt, CO_2 methanation is the key pathway for CO_2 recycling, which requires a catalyst to achieve acceptable rates and selectivities. And extensive studies have been conducted on metal-based catalytic systems in the hydrogenation of CO_2 to methane.

Noble metals (e.g., Ru, Rh, Pd) supported on oxide supports (e.g., TiO_2, Al_2O_3, CeO_2) were the most effective catalysts for CO_2 methanation under relatively mild operating conditions [12–14]; however, the high cost of the catalysts limited their practical applications [15]. Therefore, to obtain a feasible and cost-effective catalytic process, nonnoble metal catalysts (e.g., Ni, Co) were focused by many scholars [16, 17]. This review attempts to present the catalytic reactivity and reaction mechanism over the catalysts, particularly over the heterogeneous catalysts with an emphasis on the effects of supports and the second metal additives, as well as an overview regarding the challenges and opportunities for future research in the field.

2. Catalysts for CO_2 methanation

2.1. Noble metal catalysts for low-temperature methanation of CO_2

The most widely used catalysts for the CO_2 methanation are noble metals, such as Rh, Ru, and Pd, and Ni-based catalysts. The noble metals are highly active toward CO_2 methanation at lower temperature and more resistant to the carbon formation than other transition metals; however, they are expensive. In particular, the noble metals also used to promote the Ni catalysts to enhance their catalytic activities. The noble metal catalytic systems for the synthesis of methane by CO_2 hydrogenation are summarized in **Table 1**.

Catalyst	Preparation method	T/°C	TOF (10^{3} s^{-1})	Ref.
0.8 wt% Ru/TiO_2	Polygonal barrel-sputtering	160	8.5	[18]
5 wt% Ru/rutile-TiO_2	Wet-impregnation	160	6.0	[12]
3 wt% Rh/γ-Al_2O_3	Wet-impregnation	200	18.78	[13]
3 wt% Rh/TiO_2	Wet-impregnation	150	22.66	[24]
2 wt% Ru/TiO_2(101)	Hydrothermal	150	4.51	[20]

Table 1. Summarization of activities of CO_2 methanation on noble metal catalysts.

2.1.1. *Role of the support on catalyst activity*

CO_2 methanation has been studied over a series of supported Ru and Rh catalysts, which were very active for CO_2 hydrogenation [13, 14, 20, 21]. The supports, including Al_2O_3, TiO_2, and CeO_2 for these active metals, have also been investigated. To clarify the influences of the supports on the catalytic behavior of ruthenium, a FT-IR study is used to obtain more insight into the reaction mechanism [21]. Based on the FT-IR spectra of CO and CO_2 adsorbed on the catalysts, the improvement in the CO_2 methanation activity was related to a higher positive polarization of ruthenium on the zeolite, which led to a weaker Ru–CO bond on the H-ZSM-5-supported sample with a corresponding increase of the hydrogen surface coverage, which favors the transformation of the intermediate CO to methane, and which indicated that Ru/ZSM-5 exhibits more CH_4 selectivity than Ru/SiO_2 [21].

The Ru dispersion was significantly influenced by the crystal phase structure of the TiO_2 supports [19]. Rutile-type TiO_2 (r-TiO_2) was a much better support than anatase-TiO_2 (a-TiO_2) in stabilizing of RuO_2 due to the interfacial lattice matching, resulting in a higher reactivity and stability in CO_2 methanation. Owing to the highly dispersed Ru catalyst with a narrow size distribution, r-TiO_2 was a promising support [12]. There was a strong interaction between RuO_2 and r-TiO_2 during the calcination process, which prohibited the aggregation of RuO_2 in the presence of the Ru–O–Ti bond. As represented in **Figure 1**, upon calcination at 300°C, the Ru/r-TiO_2 exhibited a much higher activity and thermal stability in CO_2 methanation than Ru/a-TiO_2. Moreover, the reaction rate of the Ru/r-TiO_2 was 2.4 times higher than that of the Ru/a-TiO_2, which mainly originated from the different particle sizes of ruthenium [12].

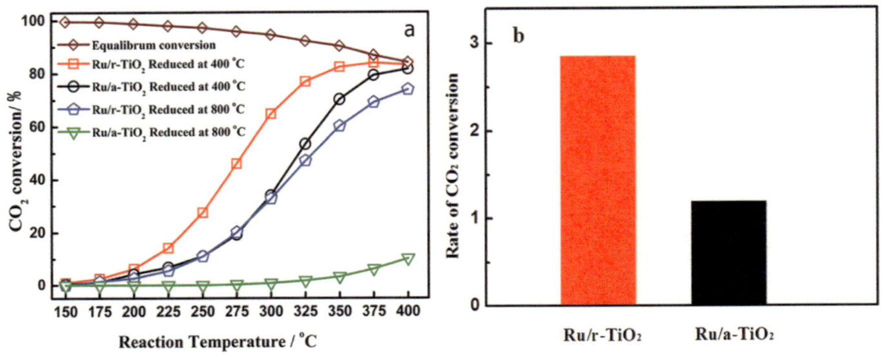

Figure 1. (a) The effects of reaction temperature on the CO_2 conversion over the Ru catalysts and (b) the specific rates of CO_2 conversion calculated at 225°C. The feed gas was 18 vol.% CO_2+ 72 vol.% H_2+ 10 vol.% N_2, and the catalyst was each 0.040 g of Ru/TiO_2 diluted with 0.400 g of SiO_2, the total space velocity was 75,000 mL·g_{cat}^{-1}·h^{-1} [12].

The Ru/TiO_2 catalysts were prepared via a spray reaction (SPR) [20, 22], and the catalytic CO_2 hydrogenation activities of the SPR fine particles were much higher than those of impregnation catalysts [20]. The high activity of the SPR catalysts was attributed to the occurrence of new active sites at the metal-support perimeters without any strong metal-support interaction phenomenon. In addition, highly dispersed Ru nanoparticle-loaded TiO_2 was prepared using a "dry" modification method [18], which markedly enhances the performance of low-temperature methanation, achieving a 100% yield at 160°C. In addition, the methanation reaction over Ru/TiO_2 proceeded at temperatures as low as room temperature with a reaction rate of 0.04 mmol·min^{-1}·g^{-1}.

Although Ru catalysts deposited on different supports, such as alumina, titanium, or silica, have been extensively studied, and the effect of the support on the catalytic properties of small Ru particles in CO_2 hydrogenation has not been fully recognized. Different supports (low and high surface area graphitized carbons, magnesia, alumina and a magnesium-aluminum spinel) were used in CO_2 methanation, and alumina was found to be the most advantageous material [23]. The catalytic properties of very small ruthenium particles are strongly affected by metal-support interactions. In the case of Ru/C, the carbon support partly covers the metal surface, lowering the number of active sites (site blocking). A sequence of the surface-based activities (TOF): Ru/Al_2O_3 > Ru/$MgAl_2O_4$ > Ru/MgO > Ru/C is almost identical to that of electron-deficiencies of the metal, determined by the Lewis acidities of the supports [23].

2.1.2. Effect of metal loading

The most likely effects caused by increasing the loading amount are the growth of the particle size, e.g., the mean particle size of surface Rh species increased with the metal loading amount, which affected the reactivity [24]. From the study over Rh/γ-Al_2O_3, varying Rh amounts show Rh particle sizes of 3.6–15.4 nm, and a 100% methane selectivity was observed over the entire temperature range and Rh amounts, and the turnover frequency for CH_4 formation depended on the Rh particle size. Larger Rh particles exhibited a catalytic activity of up to four times higher than the smaller particles at 135–150°C, whereas at higher temperatures (200°C) the turnover frequencies are similar for all particle sizes [13].

The Rh loading amount can significantly change the product selectivity of CO_2 hydrogenation over Rh/SiO_2 [25], and the main products transformed from CO_2 to CH_4 with the loading amount of Rh, as shown in **Figure 2**. To the 1 wt% Rh/SiO_2 catalyst, the concentration of surface Rh particles was low, and the Rh species were surrounded by the hydroxyl groups of SiO_2. For the 10 wt% Rh/SiO_2, 5.8 times more surface Rh particles than that of 1 wt% Rh/SiO_2 were found with accordingly less surface hydroxyl groups of SiO_2 existed around Rh particles [25]. In the Ru/Al_2O_3 catalysts with a Ru amount of 0.1–5.0%, the CH_4 selectivity in CO_2 methanation increased with the increase in the Ru loading amount [26]. In the 0.1% Ru/Al_2O_3 catalyst, Ru is mostly present in the atomic dispersion, and the agglomeration of small metal particles (and atoms) in the 3D clusters was observed, indicating a decrease in CH_4 selectivity.

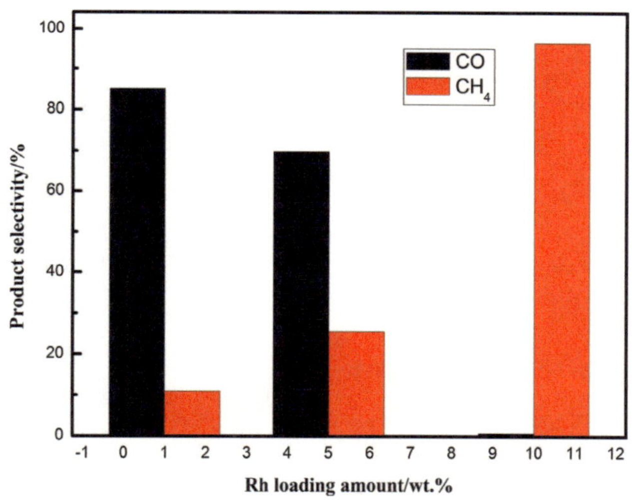

Figure 2. Effect of Rh loading on the distribution of CH_4 and CO [25]. Reaction conditions: temperature = 473 K, pressure = 5 MPa, H_2/CO_2 ratio = 3, flow rate = 100 cm^3 min^{-1}.

2.1.3. Effect of second metal

Actually, when the alkaline salts were added to Ru/Al_2O_3 catalysts, a synergetic effect can be detected, including the electron donation of an alkaline promoter modified the local electron density of the Ru metal, the formation of alkaline chlorides to neutralize the residual chlorine ions, and the removal of the depositional inactive carbon, which was formed on the catalyst surface during CO_2 hydrogenation [22]. Tests of the Ba- and K-containing Rh/Al_2O_3 and the pure Rh/Al_2O_3 in 300–700°C revealed remarkable differences in the catalytic behavior (**Figure 3**). The Ba-containing and especially the pure Rh/Al_2O_3 catalyst showed high selectivity to CH_4 below 500°C with a maximum CH_4 yield of 60% at 400°C; however, at higher temperatures, the CO formation became significant. K-containing Rh/Al_2O_3 converted CO_2 only to CO in 300–700°C and no CH_4 was found. A vastly different adsorption behavior of the Ba- and K-containing catalysts and a significant influence of these additives on the Rh(0)/Rh(I) ratio were revealed [27].

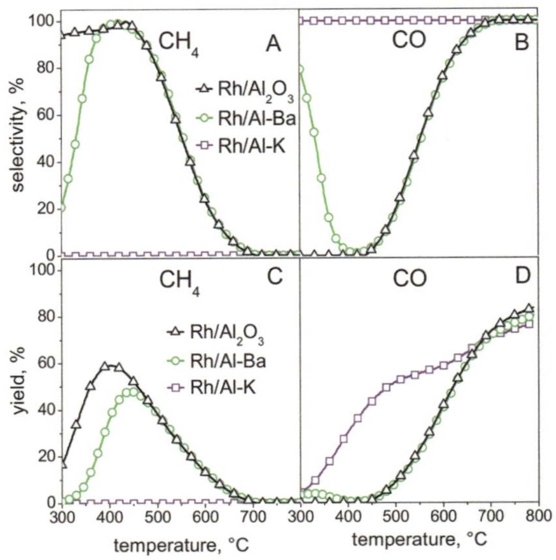

Figure 3. Comparison of selectivity and yield to CH_4 (A and C, respectively) and CO (B and D) is shown as a function of temperature for Ba-containing (circles) and K-containing (squares) Rh/Al_2O_3 catalysts, as well as for pure Rh/Al_2O_3 (triangles) [27].

2.2. Recent advances in Ni-based catalysts

2.2.1. *Effect of supports*

2.2.1.1. *Enhancement of catalytic performance*

Choosing a suitable support is mostly according to its properties to activate CO_2 and the interaction between the metal and supports, which is a key parameter for the methanation reaction [28]. The structure and properties of the support do affect the dispersity of active metals and the stability, which enhance the activity of catalysts.

Currently, various materials are used as the supports for nickel catalysts, such as γ-Al_2O_3 [29–31], SiO_2 [32, 33], $Ce_xZr_{1-x}O_2$ [33–36], and TiO_2 [37]. Because the support has a significant influence on the morphology of the active phase, adsorption, and catalytic properties [38], Ni was supported on the mesostructured silica nanoparticles (MSNs), MCM-41, HY zeolite, SiO_2, and γ-Al_2O_3. And the CO_2 methanation activity followed in the order of Ni/MSN > Ni/MCM-41 > Ni/HY > Ni/SiO_2> Ni/γ-Al_2O_3 [32]. The high activity of Ni/MSN is due to the presence of both intraparticle and interparticle porosities, which led to a high concentration of basic sites. In addition, the defect sites or oxygen vacancies in MSNs were responsible for the formation of surface carbon species, while Ni sites dissociated hydrogen to form atomic hydrogen.

An encouraging result was found in the CO methanation reaction over the zeolite supports, and the same results also found in the Ru/Y and Ru/Al_2O_3 catalysts [39], as well as the supporting Pd on the zeolites, and the catalytic activity on the supporter was in the order of HY > HZSM-5 > NaZSM-5 > NaY > SiO_2 [40]. Similarly, when CO_2 hydrogenation to methane was carried out over nickel species supported on a HNaUSY zeolite, interesting CO_2 conversions and CH_4 selectivities were achieved. CO_2 conversion increased with the Ni content from 2 to 14%, due to the higher amount of Ni^0 species after reduction [41]. Nickel particles were grafted onto SBA-15, and a chemical bond was formed between Ni and Si by O, and no bulk nickel oxides existed in the Ni-grafted SBA-15 [42]. Therefore, the Ni-grafted SBA-15 suited CO_2 methanation, resulting in the higher CO_2 conversion (TOF of 19.4 s^{-1}) and methane selectivity (92%) than a NiO dispersed SBA-15. The status of catalytic systems for the synthesis of methane by CO_2 hydrogenation is summarized in **Table 2**.

Catalyst	Preparation method	T/°C	CO_2 conversion (%)	Methane selectivity (%)	Ref.
20 wt% Ni-Al_2O_3-HT	Coprecipitation	350	82.5	99.5	[31]
20 wt% Ni/Al_2O_3	Impregnation	350	70.8	98.1	[31]
Ni/H–Al_2O_3	Hydrothermal and *in situ* reduction	300	99	99	[43]
15 wt% Ni/TiO_2	Deposition-precipitation	260	96	99	[37]
5 wt% Ni-$Ce_xZr_{1-x}O_2$	Pseudo sol-gel	350	79.7	99.3	[44]
5 wt% Ni-$Ce_xZr_{1-x}O_2$	Hydration process and impregnation	360	71.5	98.5	[48]
5 wt% Ni/MSN	Wet-impregnation	300	64.1	99.9	[32]
5 wt% Ni/MCM-41	Wet-impregnation	300	56.5	98.3	[32]
35 wt% Ni/Fe/Al_2O_3 alumina xerogel	Single step sol-gel	220	63.4	99.5	[45]
10 wt% Ni/MOF-5	Impregnation	320	75.09	100	[46]
14 wt% Ni/USY	Impregnation	400	65.5	94.2	[41]

Table 2. Summary of various Ni catalysts for CO_2 methanation.

2.2.1.2. Nickel dispersion

As a highly active catalyst for CO_2 methanation, a highly uniform dispersed active species over the support is required; therefore, a high specific surface area support is needed. In general, the support usually plays a very important role in the interaction between the Ni and the support. The nickel compounds on different support surfaces result in different "metal-support effects" [30], which implies that catalysts would exhibit different performance toward activity and selectivity for a given process.

Ni/Al_2O_3 with a high specific surface area showed an excellent controllability on the specific surface area of catalysts with the increase in the Ni amount, and increased the reducibility of the catalyst. However, a further increase in the Ni amount would cause a decrease in CO_2 conversion due to the bigger crystallite size and lower surface area of the catalyst [29, 30]. Indeed, the CO_2 conversion and CH_4 yield are strongly dependent on the Ni amount and the calcination temperature. Compared with the no pretreatment catalysts, the prereduced 16% Ni catalyst obtained 100% CH_4 selectivity with no CO detected [47]. With a higher calcination temperature, the metal nickel is in the form of $NiAl_2O_4$, which is an inactive phase for methanation [47, 48]. The existential state of Ni is usually affected by the support. Cubic metallic Ni particles are found mostly without carbon whiskers, and fast methanation occurs at the expense of the CO intermediate on the corners of nanoparticles interacting with Al_2O_3 [43].

The Ni-based catalyst prepared by coprecipitation is active for CO_2 methanation as well. Coprecipitated Ni/Al_2O_3 catalysts are found to be efficient promoters for CO_2 methanation, and Al_2O_3 is active for CO_2 adsorption [49]. A Ni-Al hydrotalcite-derived catalyst (Ni-Al_2O_3-HT) was prepared by a coprecipitation method with a narrow Ni particle-size distribution and an average particle size of 4.0 nm, a large number of Ni nanoparticles were surrounded by amorphous alumina [31]. As for the Ni amount up to 78 wt%, the average crystalline size of Ni was only 4 nm with a narrow distribution in the range of 3–9 nm. Compared with the 78 wt % Ni/Al_2O_3 catalyst using an impregnation method, the Ni-Al hydrotalcite-derived catalyst exhibited a much higher Ni dispersion than its impregnated counterpart, indicating that Ni-Al hydrotalcite is an ideal precursor for preparation of a well-dispersed Ni catalyst.

Recently, a surface defect-promoted Ni nanocatalyst with a high dispersion and high particle density embedded on a hierarchical Al_2O_3 matrix exhibits excellent activity and stability simultaneously for CO_2 methanation. The abundant surface vacancy clusters serve as the active sites, accounting for the significantly enhanced low-temperature activity of the supported Ni nanoparticles [43]. Ni/H–Al_2O_3(400) clearly possesses a significantly enhanced low-temperature activity for CO_2 methanation. The CO_2 conversion exceeded 90% at 265°C and reached the maximal value of 99% at 300°C (**Figure 4A**). The methane production rate increased along with the Ni surface area, indicating a strong correlation between the activity and the Ni surface area. The TOF value as a function of Ni dispersion for the three samples (**Figure 4B**) shows a linear correlation, indicative of a structure sensitive reaction. And the TOF values of the three catalysts toward CO_2 methanation decrease in the following order: Ni/H–Al_2O_3(400) > Ni/H –Al_2O_3(500) > Ni/Al_2O_3 [43].

The different Ni loading amount over the Ni/TiO_2 catalyst strongly affects catalytic CO_2 methanation. When the Ni loading amount was increased to 10 wt%, the selectivity switched to favor the CH_4 formation. Ni nanoparticles (NPs) immobilized on a TiO_2 support were synthesized using a deposition-precipitation method followed by a calcination-reduction process, and the CO_2 conversion and CH_4 selectivity achieved 96 and 99% with a Ni loading of 15 wt% at 260°C [37]. Due to the good dispersion of Ni NPs with large unsaturation facilitates a high exposure of active sites, the formation of surface-dissociated hydrogen and the subsequent hydrogenation removal of surface nickel carbonyl species was accelerated, accounting for the resulting enhanced low-temperature catalytic performance [37].

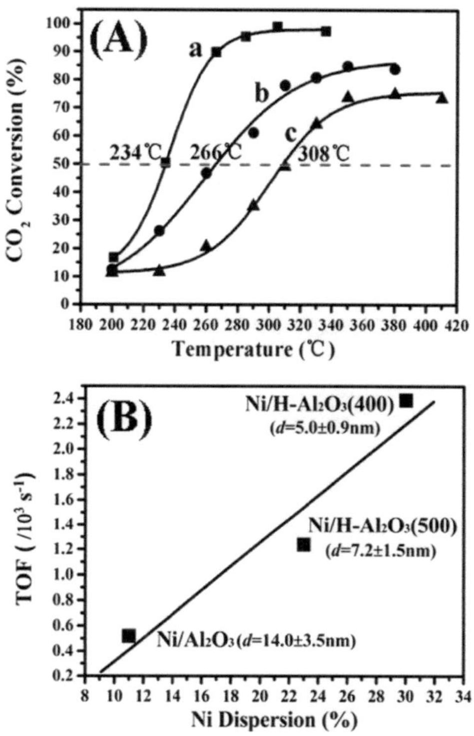

Figure 4. (A) Profiles of CO_2 conversion vs. temperature for CO_2 methanation in the presence of (a) Ni/H–Al_2O_3(400), (b) Ni/H–Al_2O_3(500), and (c) Ni/Al_2O_3 (reacted at 200–410°C and 2400 mL $g_{cat}^{-1}\cdot h^{-1}$(WHSV)). (B) The relationship between the TOF value and the Ni dispersion (reacted at 220°C, 9600 mL $g_{cat}^{-1}\cdot h^{-1}$(WHSV), and <10% CO_2 conversion) [43].

In the past few years, CeO_2–ZrO_2 solid solution ($Ce_xZr_{1-x}O_2$), an active oxygen material, has been commonly used as a support for automotive three-way catalysts because of its high oxygen storage capacity (OSC), which is important in many reactions [50, 51], and it also used as the support for CO_2 methanation. The Ni-based catalysts on $Ce_xZr_{1-x}O_2$ are greatly efficient in terms of activity and stability, which can be attributed to their high oxygen storage capacities and high Ni dispersion [34–36]. In CO_2 methanation, the Ni^{2+} ion incorporation into the Ni–$Ce_xZr_{1-x}O_2$(Ni–CZ) catalyst significantly enhances the specific catalytic activity of the CZ catalyst [44], and the global catalytic activities of CO_2 methanation on CZ catalysts depended on the surface for available metallic nickel, the composition of the support, and its modification by Ni^{2+} doping. In addition, the $Ce_xZr_{1-x}O_2$ catalyst can be synthesized by a simple hydration process, which achieved the goal of Ce and Ni enriched on the surface [34]. Meanwhile, a new NH_3 reduction method for the preparation of Ni–$Ce_{0.12}Zr_{0.88}O_2$ lead to a higher active metal reducibility, smaller Ni^0 crystallite size, and higher metal dispersion compared to the H_2-

reduction method with 100% CO and 97% CO_2 conversions and ≥ 98% CH_4 selectivity at 250°C [36]. For NH_3-treated samples, the metal dispersion is found to decrease with the increase in Ni amounts due to the formation of bulk Ni particles. However, all H_2-treated samples showed a larger NiO particle size and a lower metal dispersion than the NH_3-treated samples might owing to the H_2-reduced sample exhibits an aggregation of smaller particles and/or metal sintering [36].

Nowadays, metal-organic frameworks (MOFs) have attracted much interest as catalysts and/or supporting materials for active metals or complexes in heterogeneous catalysts [52, 53], e.g., a highly active catalyst Ni/MOF-5 showed unexpected activity at low temperature for CO_2 methanation [46]. For 10Ni/MOF-5, a very high specific surface area of 2961 $m^{-2}\cdot g^{-1}$ and a large pore volume of 1.037 $cm^{-3}\cdot g^{-1}$ led to a high dispersion of Ni of 41.8%, and the highly uniform dispersion of Ni in the framework of MOF-5 facilitates a high exposure of active sites, resulting the enhancement of the CO_2 conversion to 75.09% and CH_4 selectivity to 100% at 320°C. To further confirm the high dispersion of Ni on the MOF-5 support, the Ni dispersion on MOF-5 and SiO_2 was measured by the H_2 chemisorption. The Ni dispersion on the 10Ni/MOF-5 catalyst was 41.8% as well as that on 10Ni/SiO_2 was 33.7%, as shown in **Figure 5**, which indicated that Ni was more highly dispersed on MOF-5 [46]. In conclusion, the Ni loading amount is dependent on the type of support used, and the Ni loading amount on the support will determine its crystallite size and dispersion on the surface of the support.

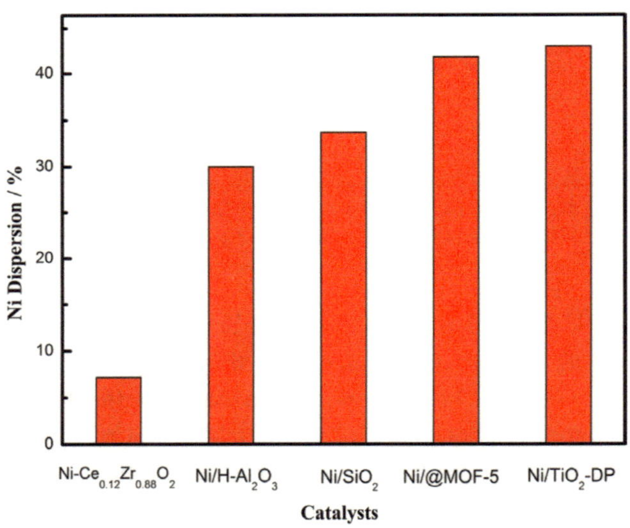

Figure 5. The relation of Ni dispersion and support [36, 37, 43, 46].

2.2.1.3. *Catalyst stability*

The stability of a catalyst is closely related to the structural destruction, coking, and metal sintering during CO_2 methanation [28, 54]. The long-term catalytic stability and thermal stability of Ni/H–Al_2O_3 was investigated, the CO_2 conversion decreases slowly in the first 180 h and then remains almost constant with a total decrease of 7% after 252 h. No obvious aggregation or sintering of Ni nanoparticles was observed for the Ni/H–Al_2O_3 catalyst after 252 h upon streaming [43]. Moreover, the control of thermal sintering is critical for maintaining the activity, which requires a stable support and an effective method to prevent particle migration and coalescence [55]. The embedding of Ni nanoparticles onto the Al_2O_3 matrix enhances the metal-support interaction, and prevents the sintering and/or the aggregation of the active nickel species, which shows that the Ni species was embedded in the hierarchical matrix by an *in situ* reduction approach, and the Ni species exhibit a high dispersion degree and high stability, guaranteeing their high activity during the long-term use.

The Ni/MOF-5 catalyst also shows the catalytic activity during 100 h of CO_2 methanation over 10Ni/MOF-5 at 280°C (**Figure 6**). The CO_2 conversion remained above 47.2% and CH_4 selectivity was almost 100% during the 80 h reaction. Obviously, the 10Ni/MOF-5 catalyst was quite stable [46]. However, on $Ce_xZr_{1-x}O_2$ support, the Ce-rich sample (5NiC4Z) showed the better stability from the CO_2 conversions (72.21–62.18%), whereas the CO_2 conversions were 51.63–36.42% and 37.64–23.19% over 5NiCZ and 5NiCZ4, respectively [34]. The higher reducibility of the Ce-rich supported highly-dispersed Ni catalyst was considered to be the important factors to ensure its long-term stability [34].

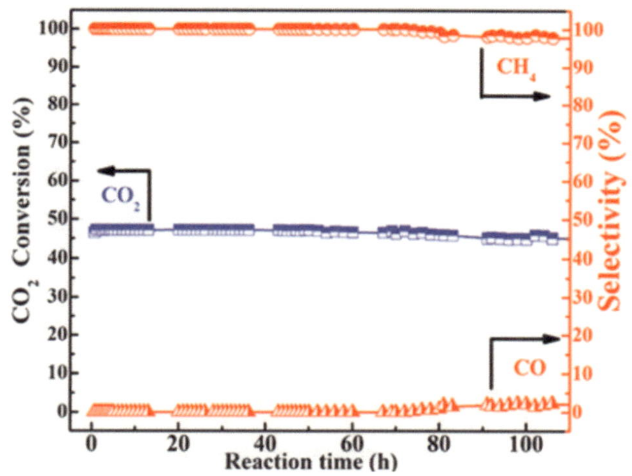

Figure 6. Long-term (100 h) stability tests using the 10Ni/MOF-5 catalyst; reaction conditions: 200 mg catalyst, H_2:CO_2= 4:1, GHSV = 2000 h^{-1}, 1 atm, 280°C [46].

As shown in **Figure 7**, the stability of different Ni supported catalysts was studied, and the rate formation of CH_4 of Ni/MCM-41, Ni/HY, Ni/SiO_2, and Ni/γ-Al_2O_3 catalysts decreases slightly with time on stream increases; however, the rate formation of CH_4 on the Ni/MSN catalyst shows no obvious decrease [32]. In particular, the Ni/MCM-41 shows a minimum percent decrease of the CH_4 formation rate of 3.4%, whereas the Ni/HY, Ni/SiO_2, and Ni/γ-Al_2O_3 is 9.0, 10.6 and 26.6%, respectively. The presence of coke deposition on the active sites is known for the catalyst deactivation; however, no coke content was observed on the Ni/MSN catalyst from the TGA result and the highest coke content was observed on the Ni/Al_2O_3 catalyst (9.1%), indicating that the Ni/MSN catalyst did not show any sign of deactivation for the methanation reaction up to 200 h of time-on-stream. Therefore, the Ni/MSN catalyst is resistant toward coke formation and presented good stabilities under the reaction conditions [32].

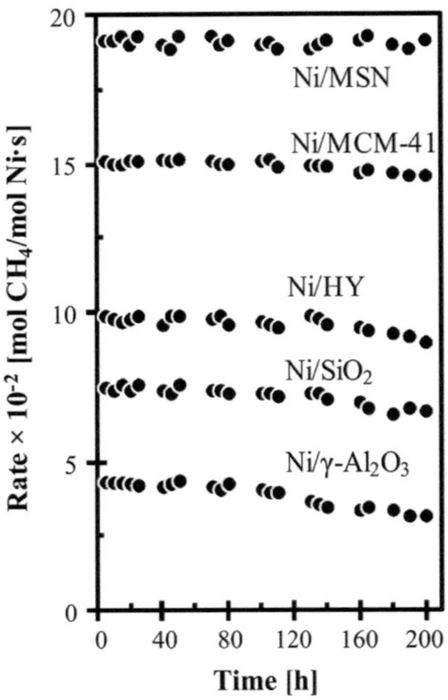

Figure 7. Long-term stability test of Ni catalysts for the CO_2 methanation reaction at a temperature of 573 K, GHSV = 50,000 $mL \cdot g^{-1} \cdot h^{-1}$ and H_2/CO_2= 4:1 [32].

2.2.2. *Effect of the second metal*

2.2.2.1. *Enhancement of catalytic performance*

Ni-based catalysts are vulnerable to sintering and coking, which may lead to their deactivation. Hence, many efforts have been made to enhance the catalytic activity, including selection of appropriate supports and addition of catalytic promoters such as Ce, Zr, La, Mg, V, and Co [45, 56, 57]. The most noticeable effect due to the promotion with these metals is a considerable increase both in the CO_2 conversion and CH_4 selectivity under steady conditions.

The catalytic performance of nickel-based catalysts supported on mesoporous nanocrystalline γ-Al_2O_3 promoted with CeO_2, MnO_2, ZrO_2, or La_2O_3 was investigated, and the Ce promoter considerably increases the CO_2 conversion in the methanation reaction (**Table 3**). The addition of the Ce promoter to Ni increased the dissociation and CO_2 hydrogenation, and weakened the C=O bond of CO_2 adsorbed on the Ni active sites. Compared with the unpromoted Ni/ Al_2O_3 catalyst, the addition of Ce strengthen the interaction between Ce and Ni, resulting in better activity of the Ce–Ni/Al_2O_3 catalyst [58]. Doping the Ni-zeolites catalysts with 3–15% of Ce would be much more enhanced the catalytic performance than the unpromoted catalysts [41]. Actually, the presence of CeO_2 after reduction might promote CO_2 activation into CO, the final catalyst properties being due to the synergetic effect between the metal active sites and the promoter.

Catalysts	CO_2 conversion (%)	CH_4 selectivity (%)
20Ni/Al_2O_3	77.2	100
2Ce-20Ni/Al_2O_3	80.3	100
2Mn-20Ni/Al_2O_3	78	100
2La-20Ni/Al_2O_3	75.4	97.6
2Zr-20Ni/Al_2O_3	74.4	99.1

Table 3. Catalytic evaluation of the Ni/Al_2O_3 catalyst with different promoters [56].

Reaction conditions: H_2/CO_2 molar ratio = 3.5, GHSV = 9000 $mL{\cdot}g_{cat}^{-1}{\cdot}h^{-1}$ and 350°C.

Some active metals, such as Co, Cu, and Fe, are also used to control the catalytic performance over the supported Ni catalyst, which behave an active aspect as the second metal. Compared with Co and Cu, iron is a suitable second metal for the Ni/ZrO_2 catalyst for low-temperature CO_2 methanation [59], which might be due to its strong electron-donating ability, and Fe^{2+} can promote the reduction of nickel and zirconia. Interestingly, similar results are verified and evaluated the catalytic performance of mesoporous nickel-alumina xerogel catalysts (denote as NiAX) with different second metal (M = Fe, Zr, Ni, Y, and Mg) in a fixed bed reactor (**Table 4**) [45]. However, the oxidized Co is more active toward the methane formation at low temperatures [59, 60], and the Co addition can remarkably change the catalytic performance when active $Ce_xZr_{1-x}O_2$ are used as a support for the Ni catalysts [61]. In addition, a homogeneous alloy of Co and Ni can be formed after H_2 reduction and remain after use for reaction in Co-Ni bimetallic catalysts, which increase the metal dispersion in the catalyst, indicating a certain amount of Co addition can considerably improve the catalytic performance [61, 62].

Catalysts	CO_2 conversion (%)	CH_4 selectivity (%)	CH_4 yield (%)
35Ni5FeAX	63.4	99.5	63.1
35Ni5ZrAX	61.6	99.1	61.0
35Ni5NiAX	61.1	99.2	60.6
35Ni5YAX	58.4	99.5	58.1
35Ni5MgAX	54.2	99.5	53.9

Table 4. Catalytic performance of 35Ni5MAX (M = Fe, Zr, Ni, Y, and Mg) catalysts for methane production from carbon dioxide and hydrogen obtained at 220°C after a 10 h-catalytic reaction [45].

2.2.2.2. *Nickel reducibility*

In general, the promotion of methanation catalysts with addition of second metals would enhance the nickel reducibility [63, 64]. The improvements in the Co reducibility may occur without any effect on the Co dispersion for the Ni-Ce/USY catalysts [41]. While the effect of promotion with Ce on the Ni reducibility is particularly pronounced with the alumina-supported Ni catalysts [63]. Compared to the unpromoted Ni/Al_2O_3, the lower reduction temperature of NiO in Ni-CeO_2/Al_2O_3 samples implies that addition of CeO_2 decreased the reduction temperature by altering the interaction between Ni and Al_2O_3, and improved the catalyst reducibility [16, 63, 64]. CNTs-supported catalysts exhibited better catalytic performance than the traditional Al_2O_3-supported catalysts [16], which attributed to the outstanding reduction properties of the CNTs-supported catalysts, which provided much more active sites for CO_2 methanation. As shown in H_2-TPR analysis (**Figure 8**), the accession of Ce could effectively promote the reduction of the nickel oxides, the high reduction peak temperature, corresponding to the highly dispersed nickel oxides in intimate contact with the exterior walls of the CNTs, decreased from 480 to 460°C for the 12Ni/CNT and 12Ni4.5Ce/CNT [16], which suggested easily reducible nickel species on the surface of the 12Ni4.5Ce/CNT catalyst, which may due to the interaction change between the metal oxides and CNTs by the addition of cerium.

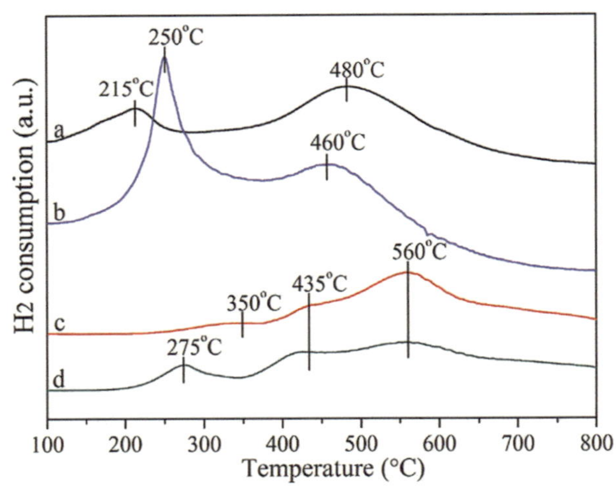

Figure 8. H_2-TPR profiles of the catalysts. (a) 12Ni/CNT, (b) 12Ni4.5Ce/CNT, (c) 12Ni/Al_2O_3, (d) 12Ni4.5Ce/Al_2O_3 [16].

Recently, a new kind of γ-Al_2O_3–ZrO_2–TiO_2–CeO_2 composite oxide supported Ni-based catalysts was synthesized for CO_2 methanation [65]. The optimal catalytic activity of the composite oxide supported Ni-based catalysts was achieved because of the improvements in the reducibility. According to the H_2-TPR profile for all the catalysts, the high temperature peak (weakly interacted with Al_2O_3, or called Ni rich phase) shifts downward for the composite oxide-supported Ni-based catalysts, suggesting a weaker interaction between NiO and the composite support. Furthermore, the reduction of the Ni rich phase would benefit the formation of large-sized Ni particles, which are active at low temperatures [66]. Therefore, increasing the fraction of Ni rich phase, i.e., NiO, the active species for the methanation reaction, would result in an increase in the CO_2 conversion at lower temperatures. Moreover, the H_2 consumed amount increased on the composite oxides support, confirming a higher reducibility of NiO on the composite oxides due to the weaker metal-support interaction [65].

2.3. Cobalt-based catalysts for low-temperature methanation of CO_2

Generally, the Co-based Fischer-Tropsch catalysts exhibit a superior catalytic performance with respect to low-temperature CO_2 methanation [17, 67, 68]. A higher CH_4 selectivity was observed in the Fischer-Tropsch synthesis when the Co catalysts were not completely reduced or when the catalysts contain smaller Co_3O_4 particles [67]. When taking the coke oven gas as feed gas and using a nanosized Co_3O_4 catalyst, CO was easily adsorbed onto the smaller nanosized Co_3O_4 surface and react with H_2, and the temperature at which CO completely converted to CH_4 was much lower than that using nanosized Co_3O_4 with large particles [67].

In addition, the Ru-doped Co_3O_4 catalyst with a relatively rough surface shows a lower light-off temperature than that of a Co catalyst [68]. The relatively rough surface morphology of Ru-doped Co_3O_4 probably results from the larger ionic radius of Ru^{3+}, which affects the dissolution-recrystallization process. Therefore, the final surface morphology of nanorods was disrupted with the addition crystalline defects. The correlation between the surface chemistry and the catalytic performances suggests that doping a noble metal to an oxide of an earth-abundant metal followed by reduction could create a chemically stable, cost-effective catalyst with a bimetallic surface, which has an equivalent or much better catalytic performance [68]. Usually, the catalytic activity affected by the catalyst composition and structure, e.g., when used the mesoporous Co/KIT-6 and Co/meso-SiO_2 in CO_2 methanation, the highly ordered bicontinuous mesoporous structure of the Co/KIT-6 catalyst exhibits higher methane selectivity than the Co/meso-SiO_2 catalyst, and the CO_2 conversion exceeds 48.9%, and the methane selectivity can be retained at 100% at 280°C [17].

3. Reaction mechanisms

According to the previous research, the reaction mechanism was difficult to establish mainly because of the different opinions on the intermediate and the methane formation process. Two feasible reaction mechanisms were proposed for CO_2 methanation in the past decades. The first one involves the CO_2 convert to CO prior to methanation, and the subsequent reaction follows the same mechanism as CO methanation [69]. Similar to the mechanism of CO_2 hydrogenation to CH_3OH, someone considered CO was an intermediate [32], and the CO hydrogenation to methane also been focused [70, 71]. The other mechanism involves the direct CO_2 hydrogenation to methane without forming CO as an intermediate [72]. However, the mechanism depends on different catalysis systems, which are still under investigation.

$$H_2 + 2^* \rightarrow 2H^* \tag{2}$$

$$CO_2 + ^* \rightarrow CO_2^* \tag{3}$$

$$CO_2^* + ^* \rightarrow CO^* + O^* \tag{4}$$

$$CO^* + ^* \rightarrow C^* + O^* \tag{5}$$

$$C^* + H^* \rightarrow CH^* \tag{6}$$

$$CH^* + 3H^* \rightarrow CH_4 + ^* \tag{7}$$

$$O^* + H^* \rightarrow OH^* \tag{8}$$

$$OH^* + H^* \rightarrow H_2O^* \rightarrow H_2O + ^* \tag{9}$$

The atomic hydrogen dissociated from Ni sites in the MSN may facilitate the formation of methane, as shown in **Figure 9**. The oxygen vacancies will be formed when H_2 react with the surface oxygen along with the water generation, which activate additional CO_2 to fill the vacancies and produce CO. During the CO_2 methanation reaction, CO was also suggested as the alternative product, which was an intermediate, as shown in Eqs. (2)–(9) [32]. Therefore, the higher CH_4 selectivity can be explained by the enhanced supply of adsorbed hydrogen to the activated adsorbed CO intermediate, which was the rate-determining step [73]. However, some researchers considered that the main mechanism for CO_2 methanation does not require CO as the reaction intermediate [28, 74], which can be explained by the importance of weak basic sites the adsorption of CO_2 [28].

Density functional theory is helpful in understanding the mechanistic aspects of the reactions. Different mechanisms of CO_2 methanation on Ni(111) surfaces were investigated, and the energy barrier of 237.4 kJ mol^{-1} is acquired for the dissociation of CO into C and O species, which support that CO_2 is converted to CO, subsequently to carbon before hydrogenation [75].

As mentioned, the CO_2 adsorption is a crucial step for methanation. Indeed, CO_2 dissociation is the rate-limiting step. CO_2 dissociation over Rh-based catalysts is influenced by the CO coverage on the surface and the strength of the bond Rh–CO, and the hydrogen adsorption at the surface is competed with CO_2 adsorption. Due to the preferential adsorption of CO_2 and the accumulation of CO on the surface, hydrogen coverage on the rhodium catalyst is very small [76]. However, CO_2 adsorption on the medium basic sites of Ni/$Ce_{0.5}Zr_{0.5}O_2$ results in monodentate carbonates, and monodentate formate derived from monodentate carbonate on medium basic sites, which could be hydrogenated more quickly than bidentate formate derived from hydrogen carbonate. The medium basic sites were proposed to promote the formation of monodentate formate species, thus to enhance the activity [77]

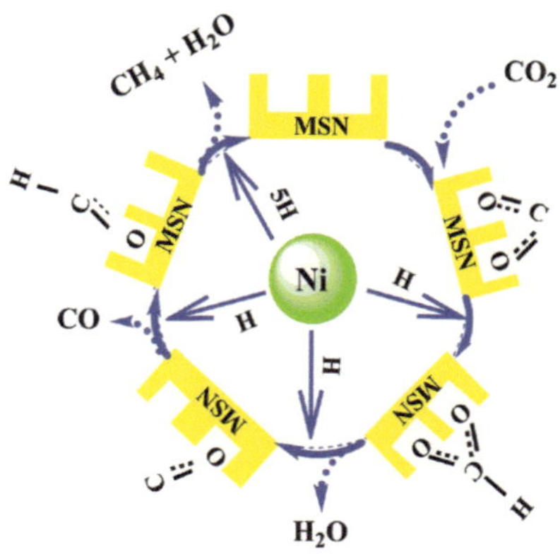

Figure 9. A probable mechanism for Ni/MSN whereby spillover of atomic hydrogen from Ni interacts with C(a) species and sequentially hydrogenates carbon until the product methane desorbs [32].

In addition, at 383 K, a reaction mechanism was proposed for the carbon dioxide methanation reaction on 2% Ru/TiO_2, which investigate the precursor existence for the adsorbed CO and reaction intermediate, and the side-product, formate was also found adsorbing on the support [78], which suggested the surface intermediate corresponding to the adsorbed formate on the metal-support interface, and the measured formate infrared bands are corresponding to the diffused formate species from the interface to the support. A pathway involving hydrogen carbonate is also presented for the formation of the interfacial formate, because the species is formed on the support during the reaction, and the transient response is consistent with the response of a CO precursor. The reaction mechanism that could account for all of these observations is presented in **Figure 10** [78].

To make a better understanding of the adsorption of possible intermediates, the reaction mechanism and factors determining the product selectivity, DFT calculations were considered to be a suitable method to investigate the hydrogenation process of CO_2 and CO on the Ru(0001) surface [79]. For CO_2 hydrogenation, the HCOO intermediate are firstly formed from the adsorbed CO_2 hydrogenation, and subsequently produces an adsorbed CHO and O species. The active C and CH species then undergo stepwise hydrogenation to CH_2, CH_3 and CH_4, or the CHx species, and further transforms to longer carbon chains. From the calculation results, CH_3 hydrogenation is considered to be the rate determining step in the sequence of C hydrogenation on the Ru(0001) surface, and the lowest barrier channel of C–C coupling occurs via the CH + CH reaction [79]. In addition, the study based on DFT calculations on a Ru nanoparticle supported on the TiO_2 catalyst further confirms the stronger electron transfer from the Ru cluster to the TiO_2(101) facet than to TiO_2(001); the Ru species supported on the

(101) plane possesses a relatively lower activation energy for the CO dissociation, resulting in the highly catalytic activity toward CO_2 methanation reaction [80].

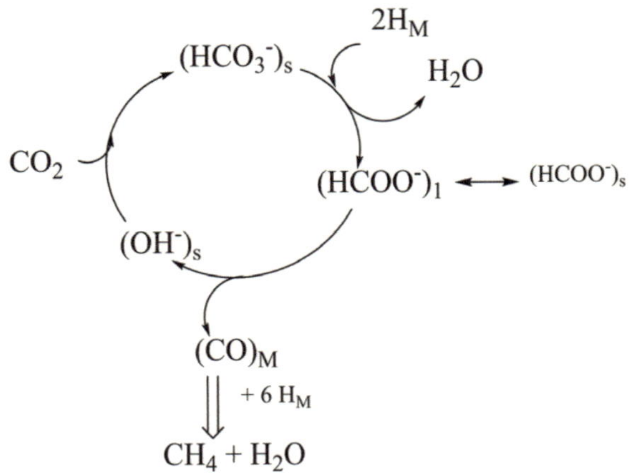

Figure 10. Reaction mechanism of CO_2 methanation [78].

Finally, the detailed mechanism was proposed for CO_2 methanation over metal-based MSNs [81]. As shown in **Figure 11**, CO_2 and H_2 were adsorbed and dissociated on the metal active sites to form CO, O, and H, followed by the migration of these atoms to the MSN surface. Subsequently, the CO dissociated from the active sites interacted with the MSN oxide surfaces to form the carbonyl, including bridged and linear carbonyl, and the H atom in the reaction facilitated the formation of bidentate formate. And the above three species were responsible for the methane formation, among them, the main route for the methane formation was due to the bidentate formate species, and the MSN support served as the sites for carbonyl species, which act as a precursor to methane formation [81].

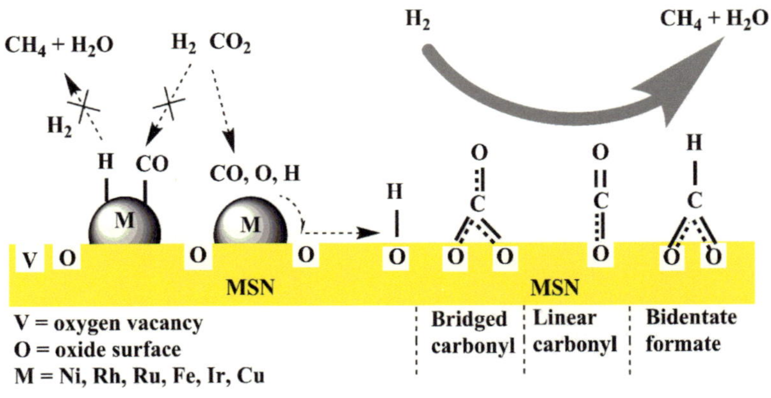

Figure 11. Plausible mechanism of CO_2 methanation on M/MSN [81].

http://dx.doi.org/10.5772/65407

4. Conclusions and perspectives

CO_2 has been promoted to an important carbon resource for conversion and utilization, and CO_2 hydrogenation is a feasible and powerful process, especially for methanation. However, CO_2 is chemically stable and thermodynamically unfavorable. To eliminate the limitations on the conversion and selectivity, various technical directions and specific research approaches on rational design of catalysts and exploration of reaction mechanisms have been presented.

Noble metal catalysts such as Ru, Rh and Pd are efficient for the formation of methane under relatively mild operating conditions, but the high cost as well as their limited availability restricts their practical applications. Therefore, researchers have paid increasing attention on the immobilization of homogenous catalysts to combine the efficient activity with the properties of separation and recyclability. Ni- and Co-based catalysts are, of course, more practical for industrial applications compared to noble metal catalysts. The catalysts with larger surface areas and higher metal dispersion can usually possess higher activity and selectivity, and longer stability in the hydrogenation of CO_2. However, the Ni-based catalysts are more resistless to carbon formation compared with noble metal catalysts. Thus, one strategy has to be proposed for pursuing high-performance catalysts with abilities of low-temperature methanation and resisting carbon formation. In addition, understanding the fundamental mechanisms of CO_2 methanation and explore its relationship with catalyst active site structures using both theoretical calculations (molecular/electronic level modeling) and experimental approaches to tailor new catalyst structures are considerably needed.

Acknowledgements

This work was supported by the National Natural Science Foundation of China (21366004, 21425627) and the Guangxi Natural Science Foundation (2015GXNSFDA139005).

Author details

Zuzeng Qin[1,2*], Yuwen Zhou[1], Yuexiu Jiang[1], Zili Liu[2*] and Hongbing Ji[1,3*]

*Address all correspondence to: qinzuzeng@gmail.com, gzdxlzl@gmail.com and jihb@mail.sysu.edu.cn

1 School of Chemistry and Chemical Engineering, Guangxi University, Nanning, China

2 School of Chemistry and Chemical Engineering, Guangzhou University, Guangzhou, China

3 School of Chemistry and Chemical Engineering, Sun Yat-sen University, Guangzhou, China

References

[1] Wang W, Wang S, Ma X, Gong J. Recent advances in catalytic hydrogenation of carbon dioxide. *Chem Soc Rev*. 2011; 40: 3703–3727. doi: 10.1039/C1CS15008A

[2] Xiaoding X, Moulijn J A. Mitigation of CO_2 by chemical conversion: plausible chemical reactions and promising products. Energy Fuels. 1996; 10: 305–325. doi: 10.1021/ef9501511

[3] Perry R J, O'Brien M J. Amino disiloxanes for CO_2 capture. *Energy Fuels*. 2011; 25: 1906–1918. doi: 10.1021/ef101564h

[4] Wang J, Huang L, Yang R, Zhang Z, Wu J, Gao Y, Wang Q, O'Hare D, Zhong Z. Recent advances in solid sorbents for CO_2 capture and new development trends. *Energy Environ Sci*. 2014; 7: 3478–3518. doi: 10.1039/c4ee01647e

[5] Jessop P G, Joo F, Tai C C. Recent advances in the homogeneous hydrogenation of carbon dioxide. *Coord Chem Rev*. 2004; 248: 2425–2442. doi: 10.1016/j.ccr.2004.05.019

[6] Bai M, Zhang Z, Bai M, Bai X, Gao H. Synthesis of ammonia using CH_4/N_2 plasmas based on micro-gap discharge under environmentally friendly condition. *Plasma Chem Plasma Process*. 2008; 28: 405–414. doi: 10.1007/s11090-008-9132-4

[7] Zeng Y, Wang G. Research and prospects of ethyne from natural gas and its downstream products. *Chem Eng Oil Gas*. 2005; 34: 89–93. doi: 10.1039/b614276a

[8] Bonura G, Cordaro M, Cannilla C, Arena F, Frusteri F. The changing nature of the active site of Cu-Zn-Zr catalysts for the CO_2 hydrogenation reaction to methanol. *Appl Catal B: Environ*. 2014; 152–153: 152–161. doi: 10.1016/j.apcatb.2014.01.035

[9] Liu R-w, Qin Z-z, Ji H-b, Su T-m. Synthesis of dimethyl ether from CO_2 and H_2 using a Cu-Fe-Zr/HZSM-5 catalyst system. *Ind Eng Chem Res*. 2013; 52: 16648–16655. doi: 10.1021/ie401763g

[10] Beuls A, Swalus C, Jacquemin M, Heyen G, Karelovic A, Ruiz P. Methanation of CO_2: Further insight into the mechanism over Rh/γ-Al_2O_3 catalyst. *Appl Catal B: Environ*. 2012; 113–114: 2–10. doi: 10.1016/j.apcatb.2011.02.033

[11] Hashimoto K, Habazaki H, Yamasaki M, Meguro S, Sasaki T, Katagiri H, Matsui T, Fujimura K, Izumiya K, Kumagai N, Akiyama E. Advanced materials for global carbon dioxide recycling. *Mater Sci Eng: A*. 2001; 304–306: 88–96. doi: 10.1016/S0921-5093(00)01457-X

[12] Lin Q, Liu X Y, Jiang Y, Wang Y, Huang Y. Crystal phase effects on the structure and performance of ruthenium nanoparticles for CO_2 hydrogenation. *Catal Sci Technol*. 2014; 4: 2058–2063. doi: 10.1039/c4cy00030g

[13] Karelovic A, Ruiz P. CO_2 hydrogenation at low temperature over Rh/gamma-Al_2O_3 catalysts: Effect of the metal particle size on catalytic performances and

reaction mechanism. *Appl Catal B: Environ*. 2012; 113: 237–249. doi: 10.1016/j.apcatb.2011.11.043

[14] Wang F, Li C, Zhang X, Wei M, Evans D G, Duan X. Catalytic behavior of supported Ru nanoparticles on the {100}, {110}, and {111} facet of CeO_2. *J Catal*. 2015; 329: 177–186. doi: 10.1016/j.jcat.2015.05.014

[15] Hetterley R D, Mackey R, Jones J T A, Khimyak Y Z, Fogg A M, Kozhevnikov I V. One-step conversion of acetone to methyl isobutyl ketone over Pd-mixed oxide catalysts prepared from novel layered double hydroxides. *J Catal*. 2008; 258 250–255. doi: 10.1016/j.jcat.2008.06.017

[16] Wang W, Chu W, Wang N, Yang W, Jiang C. Mesoporous nickel catalyst supported on multi-walled carbon nanotubes for carbon dioxide methanation. *Int J Hydrogen Energy*. 2016; 41: 967–975. doi: 10.1016/j.ijhydene.2015.11.133

[17] Zhou G, Wu T, Xie H, Zheng X. Effects of structure on the carbon dioxide methanation performance of Co-based catalysts. *Int J Hydrogen Energy*. 2013; 38: 10012–10018. doi: 10.1016/j.ijhydene.2013.05.130

[18] Abe T, Tanizawa M, Watanabe K, Taguchi A. CO_2 methanation property of Ru nano-particle-loaded TiO_2 prepared by a polygonal barrel-sputtering method. *Energy Environ Sci*. 2009; 2: 315–321. doi: 10.1039/b817740f

[19] Kondratenko E V, Amrute A P, Pohl M-M, Steinfeldt N, Mondelli C, Perez-Ramirez J. Superior activity of rutile-supported ruthenium nanoparticles for HCl oxidation. *Catal Sci Technol*. 2013; 3: 2555–2558. doi: 10.1039/c3cy00372h

[20] Li D, Ichikuni N, Shimazu S, Uematsu T. Hydrogenation of CO_2 over sprayed Ru/TiO_2 fine particles and strong metal-support interaction. *Appl Catal A: Gen*. 1999; 180: 227–235. doi: 10.1016/s0926-860x(98)00335-4

[21] Scire S, Crisafulli C, Maggiore R, Minico S, Galvagno S. Influence of the support on CO_2 methanation over Ru catalysts: An FT-IR study. *Catal Lett*. 1998; 51: 41–45. doi: 10.1023/a:1019028816154

[22] Li D, Ichikuni N, Shimazu S, Uematsu T. Catalytic properties of sprayed Ru/Al_2O_3 and promoter effects of alkali metals in CO_2 hydrogenation. *Appl Catal A: Gen*. 1998; 172: 351–358. doi: 10.1016/s0926-860x(98)00139-2

[23] Kowalczyk Z, Stolecki K, Rarog-Pilecka W, Miskiewicz E, Wilczkowska E, Karpiniski Z. Supported ruthenium catalysts for selective methanation of carbon oxides at very low CO_x/H_2 ratios. *Appl Catal A: Gen*. 2008; 342: 35–39. doi: 10.1016/j.apcata.2007.12.040

[24] Hoxha F, van Vegten N, Urakawa A, Krurneich F, Mallat T, Baiker A. Remarkable particle size effect in Rh-catalyzed enantioselective hydrogenations. *J Catal*. 2009; 261: 224–231. doi: 10.1016/j.jcat.2008.12.002

[25] Kusama H, Bando K K, Okabe K, Arakawa H. Effect of metal loading on CO_2 hydrogenation reactivity over Rh/SiO_2 catalysts. *Appl Catal A: Gen*. 2000; 197: 255–268. doi: 10.1016/s0926-860x(99)00486-x

[26] Kwak J H, Kovarik L, Szanyi J. CO_2 reduction on supported Ru/Al_2O_3 catalysts: cluster size dependence of product selectivity. *ACS Catal*. 2013; 3: 2449–2455. doi: 10.1021/cs400381f

[27] Büchel R, Baiker A, Pratsinis S E. Effect of Ba and K addition and controlled spatial deposition of Rh in Rh/Al_2O_3 catalysts for CO_2 hydrogenation. *Appl Catal A*. 2014; 477: 93–101. doi: 10.1016/j.apcata.2014.03.010

[28] Aldana P A U, Ocampo F, Kobl K, Louis B, Thibault-Starzyk F, Daturi M, Bazin P, Thomas S, Roger A C. Catalytic CO_2 valorization into CH_4 on Ni-based ceria-zirconia: Reaction mechanism by operando IR spectroscopy. *Catal Today*. 2013; 215: 201–207. doi: 10.1016/j.cattod.2013.02.019

[29] Rahmani S, Rezaei M, Meshkani F. Preparation of highly active nickel catalysts supported on mesoporous nanocrystalline γ-Al_2O_3 for CO_2 methanation. *J Ind Eng Chem*. 2014; 20: 1346–1352. doi: 10.1016/j.jiec.2013.07.017

[30] Chang F W, Kuo M S, Tsay M T, Hsieh M C. Hydrogenation of CO_2 over nickel catalysts on rice husk ash-alumina prepared by incipient wetness impregnation. *Appl Catal A: Gen*. 2003; 247: 309–320. doi: 10.1016/s0926-860x(03)00181-9

[31] He L, Lin Q, Liu Y, Huang Y. Unique catalysis of Ni-Al hydrotalcite derived catalyst in CO_2 methanation: Cooperative effect between Ni nanoparticles and a basic support. *J Energy Chem*. 2014; 23: 587–592. doi: 10.1016/s2095-4956(14)60144-3

[32] Aziz M A A, Jalil A A, Triwahyono S, Mukti R R, Taufiq-Yap Y H, Sazegar M R. Highly active Ni-promoted mesostructured silica nanoparticles for CO_2 methanation. *Appl Catal B: Environ*. 2014; 147: 359–368. doi: 10.1016/j.apcatb.2013.09.015

[33] Wu H C, Chang Y C, Wu J H, Lin J H, Lin I K, Chen C S. Methanation of CO_2 and reverse water gas shift reactions on Ni/SiO_2 catalysts: The influence of particle size on selectivity and reaction pathway. *Catal Sci Technol*. 2015; 5: 4154–4163. doi: 10.1039/c5cy00667h

[34] Cai W, Zhong Q, Zhao Y. Fractional-hydrolysis-driven formation of non-uniform dopant concentration catalyst nanoparticles of Ni/Ce_xZr_{1-x} O_2 and its catalysis in methanation of CO_2. *Catal Commun*. 2013; 39: 30–34. doi: 10.1016/j.catcom.2013.04.025

[35] Pan Q, Peng J, Sun T, Gao D, Wang S, Wang S. CO_2 methanation on Ni/$Ce_{0.5}Zr_{0.5}O_2$ catalysts for the production of synthetic natural gas. *Fuel Process Technol*. 2014; 123: 166–171. doi: 10.1016/j.fuproc.2014.01.004

[36] Razzaq R, Li C, Amin N, Zhang S, Suzuki K. Co-methanation of carbon oxides over nickel-based $Ce_xZr_{1-x}O_2$ catalysts. *Energy Fuels*. 2013; 27: 6955–6961. doi: 10.1021/ef401049v

[37] Liu J, Li C, Wang F, He S, Chen H, Zhao Y, Wei M, Evans D G, Duan X. Enhanced low-temperature activity of CO_2 methanation over highly-dispersed Ni/TiC_2 catalyst. *Catal Sci Technol*. 2013; 3: 2627–2633. doi: 10.1039/c3cy00355h

[38] Vance C K, Bartholomew C H. Hydrogenation of carbon dioxide on group viii metals. *Appl Catal*. 1983; 7: 169–177. doi: 10.1016/0166-9834(83)80005-0

[39] Elliott D J, Lunsford J H. Kinetics of the methanation reaction over Ru, Ru, Ni, Ru, Cu, and Ni clusters in zeolite Y. *J Catal*. 1979; 57: 11–26. doi: 10.1016/0021-9517(79)90039-3

[40] Saha N C, Wolf E E. CO methanation activity and XPS studies of Pd supported on ZSM-5 and Y-zeolites. *Appl Catal*. 1984; 13: 101–112. doi: 10.1016/S0166-9834(00)83331-X

[41] Graca I, Gonzalez L V, Bacariza M C, Fernandes A, Henriques C, Lopes J M, Ribeiro M F. CO_2 hydrogenation into CH_4 on NiHNaUSY zeolites. *Appl Catal B: Environ*. 2014; 147: 101–110. doi: 10.1016/j.apcatb.2013.08.010

[42] Lu B, Ju Y, Abe T, Kawamoto K. Grafting Ni particles onto SBA-15, and their enhanced performance for CO_2 methanation. *RSC Adv*. 2015; 5: 56444–56454. doi: 10.1039/c5ra07461d

[43] He S, Li C, Chen H, Su D, Zhang B, Cao X, Wang B, Wei M, Evans D G, Duan X. A surface defect-promoted Ni nanocatalyst with simultaneously enhanced activity and stability. *Chem Mater*. 2013; 25: 1040–1046. doi: 10.1021/cm303517z

[44] Ocampo F, Louis B, Kiwi-Minsker L, Roger A-C. Effect of Ce/Zr composition and noble metal promotion on nickel based $Ce_xZr_{1-x}O_2$ catalysts for carbon dioxide methanation. *Appl Catal A: Gen*. 2011; 392: 36–44. doi: 10.1016/j.apcata.2010.10.025

[45] Hwang S, Hong U G, Lee J, Baik J H, Koh D J, Lim H, Song I K. Methanation of carbon dioxide over mesoporous nickel-M-alumina (M = Fe, Zr, Ni, Y, and Mg) xerogel catalysts: Effect of second metal. *Catal Lett*. 2012; 142: 860–868. doi: 10.1007/s10562-012-0842-0

[46] Zhen W, Li B, Lu G, Ma J. Enhancing catalytic activity and stability for CO_2 methanation on Ni@MOF-5 via control of active species dispersion. *Chem Commun*. 2015; 51: 1728–1731. doi: 10.1039/c4cc08733j

[47] Garbarino G, Riani P, Magistri L, Busca G. A study of the methanation of carbon dioxide on Ni/Al_2O_3 catalysts at atmospheric pressure. *Int J Hydrogen Energy*. 2014; 39: 11557–11565. doi: 10.1016/j.ijhydene.2014.05.111

[48] Razzaq R, Zhu H, Jiang L, Muhammad U, Li C, Zhang S. Catalytic methanation of CO and CO_2 in coke oven gas over Ni-Co/ZrO_2-CeO_2. *Ind Eng Chem Res*. 2013; 52: 2247–2256. doi: 10.1021/ie301399z

[49] Aksoylu A E, Akin A N, Önsan Z İ, Trimm D L. Structure/activity relationships in coprecipitated nickel-alumina catalysts using CO_2 adsorption and methanation. *Appl Catal A: Gen*. 1996; 145: 185–193. doi: 10.1016/0926-860X(96)00143-3

[50] Wang Q, Li G, Zhao B, Shen M, Zhou R. The effect of La doping on the structure of $Ce_{0.2}Zr_{0.8}O_2$ and the catalytic performance of its supported Pd-only three-way catalyst. *Appl Catal B: Environ*. 2010; 101: 150–159. doi: 10.1016/j.apcatb.2010.09.026

[51] Zhao M, Shen M, Wang J. Effect of surface area and bulk structure on oxygen storage capacity of $Ce_{0.67}Zr_{0.33}O_2$. *J Catal*. 2007; 248: 258–267. doi: 10.1016/j.jcat.2007.03.005

[52] Mueller M, Hermes S, Kaehler K, van den Berg M W E, Muhler M, Fischer R A. Loading of MOF-5 with Cu and ZnO nanoparticles by gas-phase infiltration with organometallic precursors: Properties of Cu/ZnO@MOF-5 as catalyst for methanol synthesis. *Chem Mater*. 2008; 20: 4576–4587. doi: 10.1021/cm703339h

[53] Hermes S, Schroter M K, Schmid R, Khodeir L, Muhler M, Tissler A, Fischer R W, Fischer R A. Metal@MOF: Loading of highly porous coordination polymers host lattices by metal organic chemical vapor deposition. *Angew Chem-Int Ed*. 2005; 44: 6237–6241. doi: 10.1002/anie.200462515

[54] Razzaq R, Li C, Usman M, Suzuki K, Zhang S. A highly active and stable Co_4N/gamma-Al_2O_3 catalyst for CO and CO_2 methanation to produce synthetic natural gas (SNG). *Chem Eng J*. 2015; 262: 1090–1098. doi: 10.1016/j.cej.2014.10.073

[55] Rostrup-Nielsen J R, Pedersen K, Sehested J. High temperature methanation: Sintering and structure sensitivity. *Appl Catal A: Gen*. 2007; 330: 134–138. doi: 10.1016/j.apcata.2007.07.015

[56] Rahmani S, Rezaei M, Meshkani F. Preparation of promoted nickel catalysts supported on mesoporous nanocrystalline gamma alumina for carbon dioxide methanation reaction. *J Ind Eng Chem*. 2014; 20: 4176–4182. doi: 10.1016/j.jiec.2014.01.017

[57] Zhi G, Guo X, Wang Y, Jin G, Guo X. Effect of La_2O_3 modification on the catalytic performance of Ni/SiC for methanation of carbon dioxide. *Catal Commun*. 2011; 16: 56–59. doi: 10.1016/j.catcom.2011.08.037

[58] Enrique Daza C, Gamba O A, Hernandez Y, Centeno M A, Mondragon F, Moreno S, Molina R. High-stable mesoporous Ni-Ce/clay catalysts for syngas production. *Catal Lett*. 2011; 141: 1037–1046. doi: 10.1007/s10562-011-0579-1

[59] Ren J, Qin X, Yang J-Z, Qin Z-F, Guo H-L, Lin J-Y, Li Z. Methanation of carbon dioxide over Ni-M/ZrO_2 (M = Fe, Co, Cu) catalysts: Effect of addition of a second metal. *Fuel Process Technol*. 2015; 137: 204–211. doi: 10.1016/j.fuproc.2015.04.022

[60] Khodakov A Y, Griboval-Constant A, Bechara R, Zholobenko V L. Pore size effects in Fischer Tropsch synthesis over cobalt-supported mesoporous silicas. *J Catal*. 2002; 206: 230–241. doi: 10.1006/jcat.2001.3496

[61] Zhu H, Razzaq R, Li C, Muhmmad Y, Zhang S. Catalytic methanation of carbon dioxide by active oxygen material $Ce_xZr_{1-x}O_2$ supported Ni-Co bimetallic nanocatalysts. *AICHE J*. 2013; 59: 2567–2576. doi: 10.1002/aic.14026

[62] Guo M, Lu G. The regulating effects of cobalt addition on the catalytic properties of silica-supported Ni-Co bimetallic catalysts for CO_2 methanation. *React Kinetic Mechan Catal*. 2014; 113: 101–113. doi: 10.1007/s11144-014-0732-0

[63] Liu H, Zou X, Wang X, Lu X, Ding W. Effect of CeO_2 addition on Ni/Al_2O_3 catalysts for methanation of carbon dioxide with hydrogen. *J Nat Gas Chem*. 2012; 21: 703–707. doi: 10.1016/s1003-9953(11)60422-2

[64] Zhou L, Wang Q, Ma L, Chen J, Ma J, Zi Z. CeO_2 Promoted mesoporous Ni/gamma-Al_2O_3 catalyst and its reaction conditions for CO_2 methanation. *Catal Lett*. 2015; 145: 612–619. doi: 10.1007/s10562-014-1426-y

[65] Abate S, Mebrahtu C, Giglio E, Deorsola F, Bensaid S, Perathoner S, Pirone R, Centi G. Catalytic performance of gamma-Al_2O_3-ZrO_2-TiO_2-CeO_2 composite oxide supported Ni-based catalysts for CO_2 methanation. *Ind Eng Chem Res*. 2016; 55: 4451–4460. doi: 10.1021/acs.iecr.6b00134

[66] Zhang J, Xu H Y, Jin X L, Ge Q J, Li W Z. Characterizations and activities of the nano-sized Ni/Al_2O_3 and Ni/La-Al_2O_3 catalysts for NH_3 decomposition. *Appl Catal A: Gen*. 2005; 290: 87–96. doi: 10.1016/j.apcata.2005.05.020

[67] Zhu H, Razzaq R, Jiang L, Li C. Low-temperature methanation of CO in coke oven gas using single nanosized Co_3O_4 catalysts. *Catal Commun*. 2012; 23: 43–47. doi: 10.1016/j.catcom.2012.02.029

[68] Zhu Y, Zhang S, Ye Y, Zhang X, Wang L, Zhu W, Cheng F, Tao F. Catalytic conversion of carbon dioxide to methane on ruthenium–cobalt bimetallic nanocatalysts and correlation between surface chemistry of catalysts under reaction conditions and catalytic performances. *ACS Catal*. 2012; 2: 2403–2408. doi: 10.1021/cs3005242

[69] Lapidus A L, Gaidai N A, Nekrasov N V, Tishkova L A, Agafonov Y A, Myshenkova T N. The mechanism of carbon dioxide hydrogenation on copper and nickel catalysts. *Petrol Chem*. 2007; 47: 75–82. doi: 10.1134/s0965544107020028

[70] Fujita S I, Takezawa N. Difference in the selectivity of CO and CO_2 methanation reactions. *Chem Eng J*. 1997; 68: 63–68. doi: 10.1016/s1385-8947(97)00074-0

[71] Kraselcuk R, Isli A I, Aksoylu A E, Onsan Z I. CO hydrogenation over bimetallic nickel-vanadium catalysts. *Appl Catal A: Gen*. 2000; 192: 263–271. doi: 10.1016/S0926-860X(99)00409-3

[72] Schild C, Wokaun A, Baiker A. On the mechanism of CO and CO_2 hydrogenation reactions on zirconia-supported catalysts: A diffuse reflectance FTIR study: Part II. Surface species on copper/zirconia catalysts: Implications for methanoi synthesis selectivity. *J Mol Catal*. 1990; 63: 243–254. doi: 10.1016/0304-5102(90)85147-A

[73] Roldan L, Marco Y, Garcia-Bordeje E. Function of the support and metal loading on catalytic carbon dioxide reduction using ruthenium nanoparticles supported on carbon nanofibers. *Chemcatchem*. 2015; 7: 1347–1356. doi: 10.1002/cctc.201500016

[74] Upham D C, Derk A R, Sharma S, Metiu H, McFarland E W. CO_2 methanation by Ru-doped ceria: The role of the oxidation state of the surface. *Catal Sci Technol*. 2015; 5: 1783–1791. doi: 10.1039/c4cy01106f

[75] Ren J, Guo H, Yang J, Qin Z, Lin J, Li Z. Insights into the mechanisms of CO_2 methanation on Ni(111) surfaces by density functional theory. *Appl Surf Sci*. 2015; 351: 504–516. doi: 10.1016/j.apsusc.2015.05.173

[76] Swalus C, Jacquemin M, Poleunis C, Bertrand P, Ruiz P. CO_2 methanation on Rh/γ-Al_2O_3 catalyst at low temperature: "In situ" supply of hydrogen by Ni/activated carbon catalyst. *Appl Catal B: Environ*. 2012; 125: 41–50. doi: 10.1016/j.apcatb.2012.05.019

[77] Pan Q, Peng J, Sun T, Wang S, Wang S. Insight into the reaction route of CO_2 methanation: Promotion effect of medium basic sites. *Catal Commun*. 2014; 45: 74–78. doi: 10.1016/j.catcom.2013.10.034

[78] Marwood M, Doepper R, Renken A. In-situ surface and gas phase analysis for kinetic studies under transient conditions: The catalytic hydrogenation of CO_2. *Appl Catal A: Gen*. 1997; 151: 223–246. doi: 10.1016/S0926-860X(96)00267-0

[79] Zhang S-T, Yan H, Wei M, Evans D G, Duan X. Hydrogenation mechanism of carbon dioxide and carbon monoxide on Ru(0001) surface: A density functional theory study. *RSC Adv*. 2014; 4: 30241–30249. doi: 10.1039/c4ra01655f

[80] Wang F, Zhang S, Li C, Liu J, He S, Zhao Y, Yan H, Wei M, Evans D G, Duan X. Catalytic behavior of supported Ru nanoparticles on the (101) and (001) facets of anatase TiO_2. *RSC Adv*. 2014; 4: 10834–10840. doi: 10.1039/c3ra47076h

[81] Aziz M A A, Jalil A A, Triwahyono S, Sidika S M. Methanation of carbon dioxide on metal-promoted mesostructured silica nanoparticles. *Appl Catal A: Gen*. 2014; 486: 115–122. doi: 10.1016/j.apcata.2014.08.022

Chapter 5

Hydrogenation of Polycrystalline Silicon Thin-Film Transistors

Akito Hara and Kuninori Kitahara

Additional information is available at the end of the chapter

http://dx.doi.org/10.5772/65210

Abstract

In this chapter, the behavior of hydrogen (H) atoms in polycrystalline silicon (poly-Si) thin film is investigated in detail in order to evaluate and improve the quality of hydrogenated poly-Si thin films. Hydrogenation drastically improves the Hall effect mobility, whereas excessive hydrogenation tends to degrade it. The catalytic method is useful to inhibit excessive hydrogenation and damage suffered by the electric-field acceleration of charged particle. The H-termination of the dangling bonds at grain boundaries can be observed indirectly or directly by chemical etching and Raman microscopy. This H-termination appeared as the 2000 cm^{-1} local vibrational mode (LVM) in Raman spectra. The breaking of the Si–Si bonds by hydrogenation was detected as the 2100 cm^{-1} LVM. In addition, the defects generated in the plasma process exhibit multiple fine LVMs after hydrogenation. Moreover, we investigated the hydrogenation of low-temperature (LT) poly-Si thin-film transistors (TFTs) from the perspective of the gettering phenomenon. The most important parameter for effective hydrogenation using H gas annealing is the rate of cooling from 400°C.

Keywords: polycrystalline silicon, thin film, thin-film transistors, hydrogenation, Raman scattering

1. Introduction

Low-temperature (LT) polycrystalline silicon (poly-Si) on glass substrates is attractive for use as thin-film transistors (TFTs) on the backplane of liquid crystal displays and organic light-emitting diode displays. It is well known that hydrogen impurities improve the performance and reliability of LT poly-Si TFTs because they combine with the Si dangling bonds that affect the electrical properties of poly-Si film and TFTs. Therefore, many hydrogenation techniques such as hydrogen plasma [1–3], covering with a hydrogen-containing film [4], and hydrogen ion implantation [5–7] were evaluated. However, the behavior of hydrogen atoms and their effects on the electronic and electrochemical properties of poly-Si films are not clear yet. In this chapter, the behavior of hydrogen atoms in poly-Si film is investigated in detail. In addition, we investigated the hydrogenation of LT poly-Si TFTs from the perspective of the gettering phenomenon.

2. Hydrogenation of poly-Si thin films

2.1. Crystallization of poly-Si thin films

The fabrication of LT poly-Si was enabled by development of laser crystallization technology, where amorphous Si (a-Si) thin films on low alkali glass substrate were used as precursor. The details of the laser crystallization are provided in [8]. Here, we used quartz glass as a substrate, which was available for high-temperature heat treatment to investigate the defects in poly-Si. The thicknesses of the a-Si layers were 50–150 nm. Two techniques were employed for laser crystallization, excimer laser crystallization (ELC) [9] and continuous wave laser lateral crystallization (CLC) [10].

The poly-Si fabricated by ELC (ELC poly-Si) is already used in the industry, and the process is performed primarily by using KrF or XeCl excimer lasers, which supply intense pulsed light with durations of approximately 30 ns. A crystalline orientation map of ELC poly-Si is shown in **Figure 1(a)**, which is obtained by electron backscattering diffraction in terms of the surface normal direction. Grain growth with an average grain size of approximately 300 nm was observed. While the surface orientations of the individual grains were scattered over a wide range, they exhibited a maximum value at {001}. Most of the grain boundaries were high-energy random boundaries. The field-effect mobility μ_{FE} obtained for n-channel (n-ch) TFT increased with increase in grain size and reached 320 cm^2/Vs for an average grain size of 700 nm, where the dominant factor varied from grain boundary scattering to lattice scattering [11].

The poly-Si fabricated by CLC (CLC poly-Si) was developed by one of the authors [10]. This technique uses diode-pumped solid-state laser with a wavelength of 532 nm as the heat source. CLC poly-Si exhibits flow-shape growth as shown in **Figure 1(b)** by adjusting the scanning velocity and output power of the laser. Aligning the TFT channel in parallel with the flow effectively prevents grain boundary scattering. This alignment led to a μ_{FE} of 566 cm^2/Vs which is comparable with the μ_{FE} of 670 cm^2/Vs obtained for a MOSFET made from a single-crystalline Si-channel layer separated by ion-implanted oxygen [11].

2.2. Hydrogenation techniques of poly-Si thin films

Semiconductor devices were improved by an annealing process in a forming gas (mixture of inert gas and H_2) and a plasma hydrogenation technique [1, 12, 13]. These effects were understood as the passivation of defects with hydrogen atoms (H). However, in the case of the forming gas, something catalytic is believed to have acted in the device structure because the dissociation of H_2 is extremely small at the device process temperature. For instance, the efficiency of a multicrystalline sheet Si solar cell reportedly improved by the combination of aluminum evaporation on the back face and the forming gas annealing [14]. In this section, we will deal with plain poly-Si thin films on glass substrates and hydrogen radical (H*) intentionally generated by using plasma or hot catalyzer, which is useful for verifying the hydrogenation effects under simplified conditions.

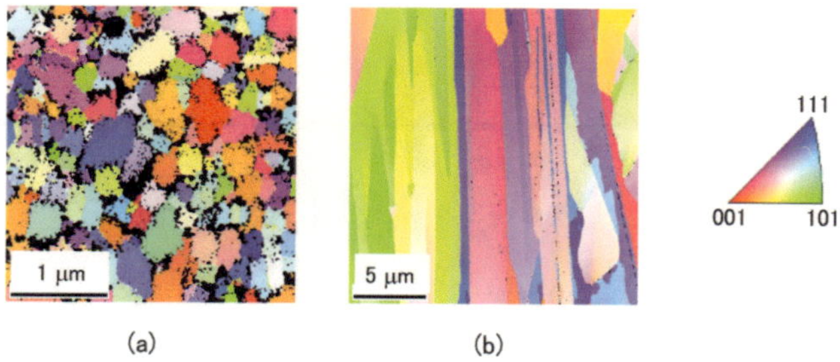

Figure 1. Crystalline orientation maps in terms of the surface normal direction for (a) ELC poly-Si and (b) CLC poly-Si observed by electron beam backscattering diffraction.

Three types of hydrogenation setups are shown in **Figure 2**. **Figure 2(a)** shows a reactor with parallel plate electrodes supplying a radio frequency of 13.56 MHz which is widely used for plasma-enhanced chemical vapor deposition (PE-CVD). Although the degree of perfection of the setup is high, electric-field-accelerated charged particles can cause damage to the semiconductors. Plasma hydrogenation was performed in a PE-CVD reactor with a power of 30 W at a substrate temperature of 350°C for 1–25 min. **Figure 2(b)** shows a simple remote plasma reactor. This setup was convenient to maintain an effective supply of H* under no electric-field acceleration in which H* was generated in a cavity supplying 2.45 GHz microwave. **Figure 2(c)** shows a reactor using an electrically heated tungsten (W) wire as the catalyzer for the dissociation of H_2. This setup was convenient for hydrogenation under no electric field. The principle was known for a long time, and catalytic CVD reactor was developed for industry [15, 16]. The W filament was mesh or coil shaped and was heated to approximately 1300°C under an H_2 pressure of 0.1–0.7 Torr [17, 18].

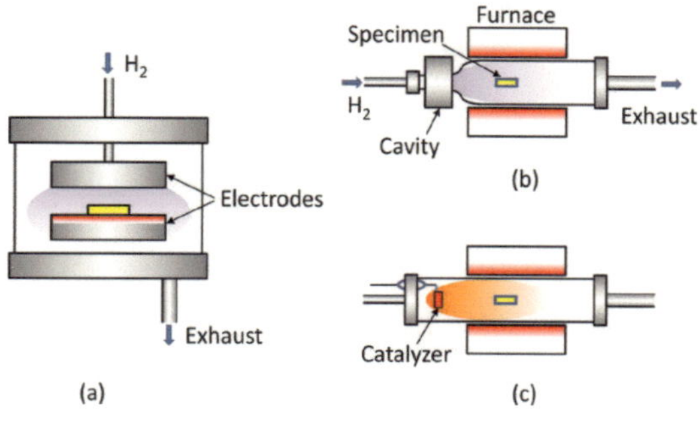

Figure 2. Setups for hydrogenation (a) PE-CVD, (b) remote plasma, and (c) catalytic reactors.

2.3. Hydrogen and grain boundaries in ELC poly-Si

The depth profile of H concentration in ELC poly-Si obtained by secondary ion mass spectroscopy (SIMS) is shown in **Figure 3**, where hydrogenation is performed in the PE-CVD reactor. The enhancement in the concentration near the surface and the Si/SiO_2 interface is apparent values. While the 1-min treatment produced a profile of H diffusing from the surface to the interface, the 10-min treatment led to an almost saturated profile, in which the average density reached was 5×10^{20} cm^{-3} that is 1 atomic percent (at.%).

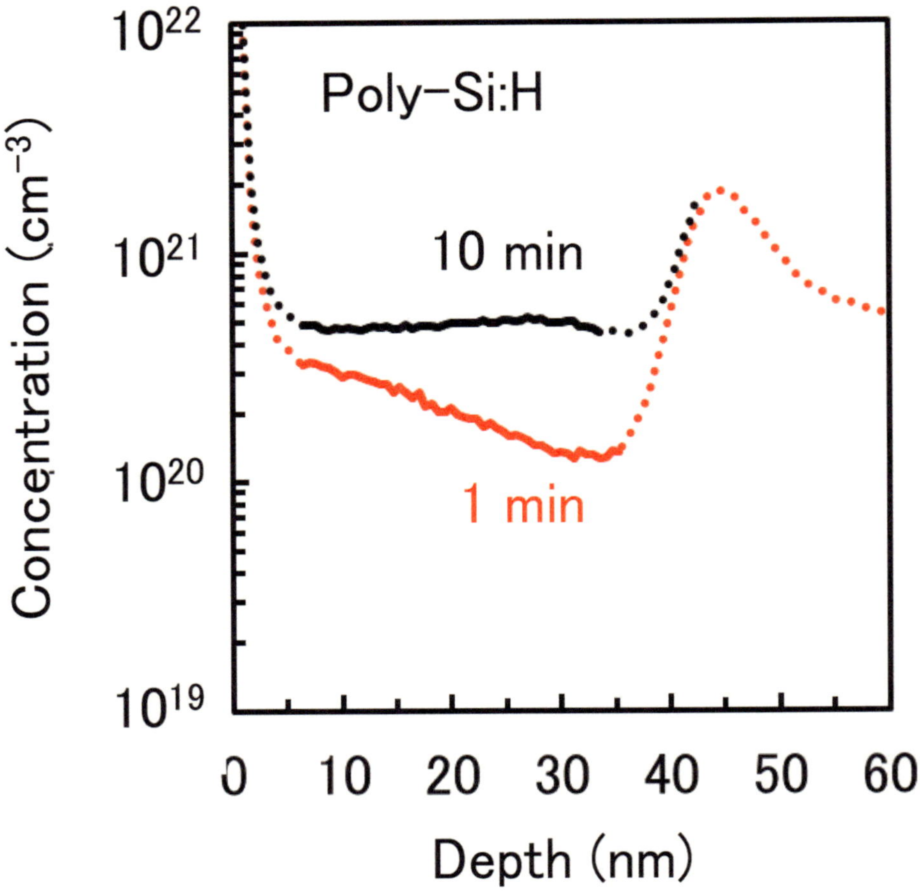

Figure 3. Depth profiles of H concentration in ELC poly-Si hydrogenated for 1 and 10 min in the PE-CVD reactor.

H in poly-Si was often related to the crystalline defects. A representative defect in poly-Si is the high-energy grain boundary. The quantitative detection of the dangling bonds at the grain boundaries was performed by using electron paramagnetic resonance (EPR), which was also used to observe effects of hydrogenation on them. It was observed that the isolated dangling bonds aggregated at the grain boundaries for ELC poly-Si. The detected electron spin density was in the order of 10^{18} cm^{-3} [19]. These dangling bonds were effectively passivated by hydrogenation. However, considering the large density of H shown in **Figure 3**, it is assumed that a majority of the H existed in the film such as interstitial instead of at the termination of dangling bonds.

Grain boundaries in poly-Si can be revealed by Secco etching, which was developed to detect dislocations in single-crystalline Si [20]. **Figure 4(a)** shows a scanning electron microscopy (SEM) image for ELC poly-Si after the etching. Grain boundaries appear as evident lines. On the other hand, almost no grain boundaries are revealed for the hydrogenated film, whereas faint twin boundaries are observed as shown in **Figure 4(b)**. This effect of hydrogenation can be understood by the electrochemical model as follows. The chemical etching proceeds with the electron transfer from the conduction band to proton in solution, which leads to production of an intermediate with a higher oxidization state. Although the electron transfer is intercepted by the energy barrier, the transfer can occur via an electron-hole recombination centers in the band gap [21]. It was shown that the localized states acting as the recombination centers are formed at the high-energy grain boundaries, and it is followed by enhancement of the etching. On the other hand, the hydrogenation relaxes the metastable states with extrinsic H-termination, which leads to suppression of the electrochemical reaction at grain boundaries.

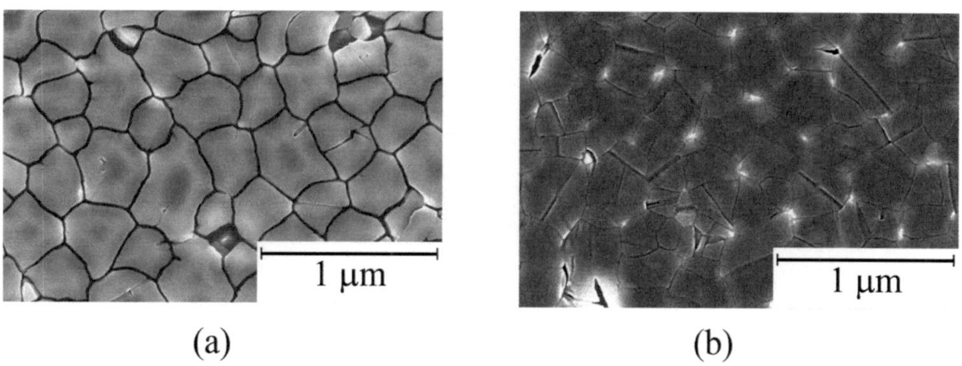

Figure 4. SEM images of ELC poly-Si after chemical etching for 40 s (a) as-crystallized and (b) hydrogenated films.

Next, hydrogenation effects on the ingrain defects were examined. The presence of defects in the grains is attributed to the large cooling velocity after the laser irradiation. The cooling velocity was estimated to be as large as ~10^{10} K/s [22]. The thermal equilibrium defects at the high temperature partially aggregate with each other during the rapid cooling and reside in the grains even at room temperature. In fact, the chemical etching rate decreased to 2/3 by reannealing at 1000°C for 10 min even with no hydrogenation, which was attributed to the annealing out of ingrain defect [23].

2.4. Hydrogenation effects on electron mobility

The mobility of poly-Si for TFT is generally expressed by μ_{FE}, which depends not only on the Si film characteristics but also on the device structures and performance of the Si/SiO_2 interface at the gate. The mobility of plain poly-Si is expressed by the value μ_0 obtained under thermal equilibrium by Hall effect. In both cases, the mobility is limited by the trapping states at the grain boundaries [24, 25].

The poly-Si layer for TFT was almost depleted because of the small thickness and high purity. Therefore, phosphorous (P) ions were doped into poly-Si followed by annealing at 600°C for 2 min in N_2 for activation. The averaged density of P obtained by SIMS was 3×10^{18} cm^{-3}. The obtained μ_0 values are treated as relative values because of the uncertainty about the effects of the charged state on the free surface and at the Si/under-layer interface in addition to the uncertain barrier height at grain boundaries depending on the impurity density. The variation in μ_0 with the hydrogenation time is shown in **Figure 5**, where the hydrogenation is performed by using plasma or catalyzer. The plasma hydrogenation enhanced the value of μ_0 in a short time. A larger value of μ_0 was obtained by catalytic hydrogenation, whereas longer time was required to reach the maximum, which was because no damage was caused by the charged particles. In both cases, excess hydrogenation decreased μ_0. The increase in defect density over a long hydrogenation time was also reported for CVD poly-Si [26].

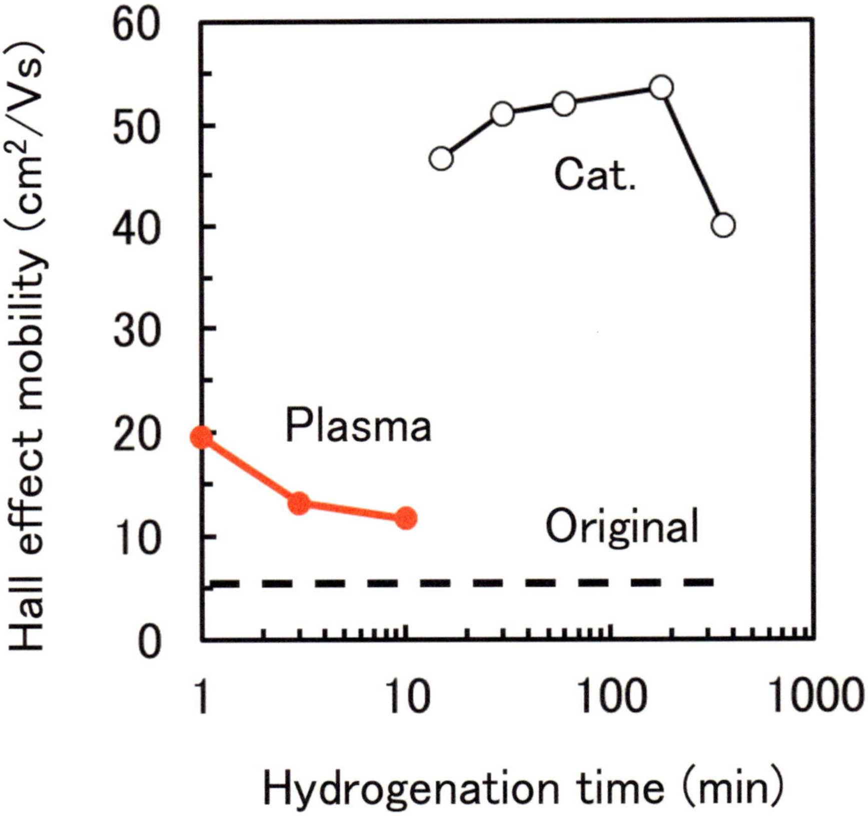

Figure 5. Variation in Hall effect mobility with hydrogenation times for ELC poly-Si hydrogenated in PE-CVD and catalytic reactors, where the broken line indicates the value at the stage of non-hydrogenation.

2.5. Si–H bonds in ELC poly-Si observed via local vibrational modes

The bonding states of H in Si were investigated via local vibrational mode (LVM) by using infrared absorption, attenuated total reflection and Raman scattering. Many literatures indicate the appearance of the LVM at 2000 and/or 2100 cm^{-1} for various crystalline phases, which were summarized in [18] and [27]. The difference in frequency was attributed to the orders of hydride, that is Si–H_x (x is a positive integer), and structures around the hydride [28]. The 2000 cm^{-1} and the 2100 cm^{-1} bands are related to the isolated SiH_x (principally Si–H) and clustered Si–H_x group (principally Si–H_2), respectively.

Raman microscopy with backscattering geometry is useful for Si thin films on glass substrates [27]. The intensity of the LVM is 10^{-3} of that of the optical phonon mode (OPM). However, it is known that Raman scattering is enhanced by roughening the surface or fabricating nanostructures, which is analyzed by using the electromagnetic model [29, 30]. High-density hillocks were generated on the surface of ELC poly-Si after crystallization, which was determined by the interference between the multiple-shot laser and the surface roughness [31]. The 20 times of enhancement was exhibited for ELC poly-Si leading to a remarkable improvement in sensitivity for LVMs [32].

In general, the defects in Si exhibited no characteristic band in the Raman scattering spectra. However, the dangling bonds at the defects were easily terminated by H. The coupling of the hydrogenation and observation of Si–H_x LVMs are expected to be useful for investigating defects in Si. The LVM spectra obtained for various H-containing Si thin films are summarized in **Figure 6**. LVMs at 2000 and 2100 cm^{-1} were observed for all the films. The device grade a-Si:H clearly exhibits a 2000 cm^{-1} band with Gaussian shape as shown in **Figure 6(a)**. Thus, the 2000 cm^{-1} band is attributed to the hydride characterizing the amorphous phase. **Figure 6(b)** shows the LVMs observed for the Si thin film deposited by catalytic CVD on glass. At least two additional LVMs are found at 2170 and 2260 cm^{-1}. The OPM spectrum indicates that the film consists of amorphous, nanocrystal, and crystal components; therefore, it is deduced that the difference in the frequency of LVM bands corresponds to the different lattice structures around the hydrides.

The LVMs for ELC poly-Si are summarized in **Figure 6(c–f)**. **Figure 6(c)** and **(d)** shows the LVMs for catalytic hydrogenated ELC poly-Si. The dominance of 2000 or 2100 cm^{-1} varies with crystallization and hydrogenation conditions. It is notable that the 2100 cm^{-1} band exhibits Lorentzian shape in some cases, which suggests that the spectral shape reflects the structural order around the hydride. The LVM bands obtained after intensive plasma hydrogenation are shown in **Figure 6(e)**. At least four additional bands are found at 2030, 2130, 2140, and 2200 cm^{-1}. The best fit was obtained when three of these bands were set to Lorentzian. These Lorentzian bands can be related to the hydrides in ordered structures such as platelets. **Figure 6(f)** shows the spectrum for oxygen plasma-irradiated ELC poly-Si followed by catalytic hydrogenation. Multiple fine bands were observed there, in which meaningful fitting is not easy. Such fine LVMs were also found for H ion-implanted Si [33]. The correlation of frequencies of multiple LVMs between the experimental and theoretical values was demonstrated for H-passivated defects arising in ion implantation and particle radiation [34]. Thus, the fine bands observed in **Figure 6(e)** and **(f)** are attributed to the damage caused by the charged particles in plasma.

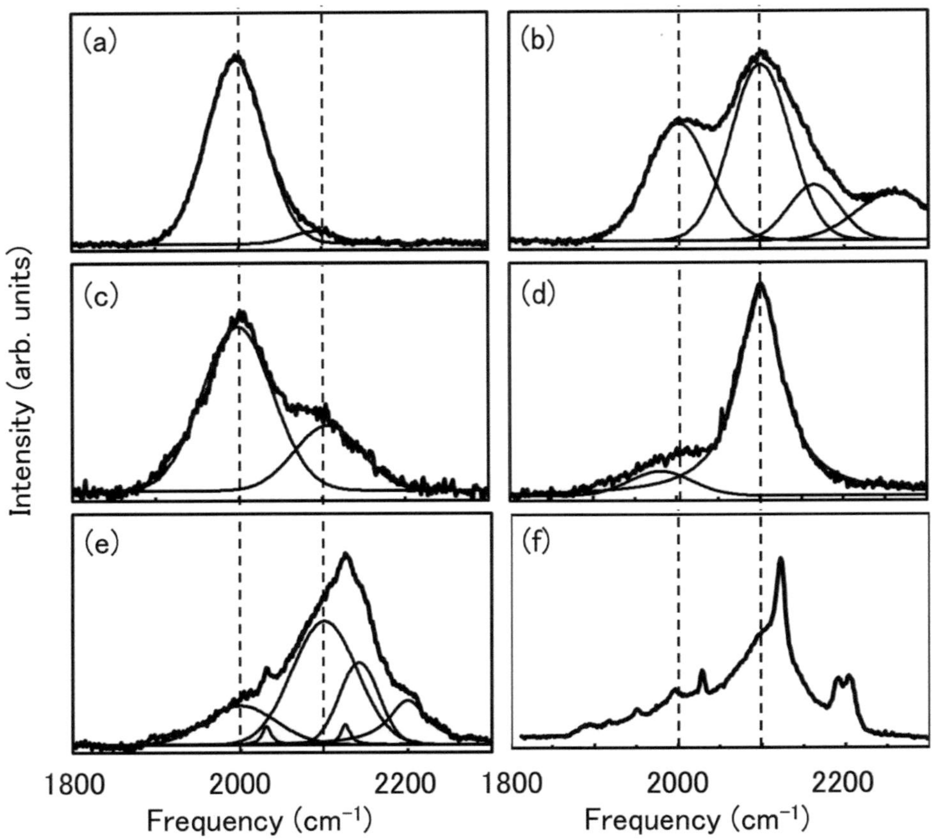

Figure 6. LVMs observed for various Si films: (a) a-Si:H, (b) catalytic CVD Si, and (c–f) hydrogenated ELC poly-Si. ELC poly-Si films were treated as follows: (c) and (d) 3 h catalytic hydrogenation, (e) 10 min plasma hydrogenation, and (f) 1 h catalytic hydrogenation after oxygen plasma exposure, where the broken lines indicate the positions of 2000 and 2100 cm^{-1}.

The location of Si–H_x bonds corresponding to the individual bands, that is, grain boundary or ingrain, was examined. The variation in LVM intensity with the etching time is shown in **Figure 7**, where hydrogenation is performed after the etching of individual time. The intensities are normalized by that at non-etching time. The 2000 cm^{-1} band disappeared in the early stage, whereas the 2100 cm^{-1} band was continuously detected until the diminishing of the Si layer. Molecular dynamics simulation predicted that a few atomic layers of amorphous components reside at the grain boundary in covalent materials [35, 36]. Thus, the 2000 cm^{-1} band was attributed to the hydride in amorphous-like structure at the grain boundary. Furthermore, the 2100 cm^{-1} band is attributed to the hydride mainly located at the ingrains. The enhancement of the 2100 cm^{-1} band in the long etching time region is due to the roughening of surface.

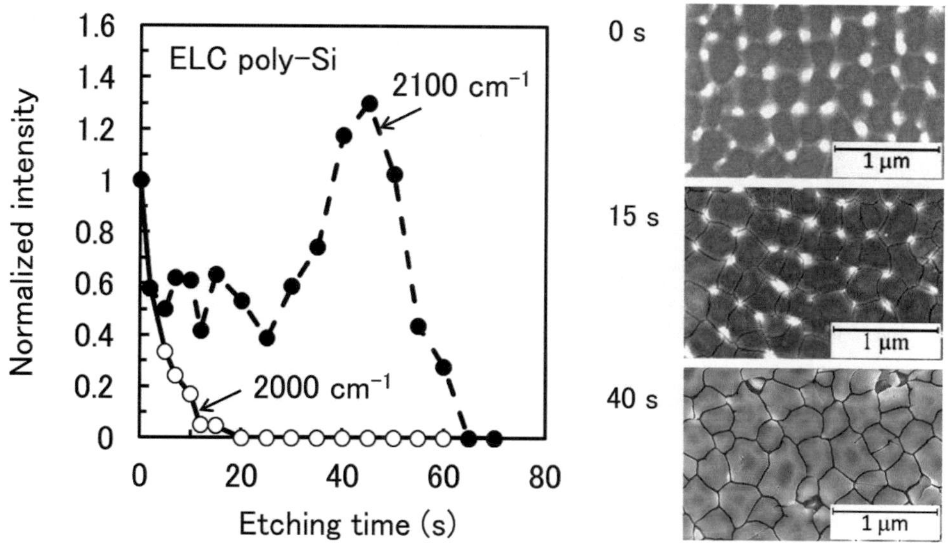

Figure 7. Variation in LVM intensity with Secco etching time, where the intensities of ~2000 and ~2100 cm^{-1} bands are normalized with the values at the surface.

The relationship between the intensity of the two dominant LVMs and hydrogenation time indicates that the 2000 cm^{-1} band intensity saturates with time, whereas the 2100 cm^{-1} band intensity monotonically increases. This implies that hydrogenation generates defects in the grains corresponding to the 2100 cm^{-1} band. A model of breaking weak bonds by H was proposed for a-Si:H for solar cells [37].

2.6. Hydrogen and defects in CLC poly-Si

CLC poly-Si is essentially different from ELC poly-Si in terms of the directionality and solidification velocity during recrystallization. Almost no LVMs were detected for CLC poly-Si even after hydrogenation, which is due to low density of H-terminated defects and/or relatively smooth surface leading to the lack of enhancement of Raman scattering.

The H-termination state in the CLC poly-Si was examined by the chemical etching in the same manner as the case of ELC poly-Si. The precursor was 150-nm-thick a-Si. At non-hydrogenation stage, high-energy grain boundaries were revealed as flow-shaped lines that extend along the laser scanning direction. The growth proceeds by repeating the generation in grains and coalescence with each other as shown by the SEM image in **Figure 8(a)**. Almost no grain boundaries were revealed in the hydrogenated film by the etching as shown in **Figure 8(b)**. This implies that the grain boundaries were inactivated by the hydrogenation as well as the case of the ELC.

2.7. Summary on the hydrogenation of poly-Si thin films

Hydrogenation has two opposite influences on the performance of poly-Si: improvement caused by the passivation of dangling bonds at the defects and degradation caused by the breaking of the Si–Si bond. Here, the hydrogenation of plane poly-Si thin films on glass with H* generated by plasma or catalyzer was described.

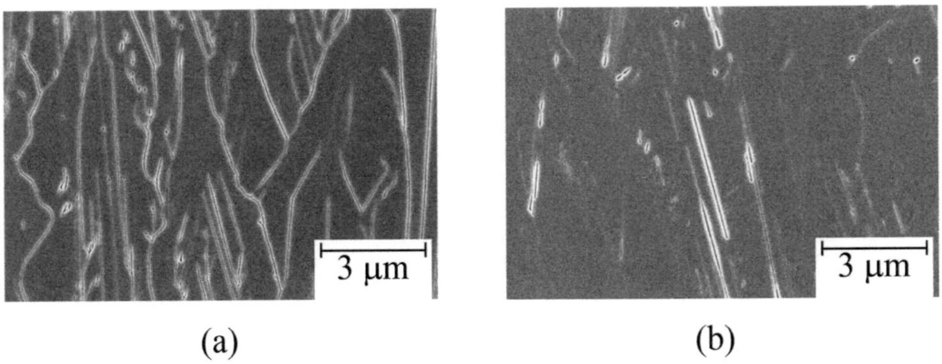

Figure 8. SEM images of CLC poly-Si after chemical etching for 25 s (a) as-crystallized and (b) hydrogenated films.

The plasma hydrogenation process introduced as much as 1 at.% H into the ELC film. Although hydrogenation drastically improves the Hall effect mobility, excessive hydrogenation tends to degrade it. The catalytic method is useful for preventing excessive hydrogenation and damage caused by the electric-field acceleration of charged particle.

The H-termination of dangling bonds at the grain boundaries can be observed indirectly or directly by chemical etching and Raman microscopy. Although preferential etching was found at the grain boundaries, hydrogenation interfered with the process because of the electrochemical inactivation of dangling bond. This H-termination appeared as 2000 cm^{-1} LVM in the Raman spectra. The breaking of Si–Si bonds by hydrogenation was determined from the appearance of the 2100 cm^{-1} LVM. In addition, the defects generated in the plasma process exhibited multiple fine LVMs after hydrogenation. The detection of extremely weak LVMs was caused by the enhancement of Raman scattering induced by the high-density hillocks that are the characteristics of ELC poly-Si.

In the case of CLC, almost no LVM was detected because of the low defect density and relatively smooth surface. Although flow-shaped grain boundaries were revealed by the chemical etching, they were protected from the etching by hydrogenation as well as ELC poly-Si. The density of defects residing in the grains was estimated to be considerably smaller than that observed for ELC, which was due to the unidirectional solidification and low cooling velocity during recrystallization.

3. Hydrogenation of poly-Si TFT

3.1. Gettering of hydrogen in poly-Si TFTs

To achieve effective passivation of electrically active defects in LT poly-Si TFTs by H, it is necessary to evaluate the interaction between H and the defects such as grain boundaries, in-

http://dx.doi.org/10.5772/65210

grain point defects, interface state at poly-Si/SiO_2, and defects in the dielectric film in a poly-Si TFT. The interaction of impurities with defects is widely known as "gettering" [38]. In this study, the evaluation of the hydrogenation of LT poly-Si TFTs in terms of the gettering of H was conducted by H_2 gas annealing [39].

3.2. Experimental procedure

A self-aligned metal double-gate (MeDG) CLC LT poly-Si TFT was used to detect the sensitivity variation in the performance of LT poly-Si TFTs in order to evaluate the interaction between H and the performance of poly-Si TFTs. **Figure 9(a)** and **(b)** shows the three-dimensional image and top view photograph of the TFT, and **Figure 9(c)** shows the transfer characteristic and mobility of the self-aligned MeDG CLC LT poly-Si TFTs [40].

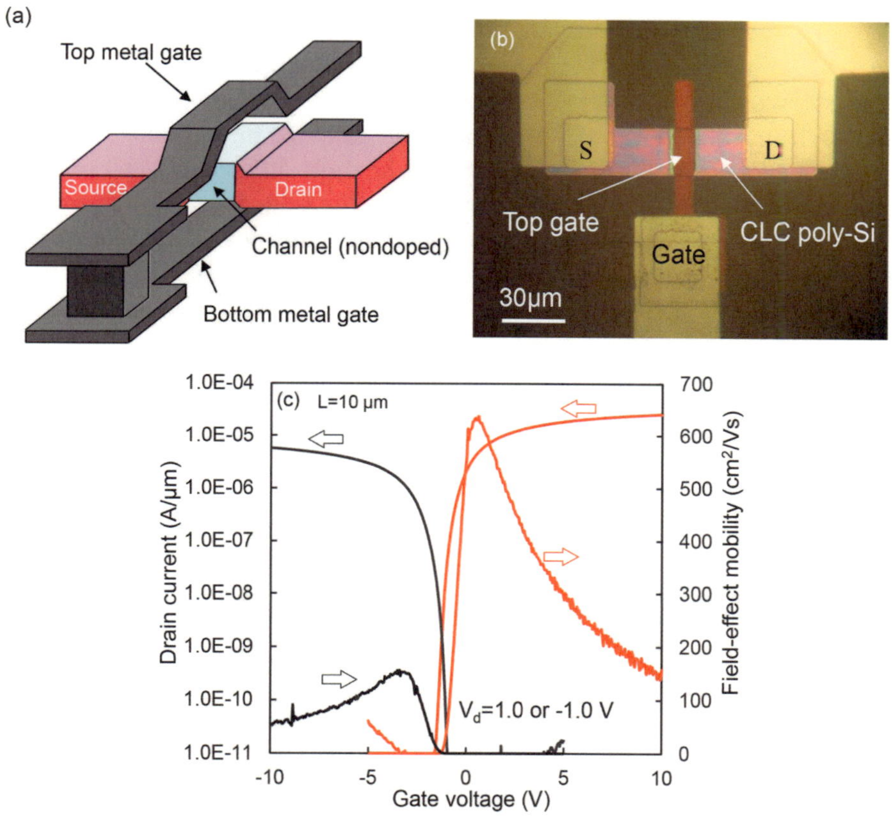

Figure 9. (a, b) Three-dimensional image and top view photograph of TFT and (c) transfer characteristic and mobility of the self-aligned MeDG CLC LT poly-Si TFT.

We used only one self-aligned MeDG CLC LT poly-Si TFT for all the evaluations in this experiment to avoid variations in the performance of poly-Si TFT. We used p-channel (p-ch) TFTs to prevent the degradation of the electrical properties of TFT under DC bias stress [41, 42]. The hydrogenation process of H_2 gas annealing (N_2:H_2 = 97:3) was applied at 400°C for 3 h in a furnace tube. The diffusion coefficient of H at 400°C is very large in Si and SiO_2, and therefore, this hydrogenation condition is sufficient for the introduction of H into the TFT. The TFT was subjected to annealing in N_2 gas at 450°C for 60 min before hydrogenation annealing to initialize its performance. This annealing process results in the initialization of the TFTs, as shown in **Figure 11(b)**.

The behavior of H in the poly-Si film after hydrogenation or dehydrogenation annealing is an indispensable bit of information for determining the mechanism of hydrogenation in LT poly-Si TFTs. The Si–H and Si–H_2 LVM in the poly-Si thin film was directly measured using an ELC poly-Si film with a grain size of 300 nm and Raman microscopy to determine this mechanism. The hydrogenated ELC poly-Si film was prepared by the tungsten (W) hot-wire catalytic technique.

3.3. TFT performance

The hydrogenation process in H_2 gas annealing follows three different cooling processes: quenching, slow cooling, and stepped slow cooling; these are shown in **Figure 10(a)**. Quenching indicates the quick removal of the TFT substrate from the furnace tube into ambient air. **Figure 10(b)** shows the transfer characteristic of the three different cooling processes. **Figure 10(c)** shows a summary of the TFT's performance. The field-effect mobility was calculated in the linear region under a drain bias V_d of -100 mV assuming a TG structure. This indicates that the stepped cooling process is attractive to improve performance of TFTs.

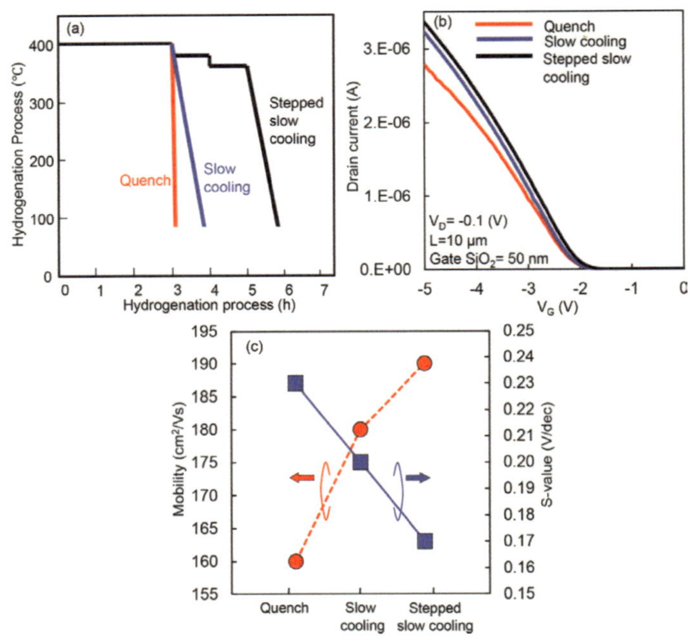

Figure 10. (a) Hydrogenation process for three different cooling processes, (b) transfer characteristic for the three different cooling processes in (a) and (c) summary of TFT performance.

The annealing procedure in **Figure 11(a)** is applied to evaluate the temperature at which the degradation of the TFT's performance begins. After hydrogenation by the stepped slow cooling, the samples were annealed in N_2 gas for 60 min and the TFT performance was evaluated. The red and blue lines in **Figure 11(b)** show the variation in the field-effect mobility and subthreshold awing (SS) value at each step. The degradation of the SS value and mobility was clearly observed after annealing at 400°C in N_2 gas.

Figure 12 shows the variations in the Si–H LVMs after W hot-wire hydrogenation followed by annealing in N_2 gas. In this experiment, hydrogenation was initially conducted under the same conditions for four samples after which N_2 annealing was performed at different temperatures for each sample. This shows that the intensities of the Si–H LVMs are reduced by N_2 annealing above 350°C. The same trends were observed for Si–H_2 LVM.

3.4. Gettering of hydrogen in poly-Si TFTs

The reduction in the Si–H and Si–H_2 LVM above 400°C is caused by dehydrogenation. Since the ELC poly-Si film is thin, it is easy to decrease its volumetric H concentration. This phenomenon also arises in poly-Si TFTs. However, in this case, out-diffusion of H starts at the surface of the SiO_2 interlayer followed by the poly-Si layer.

Figures 11 and **12** show that 400°C is a considerably high temperature for hydrogenation because the binding energy (U) between H and the defects, namely the grain boundaries, in-grain point defects, interface state of poly-Si/SiO_2 and defects in dielectric in the poly-Si TFTs, is lower than the thermal energy at 400°C. Once the H is gettered by defects, it is easily released from them at 400°C, as shown in **Figure 13(a)**. However, at temperatures below 350°C, U is greater than the thermal energy, as shown in **Figure 13(b)**. Thus, temperatures lower than 400°C are important for gettering of H in a poly-Si TFT.

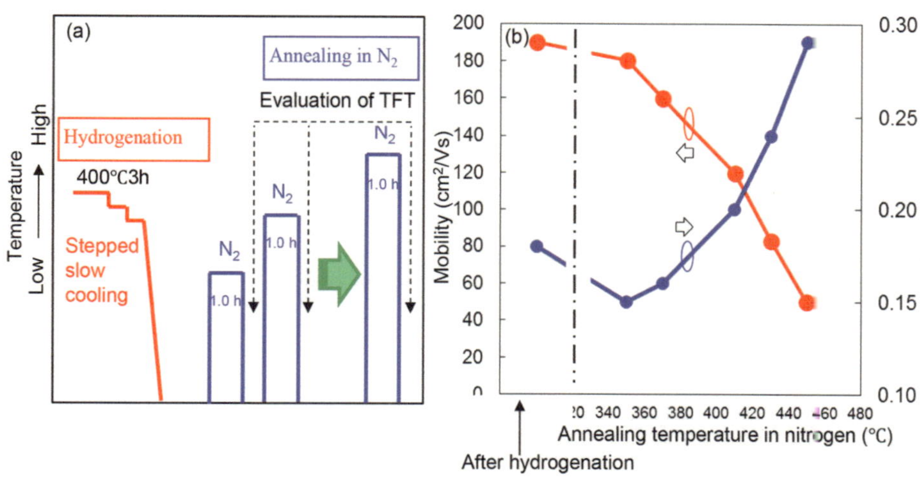

Figure 11. (a) Annealing procedure used to evaluate the temperature at which the degradation of TFT performance begins and (b) variation in the field-effect mobility and SS value at each step.

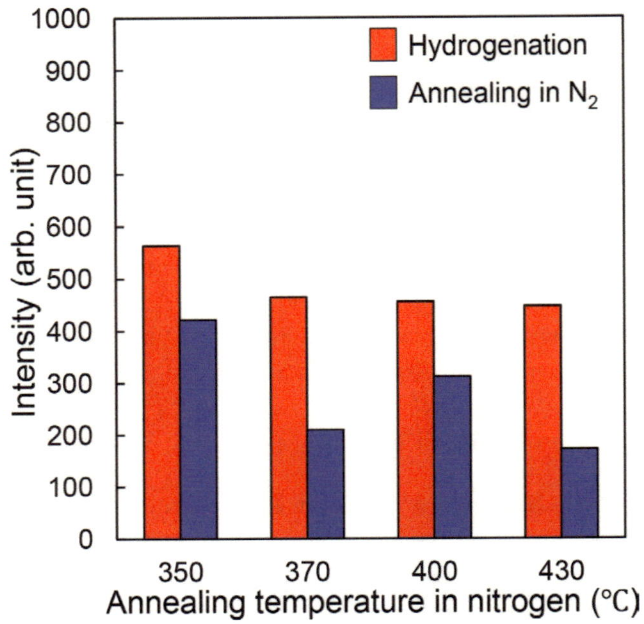

Figure 12. Variations in the Si–H LVM after W hot-wire hydrogenation followed by annealing in N_2 gas.

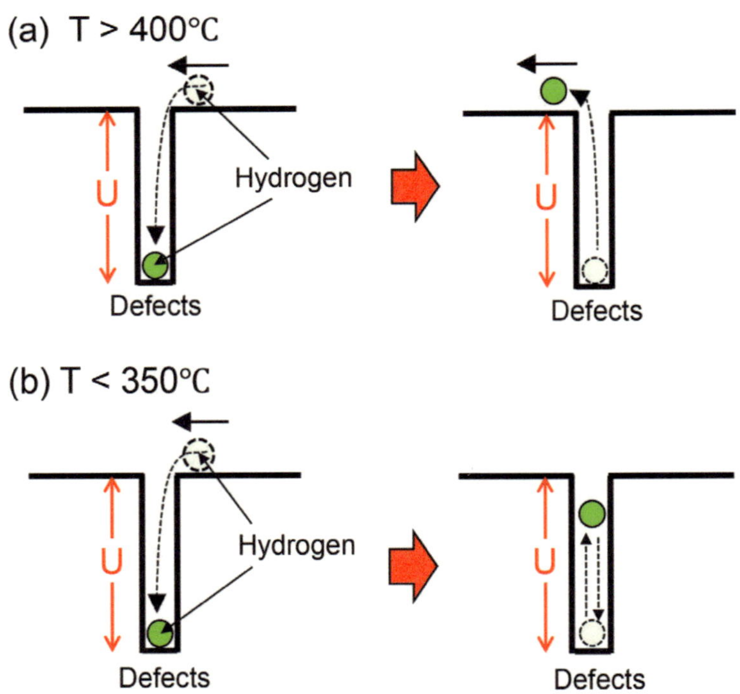

Figure 13. (a) Interaction between defects and H above 400, (b) Interaction between defects and H below 350.

Figure 14 explains the hydrogenation phenomenon from the perspective of gettering for two types of cooling processes after hydrogenation annealing at 400°C. As mentioned above, the thermal energy at 400°C is greater than the binding energy U between H and the defects in the poly-Si TFTs. Therefore, at 400°C, much of the H in the poly-Si TFTs migrates without being trapped by defects. If we used the quenching process, H diffusion motion stops because of the rapid cooling without diffusing over a long distance. Therefore, only the H in the neighborhood of the defects is gettered, which leads to a small amount of gettered H. If we used the stepped slow cooling, the diffusion length of H becomes very large. Therefore, H that is far from the defects can combine with the defects during the cooling process, and the amount of H accumulated at the defects is greater than that in the process with quenching and slow cooling. This leads to effective passivation of electrically active defects in poly-Si TFTs. Thus, the rate of cooling from 400°C is the most important parameter for hydrogenation of poly-Si TFTs by H_2 gas annealing.

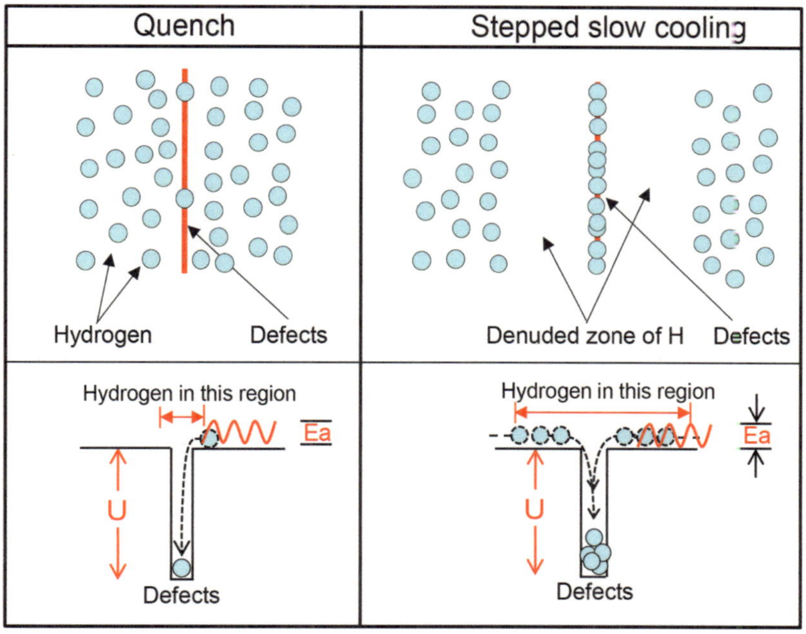

Figure 14. Hydrogenation phenomenon from the perspective of gettering for two types of cooling processes.

3.5. Summary of the hydrogenation of poly-Si TFTs

We investigated the process of hydrogenation in terms of the gettering phenomenon in a high-performance p-ch self-aligned MeDG LT poly-Si TFT. Hydrogenation was carried out by H_2 gas annealing at 400°C for 3 h. Our results show that the hydrogenation temperature of 400°C is rather high, and at this temperature, the gettered H is re-emitted. In the H_2 gas annealing, the hydrogenation of poly-Si TFTs actually occurs when it is cooled to temperatures below 400°C, but not at 400°C. The most important parameter is the rate of cooling from 400°C. In this experiment, the differences between the hydrogenation phenomena depending on the origin of defects were not considered. We need further investigation to clarify the effects of these differences.

4. Summary

In this chapter, the behavior of H atoms in poly-Si was investigated in detail in order to evaluate and improve the quality of hydrogenated poly-Si thin films. In the case of ELC, the plasma hydrogenation introduced as much as 1 at.% H into the film. Hydrogenation drastically improves the Hall effect mobility, whereas excessive hydrogenation tends to degrade it. The catalytic method is useful to inhibit the excessive hydrogenation and damage suffered by the electric-field acceleration of charged particle. The H-termination of the dangling bonds at the grain boundaries can be observed indirectly or directly by chemical etching and Raman microscopy. Although preferential etching was found at the grain boundaries, hydrogenation interfered with it because of the electrochemical inactivation of the dangling bonds. This H-termination appeared as the 2000 cm^{-1} LVM in the Raman spectra. The breaking of the Si–Si bonds by hydrogenation was detected as the 2100 cm^{-1} LVM. In addition, the defects generated in the plasma process exhibit multiple fine LVMs after hydrogenation. These extremely weak LVMs were detected due to the enhancement of Raman scattering induced by the high-density hillocks that are characteristic of ELC poly-Si. In addition, we investigated the hydrogenation of LT poly-Si TFTs from the perspective of the gettering phenomenon. The most important parameter for effective hydrogenation using forming gas annealing is the rate of cooling from 400°C, but the hydrogenation temperature is not important.

Author details

Akito Hara[1*] and Kuninori Kitahara[2]

*Address all correspondence to: akito@mail.tohoku-gakuin.ac.jp

1 Department of Engineering, Tohoku Gakuin University, Japan

2 Interdisciplinary Graduate School of Science and Engineering, Shimane University, Japan

References

[1] Kamins TI, Marcoux PJ. Hydrogenation of transistors fabricated in polycrystalline-Si films. IEEE Electron Device Lett. 1980;EDL-1:159–161. doi:10.1109/EDL.1980.25272

[2] Mimura A, Konishi N, Ono K, Ohwada J-I, Hosokawa Y, Ono YA, Suzuki T, Miyata K, Kawakami H. High performance low-temperature poly-Si n-channel TFTs for LCD. IEEE Trans. Electron Devices. 1989;36:351–359. doi:10.1109/16.19936

[3] Wu I-W, Huang T-Y, Jackson WB, Lewis AG, Chiang A. Passivation kinetics of two types of defects in polysilicon TFT by plasma hydrogenation. IEEE Electron Device Lett. 1991;12:181–183. doi:10.1109/55.75757

[4] Pollack GP, Richardson WF, Malhi SDS, Bonifield T, Shichijo H, Banerjee S, Elahy M, Shah AH, Womack R, Chatterjee PK. Hydrogen passivation of polysilicon MOSFET's from a plasma nitride source. IEEE Electron Device Lett. 1984;5:468–470. doi:10.1109/EDL.1984.25991

[5] Singh HJ, Saraswat KC, Shott JD, McVittie JP, Meindl JD, Hydrogenation by ion implantation for scaled SOI/PMOS transistors. IEEE Electron Device Lett. 1985;6:139–141. doi:10.1109/EDL.1985.26073

[6] Bernstein JD, Qin S, Chan C, King T-J. Hydrogenation of polycrystalline silicon thin film transistors by plasma ion implantation. IEEE Electron Device Lett. 1995;16:421–423. doi:10.1109/55.464804

[7] Cao M, Zhao T, Saraswat KC, Plummer JD. Study on hydrogenation of polysilicon thin film transistors by ion implantation. IEEE Trans. Electron Devices. 1995;42:1134–1140. doi:10.1109/16.387248

[8] Kitahara K, Hara A. Oriented lateral growth and defects in polycrystalline-silicon thin films on glass substrates. In: Andreeta MRB (ed) Crystallization—Science and Technology. Croatia: Intech; 2012. Chap. 19, p. 507–534. ISBN:978-953-51-07576

[9] Sameshima T, Usui S, Sekiya M. XeCl excimer laser annealing used in the fabrication of poly-Si TFT's. IEEE Electron Device Lett. 1986;EDL-7;276–278. doi:10.1109/EDL.1986.26372

[10] Hara A, Takeuchi F, Takei M, Suga K, Yoshino K, Chida M, Sano Y, Sasaki N. High performance polycrystalline silicon thin film transistors on non-alkali glass produced using continuous wave laser lateral crystallization. Jpn. J. Appl. Phys. 2002;41;L311–L313. doi:10.1143/JJAP.41.L311

[11] Hara A, Takeuchi F, Sasaki N. Mobility enhancement limit of excimer-laser-crystallized polycrystalline silicon thin film transistors. J. Appl. Phys. 2002;91;708–714. doi:10.1063/1.1420766

[12] Hickmott T-W. Annealing of surface states in polycrystalline silicon–gate capacitors. J. Appl. Phys. 1977;48;723–733. doi:10.1063/1.323662

[13] Wu I-W, Lewis AG, Huang T-Y, Chang A. Effects of trap-state density reduction by plasma hydrogenation in low-temperature polysilicon TFT. IEEE Electron Device Lett. 1989;10;123–125. doi:10.1109/55.31689

[14] Sana P, Rohatgi A, Kalejs JP, Bell RO. Gettering and hydrogen passivation of edge defined film fed grown multicrystalline silicon solar cells by Al diffusion and forming gas anneal. Appl. Phys. Lett. 1994;64;97–99. doi:10.1063/1.110880

[15] Langmuir I, Mackay GMJ. The dissociation of hydrogen into atoms. Part I. Experimental. J. Am. Chem. Soc. 1914;36;1708–1722. doi:10.1021/ja02185a011

[16] Matsumura H. Summary of research in NEDO cat-CVD project in Japan. Thin Solid Films. 2001;395;1–11. doi:10.1016/S0040-6090(01)01198-1

[17] Kitahara K, Hara A, Nalajima K, Okabe M. Silicon-hydrogen bonds in laser-crystallized polysilicon thin films and their effects on electron mobility. Jpn. J. Appl. Phys. 1999;38;1320–1325. doi:10.1143/JJAP.38.1320

[18] Kitahara K, Nogi H, Moritani A. Hydrogenation of laser crystallized poly-silicon thin films and characterization of defects using a catalytic method. Thin Solid Films. 2001;395;92–96. doi:10.1016/S0040-6090(01)01217-2

[19] Nickel NH, Anderson GB, Johnson RI. Grain-boundary defects in laser-crystallized polycrystalline silicon. Phys. Rev. B. 1997;56;12062–12068. doi:10.1103/PhysRevB.56.12065

[20] Secco d' Aragona F. Dislocation etch for (100) planes in silicon. J. Electrochem. Soc.: Solid State Sci. Technol. 1972;119;948–951. doi:10.1149/1.2404374

[21] Wijaranakula W. Characterization of crystal originated defects in czochralski silicon using nonagitated Secco etching. J. Electrochem. Soc. 1994;141;3273–3277. doi:10.1149/1.2059318

[22] Sameshima S, Usui S. Pulsed laser-induced melting followed by quenching of silicon films. J. Appl. Phys. 1993;74;6592–6598. doi:10.1063/1.355097

[23] Kitahara K, Ohashi Y, Yamamoto K, Sasaki N. Electrochemical and Raman-scattering characterizations of defects in polycrystalline silicon thin films formed by excimer-laser annealing, solid-phase crystallization, and continuous-wave lateral crystallization. Jpn. J. Appl. Phys. 2009;48;021205 (6 pages). doi:10.1143/JJAP.48.021205

[24] Seto JYW. The electrical properties of polycrystalline silicon films. J. Appl. Phys. 1975;46;5247–5254. doi:10.1063/1.321593

[25] Levinson J, Shepherd FR, Scanon PJ, Westwood WD, Este G, Rider M. Conductivity behavior in polycrystalline semiconductor thin film transistors. J. Appl. Phys. 1982;53;1193–1202. doi:10.1063/1.330583

[26] Shimizu K, Kohama N, Tani T, Hanna J. Post hydrogenation effect by hot wire method on poly-crystalline silicon based devices. J. Non-Cryst. Solids. 2004;338–340;403–407. doi:10.1016/j.jnoncrysol.2004.03.007

[27] Kitahara K, Ishii T, Suzuki J, Bessyo T, Watanabe N. Characterization of defects and stress in polycrystalline silicon thin films on glass substrates by Raman microscopy. Int. J. Spectrosc. 2011;2011;632139 (14 pages). doi:10.1155/2011/632139

[28] Beyer W. Hydrogen phenomena in hydrogenated amorphous silicon. In: Nickel NH (ed) Semiconductors and Semimetals, vol. 61. New York: Academic Press; 1999. p. 165–239. doi:10.1016/S0080-8784(08)62707-6

[29] Jayavel P, Nakamura S, Kesavamoorthy R, Srivastava GP, Tomoda W, Koyama T, Hayakawa Y. Surface morphology effects on the optical phonon modes in InAsxSb1-x epilayers on GaAs(001). Phys. Status Solidi B. 2006;243;R19–R21. doi:10.1002/pssb.200541443

[30] Cao L, Nabet B, Spanier JE. Enhanced Raman scattering from individual semiconductor nanocones and nanowires. Phys. Rev. Lett. 2006;96;157402 (4 pages). doi:10.1103/PhysRevLett.96.157402

[31] McCulloch DJ, Brotherton SD. Surface roughness effects in laser crystallized polycrystalline silicon. Appl. Phys. Lett. 1995;66;2060–2062. doi:10.1063/1.113902

[32] Kitahara K, Ishizaki A. Raman microscopy of silicon for electronic displays and solar cells: Enhanced Raman scattering observed for microstructured surface. J. Appl. Phys. 2012;112;123524. doi:10.1063/1.4769877

[33] Stein HJ. Vacancies and chemical trapping of hydrogen in silicon. Phys. Rev. Lett. 1979;43;1030–1033. doi:10.1103/PhysRevLett.43.1030

[34] Corbett JW. Hydrogen passivation of damage centers in semiconductors. In: Pankove JI (ed) Semiconductors and Semimetals, vol. 34. New York: Academic Press; 1991. p. 49–64. doi:10.1016/S0080-8784(08)62859-8

[35] Keblinski P, Phillpot SR, Wolf D. On the thermodynamic stability of amorphous interangular films in covalent materials. J. Am. Ceram. Soc. 1997;80;717–732. doi:10.1111/j.1151-2916.1997.tb02889.x

[36] Sugio K, Fukushima H. Molecular dynamics studies of crystallization and grain boundary formation in silicon. Solid State Phenomena. 2003;83;381–385. doi:10.4028/www.scientific.net/SSP.93.381

[37] Street RA, Winer K. Defect equilibria in undoped a-Si:H. Phys. Rev. B. 1989;40:6236–6249. doi:10.1103/PhysRevB.40.6236

[38] Sumino K. Basic aspects of impurity gettering. Microelectron. Eng. 2003;66:268–280. doi:10.1016/S0167-9317(02)00918-8

[39] Shika Y, Bessho T, Okabe Y, Ogata H, Kamo S, Kitahara K, Hara A. Impact of the hydrogenation process on the performance of self-aligned metal double-gate low-temperature polycrystalline-silicon thin-film transistors, Jpn. J. Appl. Phys. 2013;52:03BB01. doi:10.7567/JJAP.52.03BB01

[40] Hara A, Sato T, Kondo K, Hirose K, Kitahara K. Self-aligned metal double-gate low-temperature polycrystalline silicon thin-film transistors on glass substrate using back-surface exposure. Jpn. J. Appl. Phys. 2011;50:021401. doi:10.1143/JJAP.50.021401

[41] Wu I-W, Jackson WB, Huang T-Y, Lewis AG, Chiang A. Mechanism of device degradation in n- and p-channel polysilicon TFTs by electrical stressing. IEEE Electron Device Lett. 1990;11:167–170. doi:10.1109/55.61785

[42] Kato N, Yamada T, Yamada S, Nakamura T, Hamano T. Degradation Mechanism of Polysilicon TFTs Under DC Stress. IEDM Technical Digest; 1992, p. 677–680. doi: 10.1109/IEDM.1992.307451

Chapter 6

Catalytic Hydrogenation of Benzoic Acid

Sunil B. Shinde and Raj M. Deshpande

Additional information is available at the end of the chapter

http://dx.doi.org/10.5772/66428

Abstract

Hydrogenation of benzoic acid using mono- and bimetallic catalyst of Ru, Pd, Co, and Re yielded different products. It was observed that 5% Ru/C was an active catalyst for hydrogenation of both aromatic ring and carboxylic group, while Pd/C catalyst hydrogenated only aromatic ring. Ru-Sn/Al_2O_3 is a chemoselective catalyst for hydrogenation of –COOH group of benzoic acid.

Keywords: hydrogenation, benzoic acid, catalysts, chemoselectivity, Ru/C, Ru-Sn/Al_2O_3

1. Introduction

The hydrogenation of carboxylic acids or their esters poses major challenges, as among the carbonyl compounds, the acid/ester carbonyl is less reactive than that of ketone and aldehyde [1]. The formation of alcohols by the hydrogenation of the carbonyl group is one of the important reactions, for academia as well as industry, particularly for fatty alcohols, which have been extensively manufactured from fatty acids using hydrogenation technology [2]. Hence, there is a need for new type of catalysts that make it possible to hydrogenate carboxylic acids to alcohols under mild reaction conditions. The combination of group VIII metals with metals like rhenium, germanium, and tin has resulted in better catalysts for this reaction. The first report on the hydrogenation using these catalysts was by Louessard et al., who studied the hydrogenation of ethyl acetate to ethanol over Ru-Sn/Al_2O_3 catalyst [3]. Cheah et al. prepared Ru-Sn/Al_2O_3 catalyst by sol-gel method for the hydrogenation of carboxylic acids and esters; this catalyst was found to exhibit high catalytic performance [4]. They have also reported this catalyst for selectively hydrogenating different carboxylic acids; however, for terephthalic and isophthalic acid, the selectivity to the corresponding alcohols was 19 and 9.2%, respectively, with considerable amount of over-hydrogenated products [5].

The bimetallic Ru-Co catalyst has been reported for hydrogenation of succinic acid. The doping of ruthenium in cobalt increases the overall hydrogenation activity and the Ru/Co ratio in the Ru-Co bimetallic catalyst was found to affect the selectivity of gamma butyrolactone and tetrahydrofuran [6]. Catalysts such as Mn-K/CeO_2-Al_2O_3 are also reported for hydrogenation of benzoic acid to benzaldehyde [7].

Anderson et al. have reported that product distribution for hydrogenation of aromatic acids over Pd/C and Pd supported on carbon nanofiber in aqueous and aprotic organic solvents was almost the same [8]. Chemoselective hydrogenation of benzoic acid to cyclohexane carboxylic acid was found to be efficient over Ni-B[Ni-Zr-B-PEG(800)] amorphous alloy catalyst than noble catalyst [9].

This chapter gives information related to hydrogenation of benzoic acid, which contains two groups, i.e., aromatic ring and carboxylic group, which is not reported in detail.

2. Experimental

2.1. Catalyst preparation

The catalysts were prepared by impregnation method followed by thermal reduction. The following catalysts were prepared: Ru/C, 5% Pd/C, 5%Re/C, 5% Pd-2% Re/C, 1% Ru-6% Re/C, 5% Ru/Al_2O_3, 5% Ru/TiO_2, 5% Re/Al_2O_3, 5% Co/C, 5% Ru-29%Sn/Al_2O_3, and 5% Ru/MgO.

2.2. Reaction conditions

The reactions were carried out in a 250 ml capacity high-pressure high-temperature reactor supplied by Parr Instruments Co. USA. The reaction conditions for catalyst screening are given in **Table 1**.

Reaction parameter	Range
Benzoic acid quantity	0.82 kmol m^{-3}
Hydrogen partial pressure	6.89 MPa
Catalyst	5.00 kg m^{-3}
Solvent	1,4-Dioxane
Temperature	493 K
Agitation speed	1000 rpm
Total volume	1.0×10^{-4} m^3
Reaction time	6 h

Table 1. Reaction conditions for catalyst screening.

2.3. Analysis

The various hydrogenation products formed during the reaction were identified using GC-MS (Agilent GC 6890 with 5973 mass selective detector instrument).

For quantitative analysis, samples from the reaction mixture were filtered to remove the catalyst and analyzed by gas chromatography (Agilent Systems 6890 GC). For analysis, an HP INNOWax, polyethylene glycol capillary column (30 m × 320 μm × 0.25 μm) along with flame ionization detector was used on Agilent 6890 gas chromatograph controlled by HP ChemStation software, equipped with an autosampler unit.

3. Results and discussion

3.1. Catalyst screening study for hydrogenation of aromatic carboxylic acid

For catalyst screening study, hydrogenation of benzoic acid was carried out with different supported monometallic catalysts such as Pd, Ru, Re, and Co and bimetallic catalysts such as Ru-Re/C, Ru-Sn/Al_2O_3, and Pd-Re/C. For hydrogenation of benzoic acid, metal catalysts such as Co and Ru-Co were also used. The selectivity as well as activity for hydrogenation of benzoic acid was found to be dependent on the catalysts used. The results on catalyst screening study for hydrogenation of benzoic acid are discussed below.

3.1.1. Hydrogenation of benzoic acid with 5% Pd/C catalyst

It was observed that the aromatic ring was selectively hydrogenated to give cyclohexane carboxylic acid (CCA) as shown in **Scheme 1**. A 59.7% conversion of benzoic acid (BZA) with 100% selectivity to cyclohexane carboxylic acid was achieved.

3.1.2. Hydrogenation of benzoic acid with 5% Ru/C catalyst

5% Ru/C was an active catalyst as compared to 5% Pd/C. Conversion of benzoic acid at the end of 6 h was found to be 99%. The 5% Ru/C catalyst was found to hydrogenate the carboxylic group along with aromatic ring. The products formed in the reaction are as shown in **Scheme 2**. Under the reaction conditions, the selectivity to cyclohexane carboxylic acid was 70%, while that for cyclohexyl methanol, 30%.

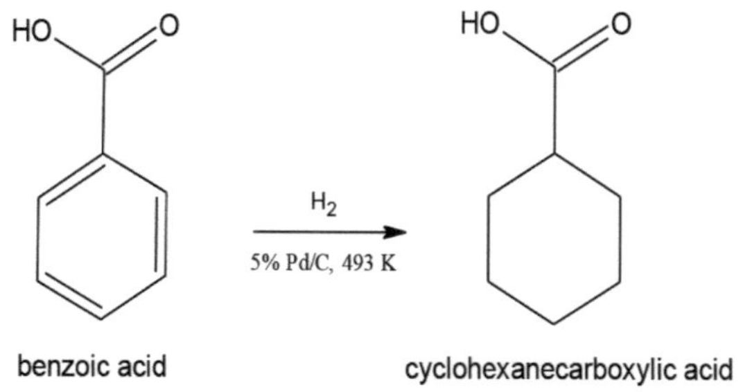

Scheme 1. Reaction scheme for hydrogenation of benzoic acid with 5% Pd/C catalyst.

3.1.3. *Hydrogenation of benzoic acid with 5% Re/C catalyst*

Hydrogenation of benzoic acid over 5% Re/C is very slow as compared to 5% Ru/C and 5% Pd/C catalysts. The reaction pathway for hydrogenation of benzoic acid over 5% Re/C is also different from that for the other two catalysts as shown in **Scheme 3**. In the literature, ReO_3 is reported to be active for selective hydrogenation of terephthalic acid to p-hydroxymethyl benzoic acid [10]. Conversion of benzoic acid was 6.1%, and selectivity to benzyl alcohol and cyclohexyl methanol was 50% each.

3.1.4. *Hydrogenation of benzoic acid with 5%Pd-2%Re/C bimetallic catalyst*

The activity of bimetallic 5%Pd-2%Re/C catalyst was higher than that for monometallic 5% Pd/C and 5% Re/C catalysts in terms of benzoic acid conversion (98.5%). But the selectivity to benzyl alcohol, cyclohexane carboxylic acid, or cyclohexyl methanol was low as compared to the other catalysts such as 5% Re/C and 5% Pd/C. This is because hydrogenation of benzoic acid with 5%Pd-2%Re/C catalyst gives benzyl alcohol which undergoes hydrogenolysis. The reaction is thus both consecutive and parallel, which gives a series of products such as CCA, cyclohexyl methanol (CHM), toluene (TOL), and methyl cyclohexane (MCH).

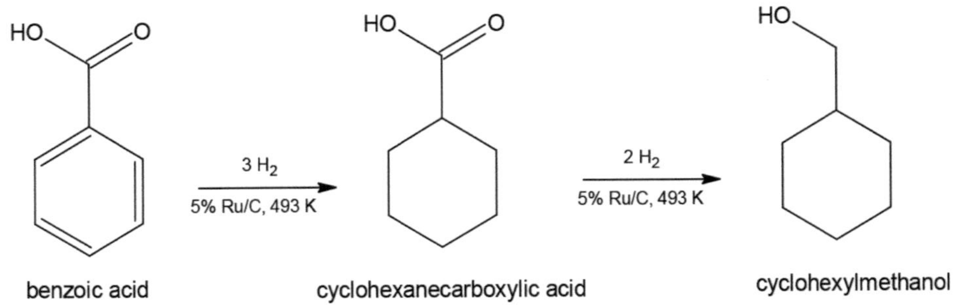

Scheme 2. Reaction scheme for hydrogenation of benzoic acid using 5%Ru/C catalyst.

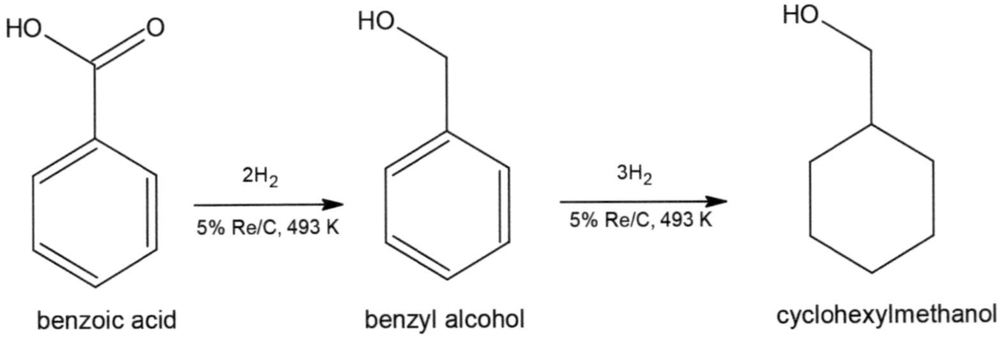

Scheme 3. Reaction scheme for hydrogenation of benzoic acid with 5% Re/C catalyst.

The selectivity for hydrogenation of BZA with 5%Pd-2%Re/C to CCA, CHM, TOL, and MCH is 19, 47, 3.3, and 30.6%, respectively (at 6 h). Such a difference in activity and selectivity for monometallic and bimetallic catalyst only confirms that Pd and Re form a synergistic combination [11].

3.1.5. *Hydrogenation of benzoic acid with 1%Ru-6%Re/C catalyst*

As compared to 5%Pd-2%Re/C catalyst, 1%Ru-6%Re/C showed lower activity in terms of BZA conversion (8.1%). This may be attributed to the low percentage of Ru in the catalyst, which alone is a good hydrogenation catalyst for benzoic acid. With 1%Ru-6%Re/C as catalyst, it was observed that –COOH group was hydrogenated in preference to the aromatic ring. The advantage of 1%Ru-6%Re/C catalyst over 5%Pd-2%Re/C and 5%Ru/C is that the benzyl alcohol (33%) is obtained at the end of reaction.

3.1.6. *Hydrogenation of benzoic acid with 5%Ru-29%Sn/Al_2O_3 catalyst*

Hydrogenation of benzoic acid with Ru catalyst on different supports such as carbon (C), alumina (Al_2O_3), titania (TiO_2), and magnesia (MgO) was studied. The activity and selectivity of the catalyst varied with support as seen in **Table 2**.

5%Ru/MgO was the least active catalyst and hydrogenated only the aromatic ring, while Ru on carbon was very active for hydrogenation of aromatic ring but relatively less active for hydrogenation of carboxylic group. With Ru on carbon and magnesia, hydrogenation of BZA to BZOH was not observed. Ru on supports like alumina, silica, and titania was active for hydrogenation but not a selective as seen in **Table 2**.

When Sn is doped in 5% Ru/Al_2O_3 catalyst, the selectivity of the 5%Ru/Al_2O_3 catalyst is changed completely. This catalyst is observed to hydrogenate only the carboxylic group of benzoic acid to give benzyl alcohol as shown in **Scheme 4**. The selectivity to benzyl alcohol is 100% with 24% conversion of benzoic acid.

3.1.7. *Hydrogenation of benzoic acid with cobalt catalysts*

Cobalt metal catalyst has been reported for the hydrogenation of carboxylic acid to alcohols [6]. Hydrogenation of benzoic acid was carried out with 5%Co/C, Co metal, and Ru-Co catalysts. These catalysts were, however, found to be inactive for the hydrogenation of both aromatic and carboxylic groups of BZA under the conditions mentioned in **Table 1**.

				Selectivity				
No.	**Catalyst**	**Conv. %**	**TOF (s^{-1})**	**CCA%**	**CHM%**	**TOL%**	**MCH%**	**BZOH%**
1	5% Ru/C	99	1.52×10^{-2}	70	30	–	–	–
2	5%Ru/Al_2O_3	96.1	1.48×10^{-2}	56	10	6	28	–
3	5%Ru/TiO_2	65	0.99×10^{-2}	13	77	7.8	2.2	–
4	5% Ru/MgO	29.2	0.45×10^{-2}	100	–	–	–	–

Table 2. Hydrogenation of benzoic acid with 5% Ru on different supports.

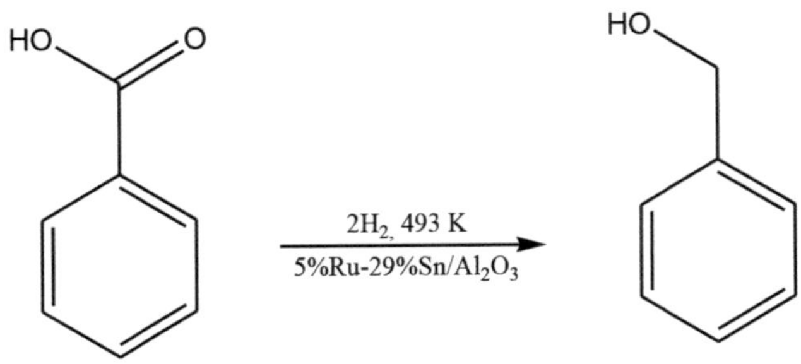

Scheme 4. Hydrogenation of benzoic acid with 5%Ru-29%Sn/Al_2O_3 catalyst.

4. Conclusion

Catalyst screening study for hydrogenation of benzoic acid suggests that 5% Ru/C catalyst was active for hydrogenation of both aromatic ring and carboxylic group, while Pd/C catalyst was observed to be active only for hydrogenation of aromatic ring. With bimetallic Ru-Sn/Al_2O_3, catalyst hydrogenation of aromatic ring was inhibited completely, and hydrogenation of carboxylic group was favored to give benzyl alcohol.

The screening of the catalysts showed that the product distribution for hydrogenation of benzoic acid with mono- and bimetallic catalyst of Ru, Pd, Co, and Re was versatile. The study also highlighted that products such as CCA, CHM, and BZOH can be selectively obtained from the same substrate (BZA).

Acknowledgements

Sunil B. Shinde thanks the Council of Scientific and Industrial Research (CSIR) India for a research fellowship.

Author details

Sunil B. Shinde[1, 2*] and Raj M. Deshpande[1, 3*]

*Address all correspondence to: sunil_b_shinde@rediffmail.com, rm.deshpande@ncl.res.in and raj.deshpande@sabic.com

1 Homogeneous Catalysis Division, NCL, Pune, India

2 Sir Parashurambhau College, Tilak Road, Pune, India

3 SABIC Technology Centre, Banglore, India

References

[1] J. A. McAlee, *J. Chem. Soc. C.* (1969) 2425 DOI: 10.1039/J39690002425

[2] T. Turek, L. D. Trimm and N. W. Cant, *Catal. Rev. Sci. Eng.* V. 36 No. 4. (1994) 645 DOI: 10.1080/01614949408013931

[3] P. Louessard, P. J. Candy, P. J. Bourneville and M. J. Basset, *Structure and Reactivity of Surface*, Amsterdam, Elsevier, (1989). DOI: 10.1163/156856791X00110

[4] Y. K. Cheath, S. T.Tang, F. Mizukami, S. Niwa, M. Toba and M. Y. Choe, *JAOCS*, V. 69. No. 5. (1992) 410 DOI: 10.1007/BF02540940

[5] M. Toba, S. Tanaka, S. Niwa, F. Mizukami, Z. Koppang, L. Guczi, K. Cheath and T. Tang, *Appl. Catal. A*, V. 189, (1999) 243 DOI: 10.1016/S0926-860X(99)00281-1

[6] R. M. Deshpande, V. V. Buwa, C. V. Rode, R. V. Chaudhari and P. L. Mills, *Catal. Commun.*, V. 3, (2002) 269 DOI: 10.1016/S1566-7367(02)00119-X

[7] D. Cheng, C. Hou and F. Chen, *React. Kinet. Catal. Lett*, V. 97. No. 2, (2009) 217 DOI: 10.1007/s11144-009-0034-0

[8] J. A. Anderson, A. Athawale, F. E. Imrie, F. M. McKenna, A. McCue, D. Molyneux, K. Power, M. Shand and R. Wells, *J. Catal.* V. 270, No. 1, (2010) 9 DOI: 10.1016/j.jcat.2009.11.028

[9] G. Bai, X. Wen, Z. Zhao, F. Li, H. Dong and M. Qiu, *Ind. Eng. Chem. Res.*, V 52, No. 6, (2013) 2266 DOI:10.1021/ie303602n

[10] L. D. Lillwitz, *US Patent 4448987*, (1984) May 15, 1984

[11] S. Zhengfeng, L. Chuang, D. Xin, X. Zihui, and L. Changhai, *Ind. Eng. Chem. Res.*, V. 53, No. 23, (2014) 9638. DOI:10.1021/ie5006405

Chapter 7

Hydrogenation of Zr-Based Quasicrystals

Devinder Singh, Radhey Shyam Tiwari and
Onkar Nath Srivastava

Additional information is available at the end of the chapter

http://dx.doi.org/10.5772/66255

Abstract

In this chapter, results of our recent investigations on the hydrogenation behavior of Zr-based quasicrystalline alloys and its effect on their structural and microhardness behavior have been discussed. The microstructural changes with respect to the addition of Ti and their correlation with hydrogen storage characteristics of $(Zr_{69.5}Al_{7.5}Cu_{12}Ni_{11})_{100-x}Ti_x$ (x = 0, 4 and 12) quasicrystalline alloys have been studied. The substitution of Ti affects the nucleation and growth characteristics of nano-quasicrystals. The grain size of quasicrystals decreases with addition of Ti. The hydrogen uptake capacity of partially quasicrystalline alloys has been improved by the addition of Ti. The alloys with x = 0, 4, and 12 absorbed 1.20 wt.%, 1.38 wt.%, and 1.56 wt.% of hydrogen, respectively. A significant effect on the structure/microstructure and mechanical behavior of $(Zr_{69.5}Al_{7.5}Cu_{12}Ni_{11})_{100-x}Ti_x$ quasicrystalline alloys due to hydrogenation has been observed. The change in the microhardness behavior has been discussed based on microstructural variation resulting to Ti addition. The study is focused on investigations of these materials to understand the structure (microstructure)-property correlations.

Keywords: hydrogenation, quasicrystalline, composites, microstructure, microhardness

1. Introduction

Storage is an important issue with hydrogen to become a renewable, clean, and climate friendly fuel. The chemical interactions between the metal and hydrogen atoms, number, type, and size of the interstitial sites for hydrogen mainly determine the storage capacity of hydrogen in metals and alloys. Metals and alloys based on transition metals are the most common materials for hydrogen storage. Hydrogen tends to occupy tetrahedral interstitial sites in these materials.

Quasicrystals have been found to exhibit certain important characteristic properties like high hardness, high oxidation resistance, low surface energy, and low thermal conductivity which make these materials attractive for technological applications [1–7]. These materials have received considerable attention as hydrogen storage materials after the discovery of thermodynamically stable icosahedral quasicrystalline phase in Ti-based alloys [8, 9]. Due to the presence of high density of interstitial voids in quasicrystalline alloys, these materials have shown to reversibly store a large amount of hydrogen. The storage capacity can reach up to 3 wt.% of hydrogen [10]. This may be better than that of the crystalline La-based, V-based, and Fe-Ti hydrogen storage alloys [11–13]. Ti- and Zr-based quasicrystalline alloys have a large number of tetrahedral coordinated sites. The Zr-Al-Cu-Ni quasicrystalline alloy is a promising material for the hydrogen storage applications due to the presence of local tetrahedral order and favorable chemical composition [14–17]. The metallic glasses and their composites are known to absorb hydrogen during electrochemical charging, up to a hydrogen per metal atom content (H/M) of 1.6 for the quasicrystalline phase [18, 19] and H/M = 1.0 for the glassy phase [20, 21], respectively. A similar hydrogen uptake from the gas phase is needed with no irreversible phase transformation in order to use these materials for hydrogen storage. It has been found that the hydrogen storage capacity is higher, and the absorption kinetics is faster for the quasicrystalline phase than for the glassy phase [15, 18, 21]. The presence of a large number of tetrahedral sites assumed for icosahedral structure has led to enhanced hydrogen storage capacity. Thus, it is relevant to study the hydrogenation characteristics of quasicrystalline phase. Though Zr-based and Ti-based quasicrystals are among the best in the group of hydrogen storage alloy systems [17, 22–26], still there are issues in regard to their structural stability with hydrogenation. The icosahedral phase (I-phase) in $Zr_{69.5}Al_{7.5}Cu_{12}Ni_{11}$ alloy is metastable. When the $Zr_{69.5}Al_{7.5}Cu_{12}Ni_{11}$ alloy is annealed for a longer time, the metastable I-phase decomposes into crystalline phases [27–31]. Hence, the hydrogen storage capacity will be deteriorated due to the lack of significant polytetrahedral order in these crystalline phases. The desorption of hydrogen was not observed to ensue at temperatures less than about 450°C, most likely due to thin oxide layers formed at the surfaces of the partially quasicrystalline $Zr_{69.5}Al_{7.5}Cu_{12}Ni_{11}$ ribbons [16]. During annealing at higher temperatures, the I-phase decomposed by a discontinuous transformation into tetragonal Zr_2Cu, tetragonal Zr_2Ni, and hexagonal Zr_6NiAl_2 starting with a precipitation reaction of Zr_2Cu [32–34]. In view of application, there are still major concerns regarding the effect of hydrogenation on the structure of these quasicrystalline alloys and consequently its influence on the mechanical properties. Thus, the study of hydrogenation effect on the structure/microstructure and mechanical behavior of $Zr_{69.5}Al_{7.5}Cu_{12}Ni_{11}$ quasicrystalline alloy is of special interest.

2. Investigation of nominal composition: $Zr_{69.5}Al_{7.5}Cu_{12}Ni_{11}$

The $Zr_{69.5}Al_{7.5}Cu_{12}Ni_{11}$ alloy with a thickness of ~40–50 μm and lengths of ~1–2 m has been synthesized using melt-spinning technique [23, 35]. During melt spinning the entire apparatus was enclosed in a steel enclosure through which argon gas was continuously flowing. This was done to prevent oxidation of the ribbons after ejection of melt from the nozzle. For annealing experiments, the ribbons were packed in a Ta foil and then sealed in a silica ampoule under an argon atmosphere. The ribbons were isothermally annealed in vacuum (10^{-6} Torr) using a Heraeus furnace with temperature control of ±1°C.

For exploring the gross structural feature, X-ray diffraction (XRD) studies of the as-formed ribbons were done. **Figure 1a** shows the representative XRD pattern of the $Zr_{69.5}Al_{7.5}Cu_{12}Ni_{11}$ as-synthesized ribbon. As can be seen, the XRD pattern shows two broad halos. This shows that the as-formed material is amorphous (glassy). However, for the as-synthesized ribbon (c.f. **Figure 1a**), broad halo within the angular range of 30–45° followed by another weaker second broad halo within the angular range of 60–70° is seen. The material was then annealed in the anticipation that the synthesized amorphous phase will transform to the quasicrystalline/ crystalline version. In order to estimate the transformation temperature, differential scanning calorimetry (DSC) technique was employed. **Figure 1b** shows the constant-rate heating DSC scan (20 K/min) of the melt-spun $Zr_{69.5}Al_{7.5}Cu_{12}Ni_{11}$ alloy. Heating the melt-spun ribbon from room temperature up to elevated temperatures reveals a distinct endothermic event characteristic of the glass transition (T_g) at 624 K. With increasing temperature the DSC curve exhibits two peaks (T_{p1} and T_{p2}) with onset temperatures T_{x1} and T_{x2} (702 K and 727 K, respectively). The supercooled liquid region, defined as $\Delta T_x = T_{x1}-T_g$ corresponds to the stability of the supercooled liquid against crystallization and is found to be 78 K. The melting temperature (T_m) of the melt-spun $Zr_{69.5}Al_{7.5}Cu_{12}Ni_{11}$ is found to be 1064 K. The first peak (T_{p1}) results from the quasicrystal (qc) formation. The second peak (T_{p2}) corresponds to the decomposition of quasicrystals as well as the transformation of the remaining amorphous matrix into a crystalline phase.

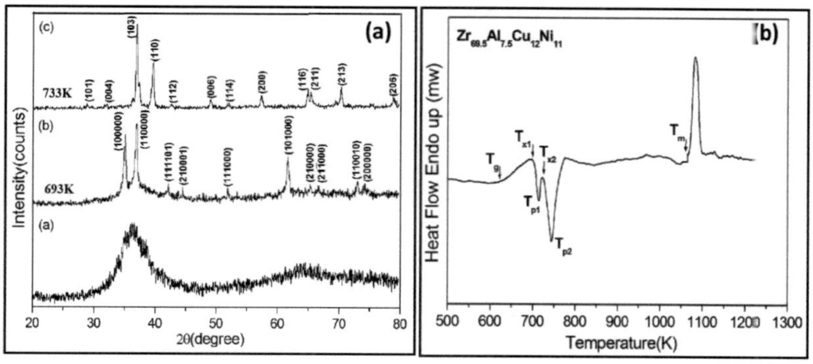

Figure 1. (a) XRD patterns of as-synthesized and annealed ribbons of $Zr_{69.5}Al_{7.5}Cu_{12}Ni_{11}$ alloy at 693 K and 733 K for 15 min and (b) DSC curve of the melt-spun $Zr_{69.5}Al_{7.5}Cu_{12}Ni_{11}$ alloy at the heating rate of 20 K/min.

The XRD patterns of the melt-spun ribbons (**Figure 1a**) exhibit the typical broad diffuse maxima characteristic of amorphous materials and no traces of crystalline phases. When the sample is isothermally heated at 693 K (i.e., near T_{x1}) for 15 min, the XRD pattern displays diffraction peaks with positions and intensity ratios that are typical for powder diffraction patterns of quasicrystals. The reflections can be indexed to I-phase. One can notice the broad overlapping diffraction peak due to amorphous phase in addition to the quasicrystalline diffraction peaks in the annealed XRD patterns. This indicates that during the first devitrification step, the phase transformation from glass to quasicrystalline is not complete and a residual glassy phase retained in the sample. This suggests that quasicrystal formation is a primary transformation. As evident from **Figure 1a**, heating the ribbon to 733 K (i.e., near T_{x2}) leads to the formation of a tetragonal Zr_2Cu-type phase (space group I4/mmm (139), a = 3.218 Å, c = 11.180 Å). No trace of quasicrystals is detected at this stage of crystallization, indicating that the quasicrystalline phase is metastable and transforms into more stable crystalline phase.

The $Zr_{69.5}Al_{7.5}Cu_{12}Ni_{11}$ melt-spun ribbon was observed to be fully amorphous by means of transmission electron microscopy (TEM). **Figure 2a** and the inset show the TEM micrograph and corresponding selected area electron diffraction (SAED) pattern displaying diffuse halos for $Zr_{69.5}Al_{7.5}Cu_{12}Ni_{11}$ alloy. The presence of two halos is suggestive of the presence of short-range order with two different correlation lengths. The TEM bright-field micrograph displaying no discernible contrast indicates the formation of glassy phase in the system. **Figure 2b** shows the typical microstructure of a partially transformed $Zr_{69.5}Al_{7.5}Cu_{12}Ni_{11}$ ribbon with spherical quasicrystals. The inset of **Figure 2b** shows the fivefold diffraction pattern of the quasicrystalline phase. Since no diffraction patterns corresponding to the other crystalline structure are seen, the initial precipitation phase in $Zr_{69.5}Al_{7.5}Cu_{12}Ni_{11}$ alloy is only quasicrystalline phase.

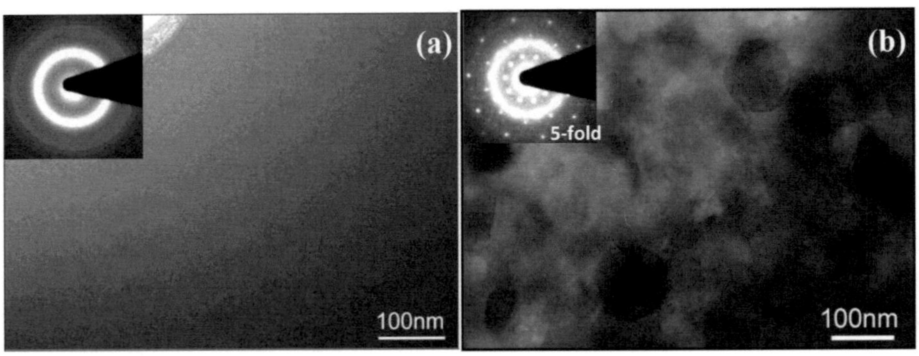

Figure 2. (a) TEM image and the corresponding diffraction pattern of as-synthesized $Zr_{69.5}Al_{7.5}Cu_{12}Ni_{11}$ alloy. (b) TEM microstructure of $Zr_{69.5}Al_{7.5}Cu_{12}Ni_{11}$ alloy showing the presence of spherical quasicrystals. Inset of (b) shows the SAED pattern of the quasicrystalline (qc) phase along the fivefold direction.

3. Effect of material tailoring on the hydrogen storage properties

The structure and phase formation of Zr-Al-Cu-Ni alloy system is strongly affected by material tailoring. Several studies reveal that the compositional changes may influence the hydrogen storage capacity of Zr-based alloys [19–21]. In the present chapter, the effect of Ti addition on the phase formation and the hydrogen storage characteristics of $(Zr_{69.5}Al_{7.5}Cu_{12}Ni_{11})_{100-x}Ti_x$ alloys with $0 \leq x \leq 12$ have been discussed. The influence of hydrogenation from the gas phase on the structure/microstructure and microhardness behavior of these melt-spun quasicrystalline alloys have been studied. The present study deals with the synthesis, phase formation, change in the microstructure with respect to Ti addition, and their correlation with hydrogen storage characteristics of Zr-Al-Cu-Ni alloys.

3.1. Microstructural and structural features

Figure 3a shows the XRD patterns of as-synthesized $(Zr_{69.5}Al_{7.5}Cu_{12}Ni_{11})_{100-x}Ti_x$ (x = 0, 4, 12 at. %) melt-spun alloys. It can be seen that all the patterns of the alloys consist of only broad diffraction maxima without a detectable sharp Bragg peak, indicating that the samples are amorphous in nature. The TEM image and its corresponding SAED pattern displaying diffuse halos for the alloy with x = 12 is shown in **Figure 3b**. The TEM microstructure for the samples with x = 12 or less displays a featureless contrast, which is the typical appearance of an amorphous phase. The DSC scans for the samples recorded at a heating rate of 20 K/min are shown in **Figure 4**. **Table 1** gives the thermal stability data for all the investigated samples. With increasing Ti content, the onset crystallization temperature (T_x) decreases gradually, whereas the change of glass transition temperature (T_g) is very subtle. The crystallization for the alloys with x = 0–12 follows a two-step process.

The primary crystallization products of these melt-spun alloys were examined after isothermal annealing. **Figure 5** compares XRD patterns for partially crystallized ribbons with x = 0, 4, and 12. Isothermal annealing of these samples was carried out for 15 min at 425°C (for x = 0 and 4) and 420°C (for x = 12). The respective annealing temperatures for these alloys have been found through DSC investigation (c.f. **Table 1**). The crystallization processes of alloys with x = 0–12 show the formation of quasicrystalline phase. The indexing of the Bragg peaks in the XRD patterns was done on the basis of I-phase [36, 37]. For the alloy with x = 12, the XRD pattern shows considerable peak broadening as compared to the alloys with x = 0 and 4. This is for the reason that the size of the precipitates is decreasing with increasing Ti content. The quasicrystalline-phase formation in these samples was further investigated by TEM. The nanometer-sized quasicrystalline grains of the annealed samples for the alloys with x = 0, 4, and 12 are shown in **Figure 6a–c**. The grain size of quasicrystals decreases with addition of Ti. The quasicrystal grains with average size of ~125 nm and ~80 nm have been observed for the alloys with x = 0 and 4, respectively. The presence of characteristic fivefold icosahedral symmetry can be seen in the corresponding SAED pattern. The alloy with x = 12 reveals the formation of 5–10 nm grains. Thus, the strong reduction in the grain size has been observed for this alloy. The corresponding SAED pattern for the alloy with x = 12 shows the presence of diffraction rings (inset in **Figure 6c**) indexed with I-phase. The diffraction patterns with diffuse ring confirm the presence of a residual glassy phase after the first devitrification step. These results show that the size of the quasicrystalline grains decreases with addition of Ti in $(Zr_{69.5}Al_{7.5}Cu_{12}Ni_{11})_{100-x}Ti_x$ alloys. The nucleation rate of the I-phase increases with increase in the concentration of Ti. Thus, in the present case, it can be said that the addition of Ti affects the nucleation and growth characteristics of quasicrystals significantly.

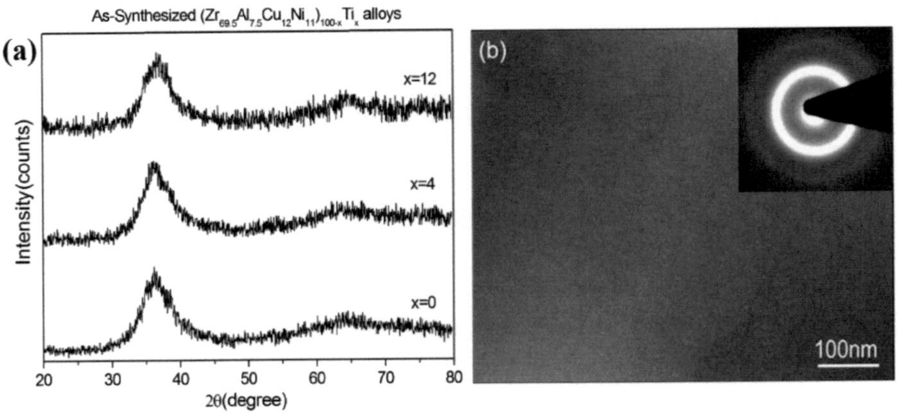

Figure 3. (a) XRD patterns of as-synthesized ribbons of $(Zr_{69.5}Al_{7.5}Cu_{12}Ni_{11})_{100-x}Ti_x$ alloys. (b) TEM image and the corresponding diffraction pattern of as-synthesized $(Zr_{69.5}Al_{7.5}Cu_{12}Ni_{11})_{88}Ti_{12}$ alloy (reprinted with kind permission from Ref. [23], Copyright 2013, Elsevier).

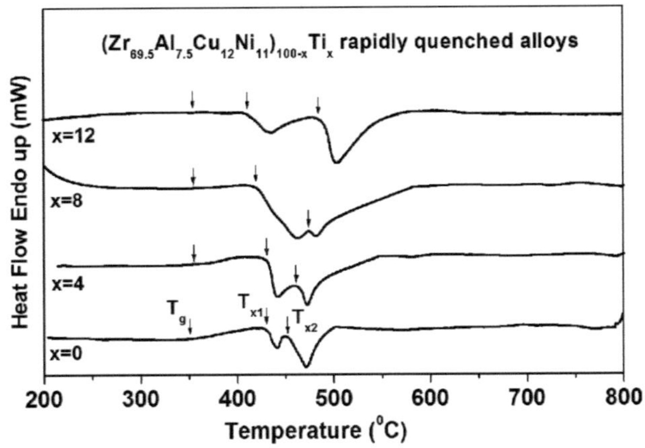

Figure 4. DSC curves of the melt-spun $(Zr_{69.5}Al_{7.5}Cu_{12}Ni_{11})_{100-x}Ti_x$ alloys [35].

x (at.%)	T_g (°C)	T_{x1} (°C)	T_{x2} (°C)	ΔT_x (°C)
0	351	429	453	78
4	355	429	461	74
8	355	420	474	65
12	355	413	480	58

T_g: glass transition temperature; T_{x1}: first-onset crystallization temperature; T_{x2}: second-onset crystallization temperature; ΔT_x: supercooled liquid region

Table 1. Thermal analysis of the melt-spun $(Zr_{69.5}Al_{7.5}Cu_{12}Ni_{11})_{100-x}Ti_x$ ribbons [35].

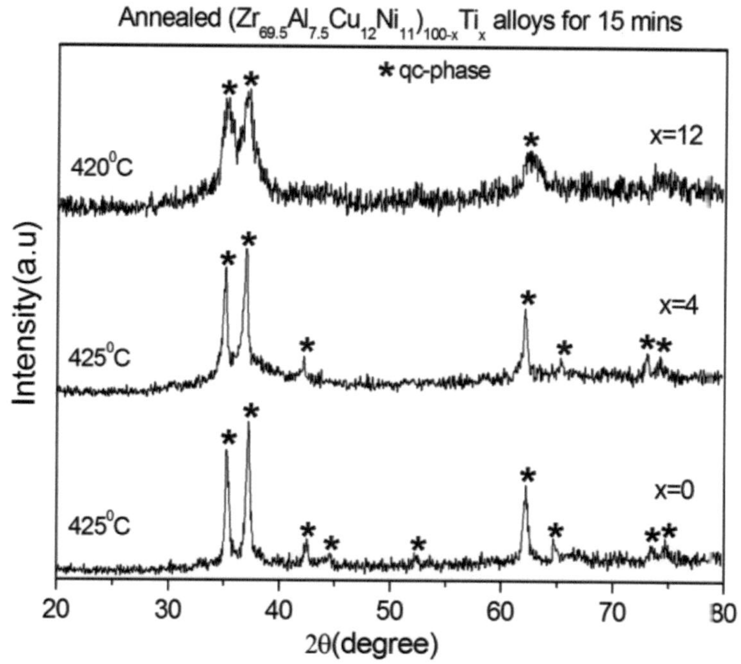

Figure 5. XRD patterns of annealed ribbons of $(Zr_{69.5}Al_{7.5}Cu_{12}Ni_{11})_{100-x}Ti_x$ alloys (reprinted with kind permission from Ref. [23], Copyright 2013, Elsevier).

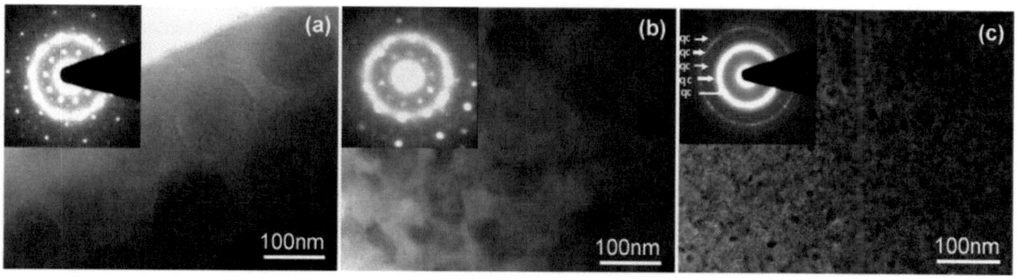

Figure 6. Bright-field TEM images of $(Zr_{69.5}Al_{7.5}Cu_{12}Ni_{11})_{100-x}Ti_x$ alloys (a) x = 0, (b) x = 4, and (c) x = 12 showing the influence of the Ti content on size of the quasicrystalline phase (reprinted with kind permission from Ref. [23], Copyright 2013, Elsevier).

A quantitative energy-dispersive X-ray (EDX) analysis was employed for the composition determination. **Figure 7a–c** shows EDX spectra of the $(Zr_{69.5}Al_{7.5}Cu_{12}Ni_{11})_{100-x}Ti_x$ alloys with x = 0, 4, and 12, respectively. **Table 2** presents EDX quantitative analysis along with deviations of the $(Zr_{69.5}Al_{7.5}Cu_{12}Ni_{11})_{100-x}Ti_x$ (x = 0–12) alloys. The semiquantitative analysis of EDX pattern at 3–5 points gives the deviations in the elemental compositions. Based on the EDX quantitative analysis, it has been found that the investigated compositions of the alloys are very close to stoichiometric proportions of nominal compositions. The presence of oxygen within the detectable limit of EDX was not found.

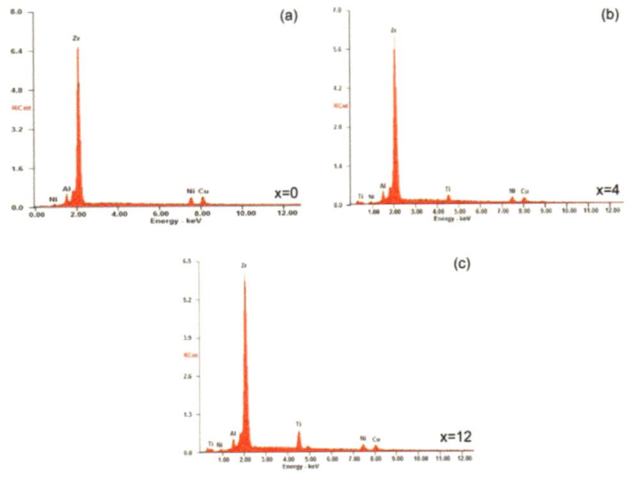

Figure 7. Energy-dispersive spectra of the melt-spun $(Zr_{69.5}Al_{7.5}Cu_{12}Ni_{11})_{100-x}Ti_x$ (x = 0, 4 and 12) alloys.

Alloy composition x (at.%)	Zr (±0.7)	Al (±0.3)	Cu (±0.5)	Ni (±0.5)	Ti (±0.5)
0	68.7	7.3	12.5	11.5	–
4	66.1	6.9	12.2	9.8	5.0
12	60.7	6.5	11.6	8.3	12.9

Table 2. Compositions of the $(Zr_{69.5}Al_{7.5}Cu_{12}Ni_{11})_{100-x}Ti_x$ alloys (in at.%) based on EDX quantitative analysis (reprinted with kind permission from Ref. [23], Copyright 2013, Elsevier).

3.2. Hydrogenation characteristics

In this section, we discuss the results of the hydrogenation characteristics of three glassy composites. The computerized pressure-concentration-temperature (PCT) apparatus supplied by Advance Materials Corporation (USA) has been used to study the hydrogen sorption characteristics of the partially crystallized ribbons. The gas reaction control-based software has been used to monitor the temperature, pressure, and gas desorbed/absorbed through the samples. The estimated error in the hydrogen storage capacity measurement for the alloys is ±0.02 wt.%. Temperature-programmed desorption (TPD) experiments at a heating rate of 5°C/min were also performed in addition to hydrogen sorption behavior. The absorption kinetic experiments were performed at fixed temperature and pressure to study the hydrogen storage characteristics of $(Zr_{69.5}Al_{7.5}Cu_{12}Ni_{11})_{100-x}Ti_x$ (x = 0, 4, and 12) quasicrystal-glass composites. It may be noted that hydrogen storage here is through dissociation of hydrogen molecule into hydrogen atoms. The catalytic activity of surface transition metal atoms is responsible for the dissociation of hydrogen molecule into hydrogen atoms. The dissociation is followed by diffusion of hydrogen atoms in the interstitial sites. Different temperatures and pressures range from 250 to 300°C and 5–10 MPa for 2–3 h have been performed for the hydrogenation experiments. The best results were found with temperature 270°C and 8 MPa of hydrogen pressure. The hydrogenation for the partially crystallized ribbons was done in a high-pressure reactor. The reactor containing the ribbons was evacuated by a rotary pump (10^{-2} Torr), heated up to 270°C, and then charged with 8 MPa of hydrogen pressure.

The hydrogen absorption characteristics for the alloys with x = 0, 4, and 12 at 270°C temperature and 8 MPa hydrogen pressure are shown in **Figure 8**. The absorption kinetics and hydrogen uptake capacity of the composites increase with increasing concentration of Ti (as shown in **Table 3**). The hydrogen storage capacity for the alloys with x = 0, 4, and 12 is found to be 1.20 wt. %, 1.38 wt.%, and 1.56 wt.%, respectively. The alloy with x = 0 having less hydrogen storage capacity of 1.20 wt.% with 100–150 nm grains whereas the alloy with x = 4 can store 1.38 wt.% of hydrogen with 60–100 nm grains at same conditions of temperature and pressure. The hydrogen storage capacity further increases to 1.56 wt.% for the alloy with x = 12 for grain size in the range 5–10 nm. It may be noted here that the experimental conditions for the synthesis of melt-spun ribbons were identical for all the alloys. In the present case, the enhancement in storage capacity is ~23% over the normal storage capacity of $Zr_{69.5}Al_{7.5}Cu_{12}Ni_{11}$ quasicrystalline phase.

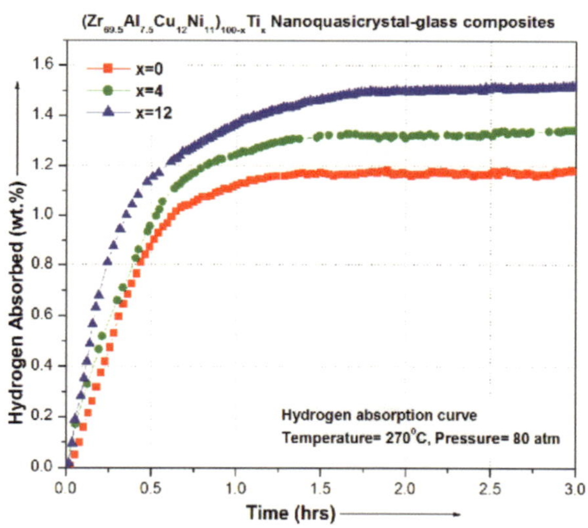

Figure 8. Hydrogen absorption kinetic curves of $(Zr_{69.5}Al_{7.5}Cu_{12}Ni_{11})_{100-x}Ti_x$ (x = 0, 4 and 12) quasicrystal-glass composites (reprinted with kind permission from Ref. [23], Copyright 2013, Elsevier).

Alloy composition	Quasi-lattice parameter		Storage capacity (wt.%) (±0.02)
x (at.%)	Before hydrogenation (a_i^0)	After hydrogenation (a_i^H)	
0	5.108 Å	5.201 Å	1.20
12	5.141 Å	5.293 Å	1.56

Table 3. Hydrogenation characteristics of $(Zr_{69.5}Al_{7.5}Cu_{12}Ni_{11})_{100-x}Ti_x$ (x = 0, 4, and 12) quasicrystal-glass composites (reprinted with kind permission from Ref. [23], Copyright 2013, Elsevier).

The probable reasons for the enhancement in the hydrogen storage capacity of $(Zr_{69.5}Al_{7.5}Cu_{12}Ni_{11})_{100-x}Ti_x$ alloys are: (i) it has been observed by XRD and TEM analysis that the grain refinement occurred with increase in the concentration of Ti. The increase of grain boundary density provides better interaction between hydrogen and I-phase and consequently increases the hydrogen storage uptake capacity. The enhancement in hydrogen storage capacity due to grain refinement has also been reported in Ti-Zr-Ni and Mg-Ni-Mm alloys [22, 38]. (ii) The other probable reason for the increase in hydrogen absorption may also be due to Ti addition. The addition of Ti in $Zr_{69.5}Al_{7.5}Cu_{12}Ni_{11}$ quasicrystalline alloy may have twofold effects on the hydrogen storage behavior. First, the catalytic effect of Ti plays an important role in the improvement of hydrogen storage behavior. The Ti addition enhances the dissociation of hydrogen molecule to hydrogen atom at the surface due to its catalytic effect [36, 37, 39]. Second, the overall affinity for hydrogen increases due to the increase in the (Zr + Ti) combined content, from 69.5 for x = 0 to 73.2 for x = 12 and might lead to an increased number of energetically favorable sites for hydrogen. Thus, it can be said that the combined effect of the above two probable reasons may lead to improved hydrogen storage characteristics of $(Zr_{69.5}Al_{7.5}Cu_{12}Ni_{11})_{100-x}Ti_x$ alloys.

Temperature-programmed desorption (TPD) experiment at heating rate 5°C/min has been done to study the hydrogen desorption behavior of $(Zr_{69.5}Al_{7.5}Cu_{12}Ni_{11})_{100-x}Ti_x$ (x = 0, 4, and 12) quasicrystal-glass composites. The desorption curves of $(Zr_{69.5}Al_{7.5}Cu_{12}Ni_{11})_{100-x}Ti_x$ (x = 0, 4, and 12) alloys are shown in **Figure 9**. The nature of desorption curves for hydrogenated ribbons changed with increasing Ti addition, and this may be attributed to the change in the microstructure of the ribbon with addition of Ti. A small decrease in desorption temperature has been observed at higher Ti content of x = 12. It can be seen from **Figure 9** that full desorption was not observed in the case of $(Zr_{69.5}Al_{7.5}Cu_{12}Ni_{11})_{100-x}Ti_x$ alloys. This behavior is in contrast to stable Ti-based quasicrystal $Ti_{45}Zr_{38}Ni_{17}$, which allows nearly full desorption [22, 40].

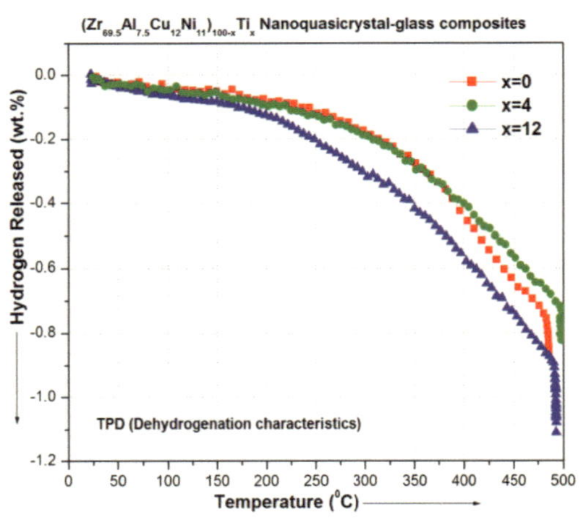

Figure 9. Temperature-programmed desorption (TPD) curves of hydrogenated $(Zr_{69.5}Al_{7.5}Cu_{12}Ni_{11})_{100-x}Ti_x$ (x = 0, 4, and 12) quasicrystal-glass composites (reprinted with kind permission from Ref. [23], Copyright 2013, Elsevier).

3.3. Influence of hydrogenation on the structural and microhardness behavior

In order to investigate the structural changes with hydrogenation, the hydrogenated samples are characterized through XRD. The XRD patterns of the hydrogenated $(Zr_{69.5}Al_{7.5}Cu_{12}Ni_{11})_{100-x}Ti_x$ (x = 0 and 12) ribbons are shown in **Figure 10**. This reveals that I-phase peaks coexist with small concentration of crystalline hydride. This hydride phase has been recognized as ZrH_2 which has a tetragonal lattice with space group: I4/mmm, a = b = 3.519 Å, and c = 4.450 Å. Several earlier studies [15, 16] have reported the formation of such hydride phase on hydrogenation of I-phase in Zr-Al-Cu-Ni alloy. The partial decomposition of I-phase may lead to the formation of hydride phase [15]. During charging from the gas phase, the volume fraction of the quasicrystalline phase exhibits a significant effect on the formation of hydrides [15, 25, 41]. The decrease in the intensity as well as the significant broadening of the I-phase peaks is evident in the XRD patterns for the hydrogenated samples. This is due to the decrease in the size of the grains after hydrogenation. It has been observed that the I-phase peaks are shifted to smaller angle, thus indicating the lattice expansion upon hydrogenation. In the present study, the X-ray diffraction patterns are not only used for the identification of phase transformations with hydrogenation but also used to estimate the quasi-lattice parameters of hydrogenated and partially crystallized ribbons (**Table 3**). The XRD patterns for the hydrogenated ribbons show the formation of crystalline hydride phase along with I-phase. We believe that the volume fraction of crystalline hydride phase is small and the measured hydrogen absorption reflects the amount of hydrogen stored in the quasicrystal.

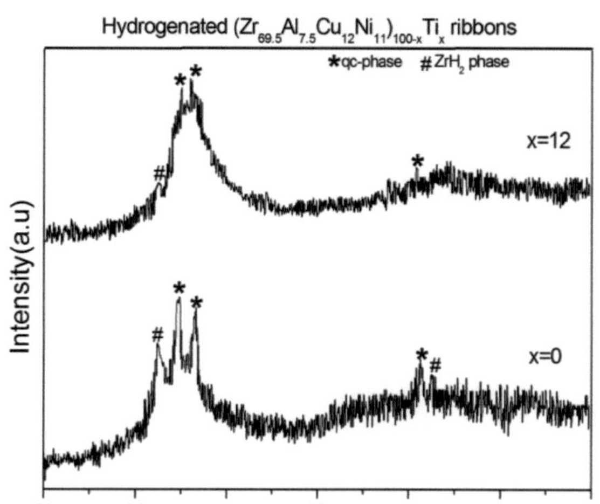

Figure 10. XRD patterns of hydrogenated $(Zr_{69.5}Al_{7.5}Cu_{12}Ni_{11})_{100-x}Ti_x$ (x = 0 and 12) ribbons (reprinted with kind permission from Ref. [23], Copyright 2013, Elsevier).

Here, we also compare the microstructural changes observed after hydrogenation. **Figure 11a** and **b** shows the TEM micrographs of the hydrogenated ribbons with x = 0 and x = 12, respectively. In comparison with **Figure 6a** and **c**, the size of the grains has decreased. The change in the morphology of the quasicrystal grains as well as weakening of the diffraction spots (marked by arrows in the inset of **Figure 11a**) has been observed for the alloy with x = 0. The weak diffraction spots superimposed on the diffuse halo ring of the amorphous phase can be seen in the SAED pattern for the alloy with x = 12 (inset of **Figure 11b**). These results are in agreement with the results obtained by Zander et al. [41]. It has been observed earlier that even weak hydrogenation may lead to the generation of defects in the icosahedral structure. These defects may be responsible for the diffuseness and domain formation in the microstructure. This would lead to a weakening of the contrast of quasicrystal grains as well as the diffraction spots [17, 41].

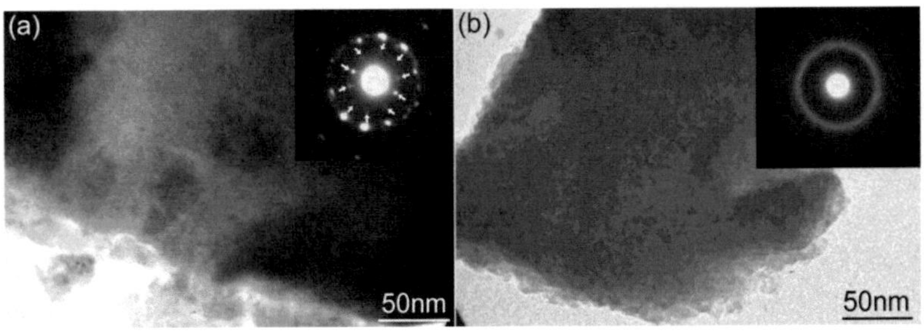

Figure 11. Bright-field TEM micrographs of hydrogenated $(Zr_{69.5}Al_{7.5}Cu_{12}Ni_{11})_{100-x}Ti_x$ ribbons with (a) x = 0 and (b) x = 12 (reprinted with kind permission from Ref. [23], Copyright 2013, Elsevier).

Indentation tests were conducted at room temperature in order to understand the effect of hydrogenation on the microhardness behavior. The SHIMADZU HMV-2T microhardness tester was used under different loads. The standard diamond-pyramid-shaped Vickers indenter with a tip size of ~0.5 μm was used. The hardness (H) was computed in GPa units by employing the following relationship [42–44]:

$$H = 1.854 \times 9.8 \times \frac{P}{d^2} \tag{1}$$

where P is the load (g) and d is the diagonal length in μm. The mean hardness values of at least five loads are reported here with deviations. A notable difference in the indentation behavior of as-synthesized, quasicrystal (qc)-glass composite and hydrogenated qc-glass composite has been observed for the alloys with x = 0, 4, and 12 (c.f. **Figure 12a–c**). Cracks around the indentation impression start to appear when the applied load goes above a certain value. The microhardness tests in the present study were carried out up to this critical load. It can be seen from the hardness versus load characteristic curves that the microhardness tests were carried out up to load of 100 g for the qc-glass composites and 50 g for the hydrogenated qc-glass composites. Hence, load to fracture decreases, and this reveals the decrease in the fracture toughness for the hydrogenated samples. The significant increase in the hardness for the qc-glass composites as compared to as-synthesized ribbons has been observed, and this is due to the precipitation of nano-quasicrystal grains in the amorphous matrix [45–53]. Minor changes in the microhardness behavior after hydrogenation have been observed for the qc-glass composites. **Table 4** gives hardness values at 50 g of load for the as-synthesized, qc-glass composites, and hydrogenated qc-glass composites. The hardness values of hydrogenated qc-glass composites of x = 0, 4, and 12 at 50 g load are found to be ~7.33 GPa, ~7.57 GPa, and ~7.93 GPa, respectively. These are slightly higher than that of qc-glass composites. The microstructural variation and the partial decomposition of I-phase into crystalline hydride phase during hydrogenation may lead to slight increase in the hardness values. The micro-hardness might change during hydrogenation by the interstitial content and microstructural changes, e.g., phase transformations as well as precipitations [54–60].

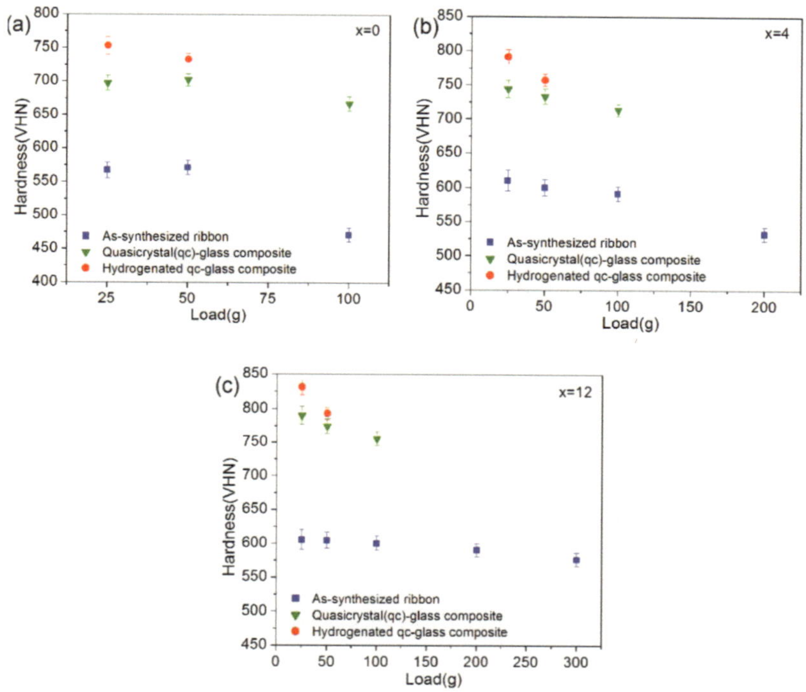

Figure 12. Variation in hardness (VHN) with respect to load for the as-synthesized, quasicrystal(qc)-glass composites, and hydrogenated qc-glass composites (a) x = 0, (b) x = 4, and (c) x = 12 (reprinted with kind permission from Ref. [23], Copyright 2013, Elsevier).

Alloy composition x (at.%)	As-synthesized (±0.10)	Quasicrystal (qc)-glass composite (±0.10)	Hydrogenated qc-glass composite (±0.10)
0	5.71	7.02	7.33
4	6.00	7.33	7.57
12	6.04	7.74	7.93

Table 4. Values of hardness (VHN) (GPa) at 50 g load of as-synthesized, quasicrystal (qc)-glass composites, and hydrogenated qc-glass composites of $(Zr_{69.5}Al_{7.5}Cu_{12}Ni_{11})_{100-x}Ti_x$ alloys (reprinted with kind permission from Ref. [23], Copyright 2013, Elsevier).

4. Conclusion

In this chapter, the hydrogenation behavior of Zr-based quasicrystalline alloy system has been discussed. Based on this study, it may be said that the hydrogen storage capacity of partially quasicrystalline $(Zr_{69.5}Al_{7.5}Cu_{12}Ni_{11})_{100-x}Ti_x$ (x = 0, 4 and 12) alloys has been improved by the addition of Ti. The hydrogen storage capacity is enhanced from ~1.20 wt.% (for x = 0) to ~1.56 wt.% (for x = 12). The enhancement in storage capacity for the alloy with x = 12 is ~23% as compared to the storage capacity of alloy with x = 0. The grain refinement of quasicrystals resulting due to Ti addition has led to the improvement in the hydrogen uptake capacity of $(Zr_{69.5}Al_{7.5}Cu_{12}Ni_{11})_{100-x}Ti_x$ alloys. The microhardness shows only minor changes after hydrogenation of quasicrystal-glass composites. The microstructural and morphological changes affect the hydrogen storage characteristics of quasicrystal-glass composites.

Acknowledgements

One of the authors (Devinder Singh) would like to sincerely acknowledge the financial support from the Department of Science and Technology (DST), New Delhi, India, in the form of INSPIRE Faculty Award [IFA12-PH-39]. Section 3 of this chapter is reproduced with kind permission from Ref. [23] (©2013 Elsevier).

Author details

Devinder Singh[1*], Radhey Shyam Tiwari[2] and Onkar Nath Srivastava[2]

*Address all correspondence to: devinderbhu@yahoo.com

1 Department of Physics, Punjab University, Chandigarh, India

2 Department of Physics, Institute of Science, Banaras Hindu University, Varanasi, India

References

[1] Somekawa H, Watanabe H, Mukai T. Damping properties in Mg-Zn-Y alloy with dispersion of quasicrystal phase particle. Mater Lett 2011;65:3251.

[2] Dubois JM. So useful, those quasicrystals. Isr J Chem 2011;51:1168.

[3] Singh D, Yun Y, Wan W, Grushko B, Hovmöller S, Zou XD. A complex pseudo-decagonal quasicrystal approximant $Al_{37}(Co,Ni)_{15.5}$ solved by the rotation electron diffraction (RED) method. J Appl Crystallogr 2014;47:215.

[4] Hu W, Wang L, Wang L. Quinary icosahedral quasicrystalline Ti–V– Ni–Mn–Cr alloy: a novel anode material for Ni-MH rechargeable batteries. Mater Lett 2011;65:2868.

[5] Singh D, Yun Y, Wan W, Grushko B, Hovmöller S, Zou XD. Structure determination of a pseudo-decagonal quasicrystal approximant by the strong-reflections approach and rotation electron diffraction. J Appl Crystallogr 2016;49:433–441 doi: 10.1107/S1600576716000042.

[6] Yadav TP, Singh D, Shahi RR, Shaz MA, Tiwari RS, Srivastava ON. Formation of quasicrystalline phase in $Al_{70-x}Ga_xPd_{17}Mn_{13}$ alloys. Philos Mag 2011;91:2474.

[7] Yadav TP, Singh D, Shaz MA, Tiwari RS, Srivastava ON. Synthesis of quasicrystalline film of Al-Ga-Pd-Mn alloy. Thin Solid Films 2013;534:265.

[8] Kelton KF, Kim WJ, Stroudb RM. A stable Ti-based quasicrystal. Appl Phys Lett 1997;70:24.

[9] Stroud RM, Viano AM, Gibbons PC, Kelton KF. Stable Ti-based quasicrystal offers prospect for improved hydrogen storage. Appl Phy Lett 1996;69:20.

[10] Takasaki A, Kelton KF. Hydrogen storage in Ti-based quasicrystal powders produced by mechanical alloying. Int J Hydrogen Energy 2006;31:183.

[11] Hudson MSL, Dubey PK, Pukazhselvan D, Pandey SK, Singh RK, Raghubanshi H, et al. Hydrogen energy in changing environmental scenario: Indian context. Int J Hydrogen Energy 2009;34:7358.

[12] Lototsky MV, Yartys VA, Zavaliy IY. Vanadium-based BCC alloys: phase-structural characteristics and hydrogen sorption properties. J Alloys Compds 2005;404:421.

[13] Sakintuna B, Lamari D, Hirscher M. Metal hydride materials for solid hydrogen storage: a review. Inter J Hydrogen Energy 2007;32:1121.

[14] Rud AD, Schmidt U, Zclinska GM, Lakhnik AM, Kolbasov GY, Danilov MO. Atomic structure and hydrogen storage properties of amorphous–quasicrystalline Zr–Cu–Ni–Al melt-spun ribbons. J Non Cryst Solids 2007;353:3434.

[15] Huett VT, Zander D, Jastrow L, Majzoub EH, Kelton KF, Köster U. Gaseous hydrogen charging of Zr–Cu–Ni–Al glasses and quasicrystals. J Alloys Compds 2004;379:16.

[16] Zander D, Tal-Gutelmacher E, Jastrow L, Köster U, Eliezer D. Hydrogenation of Pd-coated Zr–Cu–Ni–Al metallic glasses and quasicrystals. J Alloys Compds 2003;356-357:654.

[17] Apih T, Khare V, Klanjsek M, Jeglic P, Dolinsek J. Hydrogen diffusion in partially quasicrystalline Zr69.5Cu12Ni11Al7.5. Phys Rev B 2003;68:212202.

[18] Eliaz N, Eliezer D, Abramov E, Zander D, Köster U. Hydrogen evolution from Zr-based amorphous and quasicrystalline alloys. J Alloys Compds 2000;305:272.

[19] Köster U, Zander D, Meinhardt J, Eliaz N, Eliezer D. Hydrogen in quasicrystalline Zr-Cu-Ni-Al, in: Takeuchi S, Fujiwara T (Eds), Proceedings of the Sixth International Conference on Quasicrystals, Tokyo, 1997, World Scientific, Singapore, 1998, p. 313.

[20] Ismail N, El-Meligi AA, Uhlemann M, Gebert A, Eckert J, Schultz L. Hydrogenation of Zr–Cu–Al–Ni–Pd metallic glasses by electrochemical means. J Alloys Compds 2009;480:321.

[21] Köster U, Zander D, Leptien H, Eliaz N, Eliezer D. Hydrogenation and crystallization of Zr-Cu-Ni-Al metallic glasses, in: Johnson WL, Inoue A, Liu AT, (Eds), Proceedings of the MRS-Symposium on Bulk Metallic Glasses, Boston, 1998. Mater Res Soc Symp Proc, vol 554, Pittsburgh, 1999: p. 287.

[22] Shahi RR, Yadav TP, Shaz MA, Srivastava ON, Smaalen Van S. Effect of processing parameter on hydrogen storage characteristics of as quenched $Ti_{45}Zr_{38}Ni_{17}$ quasicrystalline alloys. Int J Hydrogen Energy 2011;36:592.

[23] Singh D, Shahi RR, Yadav TP, Mandal RK, Tiwari RS, Srivastava ON. Hydrogenation of $(Zr_{69.5}Al_{7.5}Cu_{12}Ni_{11})_{100-x}Ti_x$ quasicrystalline alloys and its effect on their structural and microhardness behaviour. J Non-Cryst Solids. 2013;380:11.

[24] Huang H, Li R, Yin C, Zheng S, Zhang P. Hydrogenation study of suction-cast $Ti_{40}Zr_{40}Ni_{20}$ quasicrystal. Int J Hydrogen Energy 2008;33:4607.

[25] Zander D, Köster U, Eliaz N, Eliezer D. Influence of hydrogen on formation and stability of Zr-based quasicrystals. Mater Sci Eng 2000;294-296:112.

[26] Sinning HR, Scarfone R, Golovin JS. Mechanical spectroscopy of hydrogen-absorbing quasicrystals. Mater Sci Eng A 2004;370:78.

[27] Wang YM. Structure evolution of Zr-based Glass-Forming Alloys and Composition Design Methodology for Good Glass-Forming Abilities, Ph.D. thesis, City University of Hong Kong, Hong Kong (2007).

[28] Kühn U, Eymann K, Mattern N, Eckert J, Gebert A, Bartusch B, Schultz L. Limited quasicrystalformationinZr–Ti–Cu–Ni–Albulkmetallicglasses.ActaMater2006;54:4685.

[29] Singh D, Tiwari RS, Srivastava ON, Mandal RK. Synthesis and Mechanical Properties of $Zr_{69.5}Ga_{7.5}Cu_{12}Ni_{11}$ Metallic Glass and Nanoquasicrystal-Glass Composites. Emerging Paradigms in Nanotechnology. Pearson Education, Chennai, India. 2013;81: ISBN: 978-81-317-8991-9.

[30] Singh D, Mandal RK, Tiwari RS, Srivastava ON. Role of Nano-Quasicrystals in the Formation of Shear Bands in Zr-based Glassy Alloys. Nanotechnology: Novel Perspectives and Prospects. McGraw-Hill, USA. 2015; ISBN: 978-93-392-2109-6.

[31] Singh D, Tiwari RS, Srivastava ON. Metallic Glasses, Quasicrystals and their Nanocomposites. Lap Lambert Academic Publishing, Germany. 2014;111156:165. ISBN: 978-3-659-62088-1.

[32] Hong E, Dunand DC, Choe H. Hydrogen-induced transformation superplasticity in zirconium. Int J Hydrogen Energy 2010;35:5708.

[33] Singh D, Yadav TP, Mandal RK, Tiwari RS, Srivastava ON. Effect of Ga substitution on the crystallization behaviour and glass forming ability of Zr–Al–Cu–Ni alloys. Mater Sci Eng A 2010;527:469.

[34] Zander D, Leptian H, Köster U, Eliaz N, Eliezer D. Hydrogenation of Zr-based metallic glasses and quasicrystals. J Non-Cryst Solids 1999;250-252:893.

[35] Singh D, Yadav TP, Mandal RK, Tiwari RS, Srivastava ON. Effect of Ti addition on the quasicrystalline phase formation and indentation characteristics of $Zr_{69.5}Al_{7.5}Cu_{12}Ni_{11}$ alloy. Philos Mag 2011;91:2837.

[36] Bancel PA, Heiney PA, Stephans PW, Goldmann AI, Horn PM. Structure of rapidly quenched Al-Mn. Phys Rev Lett 1985;54:2422.

[37] Saida J, Matsushita M, Li C, Inoue A. Effects of Ag and Pd on the nucleation and growth of the nano-icosahedral phase in $Zr_{65}Al_{7.5}Ni_{10}Cu_{7.5}M_{10}$ (M = Ag or Pd) metallic glasses. Philos Mag Lett 2000;80:737.

[38] Wu Y, Lototsky MV, Solberg JK, Yartys VA, Han W, Zhou SX. Microstructure and novel hydrogen storage properties of melt-spun Mg–Ni–Mm alloys. J Alloys Compds 2003;477:262.

[39] Qiang JB, Zhang W, Xie GQ, Inoue A. Effect of Ti addition on the crystallization behavior and glass-forming ability of Zr–Al–Cu alloys. J Non-Cryst Solids 2008;354:2054.

[40] Viano AM, Majzoub EH, Stroud RM, Kramer MJ, Misture ST, Gibbons PC, Kelton KF. Hydrogen absorption and storage in quasicrystalline and related Ti-Zr-Ni alloys. Philos Mag A 1998;78:131.

[41] Zander D, Köster U, Khare V. Hydrogen induced transformations in Zr–Cu–Ni–Al quasicrystals. J Non-Cryst Solids 2004;334–335:247.

[42] Mukhopadhyay NK, Weatherly GC, Embury JD. An analysis of microhardness of single-quasicrystals in the Al–Cu–Co–Si system. Mater Sci Eng A 2001;315:202.

[43] Singh D, Mandal RK, Tiwari RS, Srivastava ON. Effect of cooling rate on the crystallization and mechanical behaviour of Zr-Ga-Cu-Ni metallic glass composition. J Alloys Compds 2015;648:456.

[44] Singh D, Singh D, Yadav TP, Mandal RK, Tiwari RS, Srivastava ON. Synthesis and indentation behaviour of amorphous and nanocrystalline phases in rapidly quenched Cu-Ga-Mg-Ti and Cu-Al-Mg-Ti alloys. Metallogr Microstruct Anal 2013;2:321.

[45] Singh D, Yadav TP, Mandal RK, Tiwari RS, Srivastava ON. Indentation characteristics of metallic glass and nanoquasicrystal-glass composite in Zr–Al (Ga)–Cu–Ni alloys. Intermetallics 2010;18:2445.

[46] Singh D, Mandal RK, Tiwari RS, Srivastava ON. Nanoindentation characteristics of $Zr_{69.5}Al_{7.5-x}Ga_xCu_{12}Ni_{11}$ glasses and their nanocomposites. J Alloys Compds 2011;509:8657.

[47] Yadav TP, Singh D, Tiwari RS, Srivastava ON. Enhanced microhardness of mechanically activated carbon–quasicrystal composite. Mater Lett 2012;80:5.

[48] Singh D, Singh Manjeet, Yadav TP, Mandal RK, Tiwari RS, Srivastava ON. Nanoindentation studies of metallic glasses and nanoquasicrystal–glass composites in Zr–Al (Ga)–Cu–Ni alloys. Inter J Nanosci 2011;10:929.

[49] Singh D, Yadav TP, Tiwari RS, Srivastava ON. Phase formation in rapidly quenched Cu-based alloys. J Mater Sci 2009;44:3883.

[50] Singh D, Singh D, Mandal RK, Srivastava ON, Tiwari RS. Glass forming ability, thermal stability and indentation characteristics in $Ce_{75}Al_{25-x}Ga_x$ metallic glasses. J Alloys Compds 2014;590:15.

[51] Singh D, Singh D, Srivastava ON, Tiwari RS. Microstructural effect on the low temperature transport properties of Ce-Al (Ga) metallic glasses. Scripta Mater. 2016;118:24.

[52] Singh D, Singh D, Mandal RK, Srivastava ON, Tiwari RS. Crystallization behaviour and mechanical properties of $(Al_{90}Fe_5Ce_5)_{100-x}Ti_x$ amorphous alloys. J Alloys Compds 2016;687:990.

[53] Singh D, Singh D, Mandal RK, Srivastava ON, Tiwari RS. Effect of annealing on the devitrification behaviour and mechanical properties of rapidly quenched Ce-based glassy alloys. J Non-Cryst Solids 2016;445:53.

[54] Jayalakshmi S, Kim KB, Fleury E. Effect of hydrogenation on the structural, thermal and mechanical properties of Zr_{50}–Ni_{27}–Nb_{18}–Co_5 amorphous alloy. J Alloys Compds 2006;417:195.

[55] Yamaura SI, Hasegawa M, Kimura H, Inoue A. Effects of hydrogen on the mechanical properties of $Ti_{50}Ni_{25}Cu_{25}$ metallic glass. Mater Trans 2002;43:2543.

[56] Zander D, Olson DL, Eliezer D. Mutual effects of hydrogenation and deformation in Ti-Nb alloys. Metall Mater Trans A 2003;34A:2199.

[57] Jayalakshmi S, Ahn JP, Kim KB, Fleury E. Hydrogen-induced amorphization and embrittlement resistance in Ti-based in situ composite with bcc-phase in an amorphous matrix. J Mater Res 2007;22:428.

[58] Jayalakshmi S, Fleury E, Lee DY, Chang HJ, Kim DH. Hydrogenation of $Ti_{50}Zr_{25}Co_{25}$ amorphous ribbons and its effect on their structural and mechanical properties. Philos Mag Lett 2008;88:303.

[59] Singh D, Mandal RK, Tiwari RS, Srivastava ON. Mechanical Behavior of Zr-Based Metallic Glasses and Their Nanocomposites, Metallic Glasses - Formation and Properties, Dr. Behrooz M (Ed.), InTech, Rijeka, Croatia – European union. 2016, doi: 10.5772/64221.

[60] Mandal RK, Tiwari RS, Singh D, Singh D. Influence of Ga substitution on the mechanical behavior of $Zr_{69.5}Al_{7.5\text{-}x}Ga_xCu_{12}Ni_{11}$ and $Ce_{75}Al_{25\text{-}x}Ga_x$ metallic glass compositions. MRS Proceeding. 2015;1757. doi: 10.1557/opl.2015.45.

Chapter 8

Supported Nickel-Based Catalysts for Partial Hydrogenation of Edible Oils

Miroslav Stanković, Jugoslav Krstić,
Margarita Gabrovska, Vojkan Radonjić,
Dimitrinka Nikolova, Davor Lončarević and
Dušan Jovanović

Additional information is available at the end of the chapter

http://dx.doi.org/10.5772/66967

Abstract

Nickel-based catalysts, supported on diatomite, silica gel and perlite, with high nickel loadings, have been prepared by precipitation-deposition method. Various nickel precursor salts were used for the preparation of catalyst precursors. In the precursor state, the catalysts were characterized using nitrogen physisorption, mercury porosimetry, infrared, and X-ray diffraction spectroscopy. The reducibility of catalyst precursors was evaluated using hydrogen temperature programmed reduction. Hydrogen chemisorption and X-ray photoelectron spectroscopy measurements were performed with the aim of characterizing the chemical state of the catalyst precursors. This research was focused on the study of some major factors on the state, dispersion and reducibility of a deposited Ni^{2+} phase by the combined use of mentioned experimental techniques. We have examined the influence of the nature of support and the use of modifiers on activity of nickel-based catalysts in the partial hydrogenation of sunflower and soybean oils. Nitrogen physisorption and mercury porosimetry data showed that synthesis operating conditions and pore structure of supports have a profound effect on the textural properties of catalyst precursors. The analysis of infrared and X-ray diffraction spectra showed the existence of chemical species and phases which indicate the different extent of interaction between the support and the active metal. Temperature programmed reduction study revealed that the reduction features depend on the identity of the nickel precursor salt and its interaction with the support. A stronger interaction of the supported Ni^{2+} phase with support hinders the reduction of catalyst precursors. Hydrogen chemisorption results showed the presence of nickel crystallites varying from 5 to 47 nm in size. The X-ray photoelectron spectroscopy data confirmed the formation surface species with different strength of interaction and different nickel crystallite sizes. The hydrogenation results showed significant differences, depending on the support and the modifier, as well as structural characteristics of reduced catalyst precursors. The results show the importance of modifiers in the control of the activity and selectivity of the partial hydrogenation process. The developed kinetic models of the hydrogenation of soybean and sunflower oils over studied catalytic systems were found useful in the prediction of the rate of reactions, product selectivity and catalytic activity.

Keywords: hydrogenation, soybean oil, sunflower oil, *trans* fatty acids, nickel catalysts, diatomite, silica gel, perlite, activity, selectivity, kinetics

1. Introduction

Hydrogenation of edible oils is a process used since its development in the early 1900s, to convert ats [1]. Principal products obtained by hydrogenation include oleomargarine stock, shortening, salad and cooking oils [2]. Hydrogenation changes the melting and solidification characteristics of the oils and is usually employed to reduce the degree of unsaturation of the naturally occurring triacylglycerides (TAGs). Most of the unsaturated fatty acids in TAGs contain 18 carbon atoms and the unsaturated fatty acids are almost completely in *cis*-configuration. The degree of hydrogenation which leads to hardening of the oil depends on the application, but it is always desired to reduce the level of polyunsaturated fatty acids such as linolenic (C18:3) and linoleic (C18:2) acids due to their high sensitivity to oxidation [3, 4]. This process is commonly used for vegetable oils with significant levels of linoleic acid, such as soybean and sunflower oils [5].

The hydrogenation process is usually carried out in a three-phase slurry reactor in a semi-batch mode where hydrogen gas is bubbled with pressure in hot vegetable oil in the presence of a catalyst [6]. In the industrial practice, hydrogenation process is typically carried out using nickel-based catalysts, either in the form of nickels Raney, or supported on different materials [7–10]. Economic price, high activity and easy availability of nickel make it superior over the other metals. High nickel loading is usual in commercial supported catalysts [11]. Normal operating conditions in commercial batch reactors are temperature range: 120–200°C, pressure: 1–5 bar and catalyst loading ranging from 0.01 to 0.2 wt% depending on the properties of the final product [12, 13]. It is desired to maximize the amount of oleic acid (C18:1) in the final product, as well as to eliminate linolenic acid and to reduce the content of linoleic acid to a substantial extent, without going too far towards producing the fully saturated stearic acid (C18:0), since these are not easily digested as foodstuffs [14].

During partial hydrogenation, some of double bonds of unsaturated fatty acids in TAGs can be isomerized into *trans* fatty acids (TFAs). Their intake was convincingly associated with risk of coronary heart disease (CHD) based on epidemiologic and clinical studies and has been shown to be harmful to human health [15–17]. In the last three decades, numerous research works were published in order to explain the effect of TFAs on the cellular metabolism. The consumption of foods high in TFAs has been shown to increase LDL-C and decrease HDL-C levels, which increases the risk of developing CHD [18]. For this reason, the demand for smaller levels of TFAs content in hydrogenated oils has increased and the search for alternatives is important to improve the hydrogenation process. Improvements are required to find new types of hydrogenation catalysts [19–25], new technological solutions [26] and new directions in edible oil modification processes, involving interesterification, fractionation, or blending [27].

The major advances in finding solutions leading to the reduction of TFAs in hydrogenated oils have been achieved in the field of hydrogenation catalysts. In the last two decades, there has been a growing interest for nanosized structures in the range 1–20 nm in different fields of research [28]. This is also the size of metal particles in supported metal catalyst of new generation of nickel [19, 25, 29–31] and precious metal hydrogenation catalysts [20, 32–34]. In general, such nanoparticles of metals such as nickel, palladium, ruthenium, or platinum are used for hydrogenation, since the dissociatively adsorbed hydrogen is easily accessible on these group VIII metals. Supported metal catalysts containing both a group VIII and a group II metal [32, 35, 36], or a case where the catalysts containing both a group VIII and a group IB metal [22, 29, 37–40] although insufficiently studied, can be found in the literature. In these catalysts, the metal of group II or group IB is added as the modifier with the purpose of promoting the *cis*-isomer selectivity. Recently, systematic investigation has been performed over Pd-Mg, Pt-Mg and Ni-Mg supported on silica [36] and Ni-Mg-Ag supported on diatomite catalysts [29]. Regarding to *cis*/*trans*-selectivity, these catalysts produce less *trans* isomers, promoting the *cis*-selectivity. The results have been interpreted by implying electronic effect—modifying the local electron density of the transition metal either directly or through the support [36] or geometrical effect (two separate metal phases)—blocking or masking the effect on the active metal particles without forming the chemical bond [29].

Although many preparation methods have been developed for synthesis of a well-defined supported metal nanocatalyst, traditional precipitation methods remain widely used, especially for industrial applications, due to their relatively low cost and simplicity. These approaches typically involve precipitating of metal salts with an alkaline precipitant in the presence of suspended supports and then thermal decomposition of salt to yield a dispersion of metal particles on the support. It is often challenging to generate uniform metal dispersions, especially in the case of high metal loading in supported metal catalysts. The desired metal dispersion depends on different factors, including synthesis method, nature of the support, identity of metal precursor salt, concentration of the metal, prereduction treatment and reduction conditions [41–44].

The most common methods used for preparation of supported nickel catalysts include impregnation, co-precipitation and precipitation-deposition (PD). Among these methods, to prepare catalysts with high nickel loading, the most suitable is the PD method. However, in the synthesis of supported catalysts by this method, the interaction of the precipitating precursor with the support such as silica or alumina plays a dominant role. Nickel ions (Ni^{2+}) can react with hydroxyl ions and silica to form a bulk compound, nickel hydrosilicate, which is more stable than the bulk hydroxide or hydroxyl-carbonate and nuclei stabilized by interaction with silica surface [45, 46]. It is undoubtedly proven that the reason for the difficulty of reduction of the active phase on the supported metal catalyst lies in the strong mutual interaction between precipitating nickel precursors and the silica support, with at least partial formation of nickel hydrosilicates [46–49].

In the partial hydrogenation process of edible oils, a catalyst with the high activity and selectivity is required [50–52]. To meet these requirements, the catalyst support should provide sufficient surface area for the metal to disperse and there must be adequate metal-support interaction [35, 43, 46, 48, 49, 53–57]. The nickel phase on different support surfaces exhibits different extents of metal-support effects. This implies that the surface properties of the catalyst could be changed by the nature of the supported Ni^{2+} phase, thus acquiring different characteristics and exhibiting different performances toward activity and selectivity, which are known to vary considerably with changes in the preparation conditions [41].

To control the fatty acid composition through temperature, pressure, catalyst and reaction time it is necessary to have a kinetic equation. The kinetic equations based on complex mechanisms as Horiuti-Polanyi [58] obtained from an extensive experimental work, give good results for predicting the reaction products, but in practice, simple mechanisms are employed with approximate results. An alternative is to use empirical modeling approach, which includes mathematical and statistical techniques for chosen empirical model [59–66].

The present work contains a part of our comprehensive research that we conducted on different nickel-based supported hydrogenation catalysts for their use in partial hydrogenation of edible oils. In this work, the characteristics and the structure of high loading nickel-based catalysts supported on diatomite, silica gel and perlite of different properties are related to their activity and selectivity in the hydrogenation of sunflower and soybean oils. Nitrogen physisorption and mercury porosimetry measurements, infrared and X-ray diffraction spectroscopy analyses, temperature programmed reduction studies, quantitative hydrogen chemisorption measurements and X-ray photoelectron spectroscopy were used. The kinetic models for hydrogenation of soybean and sunflower oils were developed to obtain the related kinetic parameters. The partial results derived of each one, treated together, have allowed us to present an overall picture of the nickel-based supported catalysts and some conclusions concerning the relationship in the triad—synthesis, structure and properties.

2. Experimental

2.1. Materials used

The support materials used for the synthesis of nickel-based catalysts were diatomite, silica gel and perlite. Diatomite support (D) was prepared from local crude diatomite (Baroševac-"Kolubara" coal basin, Lazarevac, Serbia) in our laboratories (IChTM-DCCE, see [67]). The crude material was mechanically, chemically and thermally treated to obtain the desired support characteristics. Three types of commercial silica gel with different textural characteristics (silica gel-A, silica gel-B, silica gel-C, silica gel = SiG hereinafter expressed as SiG-A, SiG-B and SiG-C, see [68]), produced in Bulgaria, were used for the preparation of the nickel-based catalyst supported on silica gel. Expanded perlite (PF, commercial product Perfit PF-295) supplied with the courtesy of Termika Zrenjanin, Serbia, was used for synthesis of nickel-based catalysts on perlite. Some of their characteristics are summarized in **Table 1**.

Refined, bleached and deodorized commercial sunflower oil (Dijamant AD Company, Zrenjanin, Serbia) along with refined and bleached soybean oil provided by Factory of Oils and Vegetable Fats Vital-Vrbas, Serbia, were used in the hydrogenation experiments. The initial iodine value (IV) and the fatty acid composition of sunflower and soybean oil are listed in **Table 2**.

http://dx.doi.org/10.5772/66967

Support	Property							
	Nitrogen physisorption[a]			**Mercury porosimetry**[b]				**He-P**[c]
	S_{BET} ($m^2 g^{-1}$)	**V_{micro} ($cm^3 g^{-1}$)**	**V_{meso} ($cm^3 g^{-1}$)**	**V_p ($cm^3 g^{-1}$)**	**D_{mean} (nm)**	**ρ_{bulk} ($g cm^{-3}$)**	**P (vol %)**	**ρ_{app} ($g cm^{-3}$)**
Diatomite (D)*	29	0.010	0.063	1.40	1188	0.59	73	2.21
Silica gel (SiG)*								
SiG-A	777	0.280	0.170	0.05	13	1.21	6	1.29
SiG-B	581	0.210	0.780	0.21	15	1.41	29	1.98
SiG-C	387	0.140	0.850	0.31	13	0.73	22	0.94
Perlite (PF)*	6	0.002	0.015	2.56	5500	0.25	89	2.23

*SiO2 content (wt%): Diatomite (D)—93.07; Silica gel (SiG-A, SiG-B, Sig-C)—100; Perlite (PF)—74.10.

[a]Determining surface area, adsorbed volume and pore size distribution in the micro- and mesopore ranges: S_{BET}—BET specific surface area; V_{micro}—specific pore volume (micropore range: D < 2nm); V_{meso}: specific pore volume (mesopore range: 2 nm < D < 50 nm).

[b]Determining total pore volume and pore size distribution in the meso- and macropore ranges: V_P—specific pore volume at 200 MPa; pore diameter: D_{mean}—mean pore diameter; P—total porosity; constant surface tension for Hg was assumed: $\gamma = 480$ dyn cm^{-1}.

[c]Helium picnometry—determining apparent density: ρ_{app}—includes closed pore (IUPAC).

Table 1. Typical catalyst support characteristics.

		Soybean oil (SBO)	**Sunflower oil (SFO)**
Iodine value (IV)		130.2	131.5
Fatty acid composition—CX:Y[b] (mass%)			
Palmitic	C16:0	11.0	7.3
Stearic	C18:0	4.5	4.[illegible]
Oleic	C18:1	21.3	26.0
Elaidic	C18:1*t*	< 0.1	< 0.1
Linoleic	C18:2	53.8	62.2
Trans acids sum[c]	*trans*-C18:2 sum	< 0.1	< 0.1
Linolenic	C18:3	7.1	< 0.1
Others	C14:0-C24:0 (soybean oil)	2.2	–
	C20:0-C22:0 (sunflower oil)	–	0.4

[a]Initial unsaturation of soybean and sunflower oil was measured in terms of IV.

[b]First number represents the total carbon number of the acyl groups and the second number represents the total number of double bounds.

[c]Sum: C18:2*c,t* + C18:2*t,c* + C18:2*t,t*.

Table 2. Iodine value (IV) and fatty acid composition of soybean and sunflower oils.

Analytical grade chemicals and pure hydrogen (99.999%) and nitrogen (99.999%) were employed in the experiments and none of these gases contained catalyst-poisoning substances such as oxygen or sulfur.

2.2. Catalyst preparation

Catalyst precursor synthesis. All catalyst precursors were prepared by the precipitation-deposition method. **Figure 1** shows a schematic of the experimental setup for catalyst precursor synthesis.

Table 3 lists catalyst precursor samples designation, synthesis operating parameters and nickel loadings.

The details of the synthesis procedures for differently supported nickel-based catalyst precursors were given in our earlier works [29, 69, 57, 31].

Catalyst precursor reduction. The reduction of the catalyst precursors was performed by a dry reduction method in a laboratory apparatus shown in **Figure 2** (line: reduction). Catalyst precursors were heated up to 430°C at a rate of 1.5°C min^{-1} under flowing H_2/N_2 gas mixture (150 cm^3 min^{-1}, 50 vol% of H_2) and maintained at 430°C for 5 h. Precursor treated in this way is referred to as Rp (e.g., $(Ni\text{-}Mg/D)_{Rp}$).

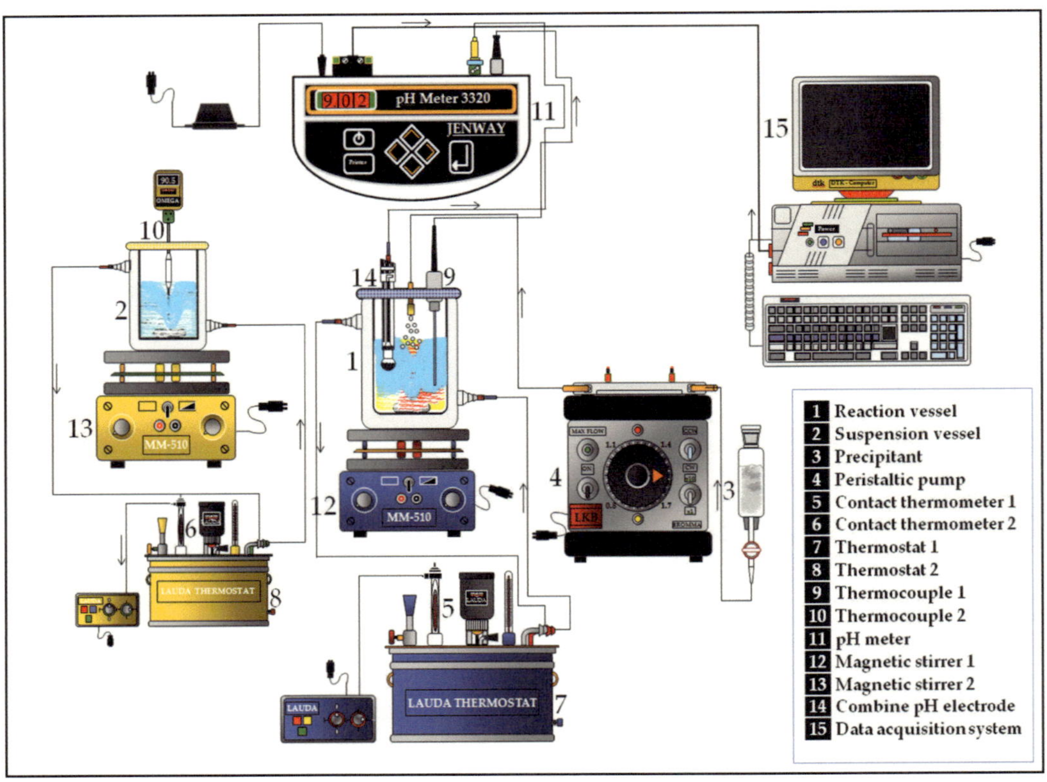

Figure 1. Schematic representation of the experimental setup—catalyst precursor synthesis.

Designation[a]		Synthesis operating parameters[b]				Ni loading
Catalyst precursor (system)	Sample (code)	Ni/SiO_2 (molar ratio)	Ni/Mg[b] (atomic ratio)	T_{PD}[b] (°C)	t_{AG}[b] (min)	Ni (wt%)
Ni-Mg/D	Ni-Mg/D*	0.93/1	10/1	90	120	36.3
	Ni_A-Mg/D	0.93/1	10/1	90	120	36.6
	Ni_C-Mg/D	0.93/1	10/1	90	120	36.2
	Ni_F-Mg/D	0.93/1	10/1	90	120	36.6
	Ni_S-Mg/D	0.93/1	10/1	90	120	35.0
Ni-Mg-Ag/D*	Ni-Mg-$Ag_{0.16}$/D*	0.93/1	10/1	90	120	35.9
	Ni-Mg-$Ag_{1.55}$/D*	0.93/1	10/1	90	120	35.2
	Ni-Mg-$Ag_{5.88}$/D*	0.93/1	10/1	90	120	33.5
Ni/SiG*	Ni/SiG-A*	1.01/1	–	90	120	43.7
	Ni/SiG-B*	1.11/1	–	90	120	45.5
	Ni/SiG-C*	1.00/1	–	90	120	43.5
Ni-Mg/PF*	Ni-Mg/PF-1*	1.00/1	10/1	90	30	30.2
	Ni-Mg/PF-2*	1.75/1	10/1	90	30	36.0

*Referred to catalyst precursors sample prepared from nitrate salts of nickel and modifier.

[a]Symbols in sample designation: A—acetate, C—chloride, F—formate, S—sulfamate; numbers refer to the Ag loading: 0.16 wt% of Ag (Ni-Mg-$Ag_{0.16}$/D), 1.55 wt% of Ag (Ni-Mg-$Ag_{1.55}$/D), 5.88 wt% of Ag (Ni-Mg-$Ag_{5.88}$/D); Ni/$Ag_{0.16}$ (atomic ratio) = 400/1, Ni/$Ag_{1.55}$ (atomic ratio) = 40/1, Ni/$Ag_{5.88}$ (atomic ratio) = 10/1.

[b]Constant parameters: T_{PD}—reaction temperature (precipitation-deposition), t_{AG}—aged temperature, precipitant: anhydrous Na_2CO_3 approx. 50% in excess with respect to a sum of Ni and Mg molar content.

Table 3. Catalyst precursor systems and samples prepared.

Reduced catalyst precursor soaking—final catalyst. After cooling down to room temperature, reduced catalyst precursors were protected by soaking in pure paraffin oil. After removing the excess of paraffin oil by filtering, the final catalyst was stored.

Reduced catalyst precursor passivation. Passivation of the reduced catalyst precursors was necessary to prevent the exceptional pyrophoricity of the metallic nickel before the reduced precursor was exposed to air for characterization and handling. This treatment was carried out in a laboratory apparatus depicted in **Figure 2** (line: passivation). Reduced catalyst precursor was passivated at room temperature in a N_2 stream (150 $cm^3\ min^{-1}$) containing 350 ppm O_2 for 30 min.

2.3. Characterization

Chemical analyses. The chemical analysis of nickel (Ni) was performed by standard test methods for quantitative analysis (dimethylglyoxime method). The silver (Ag) loading was determined using inductively coupled plasma-optical emission spectroscopy (ICP-OES) on an iCAP 6500 Duo apparatus (Thermo Fisher Scientific). Before the analysis, the precursors were subjected to microwave digestion using a Milestone Ethos 1 advanced microwave digestion system.

Nitrogen physisorption (N_2-physisorption). N_2 adsorption/desorption isotherms of the catalyst precursors at −196°C were measured on a Thermo Finnigan Sorptomatic 1990. Precursors

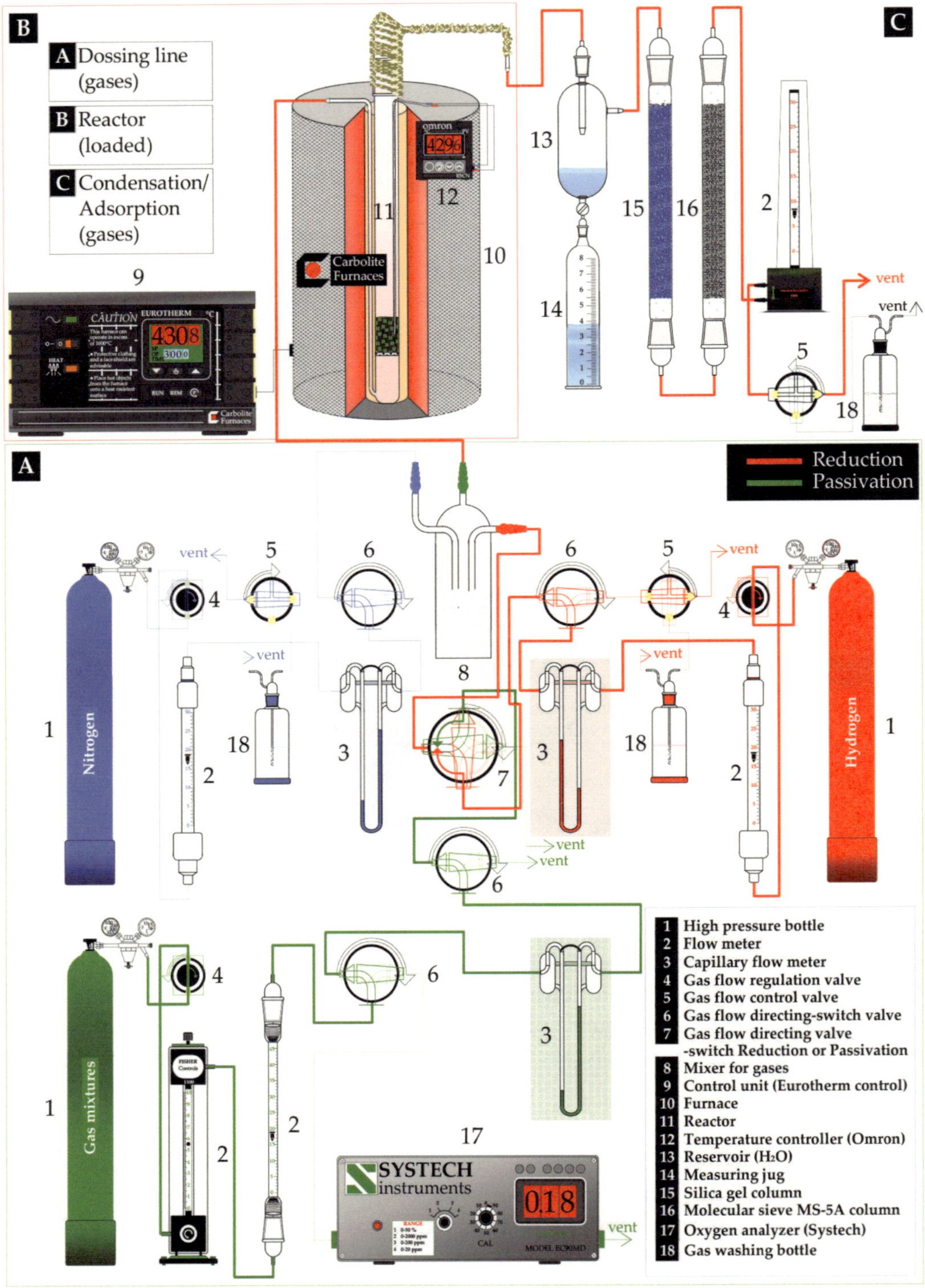

Figure 2. Schematic representation of the experimental setup—catalyst precursor reduction/passivation.

were previously degassed for 16 h at 110°C and 10^{-3} Torr (1 Torr = 133.3 Pa). The BET equation was used to calculate the specific surface area, S_{BET}. Total pore volume, V_p, was estimated at a relative pressure of 0.99.

Mercury porosimetry (Hg-porosimetry). Hg-porosimetry measurements were performed in the fully automated conventional porosimeters: 1. Thermo Scientific Pascal 140 porosimeter (pressure range: 0.01–0.1(0.4) MPa; pore size (diameter) range: 3.8–116 μm); 2. Thermo Scientific Pascal 440 porosimeter—Solid 1.3.4 software (pressure range: 0.1–400 MPa; pore size (diameter) range: 0.0036–15 μm)—data acquisition with the software package Solid 1.3.4; 3. Carlo Erba Macropore 120 porosimeter (pressure range: 0.01–0.1 MPa; pore size (diameter) range: 3.8–116 μm (or 15–150 μm); 4. Carlo Erba 2000 porosimeter (pressure range: 0.1–200 MPa; pore size (diameter) range: 0.0075–15 μm)—data acquisition with the software package Milestone 200.

Density measurements. Apparent densities were measured using a helium pycnometer, which uses a gas displacement technique to determine the volume of the solid material under test. The measurements were performed using a Pycnomatic ATC (Thermo Fisher Scientific).

IR measurements. IR spectra were recorded on a Perkin Elmer 983 G spectrometer (4000–250 cm^{-1}) and a Thermo Scientific Fourier transform Nicolet 6700 spectrometer (4000–400 cm^{-1}). KBr pellet method was used: 1 mg of precursor was well mixed into 200 mg fine KBr powder and the finely pulverized and put into a pellet-forming die.

X-ray diffraction (XRD). Powder XRD patterns were obtained with a Siemens D5005 diffractometer, equipped with a graphite monochromator and Antoan Paar chamber. Copper filtered Cu Kα radiation (λ = 0.154184 nm) was employed covering 2θ angles from 5 to 80° or from 10 to 100°. The average metal crystallite sizes were estimated by application of the Sherrer equation. The width of the Ni-(111) peak at half-maximum was corrected for Kα doublet and instrumental broadening.

Hydrogen temperature programmed reduction (H_2-TPR). Hydrogen temperature programmed reduction experiments were carried out in an automatic apparatus Thermo Scientific TPDRO, Pulse Chemisorb 1100. H_2-TPR experiments of catalyst precursors were performed in the temperature range 50-900°C, using a flow of H_2/Ar (20 cm^3 min^{-1}, 4.9 vol% of H_2) and a heating rate of 2°C min^{-1}. A cooling trap filled with 3A-MS was installed between the oven and TCD to remove the water formed during the reduction. The consumption of H_2 was monitored by a thermal conductivity detector. The detector response was calibrated by reducing a known mass of CuO. TPR profiles were normalized to the same mass of catalyst precursors.

Hydrogen chemisorption (H_2-chemisorption). H_2-chemisorption data were obtained using both the static method (standard volumetric apparatus) and the dynamic method (Thermo Scientific TPDRO, Pulse Chemisorb 1100).

Experimental setup—standard volumetric apparatus. In situ reduction of catalyst precursors at 430°C for 1 h by passing a mixture of hydrogen and nitrogen (H_2/N_2 = 1:1 v/v) over the sample. Reduction conditions were as follows: heating rate: 2°C min^{-1} and H_2/N_2 gas mixture flow of 20 cm^3 min^{-1}. After reduction, the precursors were degassed at 10^{-4} Pa for 4 h at the same temperature and cooled at 25°C to carry out chemisorption measurements. The pressure range of the isotherms was 0.1–13.3 kPa and the amounts of hydrogen chemisorbed were calculated by extrapolation of the isotherms to zero pressure. Further details about the expressions used to calculate nickel crystallite sizes can be found in our previous paper (see [29]).

Experimental setup—pulse chemisorb TPDRO 1100. In situ reduction of catalyst precursors at 430°C (2°C min^{-1}) for 1 h, under a flow of 20 cm^3 min^{-1} H_2/Ar (4.9 vol% of H_2). After reduction, the precursors were degassed by temperature programmed desorption at 425°C (carrier gas Ar) and cooled at 45°C, to carry out chemisorption of H_2. Finally, the catalyst precursors were subjected to a known number of calibrated pulses (0.353 cm^3) of pure H_2 at 45°C.

X-ray photoelectron spectroscopy (XPS). The quantitative chemical composition of surfaces of catalyst precursors and the valence state of studied ions were obtained by X-ray photoelectron spectroscopy. XPS measurements were performed using a VG Scientific Escalab Mk II spectrometer interfaced with the necessary data handling software Lab Cal 2. Spectra were recorded under ultrahigh vacuum conditions (10^{-8} Torr), using Al Kα primary radiation (1486.6 eV). Data were collected in the sequence of a survey scan (to determine the C 1s reference), followed by scans of the O 1s, S 2p, Ni 2p$_{3/2}$ and Ag 3d regions to minimize the time of exposure to X-rays.

2.4. Catalytic activity measurements

Partial hydrogenation of sunflower and soybean oil over prepared catalysts was carried out under laboratory and pilot plant conditions. Schematics of the lab- and pilot-experimental setup are shown in **Figures 3** and **4**.

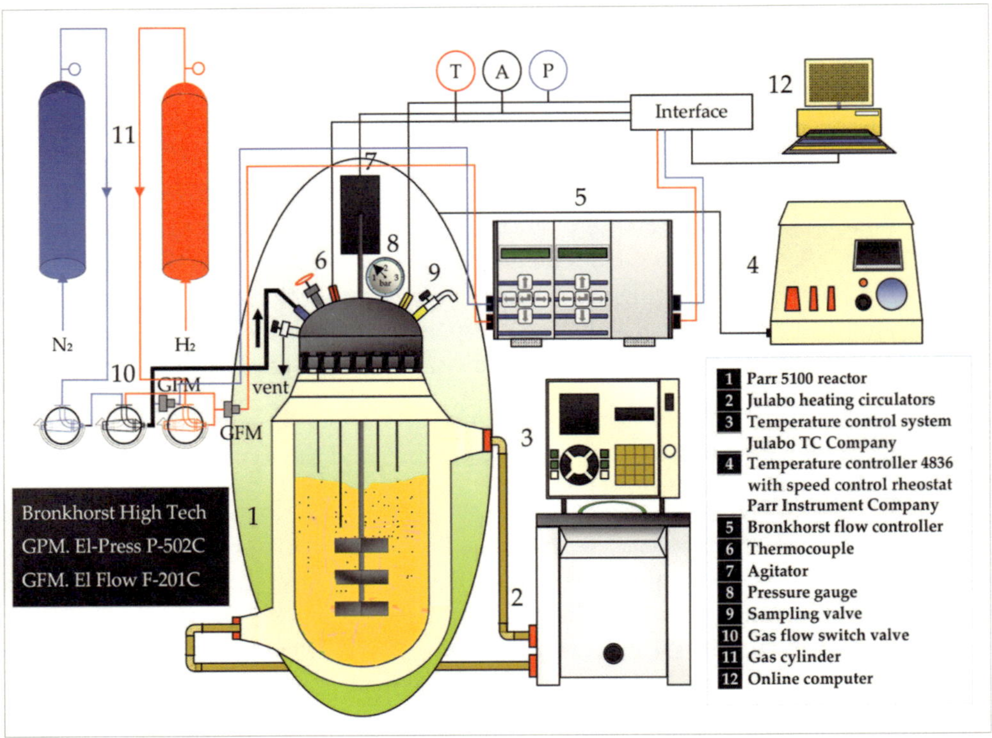

Figure 3. Schematic representation of the experimental setup—laboratory reactor system.

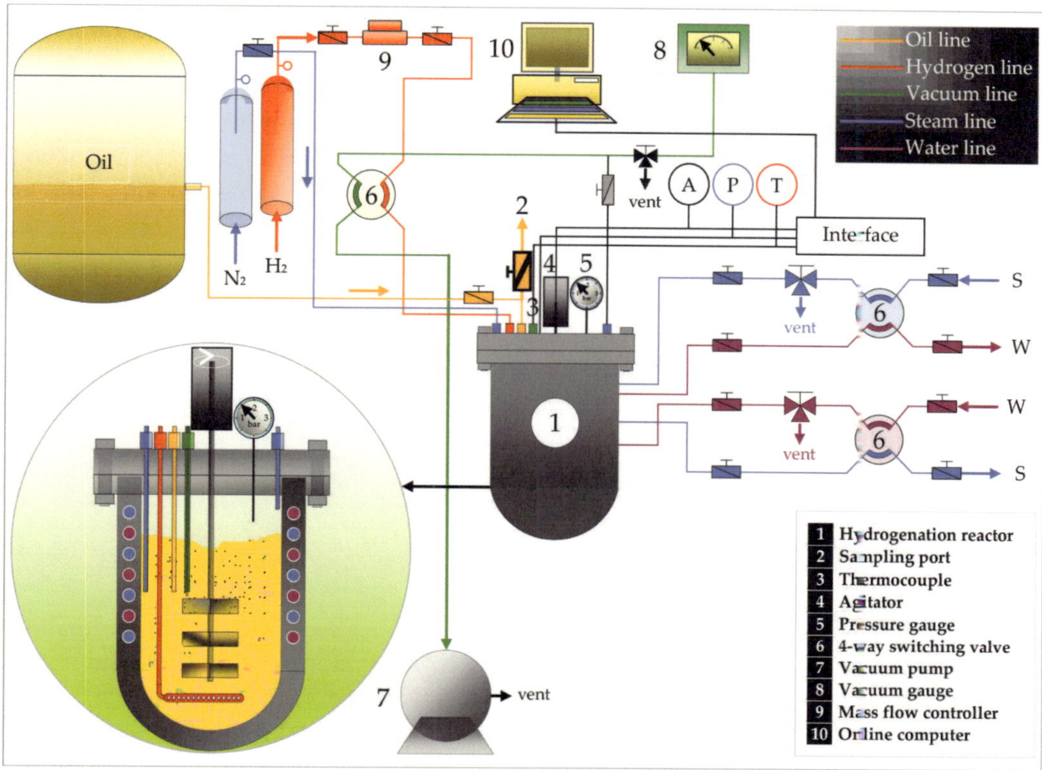

Figure 4. Schematic representation of the experimental setup—pilot plant reactor system.

Experimental setup—laboratory reactor system. The hydrogenation tests were carried out in a Parr 5100 glass 1 L jacketed reactor operated in a semibatch mode. The laboratory system comprises a catalytic reactor interfaced with a mass flow controller—Bronkhorst EL-FLOW model F-201C and a digital electronic pressure meter—Bronkhorst EL-PRESS model P-502C and a minicomputer. The reactor was connected to a hydrogen source, maintained at a constant pressure. The reduced catalyst was added to the oil when the reaction temperature was reached using a new type of catalyst feeder constructed in our laboratories (IChTM-DCCE). The catalytic tests were performed at 160°C and 0.2 MPa, using 900 g of sunflower oil. The catalyst weight was adjusted in order to keep constant oil to Ni mass ratio (0.06 wt% Ni with respect to the amount of oil). The stirring rate was 1200 rpm. An experimental procedure of hydrogenation test under laboratory conditions was described in detail in our previous paper [30].

Experimental setup—pilot plant reactor system. The catalytic experiments were carried out in a 7.5-L batch-stirred PPV (Pilot Plant Vital) reaction vessel. The PPV and its ancillaries were available at the research facilities of Oils and Vegetable Fats Factory Vital-Vrbas. **Figure 4** shows a schematic representation of the experimental setup.

Each experiment was performed at a constant liquid volume and constant oil/catalyst mass ratio (see [29]). Before the reactor was heated, the headspace was purged with nitrogen to remove oxygen. The catalyst sample was precisely weighed and added to the liquid soybean oil at working temperature (160°C), under a mixing speed set at 750 rpm. The reactor was then pressurized with pure hydrogen to the operating pressure (0.16 MPa). During the experiments, the heat flow, hydrogen uptake and reactor temperature and pressure were monitored by instruments interfaced to the reactor PPV system. For each run, the soybean oil batch was partially hydrogenated to a final IV of 90. The composition of fatty acids in the original soybean oil and hydrogenated products was analyzed by the capillary gas chromatographic method. Experiments were performed on a Schimadzu GC-9A equipped with flame ionization detector (FID). Chromatographic conditions were as follows: HP-88 capillary column (100 m × 0.25 mm, 0.20 μm film thickness, Agilent), oven temperature of 180°C, detector and injector temperature of 240°C. Injection was carried out in the splite mode at a splite ratio of 1:4. The injection volume was 2 μL. Helium was used as the carrier gas at a constant flow rate of 1.2 cm^3 min^{-1}. The IUPAC method II.D.19 [70] for preparation and CG analysis of fatty acids methyl esters was used to convert fatty acids, taken out at predetermined time intervals from the catalytic reactor, into their corresponding methyl esters.

3. Results and discussion

3.1. Catalyst precursor characterization

Nitrogen physisorption (N_2-physisorption—NP). Nitrogen adsorption-desorption measurements were performed to determine the effect of the support and preparation parameters on the BET surface area, pore volume and mean pore diameter of prepared catalyst precursor samples. The obtained results are summarized in **Table 4**.

Table 4 reveals that specific surface area (S_{BET}) was greatly enhanced after deposition Ni^{2+} precipitates onto the macroporous diatomite and/or perlite support. The prepared samples have BET surface areas, which are an order of magnitude larger than the surface area of the supports. The catalyst precursors have significantly different values of the pore volume and mean diameter of pores compared to the starting support materials (**Tables 1** and **4**). It has been shown in numerous studies that improved textural properties of the supported metal catalysts are related to the metal-support interaction (MSI) [49, 71–73]. This phenomenon was also observed in the nickel catalysts supported on diatomite and/or silica [54, 71]. The authors explained an extremely enhanced BET surface area by forming supported intermediate phases layered structure-nickel hydrosilicates as a result of nickel-support interaction. Nepouite-like (1:1 nickel hydrosilicate) is the main Ni^{2+} phase formed in Ni/SiO_2 samples prepared by the PD method on silica with a high surface area [56]. In the case of Ni-Mg/D precursors, it is reasonable to assume that the improved textural properties of the precursors are the results of MSI, as elucidated by IR and XRD patterns discussed later in this paper. Ni/SiG precursors showed different behavior. A considerable decrease in the BET surface area was observed in the Ni/SiG-A and Ni/SiG-B precursors formed by precipitating the Ni^{2+} species onto micro- (SiG-A) and mesoporous (SiG-B) supports initially having high surface areas. It is well known that the supports with low specific surface areas favor the formation of nickel hydroxide or hydroxyl-carbonate phases whatever the preparation method is, while supports with high specific surface areas allow the growth of hydrosilicate to occur [74]. It is likely that the high nickel loadings in those precursors (above 40 wt%, **Table 3**)

Sample	S^a ($m^2\ g^{-1}$)	S_{meso}/S^b	V_p ($cm^3\ g^{-1}$)	V_{meso}/V_p ^{b}	D_{mean} ^{c} (nm)
Ni-Mg/D	224	0.602	0.201	0.695	3.5
Ni_A-Mg/D	223	0.430	0.161	0.571	3.7
Ni_C-Mg/D	208	0.516	0.170	0.694	4.1
Ni_F-Mg/D	177	0.477	0.150	0.613	4.5
Ni_S-Mg/D	255	0.472	0.232	0.698	5.5
Ni-Mg-$Ag_{1.55}$/D	167 (157*)	0.655 (0.585*)	0.172 (0.157*)	0.610 (0.599*)	4.1 (4.0*)
Ni-Mg-$Ag_{5.88}$/D	124 (136*)	0.812 (0.753*)	0.149 (0.157*)	0.691 (0.662*)	4.8 (4.6*)
Ni/SiG-A	269 (146*)	0.781 (0.836*)	0.530 (0.310*)	0.604 (0.710*)	8.8 (8.9*)
Ni/SiG-B	392 (193*)	0.735 (0.642*)	0.840 (0.320*)	0.464 (0.594*)	6.7 (7.0*)
Ni/SiG-C	367 (216*)	0.757 (0.676*)	0.630 (0.340*)	0.730 (0.647*)	10.2 (7.0*)
Ni-Mg/PF-1	210	0.622	0.169	0.799	3.7
Ni-Mg/PF-2	225	0.898	0.186	0.946	3.6

*Data for reduced-passivated samples: $(Ni\text{-}Mg/D)_{Rp}$, $(Ni\text{-}Mg\text{-}Ag_{1.55}/D)_{Rp}$ and $(Ni\text{-}Mg\text{-}Ag_{5.88}/D)_{Rp}$; $(Ni/SiG\text{-}A)_{Rp}$, $(Ni/SiG\text{-}B)_{Rp}$ and $(Ni/SiG\text{-}C)_{Rp}$.

[a]S: specific surface area-BET method.

[b]S_{meso} and V_{meso}: contribution of the mesopores.

[c]D_{mean}: mean pore diameter.

Table 4. Nitrogen physisorption data for dried and selected reduced-passivated catalyst precursor samples.

mitigate the effect of MSI leading to the formation of bulk nickel hydroxyl-carbonate covering the hydrosilicate layer formed in MSI. On the other hand, the precipitation of Ni^{2+} species onto the SiG-C support does not affect the BET surface area of the Ni/SiG-C precursor (see **Tables 1** and **4**).

The nitrogen adsorption-desorption isotherms are presented in **Figure 5**. Comparative plots have been constructed for all supports and systems of catalyst precursors.

In general, the experimental N_2 adsorption-desorption curves of samples closely resemble a type IV isotherm characteristic for mesoporous solids according to the IUPAC classification, with exception of the samples of D and SiG-A that appear to be extensively macroporous (isotherm type II) and microporous (isotherm type I). In the case of catalyst precursors of type Ni-Mg/D (**Figure 5a** and **d**) after saturation of micropores, nitrogen uptake monotonically increases with p/p_0 values due to sorption in the sample larger pores. After filling of mesopores, the uptake remains essentially invariant with p/p_0 values. The slope of curves is small, indicating a low content of mesopores having a wide pore size distribution (PSDs). In contrast, the catalyst precursors prepared on SiG and PF supports had sharper slope of nitrogen adsorption curves, indicating mesopores present with narrow PSDs.

It is well known that the occurrence of the capillary hysteresis loop depends on the pore sizes. In N_2-physisorption, the isotherms of the samples with the smallest pore size do not exhibit hysteresis, while the samples with the smaller pore sizes exhibit isotherms with narrow hysteresis.

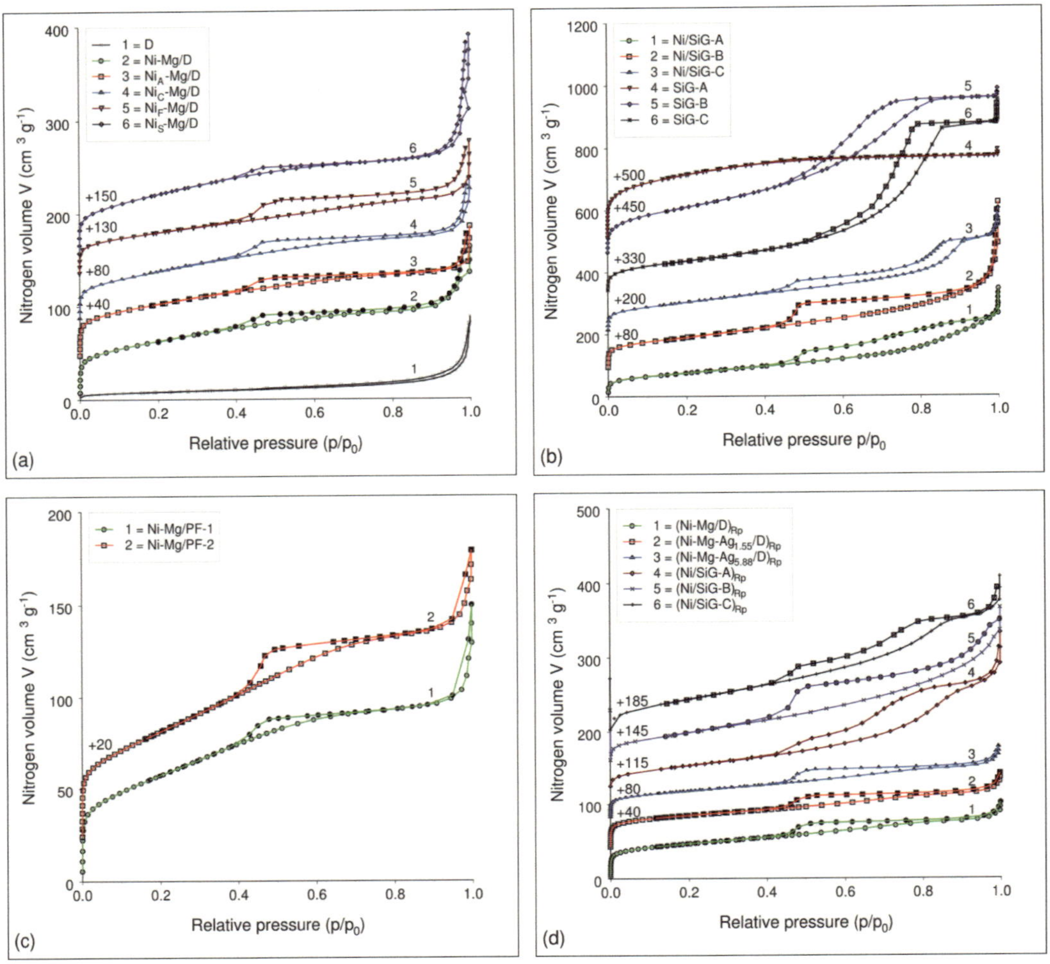

Figure 5. Experimental N_2 adsorption-desorption isotherms at -196°C for dried and reduced-passivated samples: (a) D, Ni-Mg/D; (b) SiG, Ni/SiG; (c) PF, Ni-Mg/PF; (d) $(Ni\text{-}Mg/D)_{Rp}$, $(Ni\text{-}Mg\text{-}Ag/D)_{Rp}$, $(Ni/SiG)_{Rp}$. Units and shifts along Y axis are chosen for convenience.

Wider capillary hysteresis loops are observed in the nitrogen isotherms on the samples with larger pore sizes. Besides, the shapes of the capillary hysteresis loops vary from a "triangle" to a well-pronounced "parallelogram". The results obtained for N_2-physisorption showed that samples differ in the shape and types of hysteresis. As can be observed, the precursors Ni-Mg/D, Ni-Mg-Ag/D and Ni-Mg/PF are characterized by capillary hysteresis loops with shape of a triangle (**Figure 5a** and **c**) unlike the precursor Ni/SiG having capillary hysteresis loops shape that look like a parallelogram (**Figure 5b**). The hysteresis type of the samples cannot be classified into any types of IUPAC classification and mostly resembles the H3 type corresponding to the mesoporous solids with a broad distribution of the pore sizes. Despite the expected absence of hysteresis for the diatomite support, a narrow loop of H1 type in the IUPAC classification was observed (**Figure 5a**).

http://dx.doi.org/10.5772/66967

Thermal treatment of reduced samples has not changed the pore structure significantly, preserving their mesoporosity (**Figure 5d**).

Mercury porosimetry (Hg-porosimetry—MP). The relevant mercury porosimetry experimental data are summarized in **Table 5**.

Table 5 indicates a large total pore volume and porosity for the dried catalyst precursor of Ni/SiG and Ni-Mg/PF systems. In the case of Ni-Mg/D catalyst precursor, the pore volume and porosity, as shown in **Table 5**, are obviously different from those mentioned above. The total pore volume and porosity are, in general, significantly lower and the pores are significantly smaller in diameter (D_{mean} and/or D_{av}). The differences between the prepared precursors may be associated with differences in pore structure characteristics of supports and the nature of nickel precursor salt.

Sample	V_p [b] ($cm^3\ g^{-1}$)	P[c] (vol %)	ρ_{bulk} [d] ($g\ cm^{-3}$)	D_{mean} [e] (nm)	D_{av} [f] (nm)
Ni-Mg/D	0.393	47	1.19	164	7.0
Ni_A-Mg/D	0.228	34	1.48	219	4.1
Ni_C-Mg/D	0.246	34	1.39	177	4.7
Ni_F-Mg/D	0.118	20	1.72	73	2.7
Ni_S-Mg/D	0.585	54	0.93	229	9.2
Ni-Mg-$Ag_{1.55}$/D	0.092 (0.135*)	16 (26*)	1.78 (1.89*)	29 (177*)	2.2 (3.4*)
Ni-Mg-$Ag_{5.88}$/D	0.111 (0.144*)	21 (29*)	1.84 (2.01*)	89 (189*)	3.6 (4.2*)
Ni/SIG-A	0.655	49	0.75	2331	9.7
Ni/SIG-B	1.876	72	0.38	1152	19.1
Ni/SIG-C	1.268	61	0.48	3087	13.8
Ni-Mg/PF-1	1.683	79	0.47	3690	32.0
Ni-Mg/PF-2	0.945	66	0.70	2237	16.8

*Data for reduced-passivated samples: $(Ni\text{-}Mg/D)_{Rp}$, $(Ni\text{-}Mg\text{-}Ag_{1.55}/D)_{Rp}$ and $(Ni\text{-}Mg\text{-}Ag_{5.88}/D)_{Rp}$.
[a]Acquisition data obtained by Milestone 200 and Pascal softwares.
[b]Pore volume.
[c]Porosity calculated from bulk density of dried sample as measured in porosimeter.
[d]Bulk density.
[e]Pore diameter: D_{mean}—computed from the corresponding PSD curves.
[f]Pore diameter: D_{av}—average pore diameter calculated according to the $D_{av}=4V_p/S_{BET}$ assuming cylindrical pore shape.

Table 5. Mercury porosimetry data for dried and selected reduced-passivated catalyst precursor samples[a].

The experimental PSDs data are presented in the form of cumulative pore diameters distribution curves in **Figure 6**. The data are cumulated from larger pore diameters measured to the smallest diameter limit set by the pressuring capacity of the instrument. According to the hysteresis curves (not shown), the main part of mercury remains in the pores after the measurement, indicating the presence of ink bottle-like pores.

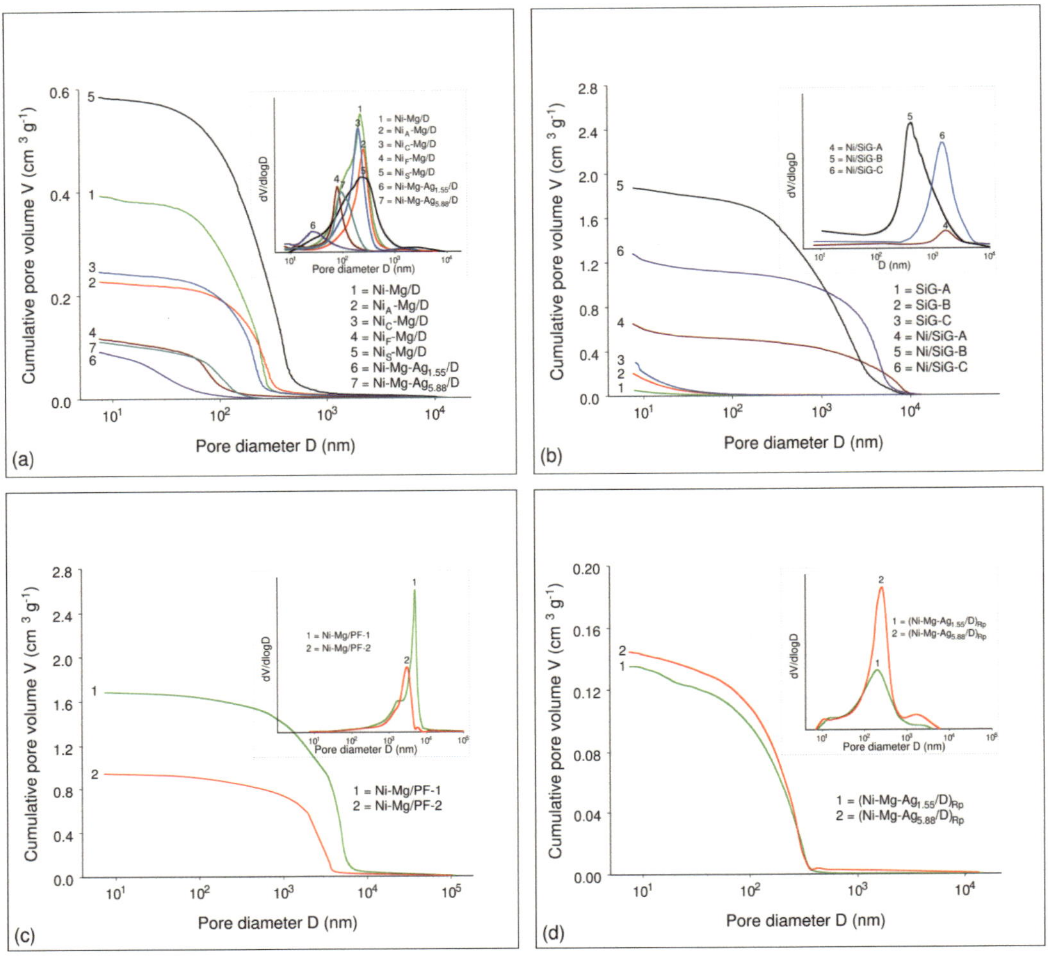

Figure 6. Cumulative and derivative PSDs for the SiG supports and catalyst precursor systems: (a) Ni-Mg/D, Ni-Mg-Ag/D; (b) SiG, Ni/SiG; (c) Ni-Mg/PF; (d) $(\text{Ni-Mg-Ag/D})_{Rp}$.

The derivate distribution function (dV/dlogD) is represented as insert (see **Figure 6a–d**). It should be noted that the pressurization data from mercury intrusion yields information about the size of the opening of pores and/or voids and does not reflect the pore size behind the "neck". It is apparent that the deposition of Ni^{2+} precipitates onto surface of SiG supports leading to a shift in the PSD curves towards the larger pore diameters. The explanation for this effect appears to lie in different microstructural arrangements of supported Ni^{2+} species in dried samples compared to the starting supports (**Figure 6b**). Thermal treatment has led to opening of smaller pores and a slight displacement of PSDs to larger pore diameters (**Table 5**, **Figure 6d**).

Combined nitrogen physisorption and mercury porosimetry studies showed that all catalyst precursors had good textural properties, namely a high specific surface area and a well-developed porous structure, containing mesopores stable to thermal treatments. Mesoporosity

http://dx.doi.org/10.5772/66967

(pore width of 2–50 nm) is preferable for application that involves the liquid phase since it provides a balance between good diffusion rates of reactants and useful in-pore effects.

IR analysis. The IR method was employed to identify the Ni^{2+} species in the precursor samples. The identification of Ni^{2+} species by IR analysis is made by the comparison with reference bulk compound—basic nickel carbonate (BNC). **Figure 7** shows infrared and Fourier transform infrared spectra of the studied samples. The 2000–400 cm^{-1} region has been presented in order to obtain a better representation.

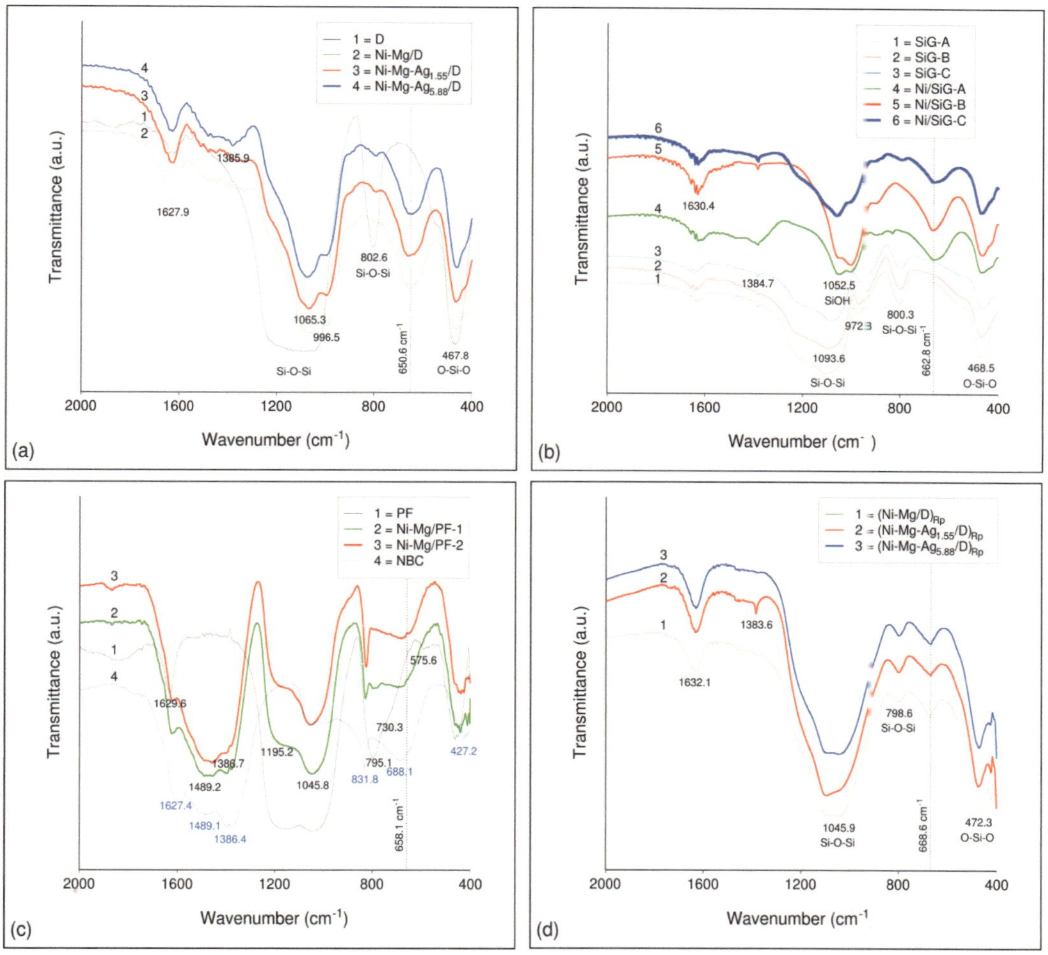

Figure 7. IR spectra of the supports and catalyst precursor samples: (a) D, Ni-Mg/D, Ni-Mg-Ag/D; (b) SiG, Ni/SiG; (c) PF, Ni-Mg/PF; (d) $(Ni\text{-}Mg\text{-}Ag/D)_{Rp}$. Units and shifts along Y-axis are chosen for convenience.

The infrared spectra of the D support and dried Ni-Mg/D catalyst precursors are shown in **Figure 7a**. The IR spectrum of the D support having silica as its main constituent ingredient shows antisymmetric stretching vibration band at around 1090 cm^{-1} and symmetric stretching vibration band at around 800 cm^{-1} characteristic for the Si-O-Si bonds [75]. The band at around 470 cm^{-1} is associated with O-Si-O bond bending vibrations. The absorption band at around 1630 cm^{-1} can be assigned to the vibration of adsorbed molecular water.

IR spectra of precursors are similar to the spectrum of diatomite in the OH-stretching region containing vibration bands of silica-free hydroxyl groups, hydrogen-bonded hydroxyl groups and adsorbed molecular water (not shown in **Figure 7a**). The main absorption broad band in the spectrum of diatomite (1090 cm^{-1}) in the IR spectra of the precursors appeared to be composed of multiple bands around 1100 cm^{-1}. It can be observed the presence of three bands: IR band characteristic of silica at 1090 cm^{-1} was reduced to a shoulder, IR band characteristic of silicate-type species connected/interacted with carbonate rich BNC species appeared at 1065 cm^{-1} and shoulder appeared around 1000 cm^{-1} which could be attributed to silicate-type species connected/interacted with hydroxide rich BNC species [54]. The presence of intercalated anionic species in the Ni^{2+} precipitates is attested by the existence of bands at around 1630 cm^{-1} and at around 1385 cm^{-1} may be attributed to adsorbed molecular water and carbonate ions. By comparing spectra of precursors and diatomite, it can be observed that IR spectra of precursors contain a band at 650.6 cm^{-1} (**Figure 7a**) which does not exist in the IR spectrum of the diatomite support. The appearance of a new phase can be attributed to nickel hydrosilicates arisen from the interaction between nickel and the diatomite support. It is apparent that this method of preparation leads to extensive interaction between Ni ions and silica, presumably because under alkaline conditions the silica have a tendency to dissolve. Characterization studies have shown that Ni^{2+} precipitates on silica as layered nickel hydrosilicate [45–49, 53–57, 68, 72, 74]. In support of this claim is the fact that samples are prepared by a precipitation-deposition method, which leads to the formation of supported nickel hydrosilicates.

The IR spectra of SiG supports and Ni/SiG precursor samples are shown in **Figure 7b**. It is obvious that the IR spectra of these samples resemble those prepared with diatomite support (**Figure 7a** and **b**). This result was expected, having in mind the proportion of the SiO_2 component in the chemical composition of diatomite and SiG supports. In addition, a band at 972 cm^{-1} can be observed in all three types of SiG support. The band at around 970 cm^{-1} has been widely used to characterize the incorporation of metal ions in the silica framework as the stretching Si-O vibration mode perturbed by the neighboring metal ions. A new band at 662.8 cm^{-1} attributable to nickel hydrosilicates can also be observed in all Ni/SiG precursor samples (**Figure 7b**).

FT-IR spectra of the perlite support and Ni-Mg/PF precursors are presented in **Figure 7c**. The spectrum of perlite support in the region from 400 to 2000 cm^{-1} shows the presence of the main absorption structures, an intense band at about 1045 cm^{-1} with a shoulder at about 1200 cm^{-1}. These two bands are attributed to Si-O-Si and Si-O-M anti-symmetric stretching vibrations, where M can be Al or Si. A further group of three bands of medium intensity is present at lower wavelengths: 795, 730 and 575 cm^{-1}. The band at 795 cm^{-1} is assigned to symmetric stretching of Si-O-Si, at 730 cm^{-1} to bending Si-O-Al and at 575 cm^{-1} to symmetric stretching of Si-O-R [76]. Water molecule deformation vibrations at around 1630 cm^{-1} are also registered. The well-expressed bands at 1387 and 1489 cm^{-1} in the IR spectrum of precursors are attributed to the presence of an additional carbonate containing phase, most probably located on the surface of the support. Comparing bands gained in synthesized precursors are apparent evidence of the created Ni^{2+} species on the support surfaces. On the reference sample spectra, existence of broad antisymmetric band at 688 cm^{-1} is evident. This band also exists in spectra of precursors, although slightly shifted towards lower wavenumbers with minimum at around 658 cm^{-1} (**Figure 7c**). Shift indicates a new type of interaction with the support, compared to the reference material and may be attributed to the Ni-O-Si vibrations [77]. It can also be stated that no evidence of structural change among the dried precursors can be acknowledged.

The IR spectra of precursors after the reduction treatment are presented in **Figure 7d**. The thermal treatment produced the elimination of adsorbed molecular water and carbon dioxide. The absence of silanol groups, as attested by disappearance of the anti-symmetric band at 980 cm^{-1} is clearly evident. Besides, a low intense band at 668 cm^{-1} indicates the presence of nonreduced silicate species in smaller amounts than in the dried samples (**Figure 7a** and **d**).

From the above results and with the available information in the literature, it could be concluded that during the deposition reaction under alkaline conditions, the silica as a constitutive component of all studied supports reacts with the basic nickel carbonate precipitate and generates the new supported nickel hydrosilicate phase containing Si-O-Ni linkages.

Powder X-ray diffraction (XRD). Powder XRD analyses were performed on the dried and reduced-passivated catalyst precursors to identify the phases present in the precursor samples at various stages of synthesis. **Figure 8** shows the XRD diffractograms obtained for dried and reduced-passivated precursor samples.

The diffractogram of the diatomite support (**Figure 8a**, curve 1) shows reflections characteristic of amorphous silica (silica halo peak centered at two-theta around 21°) and the well-crystallized quartz (Q) phase (two-theta = 26.6°; JCPDS 46-1045). Typical diffractograms of dried precursor samples exhibited only broad and asymmetrical bands attributable to ill-defined and badly crystallized nickel hydrosilicates (**Figure 8a**, curves 2–6). Besides, the XRD patterns of bulk BNC disappear in the patterns of precursors. The observed phenomenon proves that an interaction occurs between BNC and silica from the support. The formation of surface lamellar hydrosilicates in the preparation of silica supported nickel catalysts was postulated on the basis of several techniques of characterization (IR, TPR, XPS and Extended X-ray Absorption Fine Structure-EXAFS). X-ray diffraction was also employed to corroborate the presence of nickel hydrosilicate. The XRD studies showed that the nickel hydrosilicates are formed under various conditions at relatively low temperatures (under 100°C) [48, 71, 78–81]. As it is well known, when the nickel salt is precipitated with Na_2CO_3 in the presence of silica, the precipitated Ni^{2+} phase is silica supported-BNC with composition of $Ni(OH)_x(CO_3)_y/SiO_2 \times zH_2O$ that varies considerably with the changes in the precipitation conditions [41]. This interaction leads to the formation of badly ill-crystallized layered nickel hydrosilicate compounds, identified as nepouite-like (1:1 nickel hydrosilicate; $Ni_3(OH)_4(Si_2O_5)$) and/or talc-like structure (2:1 nickel hydrosilicate; $Ni_3(OH)_2(Si_2O_5)_2$). The formed phases have been identified as nickel antigorite [48, 71, 78, 81], nickel chrysotiles [82], nickel montomorillonite [48, 71], nickel palygorskite [54], serpentine [83], or orthosilicate-type [54].

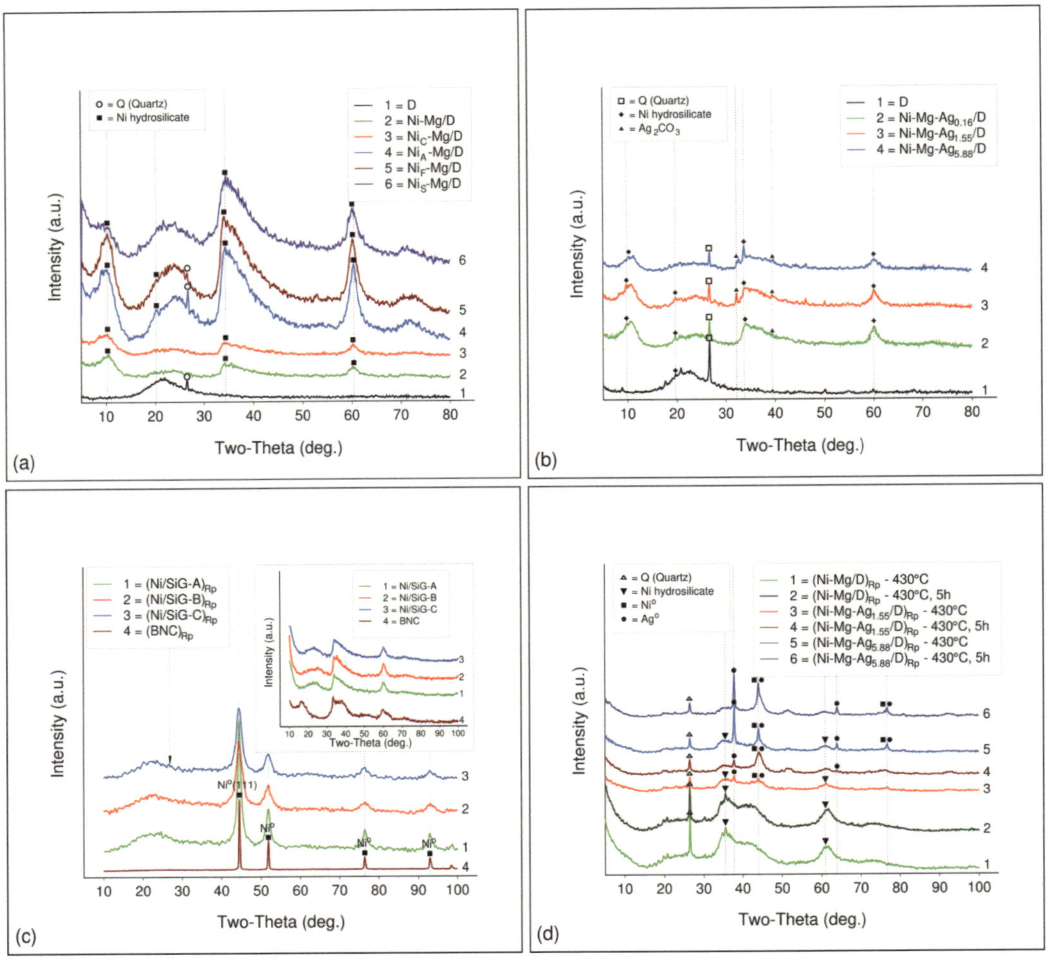

Figure 8. XRD diffractograms of supports, dried and reduced-passivated precursor samples: (a) D, Ni-Mg/D; (b) Ni-Mg-Ag/D; (c) Ni/SiG, $(Ni/SiG)_{Rp}$; (d) $(Ni\text{-}Mg/D)_{Rp}$, $(Ni\text{-}Mg\text{-}Ag/D)_{Rp}$.

Samples modified with silver (Ni-Mg-Ag/D, **Figure 8b**) have slightly altered XRD spectra. In addition to XRD peaks characteristic to the nickel hydrosilicates, two new peaks can be observed at 32.2 and 39.4° attributable to the α-Ag_2CO_3 phase (JCPDS file 31-1237). Moreover, it is observed that the modification with silver contributes to the further amorphization of the dried precursor samples [84].

The XRD patterns of the dried and reduced-passivated silica gel supported Ni catalyst precursors and reference material-BNC are presented in **Figure 8c**. The XRD patterns of three types of silica gel supports had characteristic reflections of amorphous silica (not shown). The insert in **Figure 8c** represents a comparison between the XRD patterns of the dried precursors of the Ni/SiG system and the reference BNC material. The absence of the diffraction line at 16.3° of the reference material-BNC sample and appearance of a new broad reflection at around 23° in the spectra of this catalyst precursor system represented a substantial difference between the bulk reference material and the supported Ni^{2+} phase, present in the samples of this catalyst precursor system. The turbostratic structure of nickel hydrosilicate [48, 54] predetermined the ill-organized reflections of the Ni/SiG system of the precursor. Moreover, the nickel hydrosilicate phase exhibits different degrees of crystallization, more pronounced in the Ni/SIG-B sample. It is obvious that the usage of different silica gel types affects the crystallinity of the deposited Ni containing phase. Note that, the registered high background below two-theta values of 15° in XRD diffractograms of all samples indicates advanced amorphization of the observed phase.

The XRD patterns of the reduced precursor samples at 430°C (**Figure 8c**) display reflections located at two-theta, typical for nickel metal (Ni^0) (JCPDS file 00-004-0850). The peaks of lower intensity between two-theta from 32 to 40° indicate the presence of nickel hydrosilicate in all reduced samples, but it is better represented in the sample Ni/SiG-B.

In the case of the reduced catalyst precursors of $(Ni\text{-}Mg/D)_{Rp}$ and $(Ni\text{-}Mg\text{-}Ag/D)_{Rp}$ systems, typical XRD spectra showed common peaks corresponding to nickel metal (Ni^0) and silver metal (Ag^0) (**Figure 8d**). The layered structure of the nickel hydrosilicate phase was also registered. The experimental conditions for the reduction step was selected in order to establish the relation between the reduction time and the reduction temperature (selected temperature of 430°C) was assumed to be very important. Despite the obvious reduction in intensity of peaks caused by prolonged dwell time (5h) in the selected temperature, reflections corresponding to nickel hydrosilicates are still visible. This shows that the reduction temperature of 430°C used for reduction of the dried precursor with H_2 was not sufficient to reduce all the nickel hydrosilicate species to the nickel metal (Ni^0) and silica for these two systems of catalyst precursors.

The results concerning the influence of the preparation stage and nature of the support and the modifier clearly illustrate the feature of the supported Ni^{2+} phase and demonstrate that XRD measurements may offer an effective tool to identify the nickel species and their interaction with the support in differently supported and modified nickel-based catalyst precursors.

Hydrogen temperature programmed reduction (H_2-TPR). The objective of H_2-TPR experiments was to determine the reducibility as well as the optimum reduction temperature for prepared catalyst precursor systems. In conjunction with IR and XRD data, TPR profiles were also useful in determining the type of Ni^{2+} phases present in the catalyst precursors and may be indicative of the actual activity of the final reduced metal catalyst. The H_2-TPR profiles of all prepared catalyst precursors are given in **Figure 9**.

The influence of the nature of precursor salts of nickel on the reducibility of prepared samples is reported in **Figure 9a**. These results demonstrate rather well the differences between the Ni^{2+} species formed in the case of diatomite supported nickel-based catalyst precursors. A peak due to the reduction of the Ni^{2+} phase, which corresponds to the BNC (**Figure 9b–d**—insert) was seen only in the sample prepared from the sulfamate salt of nickel (Ni_S-Mg/D). Among the prepared catalyst precursors, the smallest proportion of the Ni^{2+} phase from BNC can be seen in the Ni_A-Mg/D sample. The high reduction temperature needed for the samples prepared from the acetate nickel precursor salt is obtained by the presence of difficult to reduce nickel hydrosilicates, which is the form in which nickel precipitates are deposited during synthesis. Consequently, layered nickel hydrosilicates whose thermal decomposition starts above 450°C

[47] appear to be the main nickel species present in this sample. The stronger interaction of nickel and diatomite support hinders reduction of samples. This leads to a shift in the T_{max} value of peaks corresponding to the reduction of Ni^{2+} phases interacting with the support from 320°C over Ni_S-Mg/D to 462°C over Ni_A-Mg/D. The reduction extent (R, %) of the Ni^{2+} supported phase at 430°C (**Figure 9a**) increases with the following sequence: Ni_A-Mg/D (24.8%) < Ni_F-Mg/D (26.4%) < Ni-Mg/D (43.1%) < Ni_C-Mg/D (54.5%) < Ni_S-Mg/D (75.8%).

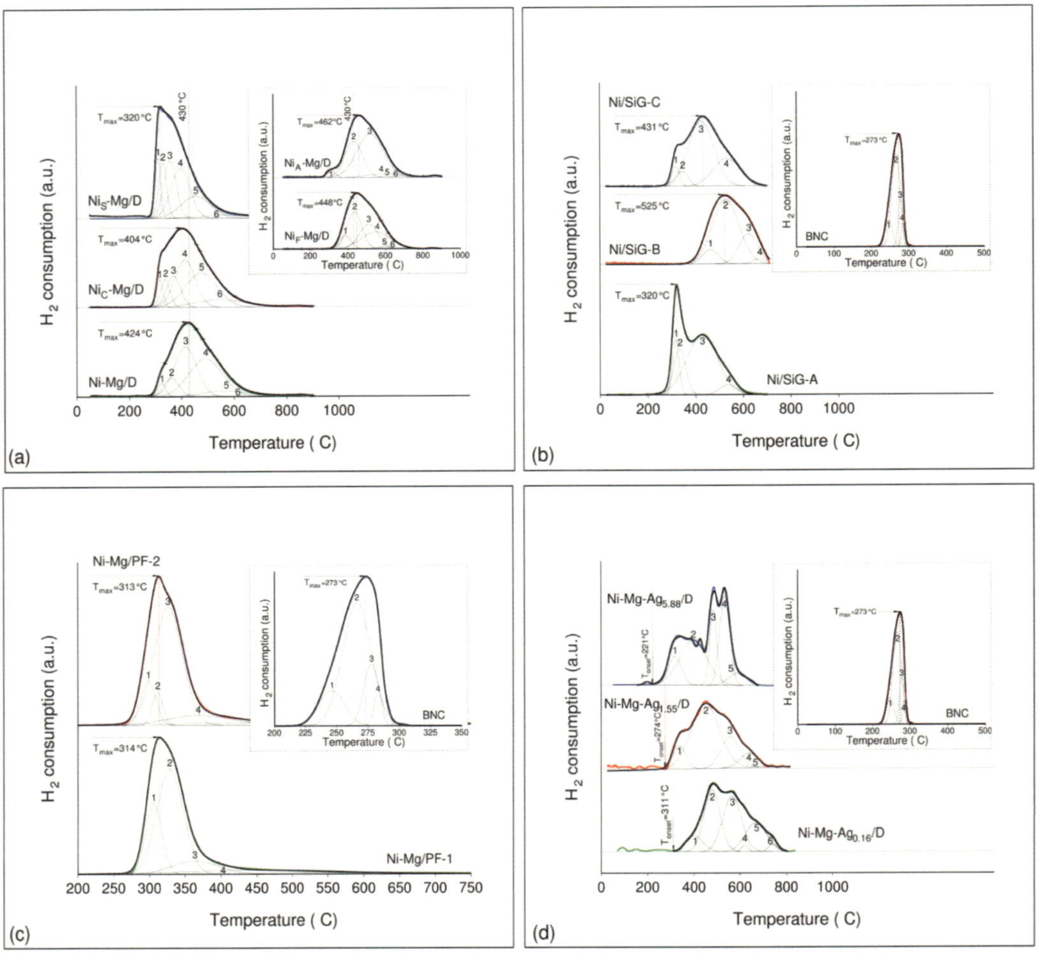

Figure 9. TPR profiles normalized to the sample weight and fitted with Gaussian deconvolution peaks for all prepared catalyst precursor systems: (a) Ni-Mg/D; (b) Ni/SiG; (c) Ni-Mg/PF; (d) Ni-Mg-Ag/D.

The reduction properties of Ni containing SiG-A, SiG-B and SiG-C catalyst precursors are shown in **Figure 9b**. The interpretation of the TPR profiles of the precursor samples is accomplished by comparing them with the profile of the reference BNC sample. The comparison is supposed to clarify the support role in the studied solids. Indeed, the experiments revealed a quite different reduction behavior of the formed Ni^{2+} species. The higher reducibility of the unsupported BNC is attested by the single low temperature peak in the region 220–310°C which is assumed to represent the full reduction of bulk Ni^{2+} ions to the nickel metal. In contrast, multiple reduction peaks with poorly resolved maxima characterize the TPR profiles of the precursors indicating a complex interaction between the Ni^{2+} species and SiG supports. Two most distinguishable peaks at 320 and 428°C can be observed in the Ni/SiG-A sample, alongside a shoulder at 540°C. The reduction peak at 320°C can be attributed to the reduction of BNC species on Ni/SiG-A. All the reduction temperatures above this temperature can be directly associated with the different type of interaction between the nickel species and the supported material. The existence of a broader peak at 428°C and the shoulder at 540°C is caused by the strong interaction between the Ni^{2+} supported phase and the SiG-A framework which points to the existence of hydrosilicate species. The TPR profile of the Ni/SiG-C sample resembles that of the Ni/SiG-A sample with a clearly observed difference in the contribution from low temperature (BNC) and high temperature Ni^{2+} hydrosilicate species. Finally, the absence peak at 320°C and no reduction at all up to 350°C in Ni/SiG-B sample shows that BNC species does not exist, while three peaks at around 462, 525 and 624°C can be attributed to Ni^{2+} hydrosilicate species [85]. It may be summarized that the TPR profiles of the Ni/SiG system evidenced a variety of interaction strength depending on the type of the SiG support resulting in the formation of varying amounts of different Ni^{2+} species.

Reducibility of the Ni-Mg/PF precursors studied by TPR is shown in **Figure 9c**. Clearly, the profiles show almost the same tendency, since both precursors have almost the same T_{max} and slightly anti-symmetric profile toward higher temperature. The low temperature reduction at 313 and/or 314°C for Ni-Mg/PF-2 and Ni-Mg/PF-1, respectively implies the presence of bulk Ni^{2+} easily reduced species and the weak interaction between nickel and perlite support. By comparing TPR profiles of precursors with reference material (**Figure 9c**—insert) the shifting of the T_{max} of about 40°C towards higher temperatures is apparent. A reasonable postulation for a lower temperature reduction of the Ni-Mg/PF system is that it contains the majority of Ni^{2+} species nucleated as BNC and anchored to the oxygen containing groups of the perlite support after deposition. Clearly, reduction of BNC proceeds easily. Based on the entire temperature range, it can be stated that all Ni^{2+} species in this system are reducible under 430°C since no temperature profiles above this temperature have been observed. A complete reduction of the Ni^{2+} species does not necessarily mean that the catalyst will be active, due to the possibility of nickel crystallite growth through aggregation which therefore decreases the number of active sites on the support, resulting in overall decrease of catalyst activity.

The influence of the silver modifier on the reducibility of Ni-Mg-Ag/D catalyst precursors is displayed in **Figure 9d**. TPR measurements clearly showed the differences in reducibility of Ni-Mg-Ag/D samples. The TPR profile of the sample Ni-Mg-$Ag_{5.88}$/D displays five peaks. It is obvious that there are two different areas where hydrogen is consumed. The first is the low temperature region (LTR) between 210 and 440°C (TPR peaks maxima at 318 and 398°C) and the second is high temperature region (HTR) from 440 to 700°C (TPR peaks maxima at 485, 535 and 591°C). In the LTR area, the reduction of easily reducible silver (Ag^+) and nickel (Ni^{2+}) phases occurs. The higher TPR peak at LTR is in agreement with XRD, which also revealed the presence of more bulk-like silver upon increasing Ag loading (**Figure 8d**). Hydrogen consumption at HTR is normally attributed to hardly reducible Ni^{2+} phases—nickel hydrosilicates. When the Ag loading is decreased, the reduction profiles become less resolved (Ni-Mg-$Ag_{1.55}$/D and Ni-Mg-$Ag_{0.16}$/D) suggesting that the reduction occurs in a single unresolved step. Obviously, increasing Ag loading in the precursors shifts the onset temperatures of the initial reduction to lower temperatures (**Figure 9d**). In addition, the reduction was completed at lower temperatures with the precursors of higher Ag loading.

Although the nickel particles of samples do not easily sinter because of the strong interaction with support [49, 72, 73], XRD and H_2-chemisorption result (discussed later in the paper) have shown that the silver loading has an impact on the reduction ability of modified catalyst precursors. The increase in the silver content leads to larger nickel particles in the Ag modified catalyst precursors, Ni-Mg-Ag/D, which displayed easier nickel reduction. In the literature, it is widely accepted as influence of reduction ability of supported metal catalysts on the particle size of the active metal: the lower the reduction ability, the smaller the nickel metal particles [56, 86]. Such a conclusion could be an explanation for better reduction ability of the catalyst precursors with higher Ag loadings.

TPR is a favorable technique for studying the impact of co-metal modifier and support effects on the ability of the reduction of supported metal catalysts [87]. The effect of adding silver on the ability of transition metals reduction is not sufficiently studied in the literature. Richardson and co-workers [88] showed the positive role of silver oxide to promote a better understanding of nickel oxide reduction. A higher degree of nickel oxide reduction in the presence of silver was interpreted by easier nucleation of the nickel clusters, which is rate determining for the reduction, according to Coenen [89]. In bimetallic silver-based catalysts, the higher the Ag loading, the deeper the reduction occurs. In the system with nickel, reduced silver forms metal particles that act as foreign nuclei for subsequent growth and reduction of nickel crystallites. The more silver cations (Ag^+) were introduced in the catalyst, the more silver nuclei formed for Ni crystallites growth and more nickel was reduced.

The TPR results demonstrate rather well the differences between Ni compound formed in the case of diatomite, silica gel and perlite. In the case of the Ni-Mg/PF system, the sharp peak at about 310°C has been identified as being due to the reduction of bulk Ni^{2+} species. The Ni-Mg/D and Ni/SiG systems are difficult to reduce and are comparable in reduction characteristics to Ni hydrosilicates. Significantly however for the Ni-Mg-Ag/D system, reduction is much more facile due to easier nucleation of the nickel crystallites in the presence of silver. In addition, it has been shown that the nature of the nickel precursor salt has a profound effect on the reducibility of Ni-Mg/D catalyst precursors.

Hydrogen chemisorption (H_2-chemisorption). Conventional supported metal catalysts are prepared by *in situ* reduction of a metal salt. The metallic surface is formed by particles of ultimate size and the rates of structure-dependent reactions depend on the size of these particles. Therefore, the catalytic activity of metal particles formed is strongly related to their size and shape. Previous work of our group showed that supported nickel nanoparticles can be obtained in the Ni/Diatomite and Ni/Water glass catalyst systems synthesized by the DP method [90]. They were found as active catalysts in partial hydrogenation of soybean oil. As to the silica support, it is known not to give rise of nickel precipitates in different oxidation states and allows a better approach of the particle size effect in the behavior of supported nickel catalysts [91].

The chemisorption of a gas on a catalyst surface, such as hydrogen chemisorption, is commonly used as a suitable method for the determination of the size of active metal surface area in supported metal catalytic systems [92]. The active metal surface may be measured under suitable conditions taking into account the peculiarity of the system being tested. Hydrogen chemisorption method consists of the use of a hydrogen molecule, which chemisorbs selectively on the metal and not on the support. Assuming a given stoichiometry for this surface reaction, it is possible to obtain an estimate of the metal surface area and of the average metal particle size. Thus, in the H_2-chemisorption studies, the measured value of the active metal surface is dependent on the stoichiometry of the hydrogen adsorption, which in turn depends on the metal-support interaction, modifiers and preparation method [93]. The estimation of the metal crystallite size from hydrogen uptake requires the assumptions to be made regarding metal crystallite morphology. However, it should be noted that the results of chemisorption for supported Ni catalysts in the literature are not always in agreement, mainly for two reasons: the first is that adsorption of H_2 on supported nickel catalysts involves simultaneous physical adsorption, chemisorption, reduction of Ni compound, activated adsorption and hydrogen spillover; the second is the existence of several forms of chemisorbed hydrogen bonded to surface, subsurface, edge and vertex Ni atoms.

Hydrogen chemisorption results for nickel dispersion, nickel surface area and nickel crystallite size are summarized in **Table 6**.

Although some discussion concerning the adequacy of this procedure can be found in the literature [55, 71, 92–95], the values thus obtained are useful from a comparative point of view. **Table 6** shows that a broad range of crystallite sizes is obtained as a consequence of the nickel salt precursors and modifiers used, metal loading and support type in each case. By using both static and dynamic methods under selected conditions of TPR and H_2-chemisorption experiments, the overall dispersion degree does not exceed 13% and crystallite size is lower than 23 nm (excepting the sample Ni/SiG-A). It is known that for the nickel loadings higher than 30 wt% Ni, an important decrease in dispersion is observed [55]. This fact is due to that higher nickel loadings favor agglomeration of particles. Moreover, this agglomeration process is also favored by the weak interaction between the metal and the surface of the support. Hydrogen chemisorption results showed that the particle sizes of nickel metal (Ni^0) in the samples from the different precursors salts may be correlated with their reducibility (see **Table 6** and **Figure 9a**) and textural properties (**Table 4**).

The addition of Ag (hydrogen does not chemisorb onto silver) to the Ni-Mg/D system decreased its chemisorption capacity. The cause of decreased hydrogen adsorption can be a result of blocking of the nickel active site by silver atoms, electronic interactions between Ni and Ag atoms that affect the hydrogen binding to the surface Ni and changes in the stoichiometry of hydrogen adsorption on Ni surfaces due to structure sensitivity [94]. The estimates of the crystallite size from hydrogen chemisorption are also compared with the values determined from X-ray diffraction methods line broadening (**Table 6**). The mean size of nickel particle deduced by the static H_2-chemisorption method was confirmed by the XRD method. The fact that the ratio of these two values is close to unity may be taken as added support for the assumed geometric model. The hydrogen chemisorption results for the samples of the Ni-SiG system are in agreement with their

NP measurements (**Tables 1** and **4**). The lowest dispersion of Ni/SiG-A samples is most likely caused by steric hindrances (microporous SiG-A support). In such a case, metal distribution on the external surface of the support is to be preferred, with consequent lower dispersion (the nickel content being almost the same in each sample (Ni/SiG-A, Ni/SiG-B and/or Ni/SiG-C, **Table 6**).

Sample code	**Ni**	**Nickel metal properties (Ni^0)**				
	(wt%)	**$H_{2\text{-chs}}$ [b] ($\mu mol\ g_{Ni}^{-1}$)**	**S_{Ni} [c] ($m^2\ g_{Ni}^{-1}$)**	**$d_{Ni\ chs}$ [d] (nm)**	**$Ni_{surf} \times 10^{-20}$ [g] ($at_{Ni\ acc}\ g_{Ni}^{-1}$)**	**D[h] (%)**
Ni-Mg/D[i]	36.3	744	58.2	7.7 (7.1[e])	9.0	8.7
Ni-Mg-$Ag_{0.16}$/D[i]	35.9	709	55.5	8.1 (8.3[e])	8.5	8.3
Ni-Mg-$Ag_{1.55}$/D[i]	35.2	538	42.1	10.7 (11.7[e])	6.5	6.3
Ni-Mg-$Ag_{5.88}$/D[i]	33.5	513	40.2	11.2 (13.6[e])	6.2	6.0
Ni/SiG-A[i]	43.7	123	9.6	46.7	1.5	1.4
Ni/SiG-B[i]	45.5	571	44.8	10.0	6.9	6.7
Ni/SiG-C[i]	43.5	1073	84.0	5.3	12.9	12.6
Ni-Mg/PF-1[i]	30.2	576	45.1	9.9 (14.9[f])	6.9	6.8
Ni_A-Mg/D[j]	36.6	259	20.3	22.2	3.1	3.0
Ni_C-Mg/D[j]	36.2	437	34.2	13.1	5.3	5.1
Ni_F-Mg/D[j]	36.6	244	19.1	23.6	2.9	2.9
Ni_S-Mg/D[j]	35.0	462	36.2	12.4	5.6	5.4

[a]Each entry represents a different run.
[b]Hydrogen chemisorbed.
[c]Ni metal surface area; S_{Ni} was calculated assuming $Ni_{surf}/H_{chs} = 1$ and mean surface area of Ni atom equal to $6.5 \times 10^{-20}\ m^2$.
[d]Mean Ni particle or crystallite size: $d_{Ni\ chs}$ — H_2-chemisorption method (surface mean value).
[e]Mean Ni particle or crystallite size: $d_{Ni\ XRD}$ – XRD line broadening method (volume mean value); Ni crystallite geometric model: hemispherical crystallites attached to support with their equatorial plane.
[f]Computed assuming spherical model.
[g]Number of accessible Ni atoms-exposed fraction.
[h]Dispersion.
[i]Chemisorption method: static.
[j]Chemisorption method: dynamic.

Table 6. Chemisorption experiments on the selected samples of Ni-Mg-Ag/D, Ni/SiG and Ni-Mg/PF systems[a].

Comparison was made between Ni-Mg/PF-1 and Ni-Mg/D samples. A smaller metal surface area and a larger Ni crystallite size can be observed in Ni-Mg/PF-1 and attributed to the rapid aggregation of nickel crystallites formed in the reduction stage. It is likely that the reason for this behavior is the weak interaction between the PF support and the Ni surface. For this sample, a crystallite size of nickel calculated assuming a spherical model which is suitable for the supported catalysts with weak metal-support interaction represents a more realistic result (**Table 6**). On the contrary, a larger nickel metal surface area and a smaller nickel particle size is observed on the Ni-Mg/D sample due to the strong anchoring effect of D support.

This anchoring restricts the migration of nickel particles hence prevents the formation of large nickel particles that did not sinter on the mild reduction at 430°C.

The hydrogen chemisorption study showed that the size of nickel nanoparticles obtained in the studied catalyst precursor systems depended on the nature of precursor nickel salt from which they are formed, the kind and loading of metal modifier and the type of support used.

X-ray photoelectron spectroscopy (XPS). In the previous part of this work, it was pointed out that two major factors that affect the reduction of Ni^{2+} supported phase are: (i) interaction between precipitating nickel precursors and the catalyst support and (ii) the dispersion nickel metal phase (Ni^0) arising by reduction of the deposited Ni^{2+} phase. These factors are in turn affected by various examined parameters including the nature of the support, the use of modifier and the choice of nickel precursor salt. Useful confirmatory evidence concerning the interaction between the Ni^{2+} supported phase and support including the structural changes occurring during preparation, drying and reduction and the dispersion of nickel metal phase can be obtained by photoelectron spectroscopy. XPS is virtually surface sensitive (a few atomic layers) and a quantitative instrumental method particularly suitable for the evaluation of surface character of supported nickel catalysts [96].

The XPS results discussed in this work will be restricted to the cases of Ni/SiG, Ni-Mg/D and Ni-Mg-Ag/D systems. The XPS spectra of the Ni 2p and Ag 3d peaks for the systems under investigation are shown in **Figure 10**.

On investigating the influence of the support characteristics on the strength of the interaction between Ni containing species and the support, we chose the Ni/SiG precursor system keeping in mind that the interaction between the Ni^{2+} species and the support is commonly accepted depending on the characteristics of the support. By choosing such a system containing the samples prepared without addition of the modifier with almost the same Ni loadings (**Table 3**), we hoped to increase the value of any comparison one may make. The 2p peaks in the nickel spectrum were used to characterize the chemical state of nickel. The intensity of the Ni 2p signal was obtained by integration over the binding energy (BE) range of 850–890 eV to include the double excitation, shake-up and shake-off peaks. It is known that the chemical forms of nickel have certain characteristics, which serve to identify their presence [97]. The shape of the peaks also contains information. The separation and intensity of the shake-up satellite of the Ni $2p_{3/2}$ core level can be helpful in identifying a particular species.

As expected for dried samples, nickel in these precursors is present in the Ni^{2+} oxidation state—Ni $2p_{3/2}$ peak is a doublet structure (splitting a few electronvolts). The nickel 2p core level, as seen in **Figure 10a**, is similar in shape for all samples, however, the binding energies of the Ni 2p level vary from each other. Since XPS is surface sensitive, the differences in binding energies of the XPS peaks indicate that the nickel species on the surface are changed. The XPS data in **Table 7** show that the binding energies of the Ni $2p_{3/2}$ level for the Ni/SiG-A sample and for the reference material (BNC) are the same. It means that the aggregates of BNC are situated on the surface of nickel hydrosilicates located on the precursor SiG-A [72, 98]. The chemical shift of the Ni $2p_{3/2}$ peak toward higher binding energy values for the Ni/SiG-B sample is assigned to the stronger interaction between the Ni^{2+} species and the SiG-B support. On the contrary, the observed shift toward lower binding energies for the Ni-SiG-C sample suggests weakening interaction between the Ni^{2+} supported phase and the SiG-C support [68]. **Table 7** reveals the variations of the Ni/Si ratio suggesting the different dispersion of the Ni^{2+} species in analyzed samples, as previously shown in the H_2-chemisorption results of the corresponding precursor samples (see **Table 6**).

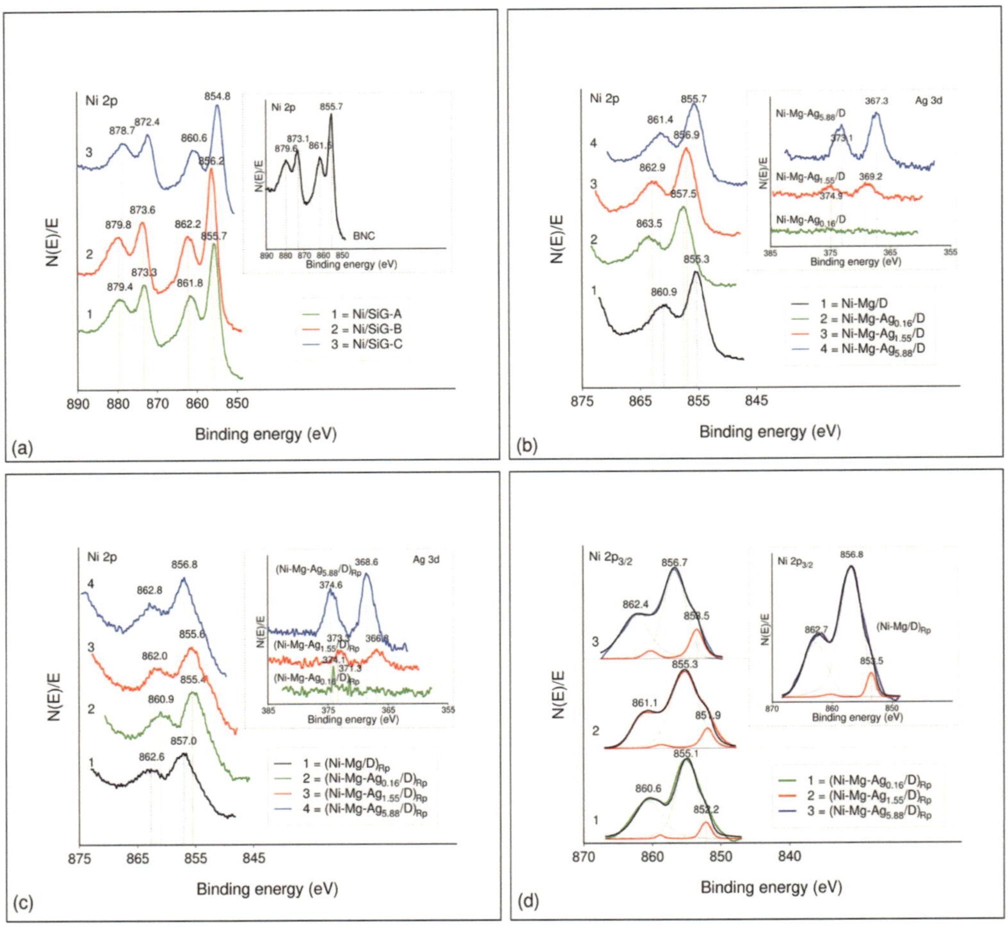

Figure 10. Ni $2p_{3/2}$ XPS spectra of supported nickel catalyst precursors: (a) dried Ni/SiG; (b) dried Ni-Mg/D, Ni-Mg-Ag/D; (c) reduced passivated $(Ni\text{-}Mg/D)_{Rp}$, $(Ni\text{-}Mg\text{-}Ag/D)_{Rp}$; (d) reduced passivated $(Ni\text{-}Mg/D)_{Rp}$, $(Ni\text{-}Mg\text{-}Ag/D)_{Rp}$ (deconvoluted).

The XPS spectra for diatomite supported nickel-based catalyst precursors are presented in **Figure 10b–d**. The Ni 2p core level signal of dried Ni-Mg/D and Ni-Mg-Ag/D precursor samples consist of a single Ni $2p_{3/2}$ peak centered at around 855 eV (Ni-Mg/D and Ni-Mg-$Ag_{5.88}$/D) assigned to Ni^{2+} (**Figure 10b**). The existence of a second component at a higher BE (860.9-863.5) could be due to the presence of hardly reducible Ni^{2+} species, as previously noted in the discussion of the corresponding TPR curves. Ag 3d XPS spectra for dried Ni-Mg-Ag/D catalyst precursors are depicted in **Figure 10b**—insert. There are two peaks, a result of spin-orbit splitting, designated Ag $3d_{5/2}$ and Ag $3d_{3/2}$, respectively, corresponding to the strongest photoelectron lines. These peaks are observed at 367.3 eV and 373.1 eV, respectively, shifted to lower BE values in relation to metallic Ag and could be assigned to the Ag^+ oxidation state [99]. Silver modification provokes a shifting of the Ni $2p_{3/2}$ peak toward higher BE values. The observed shift could be due to the interaction between the components in the dried samples, which is more intense for the sample with the lowest Ag loading.

Sample code	Binding energy (eV)	Surface concentration (at%)			Surface atomic ratio
		Ni	O	Si	Ni/Si
Ni/SiG-A	855.7	21.1	64.6	13.9	1.5
Ni/SiG-B	856.2	14.2	67.4	17.9	0.8
Ni/SiG-C	854.8	15.0	70.6	14.3	1.1
BNC	855.7	27.7	72.3	–	–

Table 7. X-ray photoelectron spectroscopy data for Ni-SiG catalyst precursors—dried samples.

After H_2 reduction, a shoulder appears on the low binding energy side of the Ni $2p_{3/2}$ peak (**Figure 10d**). The shoulder can be deconvoluted and the binding energy is that of the nickel metal (851.9 eV). The other peaks observed on the reduced-passivated samples at higher BE values assigned to Ni^{2+} also appear. The presence of Ni^{2+} species after H_2 reduction has also been confirmed by IR and XRD (**Figures 7d** and **8d**). It is worth mentioning that the increase of Ag $3d_{5/2}$ binding energy after reduction treatment with the respect to the dried precursors (**Figure 10c** and **b**—insert). Although it is known that silver oxides are quite unstable and the two silver oxides, Ag_2O and AgO, decompose below the temperature of 230°C, even in the oxygen atmosphere [100, 101] the results of the Ag 3d and O 1s (not shown here) XPS spectra seem to suggest the presence of silver (I) oxide on the surface of reduced-passivated samples.

The XPS study of Ni/SiG, Ni-Mg/D and Ni-Mg-Ag/D precursor samples confirm the formation of surface species with different strength of interaction and different dispersion of the supported nickel species.

3.2. Partial hydrogenation of soybean oil (SBO)

3.2.1. Activity of Ni-Mg/D and Ni-Mg-Ag catalysts in partial hydrogenation of SBO

Hydrogenation overall activity was monitored by the decay of the iodine value which indicates the level of unsaturation of double bonds (C=C). Activity (A) was calculated from the following equation:

$$A = \frac{H_{2c}}{t \cdot m_{oil} \cdot Ni_{lcc}} \quad (1)$$

where H_{2c} is the hydrogen consumption for decay in iodine value; t is the reaction time; m_{oil} is mass of oil; and Ni_{lcc} is the nickel loading in the catalyst charge for hydrogenation run (in grams).

The results of SBO hydrogenation over Ni-Mg/D and Ni-Mg-Ag/D catalysts in the slurry pilot-plant reactor system (**Figure 4**) are presented in **Table 8**.

Sample code	**Activity results**[a]					
	Ni_{lcc}[b] **(g)**	**IV_{final}**[c]	**t (min)**	**H_{2c}**[d] **(mol)**	**r_{H2c}**[e] **(μmol H_2 min^{-1} g_{oil}^{-1})**	**A (μmol H_2 min^{-1} g_{Ni}^{-1} g_{oil}^{-1})**
Ni-Mg/D	0.9972	89.3	75	8.018	21.3	21.4
Ni-Mg-$Ag_{0.16}$/D	0.8985	90.6	81	7.762	19.2	21.3
Ni-Mg-$Ag_{1.55}$/D	0.8790	90.0	155	7.880	10.2	11.6
Ni-Mg-$Ag_{5.88}$/D	0.8377	90.0	255	7.880	6.2	7.4

[a]Process parameters: 160°C (temperature); 0.16 MPa (H_2 pressure); 750 rpm (stirring rate).
[b]Ni loading in catalyst charge for hydrogenation run (catalyst concentration: 0.05 wt%, i.e., 2.5 g catalyst/5000 g oil).
[c]Iodine value determined by Wijs method.
[d]Hydrogen consumption for decrease of iodine value by *ca* 40–41 units.
[e]Rate of hydrogen consumption per unit mass of oil.

Table 8. Soybean oil hydrogenation over Ni-Mg/D and Ni-Mg-Ag/D catalysts in slurry pilot plant reactor system—catalytic activity test runs results.

The obtained results showed clearly the influence of the silver addition on the catalyst activity. Under the same process conditions, Ni-Mg/D and silver modified catalysts exhibited different activities toward SBO hydrogenation (see **Table 8**). By comparing the results of catalytic test runs over Ni-Mg/D and Ni-Mg-$Ag_{0.16}$/D catalysts, it can be observed that the activity of the sample without silver is slightly higher. The modification by silver inhibits hydrogenation activity, this effect being more obvious as the Ag loading is higher. From these results, the hydrogenation activity for the studied catalyst (**Table 8**) increases in the following order: Ni-Mg-$Ag_{5.88}$/D < Ni-Mg-$Ag_{1.55}$/D < Ni-Mg-$Ag_{0.16}$/D < Ni-Mg/D.

The observed differences in the activity of the studied catalysts could be attributed to nickel dispersion and different textural properties of the catalysts. From the chemisorption results, the silver-modified Ni catalyst sample with high loading (Ni-Mg-$Ag_{5.88}$/D) demonstrated 6.0% nickel dispersion and an average nickel crystallite size of ≈ 11 nm. A higher nickel dispersion and a smaller nickel crystallite size were obtained for the Ni-Mg/D catalyst sample (**Table 6**). Besides, among the studied catalysts Ni-Mg catalyst sample had the highest S_{BET} surface area (**Table 4**). This indicates that the role of catalyst texture and dispersion of the active phase is critical in assessing the catalytic efficiency. In considering an explanation for the diminished hydrogenation activity of silver modified nickel catalyst it can be also assumed a physical blocking of nickel active sites or even changes in the morphology of nickel metal particles by the silver modifier. It is difficult to discriminate between these different possibilities. Apart from the effect of co-metal blocking of the surface nickel atoms, it should also be noted that an electronic effect has been taken into account [94]. It must be noted that the electronic properties of very small particles—nanoparticles should be different from those of large particles at least for two reasons. The first relates to the differences in the fraction of the total atoms that are present on the surface. The second is the incomplete coordination of the surface atoms from those in the bulk. Based on the assumption that catalytic activity of a metal is related to its electronic properties, it seems reasonable that the activity would vary with the crystallite size. However, a clearer understanding of the factors responsible for the crystallite size effects will require more information on the properties of nanoparticles.

3.2.2. *Cis/trans isomerization*

The factors influencing the *cis/trans* composition are catalyst activity and loading, as well as process conditions that include the hydrogen pressure, temperature and stirring rate. These parameters determine the hydrogen concentration on the catalyst surface, which is crucial not only for hydrogenation of double bonds but for *cis/trans* isomerization as well.

The hydrogenation of SBO is a complex network of chemical reactions involving consecutive saturation of C18:3*c* to C18:2*c*, C18:2*c* to C18:1*c* and subsequent saturation of C18:1*c* to C18:0 as well as parallel reversible isomerization of C18:2*c* to C18:2*t* and C18:1*c* to C18:1*t*. The overall hydrogenation reaction scheme may also involve a simple step hydrogenation of C18:1*t* to C18:0 and C18:2*t* to C18:1*t*.

The fatty acid compositions of hydrogenated SBO at conversion of 30.8 ± 0.5% are shown in **Table 9**. The experimental data of the fatty acid compositions in hydrogenated SBO summarized in **Table 9** showed that in all the cases, there was an increase in the concentration of stearic acid (C18:0). On the contrary, a decrease in linoleic acid (C18:2*c*) was also observed in all the cases. It should be noted that linolenic acid (C18:3*c*) conversion after 30 min of reaction was 100% when the hydrogenation was performed over any of the studied catalysts. Taking into account a known fact that silver modified catalysts show high saturation selectivity toward linoleic acid (C18:2*c*) [102], disappearance of linolenic acid in the early stage of the hydrogenation was an expected result. As the present work is focused on the control of the TFAs, the geometric isomers are presented in **Table 9** without taking into account the position of double bond in the fatty acid chain, only distinguishing between *cis* fatty acids (CFAs) and TFAs.

From these results, it is evident that the Ni-Mg-$Ag_{5.88}$/D catalyst formed the least TFAs of all catalysts (23%) at the same conversion level. On the contrary, the Ni-Mg/D catalyst was demonstrated to have the highest content of TFAs (62.1%), which could be associated with its activity manifested in SBO hydrogenation and a higher total surface area compared to the catalyst with a silver modifier. It is well known that a large surface area encourages the isomerization reactions, due to the greater accessibility to the nickel active sites [103]. A small increase of stearic acid in the order: Ni-Mg-$Ag_{5.88}$/D < Ni-Mg-$Ag_{1.55}$/D < Ni-Mg-$Ag_{0.16}$/D < Ni-Mg/D could be explained by the differences observed in their textural properties (**Tables 4** and **5**). According to Balakos and Hernandez [13], small pores favor fatty acid saturation, since the successive hydrogenation is made easier by the mobility difficulties of the bulky molecule. The catalytic test results clearly show that adding silver to the Ni-Mg/D system have a considerable effect on the distribution of CFAs and TFAs in hydrogenated oil. A higher ratio of unsaturated *cis* fatty acids to *trans* fatty acids over the Ni-Mg-$Ag_{5.88}$/D catalyst can be attributed to the smaller formation of TFAs, suggesting a more selective hydrogenation reaction. The modification with silver is beneficial in limiting the C18:1 *cis*/*trans* isomerization during the SBO hydrogenation. On the other hand, the addition of silver did not have significant effect on the formation and distribution of C18:2*t* isomers (C18:2*c,t* + C18:2*t,c* + C18:2*t,t*), indicating that the main reason for the difference in the specific isomerization selectivity (S_i) was favored isomerization of elaidic fatty acid (C18;1*t*) over the Ni-Mg/D catalyst sample. Concerning the isomerization selectivity (S_i), the addition of silver provoked the reduction of TFAs (R_{trans}) in the range of 4–57% during SBO hydrogenation with respect to the Ni-Mg/D catalyst (see **Table 9**).

Fatty acid	**Catalyst sample**			
	Ni-Mg/D	**Ni-Mg-$Ag_{0.16}$/D**	**Ni-Mg-$Ag_{1.55}$/D**	**Ni-Mg-$Ag_{5.88}$/D**
C18:0	8.6	8.1	6.2	5.8
C18:1*c*	15.9	18.7	22.5	22.7
C18:1*t*	53.1	49.1	16.5	10.8
C18:2*cc*	0.9	1.2	22.7	31.7
C18:2*tt*	1.7	1.4	2.5	2.0
C18:2*ct*	2.9	3.9	7.6	6.9
C18:2*tc*	3.5	4.2	8.3	6.6
C18:2*t*[b]	8.1	9.5	18.4	15.5
C18:3*c*	none	none	none	none
Others[c]	13.3	13.6	13.7	13.5
CFAs/TFAs	0.3	0.3	1.3	2.1
S_i[d1]	1.5	1.5	0.9	0.6
S1[d2]	1.1	1.1	3.1	1.8
R_{trans} (%)[e]	–	4.2	43.0	57.0

[a]Conversion (%) = $[(IV_0–IV_f)/IV_0] \times 100 = 30.8 \pm 0.5$.

[b]Sum of *ct*, *tc* and *tt* fatty acid isomers.

[c]Sum of C14:0, C16:0, C16:1, C20:0, C20:1 and C22:0 fatty acids.

[d]Selectivity basically means that the reaction is guided toward a particular unsaturated bond in preference to others; [d1]Si—specific isomerization selectivity, defined as quotient of TFAs (%) in hydrogenated oil product and change in iodine value between the starting oil and hydrogenated product; [d2]S1—selectivity 1, defined as the amount of monounsaturated fatty acids (C18:1) formed with respect to the amount of diunsaturated (C18:2) converted: S1 = $(C18:1–C18:1_{(0)})/(C18:2_{(0)}–C18:2)$.

[e]R_{trans}—reduction TFAs, defined as: $R_{trans} = \{1 - [\Delta(C18:1t+C18:2t)_{catalyst2}/\Delta(C18:1t+C18:2t)_{catalyst1}]\} \times 100$, $\Delta(C18:1t+C18:2t)_{catalyst1} > \Delta(C18:1t+C18:2t)_{catalyst2}$, C18:2*t*—sum of *ct*, *tc* and *tt* fatty acid isomers.

Table 9. Soybean oil hydrogenation over Ni-Mg/D and Ni-Mg-Ag/D catalysts in slurry pilot plant reactor system—fatty acids composition[a] and selectivities[d, d1, d2].

In general, the overall hydrogenation selectivity decreased while the isomerization increased with conversion. The mechanisms of the hydrogenation and *cis/trans* isomerization are strongly interrelated. An addition-elimination mechanism according to Horiuti-Polanyi is often assumed to describe the formation TFAs in the hydrogenation processes [58]. The original concept of Horiuti-Polanyi mechanism provides that hydrogenation and isomerization should both be described by a half-hydrogenated state mechanism. Since hydrogenation is accompanied by isomerization, it can be proposed that the electron donor characteristics of the silver modified nickel catalyst would also affect this reaction. If the chemisorbed half-hydrogenated intermediate is removed quickly enough, it may not have time to isomerize to *trans* or in the case of linoleic acid, when hydrogenation of one of the double bonds is complete, the fatty acid molecule is released before hydrogenation of the second double bond can occur. The limited formation of C18:1*t* over silver modified Ni-Mg-Ag/D catalytic system compared to that of the Ni-Mg/D system may be explained by both, the changing of electronic properties and the presence of geometric effect. The promoting effect of adding silver to the Ni-Mg/D system for SBO hydrogenation is manifested for all the catalysts tested in this work when considering the reduced TFA formation.

3.2.3. *Kinetic study of SBO hydrogenation*

Several kinetics models of the hydrogenation of fatty oils containing polyunsaturated fatty acids were devised previously and reaction rate constants were evaluated for the various reactions [104–107]. All of the proposed kinetic models including various reaction pathways were incomplete. From a practical standpoint, it is justified because of the extreme complexity of the complete kinetic model, which would have to include all possible consecutive and isomerization reactions.

A mathematical model has been developed to describe the kinetics of both the hydrogenation and the *cis/trans* isomerization of SBO. A simple approach to model the rate constants for SBO hydrogenation over studied catalysts is presented. The rate constant model is constructed assuming the first order rate equations with respect to the compositions of the various *cis* and *trans* fatty acids in the hydrogenated SBO. In this model, the fatty acids are divided into five groups: (i) C18:2*c*—*cis* diunsaturated; (ii) C18:2*t*—*trans* diunsaturated; (iii) C18:1*c*—*cis* monounsaturated; (iv) C18:1*t*—*trans* monounsaturated; (v) C18:0—saturated (stearic acid). Mathematical equations were developed for all groups of fatty acids as a function of reaction time. The k_i's ($i = 1$–8) are the respective first order reaction rate constants. Numerical solutions for the set of ordinary differential equations corresponding to the kinetic model were obtained through the Gear algorithm [108]. The rate constants were computed from the kinetic experiments by minimizing the sums of squares for deviations between the computed and experimental concentrations of studying fatty acids. The minimization was performed by the simplex method [109].

A reaction scheme, time dependence concentration of fatty acids and the estimated rate constants are presented in **Figure 11** and **Table 10**.

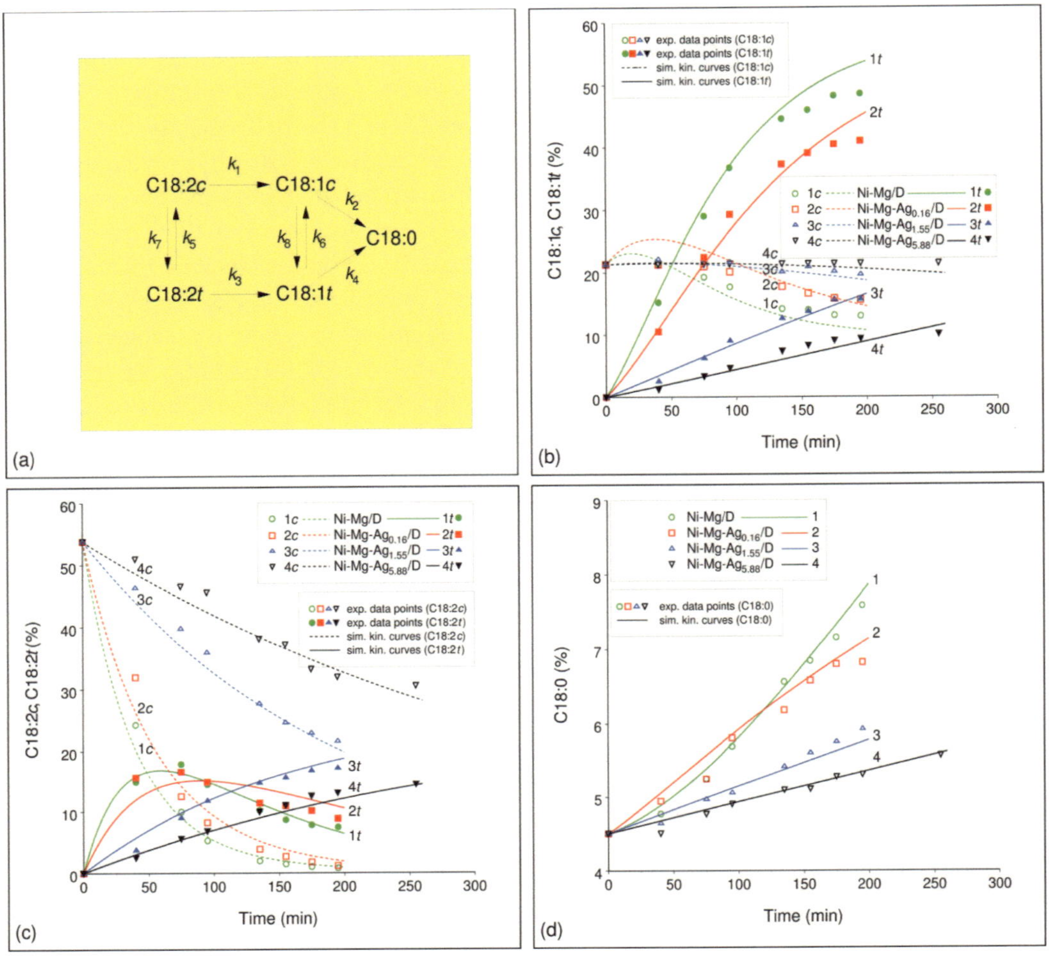

Figure 11. Hydrogenation of SBO: (a) reaction scheme; (b–d) correlation between experimental data and model predictions for SBO hydrogenation over Ni-Mg/D and Ni-Mg-Ag/D catalytic systems under operating conditions used.

Sample code	Rate constants[a]							
	k_1 (min^{-1})	k_2 (min^{-1})	k_3 (min^{-1})	k_4 (min^{-1})	k_5 (min^{-1})	k_6 (min^{-1})	k_7 (min^{-1})	k_8 (min$_{-1}$)
Ni-Mg/D	1.0×10^{-2}	3.0×10^{-4}	1.0×10^{-2}	3.5×10^{-4}	1.5×10^{-3}	2.5×10^{-3}	1.5×10^{-2}	1.5×10^{-2}
Ni-Mg-Ag$_{0.16}$/D	0.9×10^{-2}	5.6×10^{-4}	6.0×10^{-3}	6.0×10^{-5}	6.0×10^{-4}	2.0×10^{-3}	9.0×10^{-3}	1.0×10^{-2}
Ni-Mg-Ag$_{1.55}$/D	0.2×10^{-2}	3.0×10^{-4}	0.7×10^{-3}	2.5×10^{-5}	0.1×10^{-4}	0.9×10^{-3}	3.0×10^{-3}	4.0×10^{-3}
Ni-Mg-Ag$_{5.88}$/D	0.1×10^{-2}	2.0×10^{-4}	0.4×10^{-3}	2.0×10^{-5}	0.1×10^{-4}	0.1×10^{-3}	1.5×10^{-3}	2.0×10^{-3}

[a]Refers to the first order kinetic rate constants.

Table 10. Hydrogenation of soybean oil over Ni-Mg/D and Ni-Mg-Ag/D catalysts—computed values of the rate constants (T = 160°C).

Table 10 reveals that the rate constants of isomerization reaction k_7 and k_8 and constant of saturation reaction k_4, corresponding to the stearic acid formation, are higher in the case of Ni-Mg/D and Ni-Mg-$Ag_{0.16}$/D catalyst samples. These observations are in agreement with experimental data, shown in **Table 9** related to the formed C18:1*t* and C18:0 during the hydrogenation of SBO.

Figure 11b–d shows a comparison between experimentally measured and simulated modeling kinetics curves (parity plots for C18:1*c* and C18:1*t*—**Figure 11b**; C18:2*c* and C18:2*t*—**Figure 11c**; C18:0—**Figure 11d**). As it can be seen, generally, the difference between experimental results and model estimation is within 10%, which confirms the accuracy of the results. A model agrees with the general knowledge in hydrogenation and the data were fitted fairly well by the model. The proposed kinetic model could be applicable for the hydrogenation of SBO under studied operating conditions.

3.3. Partial hydrogenation of sunflower oil (SFO)

3.3.1. Activity of Ni/SiG and Ni-Mg/PF catalysts in partial hydrogenation of SFO

The Ni/SiG and Ni-Mg/PF catalysts were tested for comparison, in order to determine their activity in the sunflower oil hydrogenation. The obtained results in the laboratory reactor system (**Figure 3**) are shown in **Table 11** and **Figure 12**.

Sample code	Activity results[a]					
	Ni_{lcc}[b] (g)	$IV_{selected}$[c]	t (min)	H_{2c}[d] (mol)	r_{H2c} (μmol H_2 min^{-1} g_{oil}^{-1})	A (μmol H_2 min^{-1} g_{Ni}^{-1} g_{oil}^{-1})
Ni/SiG-A	0.5510	108.7	18.55	0.894	53.5	97.2
Ni/SiG-B	0.5510	108.7	180.00	0.894	5.5	10.0
Ni/SiG-C	0.5511	108.7	47.34	0.894	21.0	38.1
Ni-Mg/PF-1	0.5400	108.7	6.92	0.894	143.5	265.8

[a]Process parameters: 160°C (temperature); 0.20 MPa (H_2 pressure); 1200 rpm (stirring rate).
[b]Ni loading in catalyst charge for hydrogenation run; Ni concentration: 0.06 wt% with respect to oil.
[c]Selected value of IV to compare catalyst activity corresponds to the final iodine value (IV_f) in the hydrogenation of sunflower oil over the catalyst with the lowest activity (Ni/SiG-B).
[d]Hydrogen consumption for decrease of iodine value from starting (131.5) to selected value.

Table 11. Sunflower oil hydrogenation over Ni/SiG and Ni-Mg/PF catalysts in laboratory reactor system—catalytic activity test runs results.

A comparative study of SFO hydrogenation over Ni/SiG catalyst samples was performed at the same level of conversion (17.3%) in order to obtain more accurate comparative results. Activity was calculated according to Eq. (1) as hydrogenation overall activity, referred to the hydrogen consumption for a target IV value of 108.7. Analyzing the activity presented in **Table 11**, a significant variation of the values obtained for the different catalysts can be observed. As all catalysts are expected to present the same kind of active sites (metallic nickel), an explanation for this behavior should be sought in different structural and textural properties of the studied catalysts. Considering the structure of the Ni/SiG-A, Ni/SiG-B and Ni/SiG-C samples and dispersion of the nickel metal, no clear correlation of the experimental data was found. It is likely that nonuniform distribution of nickel is the main reason for this behavior of the studied catalysts. To verify this assumption, it is necessary to establish a functional relationship between the concentration of the available nickel surface area in the reaction mixture and the initial global hydrogenation rate [110]. Analyzing the results of NP measurements for the Ni/SiG system (**Table 4**), it appears that the activity of the samples is associated with their mesoporosity. The hydrogenation overall activity Ni/SiG-B < Ni/SiG-C < Ni/SiG-A follows the same order as surface area in the mesopore range (available for the hydrogenation, see **Tables 4** and **11**).

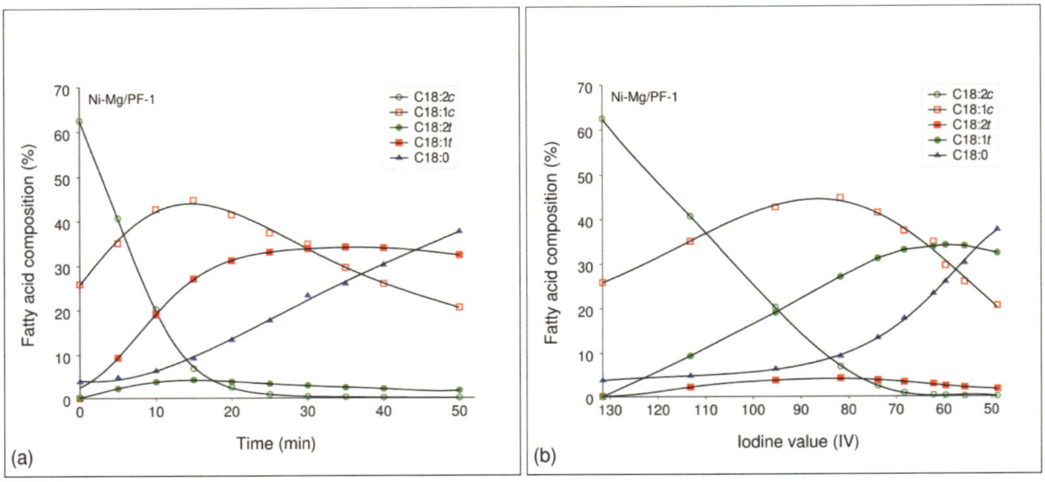

Figure 12. Time course (a) and IV decay (b) profiles of reactants and products during the hydrogenation of SFO over the Ni-Mg/PF-1 catalyst.

Regarding the performance of the Ni-Mg catalyst supported on perlite (Ni-Mg/PF-1), the hydrogenation activity was found to be very high. In addition, the Ni-Mg/PF-1 catalyst demonstrated a high activity in SFO deep hydrogenation (decrease in IV of 82.8, **Figure 12**) [31].

In **Figure 12a**, linoleic acid (C18:2*c*) was almost depleted within 30 min of reaction. This may be attributed to complex reactions evolving concomitantly. Such reactions may include hydrogenation of C18:2*c* to C18:1*t*; isomerization of C18:2*c* to C18:2*t* then hydrogenation to C18:1*t* as well as isomerization of C18:12*c* to C18:1*t*. The level of oleic acid initially increased, then leveled off; while the stearic acid rapidly increased. The monoenic *trans* fatty acid content (C18:1*t*) monotonically increased up to 33.8% after 30 min of reaction and continued to decrease slowly reaching a final level of 32.3%. Dienic *trans* fatty acid profiles (C18:12*t*) exhibited a different behavior characterized by an increase up to 4% followed by a decrease due to their conversion to stearic acid and possibly also to *trans* monoenic acid (C18:1*t*). From the standpoint of health and technical functional properties, taking into account unsatisfactory product distribution (34.3% TFAs and 37.6% stearic acid), it is necessary to optimize the properties of this catalyst with the aim of finding a good compromise between the activity and capacity to produce undesirable TFAs and stearic acid.

3.3.2. *Kinetic study of SFO hydrogenation over Ni/SiG catalysts*

A lumped kinetic model was developed to describe the evolution of the products during the SFO hydrogenation over the Ni/SiG system. This model considers the saturation of double bonds along the fatty acid chains and *cis*/*trans* isomerization, which take place simultaneously with the hydrogenation of fatty acids. The assumed reaction pathway is described in **Figure 13a**. The reaction pathway contains 12 reactions that include all possible reactions. The reaction scheme does not exclude reverse isomerization reactions (*trans* to *cis*), nor does it exclude hydrogenation with isomerization (C18:2*t* to C18:1*c*), but their low probability is emphasized with dashed arrows (see **Figure 13a**). Due to the fact that there are 12 reactions we considered all 12 reaction rate constants. For calculation of kinetic parameters, the same method as for the hydrogenation of SBO was used. Lines in **Figure 13b–c** depict the predicted data of fatty acid composition for the catalysts with the highest (Ni/SiG-A) and the lowest (Ni/SiG-B) activity. **Table 12** shows the best fitting kinetic constant values for all three catalysts.

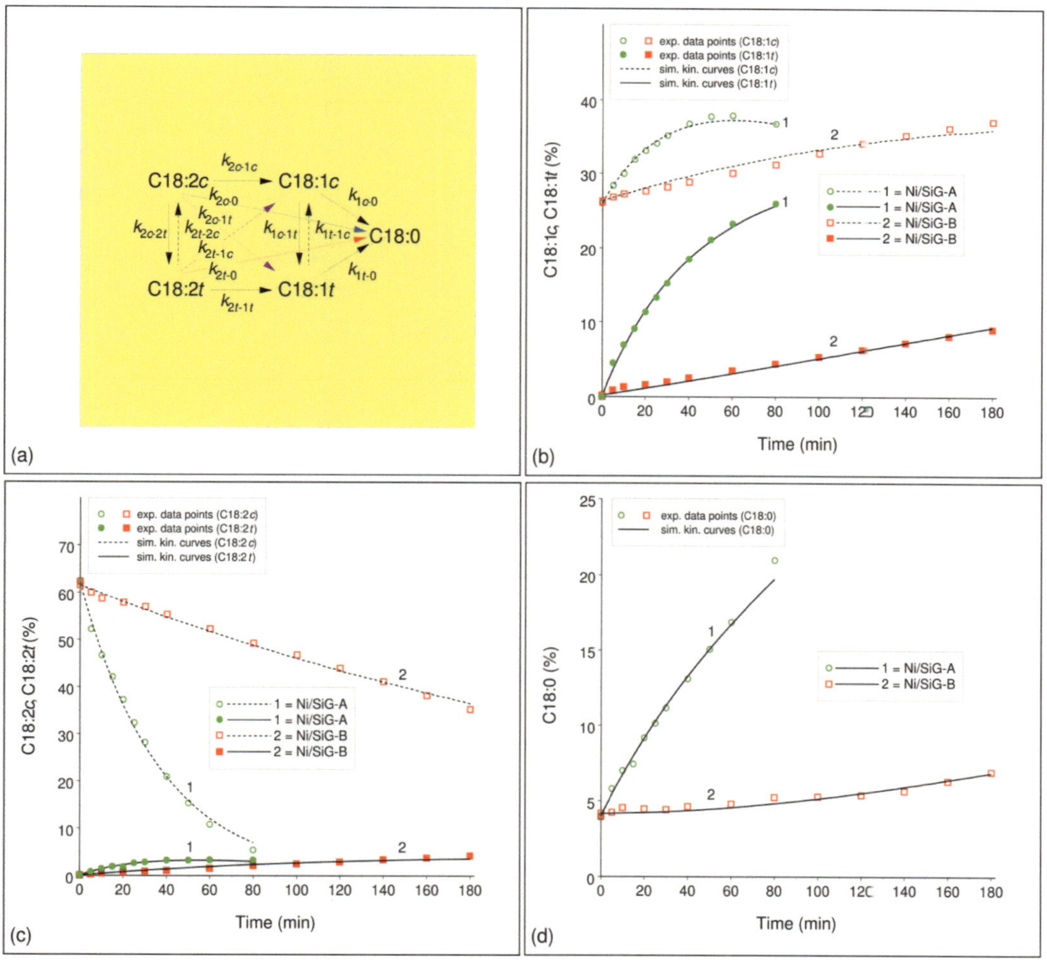

Figure 13. Hydrogenation of SFO: (a) reaction pathway; (b–d) correlation between experimental data and model predictions for SFO hydrogenation over Ni/SiG catalysts under the experimental conditions used.

The results in **Table 12** indicate that the values of the rate constants show a significant difference, regardless of the catalyst activity and hence the contribution of the individual reactions in the reaction mechanism can be recognized.

Rate constants[a]	**Sample**		
	Ni/SiG-A	**Ni/SiG-B**	**Ni/SiG-C**
$k_{2c\text{-}1c}$ (min^{-1})	**9.6×10^{-3}**	**2.3×10^{-3}**	**2.8×10^{-3}**
$k_{2c\text{-}1t}$ (min^{-1})	**1.1×10^{-2}**	1.1×10^{-7}	**2.2×10^{-3}**
$k_{2c\text{-}0}$ (min^{-1})	**4.3×10^{-3}**	8.0×10^{-9}	**2.7×10^{-3}**
$k_{2t\text{-}1t}$ (min^{-1})	8.5×10^{-7}	1.0×10^{-7}	2.4×10^{-7}
$k_{2t\text{-}1c}$ (min^{-1})	2.0×10^{-7}	6.7×10^{-7}	1.4×10^{-6}
$k_{2t\text{-}0}$ (min^{-1})	**1.1×10^{-2}**	**4.7×10^{-3}**	**9.2×10^{-3}**
$k_{2c\text{-}2t}$ (min^{-1})	**2.6×10^{-3}**	**6.3×10^{-4}**	**8.0×10^{-4}**
$k_{2t\text{-}2c}$ (min^{-1})	1.2×10^{-5}	5.3×10^{-7}	2.6×10^{-7}
$k_{1c\text{-}0}$ (min^{-1})	**8.2×10^{-4}**	2.0×10^{-7}	9.2×10^{-8}
$k_{1t\text{-}0}$ (min^{-1})	**1.8×10^{-3}**	**8.7×10^{-4}**	**3.6×10^{-3}**
$k_{1c\text{-}1t}$ (min^{-1})	**2.2×10^{-3}**	**1.7×10^{-3}**	**2.0×10^{-3}**
$k_{1t\text{-}1c}$ (min^{-1})	2.5×10^{-8}	9.3×10^{-10}	4.6×10^{-10}

[a]Values in bold show significant rate constants.

Table 12. Kinetic constant values for Ni/SiG catalysts at 160°C and 0.2 MPa.

In order to simplify the reaction pathway shown in **Figure 13a**, without compromising the accuracy of predicting the concentration of fatty acid change as criteria of importance for some rate constants, we used a value of 1% of the highest rate constant for the particular catalyst (1% k_{max}). Rate constants higher than 1% k_{max} were declared as significant and marked bold in **Table 12**. On the contrary, rate constants less than 1% k_{max} are associated as insignificant. It was shown that the number of significant rate constants increases as the activity of the catalysts grows higher. Significant constants, higher than 1% k_{max}, show that Ni/SiG-A has 8, Ni/SiG-C has 7, while the least active catalyst, Ni/SiG-B, has the lowest value 5. Using the criteria of 1% k_{max}, we can rewrite a reaction pathway (**Figure 13a**), which will include only significant rate constants for each individual catalyst and are given in **Figure 14**.

Using the reduced reaction pathways, reaction rates were rewritten, using only significant rate constants and the process of hydrogenation was simulated with only those reaction rates. Values of rate constants then obtained were the same as the ones obtained by the initial reaction mechanism, which indicates that the reduction of the initial reaction pathway changes as a function of catalyst activity.

http://dx.doi.org/10.5772/66967

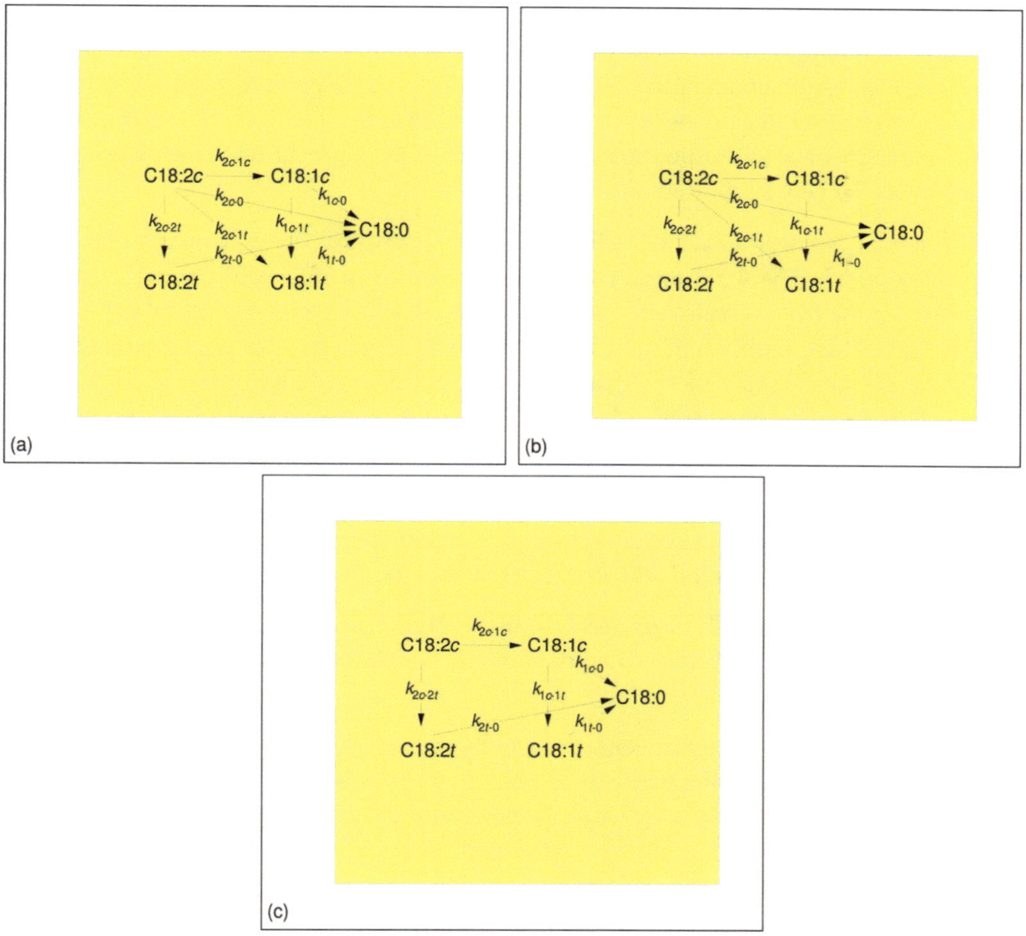

Figure 14. Reduced reaction pathways for three different catalysts: (a) Ni/SiG-A; (b) Ni/SiG-C; (c) Ni/SiG-B.

4. Conclusions

The characteristics, structure and catalytic behavior of high loading nickel-based catalysts supported on diatomite, silica gel and perlite have been analyzed. Nickel-based supported catalyst precursors were prepared by the precipitation-deposition method. The results show that the state, reducibility and dispersion of nickel in supported nickel-based catalysts vary depending on the nature of support and the preparation parameters.

Combined nitrogen physisorption and mercury porosimetry studies showed that the studied nickel-based supported systems had a high specific surface area and a well-developed porous structure, containing mesopores stable to thermal reduction treatments.

The results concerning the influence of the preparation stage and nature of the support and the modifier clearly illustrate the features of the supported Ni^{2+} phase and demonstrate that

IR and XRD measurements may offer as an effective tool to identify nickel species and their interaction with support in differently supported and modified nickel-based catalyst precursors. From the results obtained by both IR and XRD instrumental techniques, it could be concluded that during the deposition reaction under alkaline conditions, the silica as the constitutive component of all studied supports reacts with the basic nickel carbonate precipitate and generates the new supported nickel hydrosilicate phase.

The TPR results demonstrate rather well the differences between Ni compounds formed on the surface of supports. The weak metal-support interaction in the Ni-Mg/PF system is probably responsible for the hydrosilicate formation at a low level, which could decrease the difficulty in the system reduction. The Ni-Mg/D and Ni/SiG systems are difficult to reduce and are comparable in reduction characteristics to nickel hydrosilicates. The addition of silver to the Ni-Mg/D system significantly affected reducibility of nickel-based catalysts. Larger nickel crystallites in silver modified nickel catalysts displayed easier nickel reduction than smaller ones in the Ni-Mg/D catalyst.

The hydrogen chemisorption study showed that the size of nickel nanoparticles obtained in the studied catalyst precursor systems depended on the nature of precursor nickel salt from which they are formed, the kind and loading of metal modifier and the type of support used.

The XPS study of Ni/SiG, Ni-Mg/D and Ni-Mg-Ag/D precursor samples confirm the formation surface species with different strength of interaction and different dispersion of the supported nickel species.

The silver modifier inhibits hydrogenation activity, this effect being more obvious as the Ag loading is higher. Modification by silver allowed us to promote the selectivity toward the *cis* isomers, but the catalyst is less active than the non-modified catalyst in the partial hydrogenation of soybean oil.

Among the catalyst samples studied, the highest activity in the sunflower oil hydrogenation was observed over the Ni-Mg/PF-1 catalyst suggesting that the Ni-Mg/PF-1 catalyst is a promising catalyst for SFO hydrogenation. Although Ni/SiG catalysts show a lower overall activity, this system also could be considered as good, since they produced less amount of stearic acid compared to the Ni-Mg/PF system.

The kinetic models include the saturation of double bonds along the fatty acids chains and *cis/trans* isomerization. Under studied operating conditions models proved to adequately fit the experimental data for the evolution of product distribution with reaction time. It was shown that the catalysts of different activities had different reaction pathways. The more active catalysts, the reaction pathways multiply and require more complex reaction scheme to describe the results of the catalytic tests.

Acknowledgments

The authors are grateful to the Ministry of Education, Science and Technological Development of the Republic of Serbia (Project No. III 45001) for providing funding support. The support by Serbian Academy of Sciences and Arts and by Bulgarian Academy of Sciences (Joint Research Project: New nanosized hydrogenation catalysts based on metals of VIII group) is also appreciated.

Author details

Miroslav Stanković[1]*, Jugoslav Krstić[1], Margarita Gabrovska[2], Vojkan Radonjić[1], Dimitrinka Nikolova[2], Davor Lončarević[1] and Dušan Jovanović[1]

*Address all correspondence to: mikastan@nanosys.ihtm.bg.ac.rs

1 Department of Catalysis and Chemical Engineering, Institute of Chemistry, Technology and Metallurgy, University of Belgrade, Belgrade, Serbia

2 Institute of Catalysis, Bulgarian Academy of Sciences, Sofia, Bulgaria

References

[1] Shahidi F, editor. Bailey's Industrial Oil and Fat Products, Volume 4. Edible Oil and Fat Products: Products and Applications. 6th ed. Chichester: Wiley; 2005. 455 p. ISBN: 978-0-471-38549-3

[2] Gunstone FD. Oils and fat in the market place. Commodity oils and fats. Soybean Oil. The AOCS Lipid Library. [Internet]. 2012. Available from: http://lipidlibrary.aocs.org/market/soybean.htm. [Accessed: 2016-05-27]

[3] Hamilton RJ. The chemistry of rancidity in foods. In: Allen JC, Hamilton RJ, editors. Rancidity in Foods. 3rd ed. New York: Chapman and Hall; 1994. p. 1–21.ch1. ISBN: 0-7514-0219-2.

[4] Choe E, Min DB. Mint: Mechanisms and factors for edible oil oxidation. Comprehensive Reviews in Food Science and Food Safety. 2006;**5**:169–186. DOI: 10.1111/j.1541-4337.2006.00009.x

[5] Gunstone FD. Extraction, refining and modification processes. In: Early R, series editor. Oils and Fats in the Food Industry. Chichester: Wiley-Blackwell; 2008. pp. 26–36. DOI: 10.1002/9781444302424.ch3

[6] Chaudhari VR, Ramachandran AP. Mint: Three phase slurry reactors. AIChE Journal. 1980;**26**:177–201. DOI: 10.1002/aic.690260202

[7] Fernández MB, Tonetto GM, Crapiste GH, Damiani DE. Mint: Revisiting the hydrogenation of sunflower oil over a Ni catalyst. Journal of Food Engineering. 2007;**82**:199–208. DOI: 10.1016/j.foodeng.2007.02.010

[8] Ahmad I, Ali A. Mint: Industrial hydrogenation of rapeseed oil with nickel catalyst. Journal of the American Oil Chemists' Society. 1981;**58**:87–88. DOI: 10.1007/BF02672186

[9] Chu YH, Lin LH. Mint: An evaluation of commercial nickel catalysts during hydrogenation of soybean oil. Journal of the American Oil Chemists' Society. 1991;**68**:680–683. DOI: 10.1007/BF02662296

[10] Cizmeci M, Musavi A, Kayahan M, Tekin A. Mint: Monitoring of hydrogenation with various catalyst ratios. Journal of the American Oil Chemists' Society. 2005;**82**:925–929. DOI: 10.1007/s11746-005-1166-3

[11] Babaee Z, Nikoopour H, Safafar H. Mint: A comparison of commercial nickel catalysts effect on hydrogenation of soybean oil. World Applied Sciences Journal. 2007;**2**:621–626. ISSN: 1818-4952. http://www.idosi.org/wasj/wasj2(6)/10.pdf [Accessed: 2016-06-23]

[12] Coenen JWE. Mint: Hydrogenation of edible oils. Journal of the American Oil Chemists' Society. 1976;**53**:382-389. DOI: 10.1007/BF02605727

[13] Balakos MW, Hernandez EE. Mint: Catalyst characteristics and performance in edible oil hydrogenation. Catalysis Today. 1997;**35**:415–425. DOI: 10.1016/S0920-5861(96)00212-X

[14] Sohail A, Chandrappa KG, Bakhtiar M, Jayant I. Mint: Catalytic hydrogenation of soybean oil promoted by synthesized Ni/SiO_2 nanocatalyst. Der Pharma Chemica. 2013;**5**:118–126. http://derpharmachemica.com/vol5-iss2/DPC-2013-5-2-118-126.pdf [Accessed: 2016-06-03]

[15] Stender S, Dyerberg J. Mint: Influence of *trans* fatty acids on health. Annals of Nutrition and Metabolism. 2004;**48**:61–66. DOI: 10.1159/000075591

[16] Tarrago-Trani MT, Phillips KM, Lemar LE, Holden JM. Mint: New and existing oils and fats used in products with reduced *trans*-fatty acid content. Journal of the American Dietetic Association. 2006;**106**:867–880. DOI: 10.1016/j.jada.2006.03.010

[17] Kummerow FA. Mint: The negative effects of hydrogenated *trans* fats and what to do about them. Atherosclerosis. 2009;**205**:458–465. DOI: 10.1016/j.atherosclerosis.2009.03.009

[18] Mensink RP, Katan MB. Mint: Effect of dietary *trans* fatty acids on high-density and low-density lipoprotein cholesterol levels in healthy subjects. The New England Journal of Medicine. 1990;**323**:439–445. DOI: 10.1056/NEJM199008163230703

[19] Wright AJ, Wong A, Diosady LL. Mint: Ni catalysts promotion of a *cis*-selective Pd catalyst for canola oil hydrogenation. Food Research International. 2003;**36**:1069–1072. DOI: 10.1016/j.foodres.2003.08005

[20] Belkacemi K, Boulmerka A, Arul J, Hamoudi S. Mint: Hydrogenation of vegetable oils with minimum *trans* and saturated fatty acid formation over a new generation of Pd-catalyst. Topics in Catalysis. 2006;**37**:113–120. DOI: 10.1007/s11244-006-0012-y

[21] Li T, Zhang W, Lee RZ, Zhong Q. Mint: Nickel-boron alloy catalysts reduce the formation of *trans* fatty acids in hydrogenated soybean oil. Food Chemistry. 2009;**114**:447–452. DOI: 10.1016/j.foodchem.2008.09.068

[22] Stanković M, Gabrovska M, Krstić J, Tzvetkov P, Shopska M, Tsacheva T, Banković P, Edreva-Kardjieva R, Jovanović D. Mint: Effect of silver modification on structure and catalytic performance of Ni-Mg/diatomite catalysts for edible oil hydrogenation. Journal of Molecular Catalysis A: Chemical. 2009;**297**:54–62. DOI: 10.1016/j.molcata.2008.09.004

[23] Hussain ST, Zia F, Mazhar M. Mint: Modified nano supported catalyst for selective catalytic hydrogenation of edible oils. European Food Research and Technology. 2009;**228**:799–806. DOI: 10.1007/s00217-008-0991-y

[24] Alshaibani AM, Yaakob Z, Alsobaai AM, Sahri M. Mint: Palladium-boron catalyst for vegetable oils hydrogenation. RASĀYAN Journal of Chemistry. 2012;**5**:463–467. e-ISSN: 0976-0083

[25] McArdle S, Leahy JJ, Curtin T, Tanner D. Mint: Hydrogenation of sunflower oil over Pt-Ni bimetallic supported catalysts: preparation, characterization and catalytic activity. Applied Catalysis A: General. 2014;**474**:78–86. DOI: 10.1016/j.apcata.2013.08.033

[26] Jang ES, Jung MY, Min DB. Mint: Hydrogenation for *low* trans and high conjugated fatty acids. Comprehensive Reviews in Food Science and Food Safety. 2005;**4**:22–30. DOI: 10.1111/j.1541-4337.2005.tb00069.x

[27] Menaa F, Menaa A, Tréton J, Menaa B. Mint: Technological approaches to minimize industrial *trans* fatty acids in foods. Journal of Food Science. 2013;**78**:377–386. DOI: 10.1111/1750-3841.12055

[28] Mohr C, Claus P. Mint: Hydrogenation properties of supported nanosized gold particles. Since Progress. 2001;**84**:331–334. DOI: 10.3184/003685001783238925

[29] Stanković M, Čupić Ž, Gabrovska M, Banković P, Nikolova D, Jovanović D. Mint: Characteristics and catalytic behaviour of supported NiMgAg/D catalysts in the partial hydrogenation of soybean oil. Reaction Kinetics, Mechanisms and Catalysis. 2015;**115**:105–127. DOI: 10.1007/s11144-014-0829-5

[30] Krstić J, Gabrovska M, Lončarević D, Nikolova D, Radonjić V, Vukelić N, Jovanović D. Mint: Influence of Ni/SiO_2 activity on the reaction pathway in sunflower oil hydrogenation. Chemical Engineering Research and Design. 2015;**100**:72-80. DOI: 10.1016/j.cherd.2015.05.001

[31] Radonjić V, Krstić J, Lončarević D, Jovanović D, Vukelić N, Stanković M, Nikolova D, Gabrovska M. Mint: Perlite as a potential support for nickel catalyst in the process of sunflower oil hydrogenation. Russian Journal of Physical Chemistry A. 2015;**89**:2359–2366. DOI: 10.1134/S0036024415130294

[32] Nohair B, Especel C, Marécot P, Montassier C, Hoang LC, Barbier J, Mint: Selective hydrogenation of sunflower oil over supported precious metals. Comptes Rendus Chimie. 2004;**7**:113–118. DOI: 10.1016/j.crci.2003.10.012

[33] Simakova IL, Simakova OA, Romanenko AV, Murzin DYu. Mint: Hydrogenation of vegetable oils over Pd on nanocomposite carbon catalysts. Industrial and Engineering Chemistry Research. 2008;**47**:7219–7225. DOI: 10.1021/ie800663j

[34] Iida H, Itoh D, Minowa S, Yanagisawa A, Igarashi A: Hydrogenation of soybean oil over various platinum catalysts: effect of support materials on *trans* fatty acid levels. Catalysis Communications. 2015;**62**:1–5. DOI: 10.1016/j.catcom.2014.12.025

[35] Babu GP, Ghuge KD, Rammohan SV, Krishnan V, Bhat AN. Mint: Characterization of silica supported Ni and Mg basic carbonate catalysts precursors. Catalysis Letters. 1992;**15**:95–103. DOI: 10.1007/BF00770902

[36] Thunyaratchatanon C, Jitjamnong J, Luengnaruemitchai A, Numwong N, Chollacoop N, Yoshimura Y. Mint: Influence of Mg modifier on *cis-trans* selectivity in partial hydrogenation of biodiesel using different metal types. Applied Catalysis A: General. 2016;**520**:170–177. DOI: 10.1016/j.apcata.2016.04.019

[37] Parry JD, Winterbottom JM. Mint: The hydrogenation of triglycerides using supported alloy catalysts. I. Silica-supported Ni-Ag catalysts. Journal of Chemical Technology and Biotechnology. 1991;**50**:67–80. DOI: 10.1002/jctb.280500108

[38] Parry JD, Winterbottom JM. Mint: The hydrogenation of triglycerides using supported alloy catalysts. II. Silica-supported Pd-Cu catalysts. Journal of Chemical Technology and Biotechnology. 1991;**50**:81–90. DOI: 10.1002/jctb.280500109

[39] Ksibi Z, Ghorbel A, Bellamy B, Masson A. Mint: Size particle effect and copper or silver addition effect on catalytic properties of rhodium supported onto amorphous silica. Studies in Surface Science and Catalysis. 2000;**130**:2045–2050. DOI: 10.1016/S0167-2991(00)80769-7

[40] Fernández MB, Piqueras CM, Tonetto GM, Crapiste G, Damiani DE. Mint: Partial hydrogenation of sunflower oil: use of edible modifiers of the *cis/trans*-selectivity. Journal of Molecular Catalysis A: Chemical. 2005;**233**:133–139. DOI: 10.1016/j.molcata.2005.02.012

[41] Nitta Y, Imanaka T, Teranishi S. Mint: Preparation chemistry of precipitated Ni-SiO_2 catalysts for enantioselective hydrogenation. Journal of Catalysis. 1985;**96**:429–438. DOI: 10.1016/0021-9517(85)90312-4

[42] Nele M, Vidal A, Bhering DL, Pinto JC, Salim VMM. Mint: Preparation of high loading silica supported nickel catalysts: simultaneous analysis of the precipitation and aging steps. Applied Catalysis A: General. 1999;**178**:177–189. DOI: 10.1016/S0926-860X(98)00285-3

[43] Bhering DL, Nele M, Pinto JC, Salim VMM. Mint: Preparation of high loading silica supported nickel catalysts: analysis of the reduction step. Applied Catalysis A: General. 2002;**234**:55–64. DOI: 10.1016/S0926-860X(02)00198-9

[44] Song CJ, Park TJ, Moon SH. Mint: Properties of the Ni/kieselguhr catalysts prepared by precipitation method. Korean Journal of Chemical Engineering. 1985;**9**:159–163. DOI: 10.1007/BF02705133

[45] Poncelet G, Grange P, Jacobs PA editors. Volume 16. Preparation of Catalysts III. 1st ed. Scientific Bases for the Preparation of Heterogeneous Catalysts. Amsterdam: Elsevier; 1983. 852 p. Electronic ISBN: 9780080960517. Print ISBN: 978044421845.

[46] Burattin P, Che M, Louis C. Mint: Molecular approach to the mechanism of deposition-precipitation of the Ni(II) phase on silica. The Journal of Physical Chemistry B. 1998;**102**:2722–2732. DOI: 10.1021/jp980018k

[47] Coenen JWE. Mint: Characterization of the standard nickel/silica catalyst EuroNi-1 II. Chemical aspects: precipitation, reduction and chemical analysis. Applied Catalysis. 1989;**54**:65–78. DOI: 10.1016/S0166-9834(00)82355-6

[48] van Voorthuijsen van Eijk JJB, Franzen P. Mint: Structure and properties of compounds formed during the preparation of nickel-on-silica catalysts. Recueil des Travaux Chimiques des Pays-Bas. 1951;**70**: 793–812. DOI: 10.1002/recl.19510700906

[49] Montes M, Penneman de Bosscheyde Ch, Hodnett BK, Delannay, Grange P, Delmon B. Mint: Influence of metal-support interactions on the dispersion, distribution, reducibility and catalytic activity of Ni/SiO_2 catalysts. Applied Catalysis. 1984;**12**:309–330. DOI: 10.1016/S0166-9834(00)81670-X

[50] El-Shatorry Y, deMan L, deMan JM. Mint: Evaluation of hydrogenation catalyst activity. Journal of the American Oil Chemists' Society. 1980;**57**:402–404. DOI: 10.1007/BF02678924

[51] Dafler JR. Mint: Nickel crystallite size and net activity of hydrogenation catalysts. Journal of the American Oil Chemists' Society. 1977;**54**:17–23. DOI: 10.1007/BF02671367

[52] Gray JI, Russel LF. Mint: Hydrogenation catalysts—Their effect on selectivity. Journal of the American Oil Chemists' Society. 1979;**56**:36–44. DOI: 10.1007/BF02671758

[53] Martin GA, Mirodatos C, Praliaud H. Mint: Chemistry of silica-supported catalyst: preparation, activation and reduction. Applied Catalysis. 1981;**1**:367–382. DOI: 10.1016/0166-9834(81)80054-1

[54] Ghuge KD, Bhat AN, Babu GP. Mint: Identification of nickel species and their interaction with the support in Ni-silica catalyst precursor. Applied Catalysis A: General. 1993;**103**:183–204. DOI: 10.1016/0926-860X(93)85051-P

[55] Pilar Gonzáles-Marcos M, Gutiérrez-Ortiz José I, Gonzáles-Ortiz de Elguea C, Delgado José A, Gonzáles-Velasko Juan R. Mint: Nickel on silica systems. Surface features and their relationship with support, preparation procedure and nickel content. Applied Catalysis A: General. 1997;**162**:269–280. DOI: 10.1016/S0926-860X(97)00111-7

[56] Burattin P, Che M, Louis C. Mint: Metal particle size in Ni/SiO_2 prepared by deposition-precipitation: influence of the nature of the Ni(II) phase and of its interaction with the support. The Journal of Physical Chemistry B. 1999;**103**:6171–6178. DOI: 10.1021/jp990115t

[57] Gabrovska M, Krstić J, Tzvetkov P, Tenchev K, Shopska M, Vukelić N, Jovanović D. Mint: Effect of the support and the reduction temperature on the formation of metallic nickel phase in Ni/silica gel precursors of vegetable oil hydrogenation catalysts. Russian Journal of Physical Chemistry A. 2011;**85**:2392–2398. DOI: 10.1134/S0036024411130073

[58] Horiuti I, Polanyi M. Mint: Exchange reactions of hydrogen on metallic catalysts. Transactions of the Faraday Society. 1934;**30**:1164–1172. DOI: 10.1039/TF9343001164

[59] Albright LF, Wisniak J. Mint: Selectivity and isomerization during partial hydrogenation of cottonseed oil and methyl oleate: Effect of operating variables. Journal of the American Oil Chemists' Society. 1962;**39**:14–19. DOI: 10.1007/BF02633340

[60] Litchfield C, Lord JE, Isbell AF, Reiser R. Mint: *Cis-trans* isomerization of oleic, linoleic and linolenic acids. Journal of the American Oil Chemists' Society. 1963;**40**:553–557. DOI: 10.1007/BF02822465

[61] Gut G, Kosinka J, Prabucki A, Schuerch A. Mint: Kinetics of the liquid-phase hydrogenation and isomerization of sunflower seed oil with nickel catalysts. Chemical Engineering Science. 1979;**34**:1051–1056. DOI: 10.1016/0009-2509(79)80005-6

[62] Ray JD, Carr BT. Mint: Empirical modeling of soybean oil hydrogenation. Journal of the American Oil Chemists' Society. 1985;**62**:1218–1222. DOI: 10.1007/BF02541831

[63] Niklasson C andersson B, Schöön NH. Mint: Influence of hydrogen pressure on selectivity in consecutive hydrogenation reactions. Industrial and Engineering Chemistry Research. 1987;**26**:1459–1463. DOI: 10.1021/ie00067a031

[64] Grau RJ, Cassano AE, Baltanás MA. Mint: Catalysts and network modeling in vegetable oil hydrogenation processes. Catalysis Reviews Science and Engineering. 1988;**30**:1–48. DOI: 10.1080/01614948808078615

[65] Machado RM, Gaumer Freidl K, Achenbach ML. Impact of catalysts on selectivity and reaction rates for the hydrogenation of soybean oil. In: 9th RC User Forum Europe, October 1999; Berne: EU99_11; 1999. pp. 1–12

[66] Santana A, Larrayoz MA, Ramírez E, Nistal J, Recasens F. Mint: Sunflower oil hydrogenation on Pd in supercritical solvents: kinetics and selectivities. The Journal of Supercritical Fluids. 2007;**41**:391–403. DOI: 10.1016/j.supflu.2006.12.009

[67] Jovanovic D, Radovic R, Mares Lj, Stankovic M, Markovic B. Mint: Nickel hydrogenation catalyst for tallow hydrogenation and for the selective hydrogenation of sunflower seed oil and soybean oil. Catalysis Today. 1998;**43**:21–28. DOI: 10.1016/S0920-5861(98)00133-3

[68] Nikolova D, Krstić J, Spasov L, Simeonov D, Lončarević D, Stefanov Pl, Jovanović D. Mint: Surface properties of the Ni-silica gel catalyst precursors for the vegetable oil hydrogenation process: N_2 sorption and XPS studies. Russian Journal of Physical Chemistry A. 2011;**85**:2380–2385. DOI: 10.1134/S0036024411130188

[69] Stanković M, Radonjić V, Krstić J, Marinković D. Preparation of Ni/diatomite hydrogenation catalyst precursors: Effect of counter ions on textural characteristics. In: Proceedings of 12th International Conference on Fundamental and Applied Aspects of Physical Chemistry (Physical Chemistry 2014); 22–26 September 2014; Belgrade: SPCS; 2014. pp. 284–287.

[70] Diefenbacker A, Pocklington WD, editors. Standard Methods for the Analysis of Oils, Fats and Derivatives. 7th ed. 1st Supplement Section 2, 2.001. IUPAC. Oxford: Alden Press; 1992. 151 p. ISBN 0-632-03337-I. [internet]. 2012 Available from: http://old.iupac.org/publications/books/ISBN0632033371_compress.pdf. [Accessed: 2016-06-03]

[71] Coenen JWE, Linsen BG. Structure and activity of silica-supported nickel catalysts. In Linsen BG, Fortuin JMH, Okkerse C, Steggerda JJ editors. Physical and Chemical Aspects of Adsorbents and Catalysts. London and New York: Academic Press; 1970. Ch. 10. pp. 471–527.

[72] Echeverria SM andres VM. Mint: Effect of the method of preparation on the activity of nickel-kieselguhr catalyst for vegetable oil hydrogenation. Applied Catalysis. 1990;**66**:73–90. DOI: 10.1016/S0166-9834(00)81628-0

[73] Burattin P, Che M, Louis C. Mint: Characterization of the Ni(II) phase formed on silica upon deposition-precipitation. The Journal of Physical Chemistry B. 1997;**101**:7060–7074. DOI: 10.1021/jp970194d

[74] Kermarec M, Carriat JY, Burattin P, Che M, Decarreau A. Mint: FTIR identification of the supported phases produced in the preparation of silica-supported nickel catalysts. The Journal of Physical Chemistry. 1994;**98**:12008–12017. DOI: 10.1021/j100097a029

[75] Ikuhara YH, Saito T, Takahashi S, Sasaki Y, Hirayama T. Mint: Synthesis and microstructural analysis of homogeneously dispersed nickel nanoparticles in amorphous silica. Journal of the American Ceramic Society. 2012;**95**:524–529. DOI: 10.1111/j.1551-2916.2011.04897.x

[76] Sodeyama K, Sakka Y, Kamino. Mint: Preparation of fine expanded perlite. Journal of Materials Science. 1999;**34**:2461–2468. DOI: 10.1023/A:1004579120164

[77] Sohn JR. Mint: Catalytic activities of nickel-containing catalysts for ethylene dimerization and butane isomerization and their relationship to acidic properties. Catalysis Today. 2002;**73**:197–209. DOI: 10.1016/S0920-5861(01)00513-2

[78] van Dillen AJ, Geus JW, Hermans LAM, van der Meijden J. Production of supported copper and nickel catalysts by deposition-precipitation. In: Proceedings of the Sixth International Congress on Catalysis, Imperial College; 12–16 July 1976; London. Letchworth, England: Chemical Society; 1977. pp. 667–685.

[79] Geus JW. Mint: Production and thermal pre-treatment of supported catalysts. Studies in Surface Science and Catalysis. Elsevier Science. eBook 1983;**16**:1–33. ISBN: 9780080960517, Print ISBN: 9780444421845.

[80] Houalla M. Mint: An assessment of the influence of the preparation method, the nature of the carrier and the use of additives on the state, dispersion and reducibility of a deposited "nickel oxide" phase. Studies in Surface Science and Catalysis. Elsevier Science. eBook 1983;**16**:273–289. ISBN: 9780080960517, Print ISBN: 9780444421845.

[81] Anderson JA, Rodrigo MT, Daza L, Mendioroz S. Mint: Influence of the support in the selectivity of nickel/clay catalysts for vegetable oil hydrogenation. Langmuir. 1993;**9**:2485–2490. DOI: 10.1021/la00034a001

[82] Kibby CL, Massoth FE, Swift HE. Mint: Surface properties of hydrogen-reduced nickel chrysotiles. Journal of Catalysis. 1976;**42**:350–359. DOI: 10.1016/0021-9517(76)90109-3

[83] Jacobs PA, Nijs HH, Poncelet G. Mint: Characterization of reduced natural garnierite and its catalytic activity for carbon monoxide hydrogenation. Journal of Catalysis. 1980;**64**:251–259. DOI: 10.1016/0021-9517(80)90500-X

[84] Gabrovska M, D. Nikolova, Krstić J, Stanković M, Stefanov P, Edreva-Kardjieva, Jovanović D. Mint: The state of nickel in the silver modified NiMg/SiO_2 vegetable oil hydrogenation catalysts. Russian Journal of Physical Chemistry A. 2009;**83**:1461–1467. DOI: 10.1134/S0036024409090088

[85] Wang W. High nickel- and titania-containing mesoporous silicas: synthesis and characterisation [doctoral thesis]. Leicestershire, UK: Loughborough University, Institute of Polymer Technology and Materials Engineering; 2005.

[86] Turlier P, Praliaud H, Moral P, Martin GA, Dalmon JA. Mint: Influence of the nature of the support on the reducibility and catalytic properties of nickel: evidence for a new type of metal support interaction. Applied Catalysis. 1985;**19**:287–300. DOI: 10.1016/S0166-9834(00)81751-0

[87] Roman A, Delmon B. Mint: Promoter and carrier effects in the reduction of NiO/SiO_2. Journal of Catalysis. 1973;**30**:333–342. DOI: 10.1016/0021-9517(73)90148-6

[88] Richardson JT, Turk B, Lei M, Forster K, Twigg MV. Mint: Effects of promoter oxides on the reduction of nickel oxide. Applied Catalysis A: General. 1992;**83**:87–101. DOI: 10.1016/0926-860X(92)80028-B

[89] Coenen JWE. Mint: Reduction of silica supported nickel catalysts. Studies in Surface Science and Catalysis. Elsevier Science. 1979;**3**:89–108. ISBN: 0-444-41733-8 (Vol 3), ISBN: 0-444-41801-6 (Series).

[90] Gabrovska M, Krstić J, Edreva-Kardjieva R, Stanković M, Jovanović D. Mint: The influence of the support on the properties of nickel catalysts for edible oil hydrogenation. Applied Catalysis A: General. 2006;**299**:73–83. DOI: 10.1016/j.apcata.2005.10.011

[91] Boudjahem A-G, Monteverdi S, Mercy M, Ghanbaja D, Bettahar MM. Mint: Nickel nanoparticles supported on silica of low surface area. Hydrogen chemisorption and TPD catalytic properties. Catalysis Letters. 2002;**84**:115–122. DOI: 10.1023/A:1021093005287

[92] Mustard DG, Bartholomew CH. Mint: Determination of metal crystallite size and morphology in supported nickel catalysts. Journal of Catalysis. 1981;**67**:186–206. DOI: 10.1016/0021-9517(81)90271-2

[93] Bartholomew CH. Mint: Hydrogen adsorption on supported cobalt, iron and nickel. Catalysis Letters. 1990;**7**:27–51. DOI: 10.1007/BF00764490

[94] Kumar N, King TS, Vigil RD. Mint: A portal model for structure sensitive hydrogen adsorption on Ru-Ag/SiO_2 catalysts. Chemical Engineering Science. 2000;**55**:4973–4979. DOI: 10.1016/S0009-2509(00)00132-9

[95] Gubitosa G, Berton A, Camia M, Pernicone N. Mint: Influence of the preparation technique of Pd-silica catalysts on metal dispersion and catalytic activity. Studies in Surface Science and Catalysis. Elsevier Science. eBook 1983;**16**:431–438. ISBN: 9780080960517, Print ISBN: 9780444421845.

[96] Wu M, Hercules DM. Mint: Studies of supported nickel catalysts by X-ray photoelectron and ion scattering spectroscopies. The Journal of Physical Chemistry. 1979;**83**:2003–2008. DOI: 10.1021/j100478a015

[97] Shalvoy RB, Reucroft PJ, Davis BH. Mint: Characterization of coprecipitated nickel on silica methanation catalysts by X-ray photoelectron spectroscopy. Journal of Catalysis. 1979;**56**:336–348. DOI: 10.1016/0021-9517(79)90126-X

[98] Rodrigo MT, Daza L, Mendioroz S. Mint: Nickel supported on natural silicates: Activity and selectivity in sunflower seed oil hydrogenation. Applied Catalysis A: General. 1992;**88**:101–114. DOI: 10.1016/0926-860X(92)80198-L

[99] Briggs D. Handbook of X-ray Photoelectron Spectroscopy. Eden Prairie, Minnesota: Perkin Elmer Corp., Physical Electronics Division; 1979. 190 p. DOI: 10.1002/sia.740030412

[100] Venezia AM. Mint: X-ray photoelectron spectroscopy (XPS) for catalysts characterization. Catalysis Today. 2003;**77**:359–370. DOI: 10.1016/S0920-5861(02)00380-2

[101] Gaarenstroom SW, Winograd N. Mint: Initial and final state effects in the ESCA spectra of cadmium and silver oxides. The Journal of Chemical Physics. 1977;**67**:3500–3506. DOI: 10.1063/1.435347

[102] Lefèbvre VJ, Baltes J. Mint: Nickel/Silber-hydrierkatalysatoren und ihre verwendung zur selektiven härtung von fetten. European Journal of Lipid Science and Technology. First published: Fette, Seifen, Ansitrichmittel. 1975;**77**:125–131. DOI: 10.1002/lipi.19750770402. [Internet]. 2006. Available from: http://onlinelibrary.wiley.com/doi/10.1002/lipi.v77:4/issuetoc [Accessed: 2016-06-08]

[103] Macher M. Supercritical hydrogenation of vegetable oils [PhD thesis]. Göteborg: Chalmer University of Technology; 2001.

[104] Albright LF, Allen RR, Moore MC. Mint: Quantitative measure of geometrical isomerization during the partial hydrogenation of triglyceride oils. Journal of the American Oil Chemists' Society. 1970;**47**:295–298. DOI: 10.1007/BF02609498

[105] Chen AH, McIntire DD, Allen RR. Mint: Modeling of reaction rate constants and selectivities in soybean oil hydrogenation. Journal of the American Oil Chemists' Society. 1981;**58**:816–818. DOI: 10.1007/BF02665587

[106] Jovanović D, Čupić Ž, Stanković M, Rožić Lj, Marković B. Mint: The influence of the isomerization reactions on the soybean oil hydrogenation process. Journal of Molecular Catalysis A: Chemical. 2000;**159**:353–357. DOI: 10.1016/S1381-1169(00)00154-0

[107] Belkacemi K, Boulmerka A, Hamoudi S, Arul J. Mint: Activity and selectivity of novel structured Pd catalysts: kinetics modelling of vegetable oils hydrogenation. International Journal of Chemical Reactor Engineering. 2005;**3**:A59. 1–25. ISSN: 2194-5748. http://www.bepress.com/ijcre/vol3/A59 [Accessed: 2016-07-18]

[108] Gear CW. Numerical initial value problems in ordinary differential equations. Englewood Cliffs, New Jersey: Prentice-Hall; 1971. 253 p.

[109] Stone RE, Tovey CA. Mint: The simplex and projective scaling algorithms as iteratively reweighted least squares methods. SIAM Review. 1991;**33**:220–237. DOI: 10.1137/1033049

[110] Pilar Gonzáles-Marcos M, Gutiérrez-Ortiz José I, Gonzáles-Ortiz de Elguea C, Alvarez Jon I, Gonzáles-Velasko Juan R. Mint: Control of the product distribution in the hydrogenation of vegetable oils over nickel on silica catalysts. The Canadian Journal of Chemical Engineering. 1998;**76**:927–935. DOI: 10.1002/cjce.5450760510

Chapter 9

New Strategies for Obtaining Inorganic-Organic Composite Catalysts for Selective Hydrogenation

Nicolás Carrara, Juan Manuel Badano,
Carolina Betti, Cecilia Lederhos, Mariana Busto,
Carlos Vera and Mónica Quiroga

Additional information is available at the end of the chapter

http://dx.doi.org/10.5772/65959

Abstract

Pd/composites are catalysts in which the catalytically active metal phase is located in a small layer on the support surface. The metal distribution obtained on these supports corresponds to a structure commonly known as "egg-shell". This distribution is optimal for reactions that have heat or mass transfer resistances. The small thickness of the metal layer is mainly attributed to the use of support pellets of mixed organic–inorganic composition, the special interaction between the hydrophobic–hydrophillic support surface and the metal species (Pd or Pt) present in the impregnating aqueous solution. Pd/composite catalysts are active and selective for the reactions of selective hydrogenation of styrene, 1-heptyne, 3-hexyne and 2,3-butanone, and enantioselective hydrogenation of ethyl pyruvate. These reactions are of interest in both academic and industrial aspects.

Keywords: composite catalyst, inorganic–organic supports, selective hydrogenation, improved mechanical resistance, egg-shell catalysts

1. Introduction

Industrial selective hydrogenation processes mostly make use of solid catalysts and three-phase reactors (slurry or packed bed). Most catalysts are of the supported metal kind,and the metal carriers are either completely inorganic (e.g., alumina, titania, silica) or organic (e.g., resins, activated carbons).

Nowadays, a great variety of materials are being used as support, such as alumina, mordenite, silica, titania, magnesia, zeolites, refractory metal oxides of groups III, IV, V, VI, VIII, activated carbons, and polymers.

The preparation of the support is an important step in the manufacture of commercial catalysts. The size and shape of the supports are determined by the kind of reactor and the set of reaction conditions [1].

Supported metal catalysts for industrial reactors are preferably available in the physical form of cylinders, tablets, spheres, rings, fibers, cloths, etc. The form of the support in the catalyst preparation step is important since it is closely related to the structure of the final catalyst.

At the time of choosing a catalyst, its mechanical properties cannot be disregarded. Catalysts are generally placed inside the packed beds, commonly 10 m or more in height. This arrangement makes the bottom pellets support the weight of the bed above and thus need a high crushing resistance [1]. If the resistance is not high enough, the pelletsget broken and smaller particles and dust are generated. As a result the pressure drop in the packed bed gradually increases until the bed is plugged; the unit must then be shut down for maintenance. Thus, the smallest particles of catalyst could be eluted from reactor with solvents and reaction products. This leads to an activity loss and a decrease of catalyst life.

In the preparation of the inorganic carriers employed in the fabrication of supported metal catalysts, the support is first obtained in the form of a powder or tiny particles, which are mixed with water and/or organic compounds. The mixture is then extruded, granulated, or pelletized to give it the final shape [1, 2]. Once the material has the desired shape, it is dried at temperatures between 353 and 673 K for one or more hours. Then, depending on its nature and final use, the support is calcined at higher temperatures to give mechanical resistance [1–9]. This procedure is long and energy consuming. However, the preparation of organic materials is energetically less costly.

Carbons and polymers are commonly used among the organic supports. Carbons are obtained by controlled combustion in oxygen-poor atmosphere and the final product can be surface treated to modify its catalytic properties. When carbon particles of a certain shape and size are needed, the carbon material must be agglomerated. Polymeric supports are mainly obtained by reactions in emulsions and suspensions. Carbons and polymeric supports have the common disadvantage of low mechanical resistance.

Among the supported metal catalysts we prefer the catalysts with metal "profile" that comprise an inert support and a metal phase of specially tailored distribution. Depending on this distribution catalysts are classified as "egg-yolk," "egg-white," or "egg-shell". They are usually available in the form of rings, spheres, tablets, or pellets [10].

Metal-supported catalysts of the egg-shell type are mainly used in heterogeneous catalytic reactions in which the mass or heat transfer limitations have an important effect on the activity and selectivity to the desired products. They also tend to prevent or minimize the deactivation phenomena. Their main advantages are low mass transfer resistance, support-independent pore structure, and high heat transfer rate at the catalyst surface [2].

It is known that a uniform access of reactants to active sites is important for getting high activity and selectivity in chemical processes [11]. For example, consecutive reactions of the form A→B→C in which product C should be suppressed or minimized are of great industrial interest. Due to the interaction of intraparticle diffusion and reaction, the occurrence of C can be minimized in catalysts where the active sites are preferentially located on a surface layer [11–13].

Egg-shell catalysts are used with great benefits in reactions of selective hydrogenation [14–16], Fischer-Tropsch synthesis [17], methane reforming [18, 19], partial oxidation of methane [20], etc.

It is also known that diffusional restrictions decrease the chemical reaction rate and negatively affect the selectivity to the desired products. These phenomena are mainly enhanced in processes using pelletized catalysts in packed beds, such as Fischer-Tropsch synthesis [17] or selective hydrogenation reactions [14–16]. In these reactions the maximum yields and selectivities are obtained with egg-shell catalysts in which the active phase is located on the external surface of the pellet, where the thickness of the active phase plays an important role [17].

Making egg-shell catalysts with common supports is not a simple process because there are many preparation variables that have to be carefully controlled: pH and viscosity of the impregnating solution, concentration of the metal salts in solution, time of contact of the impregnating solution with the support, temperature during impregnation, drying temperature, and calcination temperature. Other complexities arise from the phenomena involved in the grafting of the active phase, the variable nature of the impregnating solution (aqueous, organic, mixed phases), etc. [12, 13, 17, 20–30].

The field of materials science has grown significantly in the past few decades. This growth has been partly due to the development of complex composite materials that combine different properties of the individual components, sometimes with convenient synergysm of physical and chemical features, thus leading to materials with unique properties. Composite-based catalysts can have enhanced mechanical resistance, selective adsorption properties, and unique selectivity for regio, stereo, or enantio hydrogenation reactions, etc.

Composite supports using organic (polymer) and inorganic components are of special interest for modern chemistry and material science because of their potential use in photochemistry, nanoelectronics, optics, and catalysis. This chapter focuses on composite supports for heterogeneous catalysts. In these supports the inorganic phase can be supplied by metal oxide or metal particles or their mixture. The organic phase is supplied by polymers of different chemical nature, with chosen surface functional groups available for acting as active sites for adsorption and catalysis.

Synthesis and use of metal catalysts supported over functional polymers or their mixture with other inorganic materials are the areas of focus. Organometallic or metal complex catalysts are not included in this work, though they can be mentioned for the sake of discussion when reviewing the chemistry of the active sites.

Two main advantages of functional polymers when used as supports of metal particles with catalytic properties are described below:

1. Metal supported particles with controlled size and distribution can be synthesized. Particle size may range from a few nanometers to hundreds of nanometers.

2. Chemical performance advantages of the catalyst can be improved because of an intimate interaction between the metal particle and the different functional groups of the polymer. The used polymers can be either soluble or insoluble in the reaction medium. The focus here is on insoluble polymers, involving a bundle of physically and/or chemically cross-linked polymer chains in which the metal nanoparticles are embedded.

Pd/composite catalysts with egg-shell structure can be easily obtained because the hydrophil-lic-hydrophobic action of the support regulates the penetration of the metal (active phase). The good mechanical properties of these materials give them an advantage over traditional supports where their use is intended for packed bed or basket reactors.

To evaluate the properties of the metal/composite egg-shell catalysts, the test reactions of selective hydrogenation of styrene to ethylbenzene, 1-heptyne to 1-heptene, 3-hexyne to 3-hexene, 2,3-butanodione to 3-hydroxy-2-butanone and ethyl piruvate to (R)-ethyl lactate are used.

2. Preparation of composite supports and catalysts

2.1. Preparation of composite supports

A brief summary is presented about the preparation and use of different organic-inorganic supports and catalysts. The materials are divided into three classes and each one is studied separately:

A. Metals supported over functional polymers

B. Metals supported over inorganic materials

C. Metals supported over new inorganic-organic composite materials

2.1.1. Metals supported over functional polymers

The different strategies of synthesis of these metal/polymer catalysts are shown in **Figure 1**.

For the preparation of these catalysts, the organic support is commonly prepared from acrylic or vinylic monomers. Polymerization cross-linking is adjusted by the use of additives, with a minimum value for preventing the dissolution in the solvents used in the subsequent reactions. Metals with catalytic properties are placed over these supports. Oxidized metal precursors are reduced to the metal state with the help of reducing chemical agents in solution or by the use of molecular hydrogen. For these kind of supports styrene and divinylbenzene N,N-dimethy-lacrylamide, 2-acrylamido-2-methylpropane sulfonic acid, methacrylic acid , 4-vinylpyridine, glycidyl methacrylate ethylene glycol dimethacrylate, etc. are the monomers of choice.

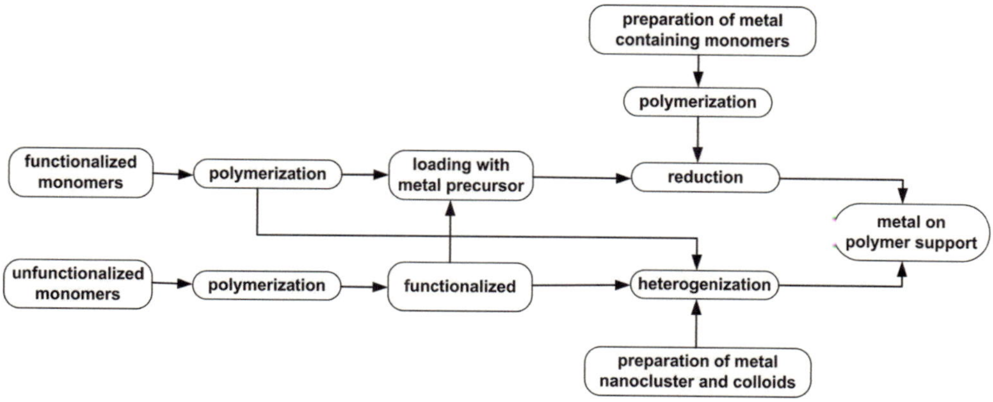

Figure 1. Routes for the preparation of metal catalysts supported on functional polymers.

2.1.2. *Metals supported over inorganic materials*

Here a functionalized polymer is used. This is generally soluble or slightly soluble in organic or aqueous solvents. Organometallic compounds or metal particles are grafted over the polymer matrix. These compounds act as metal active sites. Once the metal is firmly supported on the polymer surface, particles of this polymer are deposited over inorganic supports.

2.1.3. *Metals supported over new inorganic-organic composite materials*

Lately our research group developed new composite organic-inorganic supports in which the organic phase is made out of polymers of functionality 4. In these composites the polymeric phase has a high cross-linking degree and the inorganic phase is supplied by particulated metal oxides. The method of preparation of these composite supports is described in **Figure 2**.

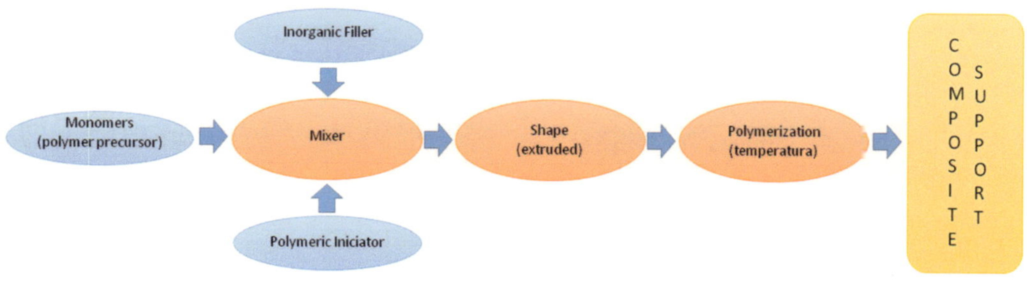

Figure 2. Scheme of preparation of a composite support.

Structured materials in the form of rings, spheres, tablets, or pellets can be prepared with this preparation technique. Metals can be deposited on them, preferentially with an egg-shell structure of small thickness of the active phase. These materials have improved mechanical resistance in comparison with other existing commercial supports. They also combine dual hydrophilic/hydrophobic, inorganic/organic surface properties; this is the main property behind the preparation of egg-shell catalysts.

The process of preparation of the composite supports is described in **Figure 2**. It comprises the following steps:

1. Weighing of monomers, polimerization starters, and inorganic material.
2. Mixing of the monomers, polimerization starters, and inorganic material.
3. Shaping of the support by means of extrusion, pelletizing, etc.
4. Polymerization of the organic phase by means of the thermal activation of the polymerization starter.
5. Making the support ready for use in the preparation of the metal catalyst.

In the preparation of the composite precursors the following polymeric phase precursors were used: bisphenol A glycerolate dimethacrylate (BGMA), triethylene glycol dimethacrylate (TEGDMA), and diurethane dimethacrylate (UDMA). The thermally activated polymerization starter used was benzoyl peroxide (BPO). **Figure 3** shows the chemical structure of the monomers and the polymerization starter used during the preparation of the UTAl and BTAl supports.

Two different composites, that is, UTAl and BTAl used in this work were made by mixing γ-alumina powder, two monomers, and BPO. The main difference between composites supports was the type of monomers used; UDMA and TEGMA were used for composite called UTAl, and BGMA and triethylene glycol dimethacrylate(TEGMA) were used for composite called BTAl. The alumina used was γ-alumina with a particle size of 0.074 mm.

UTAl and BTAl were prepared as describe by Badano et al. [31], monomers joined BPO were intimately mixed with 45 wt% γ-alumina. Then, mixture was desgassed and extruded into a cylinder of 2 mm diameter. Polymerization of the paste was carried on a stove at 393 K for 1 h. The final composite were pellets of 2 mm diameter and 2–3 mm long.

2.2. Catalysts preparation

Gamma alumina (CK-300, Ketjen), activated carbon (NORIT RX3 Extra and CNR), and composite support, BTAl and UTAl were used as supports to prepared metal catalyst. Platinum was loaded with $PtCl_2$ and $H_2PtC_{16}H_2O$ by incipient wetness technique. After impregnation, catalysts were dried at 393 K in an oven for 24 h and were maintained in desiccator for further use. The catalysts were reduced in a H_2 flow at 503 K for 1 h after catalytic test.

http://dx.doi.org/10.5772/65959

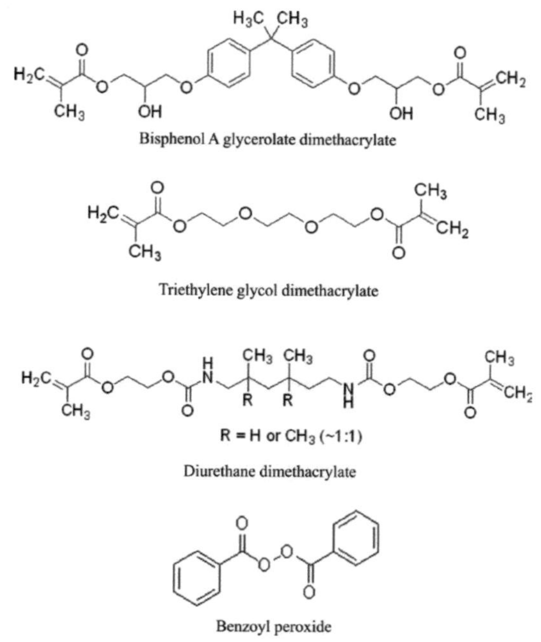

Figure 3. Monomer precursors of the polymeric phase and polymerization starter.

3. Catalyst activity test

The catalytic activity and selectivity during the hydrogenation reactions were measured in a batch, PTFE-coated, stainless steel, stirred tank reactor, using pelletized or powder catalysts. The procedure used for starting the reaction comprised of the following steps. The reactor was charged with a solution of the substrate (styrene, 1-heptyne, 3-hexyne, 2,3-butanodione, ethyl pyruvate). The stirring rate used in each catalytic test was 1200 rpm. Pressure, temperature, and catalyst concentration values were different for each system.

4. Results

4.1. Catalysts characterization

Values of the axial and radial mechanical resistance of the pellets were obtained in an universal rehearsals equipment. A compression rate of 1 mm min^{-1} was used. The attrition resistance was assessed by means of the method depicted in the ASTM D 4058 norm.

The specific surface area (S_{BET}) values were obtained from nitrogen physical adsorption-desorption isotherms at 77 K. In order to get a surface free of water and adsorbed compounds, the samples were degassed overnight at 523 K in vacuum (<10^{-4} Pa) before adsorption.

The palladium and platinum content of the catalysts was obtained by digesting the sample and then analyzing the liquors in ICP equipment. Hydrogen chemisorption measurements were performed in a pulse apparatus at atmospheric pressure. Before the test the samples were reduced in a gas mixture containing 5% v/v H_2/Ar for 1 h and degassed in an Ar flow at the same temperature of reduction for 3 h. After that, the samples were cooled in an argon flow, and the chemisorption test was done by injecting hydrogen pulses to the flowing stream until the samples became saturated.

X-ray photoelectron spectroscopy (XPS) was done over the reduced catalysts with a VG-MicrotechMultilab equipment.

Samples were analyzed using scanning electron microscopy (SEM) equipped with an energy dispersion system (EDAX) that enablesto analyze elementary chemical with an X-ray microbeam. This technique is known as electron probe microanalyzer (EPMA) and it was allowed to elucidate metal distribution and create elemental mappings. Before measure, catalyst pellets were coated with thin carbon film in order to avoid influence of charge effect during the SEM operation. The scanning speed was 0.02 mm min^{-1} and the acceleration voltage of electron beam was 20kV.

The thickness of the metal surface shell was also determined optically with the aid of micrographs of the cross section of the catalysts. Micrographs were obtained with a microscope equipped with a color video printer. To analyze the samples, this technique required to be encapsulated with a thermoplastic resin and then polished with sandpaper. Sanding encapsulated samples were done with finer paper (down 500 grit) until exposing the cross section of catalyst particles. The finest paper used was 2500 grit.

The XRD measurements were performed in a diffractometer with CuKα radiation filtered with Ni. Spectra were scanned at a rate of 0.25 min^{-1} in range between 25 and 80°. Catalyst required different preparation before analyzing. For alumina and carbon samples, material was grinded and reduced in hydrogen flow. In the case of composites, a slab of 1 cm^2 of area was impregnated with Pt or Pd, then was grinded and reduced as previous samples thermo gravimetric analysis (TGA) traces were also obtained. These traces show the dependence of the thermal weight loss of a sample as a function of the temperature. In the experiments catalyst samples (about 10 mg) were heated in an air flow of 40 mL min^{-1} from room temperature up to 1173 K at a heating rate of 5 K min^{-1}.

To measure the acidity of the supports, the reaction of dehydration of 1,4-butanediol to tetrahydrofurane (THF) was followed.

5. Results

5.1. Catalyst characterizations

Table 1 shows the results of the tests of mechanical properties, i.e., resistance to diametrical compression (SD), resistance to longitudinal compression (SL), and attrition loss obtained for the composite supports UTAl, BTAl and for the commercial supports of alumina and carbon. It can be seen that the composite supports have better mechanical properties than the commercial supports. This is especially important when γ-alumina is taken as a reference.

Support	SD[a] (Kgf/cm²)	SL[a] (Kgf/cm²)	Attrition loss (%)[b] 1800 Turns	Attrition loss (%)[b] 18,000 Turns
Carbon RX3	<100	217 ± 101	1.659	6.859
γ-Al_2O_3	<100	703 ± 199	0.167	3.103
α-Al_2O_3	282 ± 43	2053 ± 398	1.124	6.002
UTAl	392 ± 57	4708 ± 690	0.028	0.136
BTAl	348 ± 41	3741 ± 900	0.046	0.189

[a]From Ref. [31].
[b]From Ref. [32].

Table 1. Supports mechanical properties.

Table 2 contains the results of metal concentration, metal dispersion, specific surface area, binding energy of $Pd3d_{5/2}$ and Pt $4d_{5/2}$ and Cl/metal atomic ratios obtained by XPS, and metal penetration depth, as determined by different techniques.

Catalyst	Metal (wt%)	D % (at_{ad}/at_{total})	S_{BET}	XPS BE (eV) Pd $3d_{5/2}$ and Pt $4d_{5/2}$	Cl/Metal (at/at)	EMPA e (μm)
0.3PdUTAl	0.23	5	nd	335.0$^{(74\%)}$ 337.0$^{(26\%)}$	0.60	60
0.3PdBTAl	0.26	3	nd	335.6$^{(76\%)}$ 336.6$^{(24\%)}$	0.84	90
0.3PdCNR	0.49	62	nd	335.2$^{(100\%)}$	-	nd
1PdAl	1.1	19	234	335.3$^{(62\%)}$ 337.2$^{(38\%)}$	3.76	Up to Pellet center
1PdRX	1.31	4	1073	335.2$^{(100\%)}$	-	600
1PdUTAl	1.00	7	0.62	335.0$^{(89\%)}$ 337.0$^{(11\%)}$	0.20	60
1PdBTAl	1.10	4	1.13	335.2$^{(75\%)}$ 336.7$^{(25\%)}$	0.68	90
1PtUTAl	1.00	nd	nd	313.5$^{(48\%)}$ 315.5$^{(52\%)}$	1,23	nd
1PtBTAl	1.10	nd	nd	313.5$^{(55\%)}$ 315.5$^{(45\%)}$	0,87	nd

Table 2. Metal catalyst loadings, dispersion values, XPS results, and metal penetration to EMPA.

Regarding the Pd surface species, except for the 0.3PdCNR and 1PdRX catalysts, only two different species were detected. The PdUTAl catalysts had a peak with a binding energy (BE) of 335.0 eV, attributed to Pd^0, while the PdBTAl catalysts had peaks at 335.6 (0.3PdBTAl) and 335.2 eV (1PdBTAl), and attributed to $Pd\delta^+$ electrodeficient Pd species [33]. All catalysts had a second signal with BE values between 336.3 and 337.0 eV that would correspond to $Pd^{\eta+}$ species with $\delta^+<\eta^+<2$.This could possibly be related to the incomplete removal of chlorine ligands of the metal salt during the thermal treatment, leading to the presence of surface nonreduced Pd oxychloride species [34]. The same behavior can be seen for the Pt catalysts. In this case the $4d_{5/2}$ signals at 313.5 and 315.5 eV would be attributed to $Pt^{\eta+}$, oxychlorided platinum species [34].

In **Figure 4** optical microscopies of the catalysts 0.3PdBTAl, 0.3PdUTAl, 1PdBTAl, and 1PdUTAl are shown. In this figure it can be observed that the pellets images form before the catalytic tests. In the cross-sectional images of these catalysts it can be observed in the clearer region, the presence of support on the inside of the pellet, and on the outer surface, the darker region, the metal is located. Identical behavior can be seen in **Figure 5** where the microscopies for 1PtUTAl and 1PtBTAl catalysts were obtained.

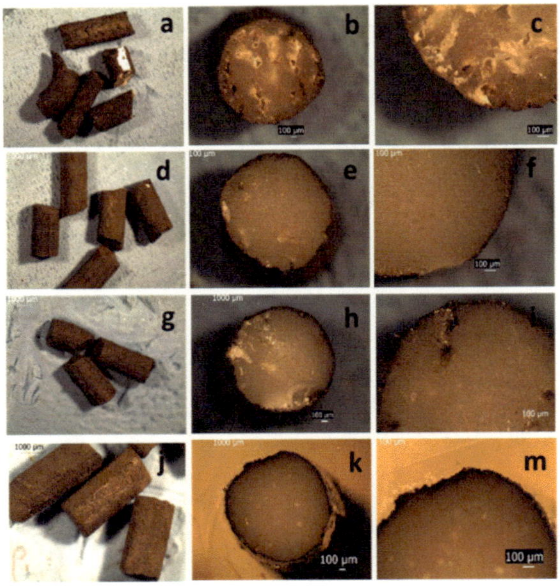

Figure 4. Optical micrographs images for transversal section of catalysts: (a)–(c): 0.3PdBTAl; (d)–(f): 0.3PdUTAl; (g)–(i): 1PdBTAl; (j)–(m): 1PdUTAl.

Figure 6 presents SEM images obtained for the catalysts 0.3PdBTAl, 0.3UTAl, 1PdBTAl, and 1PdUTAl. In the SEM images, **Figure 6(a)** and **(b)**, the topography of the surface catalysts 1PdBTAl and 1PdUTAl can be seen. Images from **Figure 6(c)–(h)** were obtained using the detector in backscattering mode in order to see the structural composition of the composite catalysts. From such images it is possible to see Pd particles as darker region on outer surface of the support. In the center of the pellet light color spots surrounded by a structure of gray colorcan be seen, which corresponds to γ-alumina particles gird for a continuous network of polymeric phase of the composite catalyst.

The egg-shell profile of the catalysts was confirmed by EPMA obtained by the penetration of Pd on the support of 60 and 90μm for UTAl and BTAl supports, respectively.

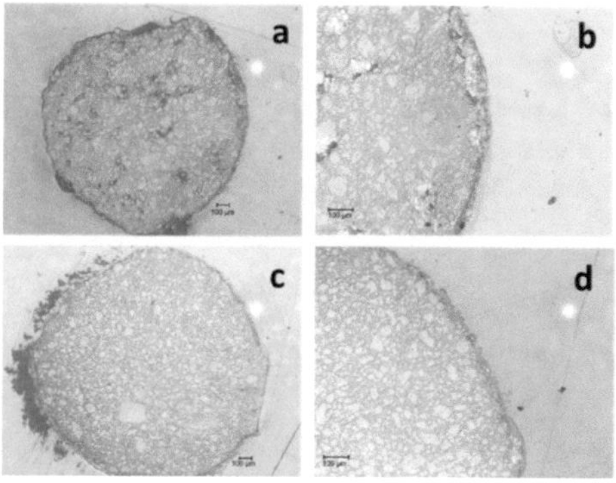

Figure 5. Optical micrograph images of the cross section of the pelletized catalysts: 1PtBTAl (a and b); 1PtUTAl (c and d).

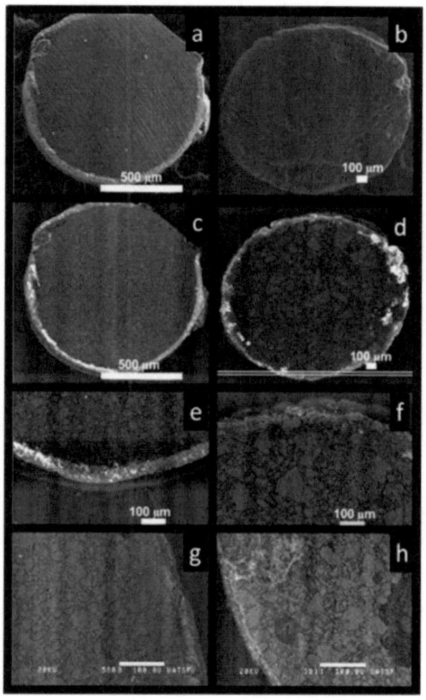

Figure 6. Catalysts transversal section SEM images. (a), (c), and (e) 1PdUTAl; (b), (d), and (f) 1PdBTAl; (g) 0.3PdUTAl, (f) 0.3BTAl. From Ref. [32].

Figure 7 shows the X-ray diffractograms of the composite supports, γ-alumina, Pt/composite, and Pd/composite catalysts. Comparing the results of gamma alumina with those of the composite supports, a close similarity can be seen. This is due to the fact that the inorganic phase used for the preparation of the composite supports is also γ-alumina and that the polymeric phase is amorphous. For all catalysts a peak at 2θ–39.9° would correspond to the reflections of the (111) planes of Pd^0 and Pt^0.

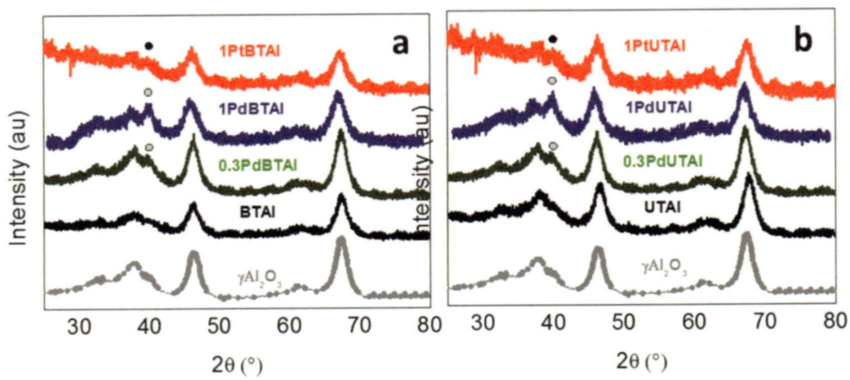

Figure 7. X-ray diffractograms of the composite BTAl (a) and UTAl (b) supports and the corresponding Pd and Pt catalysts.

Figure 8 contains the TGA results of the UTAl and BTAl composites, and 1 wt% Pd and Pt composite catalysts.

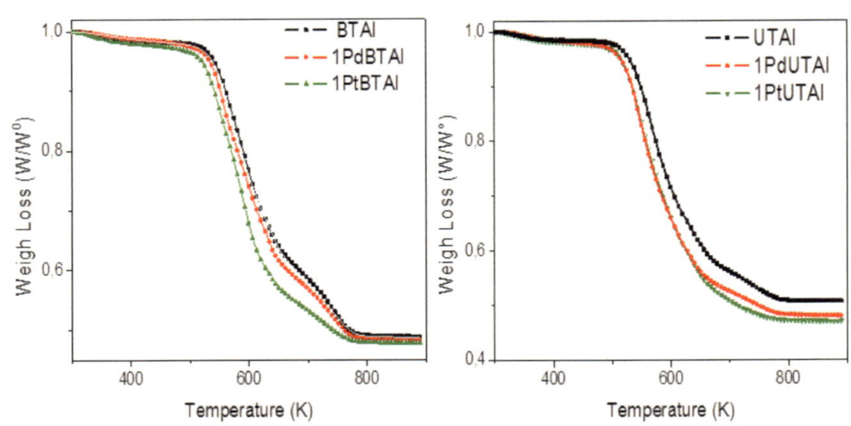

Figure 8. TGA traces of the composite supports and the corresponding Pt and Pd catalysts using air as carrier gas.

Inspection of the TGA traces indicates that for both composites the presence of the metal enhances the rate of decomposition of the organic phase.This effect is more important in the case of Pt. This behavior correlates with the nature of these metals that act as combustion catalysts in the presence of oxygen.

In order to assess the acidity of the support the reaction test of dehydration of 1,4-butanodiol to THF was used. The results of the test are shown in **Figure 9**. As expected the catalyst with the highest activity is the Amberlyst 15 resin. γ-Al_2O_3 alumina has a catalytic activity lower than the resin but higher than other supports. The composite and the Norit RX3 carbon supports display alcohol conversion values much lower than those of the resin or the alumina. There are only slight differences in activity between the activated carbon and UTAl composite while BTAl displays the lowest values of conversion as a function of time. From these results it could be inferred that the acid resin Amberlyst 15 has the highest concentration of acid sites and the BTAl composite the lowest.

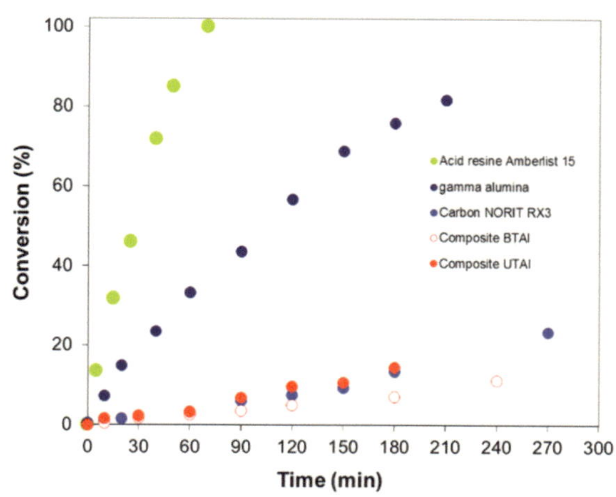

Figure 9. Acidity test. Dehydration of 1,4-butanediol to THF. Reaction conditions: 1,4-dioxane solvent, $C^0_{1,4\text{-butanediol}} = 1$ M, 0.1 MPa, 473 K, $W_{cat} = 0.2$ g.

5.2. Hydrogenation tests

5.2.1. Styrene hydrogenation

Figure 10 shows a scheme of the reaction of selective hydrogenation of styrene to ethylbenzene. In this reaction the hydrogenation of the aromatic nucleus is undesired.

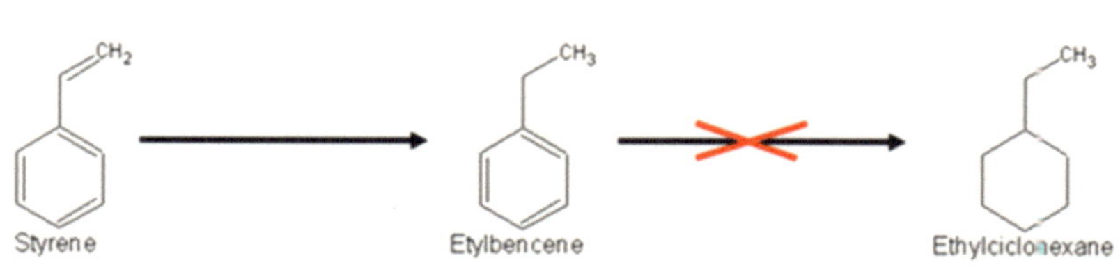

Figure 10. Scheme of the reaction of selective hydrogenation of styrene to ethylbenzene.

The reaction of hydrogenation of styrene to ethyl benzene is used as a reaction test for the process of purification of some refinery hydrocarbon streams. In this reaction the interest is to hydrogenate the vynillic external C=C bond, while keeping unaltered the bonds of the aromatic ring [34, 35].

In order to determine the mass transfer limitations in the inside of the catalyst particles, the effectiveness factor (η) was used:

$$\eta = \frac{\text{real reaction rate}}{\text{unrestricted reaction rate}} \tag{1}$$

The real reaction rate corresponds to the measured rate of the reaction at the conditions of temperature and concentration of the bulk reaction medium.The unrestricted reaction rate corresponds to the ideal reaction rate of a catalyst particle with no intraparticle or fluidparticle gradients of concentration or temperature that has the temperature and concentration of the bulk fluid phase.The unrestricted reaction rate was measured by grinding the catalyst pellets to a particle size smaller than 100 μm. For this particle size and the reaction conditions used it was estimated that all mass and heat transfer gradients were negligible.

The results of styrene conversion as a function of reaction time for the pelletized and powder catalysts are shown in **Figure 11**. The catalysts prepared with the composite supports UTAl and BTAl have higher activities (higher conversion values) than the catalysts prepared with commercial supports.

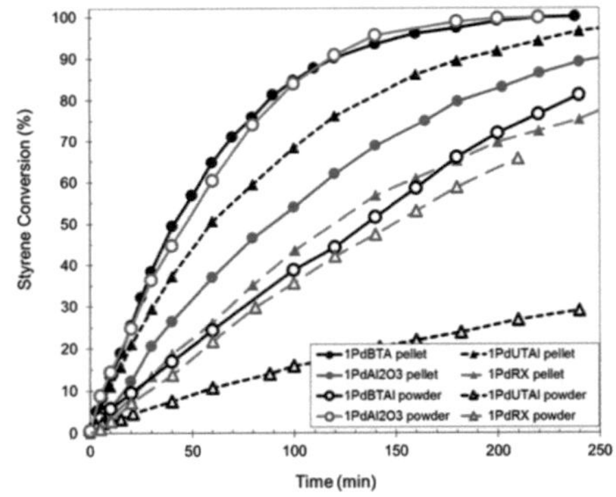

Figure 11. Styrene conversion as a function of reaction time.Catalysts with 1% Pd. Reaction conditions: toluene solvent, $C^0_{styrene}$ = 0.445 M, 2.0 MPa H_2, 353 K, W_{cat} = 0.3 g (pellet) or 0.01 g (powder).

In all the tests, for all catalysts, at all reaction conditions used, the obtained values of selectivity to ethylbenzene were higher than 99%.

Initial reaction rates were calculated by calculating the conversion rate as a function of time and extrapolating to zero conversion. With the values corresponding to the pellet and powder catalyst the effectiveness fact or could thus be estimated using the following equation:

$$\eta = \frac{\text{Initial reaction rate of pellet catalyst}}{\text{Initial reaction rate of powder catalyst}} = \frac{r^0_{pellet}}{r^0_{powder}} \tag{2}$$

Table 3 contains the values of r^0_{pellet}, r^0_{powder}, and for the different catalysts. It can be seen that the composite supports have highervalues than PdRX and PdAl catalysts. The higher effectiveness values are directly related to the lower mass transfer limitations when using the composite catalysts.

Catalyst	r^0_{pellet} (n°/gcat min)	r^0_{powder} (n°/gcat min)	η
PdRX	0.0014	0.0295	0.048
1PdAl	0.0019	0.1183	0.016
1PdUTAl	0.0029	0.0097	0.299
1PdBTAl	0.0038	0.0300	0.127

Table 3. Styrene hydrogenation initial reaction rate for catalyst pellets (r^0_{pellet}) and catalysts powders (r^0_{powder}). Effectiveness factor (η) [31].

Catalysts with a metal loading lower than 0.5 Pd wt% were also tried with this reaction test. The catalysts are named 0.3PdUTAl and 0.3PdBTAl in the case of the composites (with 0.3% Pd). LD265 is a procatalyse Pd/Al_2O_3 catalyst (0.3% Pd) and Engelhard is a BASF Pd/Al_2O_3 catalyst (0.5% Pd).The results are plotted in **Figure 12**.

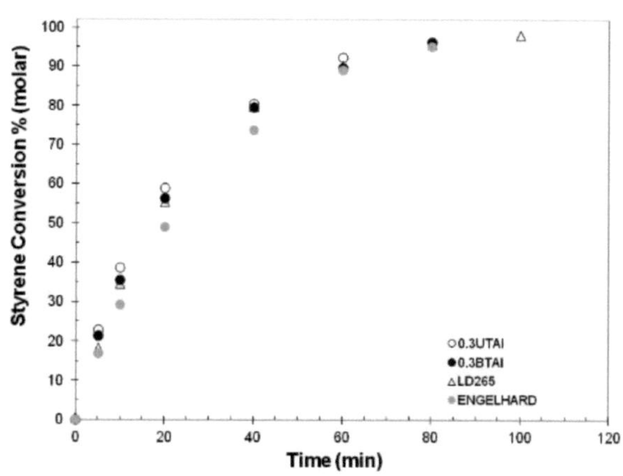

Figure 12. Styrene hydrogenation catalysts with less of 0.5 wt % of Pd. Reaction conditions: toluene solvent, $C^0_{styrene}$ = 0.445 M, 2.0 MPa H_2, 353 K, W_{cat} = 2 g pellet.

It can be seen that for all catalysts the pattern of conversion as a function of time are quite similar, though the catalytic activity is slightly better for the composite 0.3PdUTAl and 0.3PdBTAl catalysts.

5.2.2. *Alkyne hydrogenation*

The partial hydrogenation of alkynes to obtain olefinic compounds is a reaction of industrial and academic interest. It is of big importance for the petrochemical, pharmaceutical, and agrochemical industries. Most of the times the alkyne compounds to be hydrogenated are not raw materials for synthesizing new products but impurities to be removed from a certain process feedstock. The conversion of these alkynes, when the feedstock has a high concentration of alkenes, demands specialized catalysts that can react with the triple bond while keeping intact the double bonds of the alkenes. Moreover, the conversion of alkynes to valuable alkenes is usually desired [36–40].

Figure 13 shows the scheme of reaction of the selective hydrogenation of 1-heptyne, a terminal alkyne, to 1-heptene.The objective of this reaction is to hydrogenate 1-heptyne with a maximum selectivity to the intermediate product 1-heptene. The product of deep hydrogenation, n-heptane, should be completely disfavored.

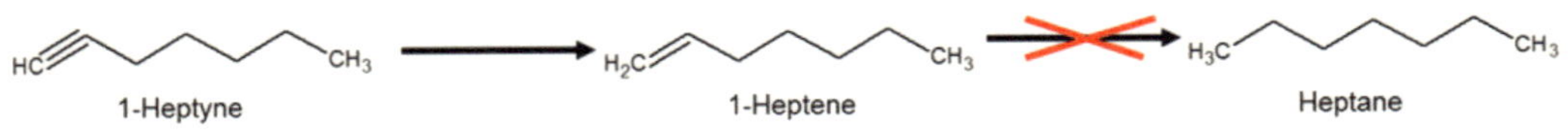

Figure 13. Scheme of the reaction for the selective hydrogenation of 1-heptyne to 1-heptene.

Figure 14 shows the values of conversion of 1-heptyne and selectivities to 1-heptene and n-heptane as a function of time for the catalysts 0.3PdAl, 0.3PdCNR, Lindlar, and 0.3PdUTAl. It can be seen that the egg-shell Pd supported catalyst base on the UTAl support is the most active for hydrogenation. At 3 h,95% conversion of heptyne is seen while the other tested pellet catalysts did not reach 30%.The Lindlar catalysts (0.75 g) with a Pd content of 5% had at 180 min values of conversion similar to 0.3PdUTAl but with selectivities toheptene close to 85%, lower than the selectivity of 0.3PdUTAl. When compared to 0.3PdUTAl, the Lindlar catalyst is disadvantaged as it is supplied in powder form. Once the reaction is finished it must be filtered from the reaction medium with a process that is lengthy and costly due to the high efficiency required for recovering the high priced, high Pd content (5%) Lindlar catalyst. 0.3PdUTAl has a much lower Pd content (0.22%), and is cheaper and has no filtering problems.

Figure 15 shows the scheme of reaction of selective hydrogenation of 3-hexyne, a nonterminal alkyne, to 3-hexene. In this reaction the desired compound is the product of mild hydrogenation, the intermediate 3-hexene, while the product of deep hydrogenation, n-hexane, is undesired.

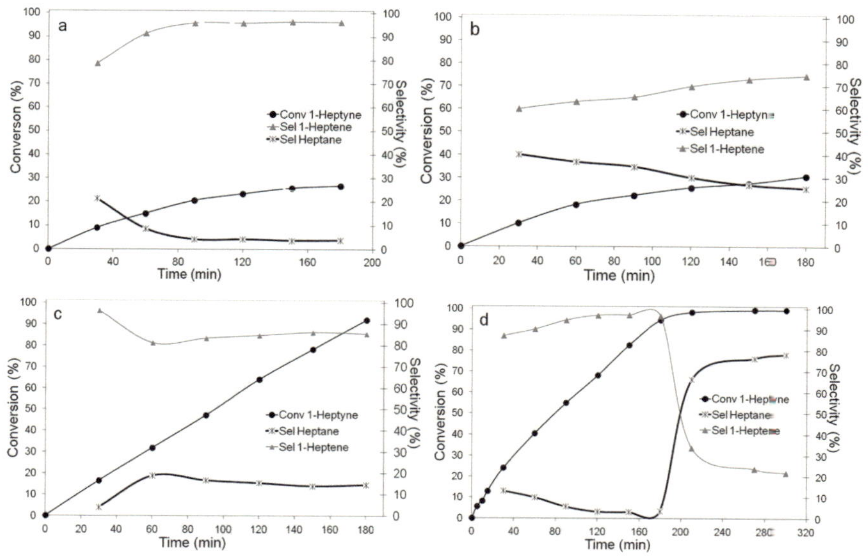

Figure 14. 1-heptyne hydrogenation. (a) 0.3PdAl, (b) 0.3PdCNR, (c) Lindlar, and (d) 0.3PdUTAl.Reaction conditions:toluene solvent, $C^0_{1\text{-heptyne}}$ = 0.350 M, 0.15 MPa H_2, 303 K, W_{Cat} = 0.75 g.

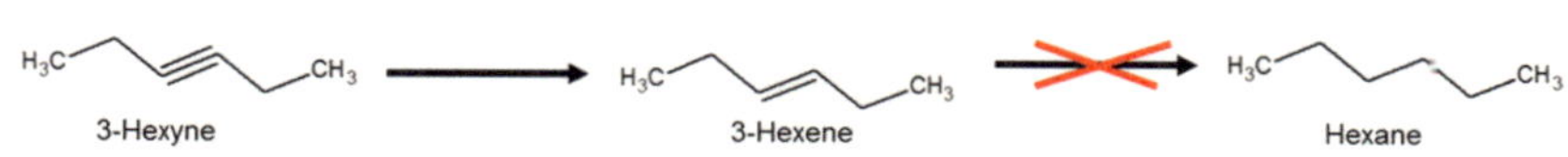

Figure 15. Scheme of the reaction of selective hydrogenation of 3-hexyne to 3-hexene.

The 0.3PdUTAl catalyst was also tried in the hydrogenation of 3-hexyne to 3-hexene. The results of total conversion of 3-hexyne (X_{3HI}) and selectivity to 3-hexene (S_{HE}) and n-hexane (S_{HA}) are plotted in **Figure 16**. Values of 99% and 94% selectivity at 180 min reaction time can be seen. Also, no deactivation of the catalyst was detected.

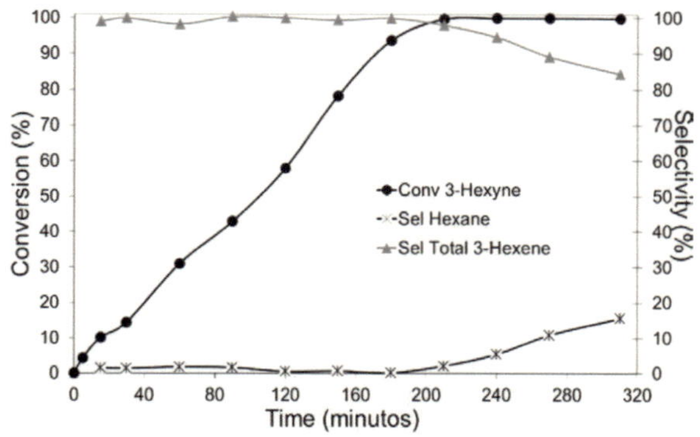

Figure 16. 3-hexyne hydrogenation.0.3PdUTAl catalyst. Toluene solvent, $C^0_{3\text{-hexyne}}$ = 0.350 M, 0.15 MPa H_2, 303 K, W_{Cat} =0.75 g.

5.2.3. 2,3-butanedione hydrogenation

Figure 17 shows the scheme of reaction for the selective hydrogenation of 2,3-butanedione to 3-hydroxybutan-2-one. In this reaction one of the carbonyl groups requires to be hydrogenated with an alcohol group.

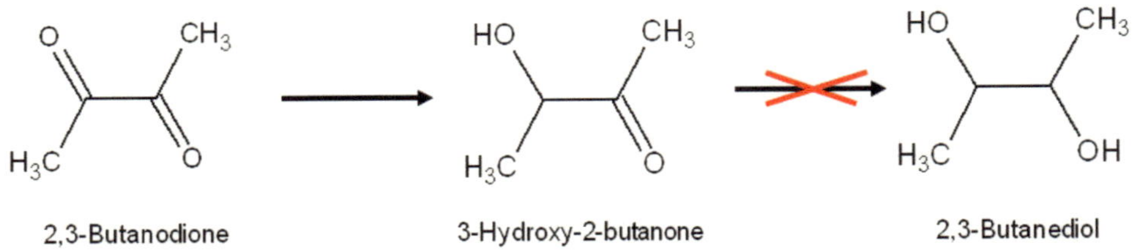

Figure 17. Scheme of reaction for the selective hydrogenation of 2,3-butanedione to 3-hydroxybutan-2-one.

The reaction of hydrogenation of 2,3-butanedione to 3-hydroxybutane-2-one is a reaction of interest for the industry of scents and fragrances [41–43]. **Figure 18** shows the graphs of total conversion of 2,3-butanedione ($X_{2,3BD}$) as a function of time for the catalysts 0.3PdUTAl, 0.3PdBTAl, 1.3PdUTAl, and 1.3PdBTAl. In all cases the selectivity to 3-hydroxybutane-2-one was higher than 98%.

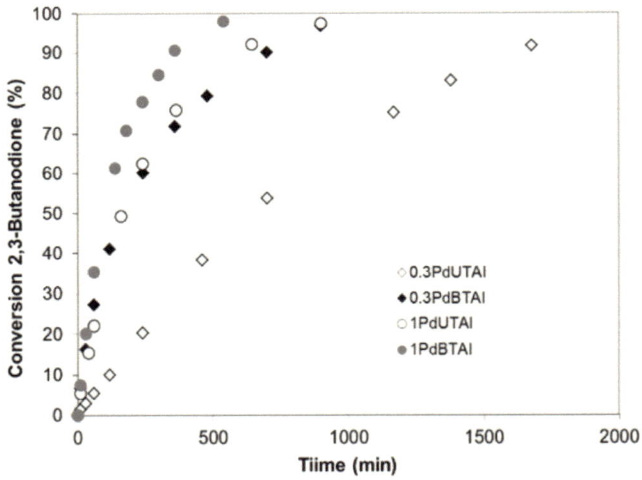

Figure 18. 2,3-butanedione hydrogenation. Reaction conditions: isopropyl alcohol solvent, $C^0_{2,3\text{-butanodione}} = 0.057$ M, 4.0 MPa H_2, 368 K, $W_{Cat} = 2$ g.

All catalysts of Pd supported over composite supports, independent of the kind of polymer used and of the metal content, were active and highly selective for the reaction of interest. Among the composite catalysts series, those based on the BTAl support were more active than those based on the UTAl. This was more evident in the case of the catalyst of low metal concentration.

The catalysts 1PtBTAl and 1PtUTAl (see **Figure 19**) were also tried in this reaction. In these catalysts a decrease is seen in the selectivity to 3-hydroxybutane-2-one due to the appearance of the product of consecutive hydrogenation of 2,3-butanediol. This pattern of lower selectivity of Pt as compared to Pd has already been reported in the literature [44].

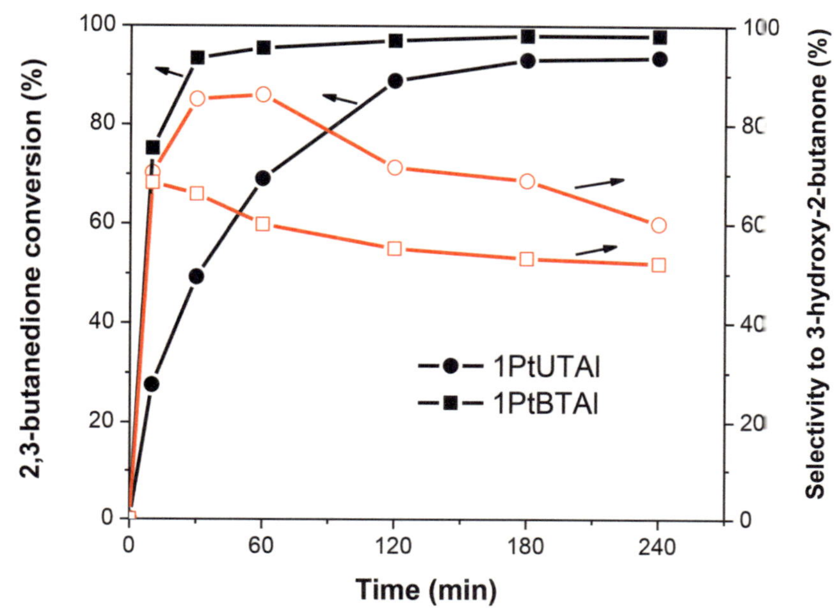

Figure 19. 2,3-butanedione hydrogenation. Reaction conditions: isopropyl alcohol solvent, $C^0_{2,3\text{-butanedione}} = 0.057$ M, 4.0 MPa H_2, 368 K, $W_{Cat} = 2$ g.

5.2.4. Ethyl pyruvate enantioselectivity hydrogenation

The enantioselective hydrogenation of ethyl pyruvate has been an intensively studied research subject. It has been found that the Pt catalysts are the most appropriate for performing this reaction [45, 46], while the Ir, Ru, and Pdcatalysts display lower yields [45, 47]. The Pt-cinchonidine catalysts favor the formation of (R)-ethyl lactate.

Figure 20 shows the scheme reaction for the selective hydrogenation of ethyl pyruvate to (R)-ethyl lactate (using cinchonidine as modifier).In this reaction the objective is to get an enantiomeric excess (ee) of the (R) form over the (S) form.

Figure 21 shows the results of catalytic activity obtained with the catalysts 1PtBTAl and 1PtUTAl employing cinchonidine as quiral modifier. The composite catalysts prepared were found to be active, showing conversion values higher than 90%. In both cases it can be seen that 1PtBTAl is more active than 1PtUTAl.This could be related to the higher availability of active sites. Also, when cinchonidine is used as chiral modifier the enantiomeric excess (ee) of (R)-ethyl lactate achieved was greater for the case of the 1PtBTAl catalyst. This could be related to the size of the metal particles. A big size would permit a better interaction with the chiral modifier.

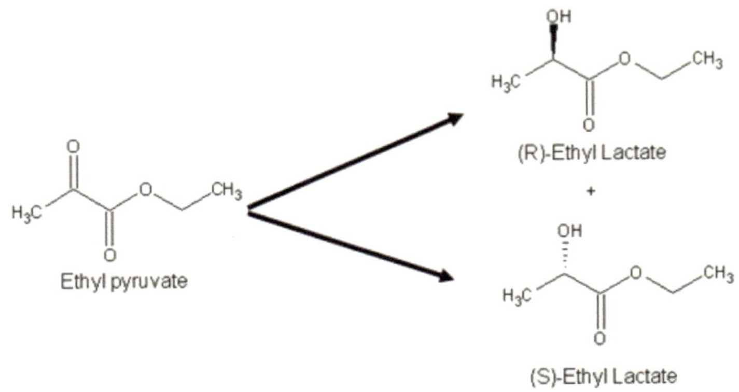

Figure 20. Scheme of reaction for the enantioselective hydrogenation of ethyl pyruvate to (R) or (S) ethyl lactate.

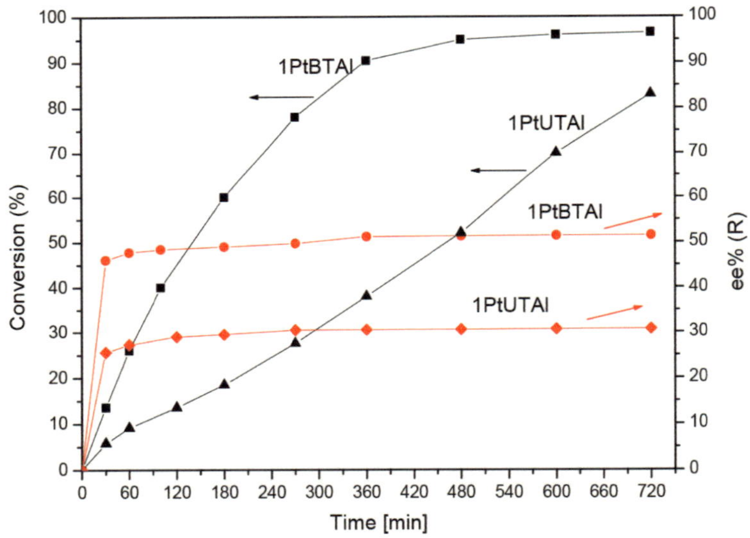

Figure 21. Ethyl pyruvate hydrogenation. Reaction conditions: isopropyl alcohol solvent, $C^0_{Ethyl\ pyruvate}$ = 0.346 M, $C^0_{Cinchonidine}$ = 0.002 M, 2.0 MPa H_2, 298 K, W_{cat} = 0.5g.

6. Conclusions

In the present work a method was presented for the preparation of new organic-inorganic hybrid materials for being used as catalyst support, generally referred to in this work as "composites."BTAl and UTAl composites are formed by a combination of inorganic material (alumina) and an organic one (polymer).The organic phase was obtained by polymerization of organic molecules of functionality 2: bisphenol Aglycerolatedimethacrylate, diurethanedimethacrylate, and triethylene glycol dimethacrylate. The inorganic phase in the form of a powder was mixed with the monomers and a heat starter of polymerization (benzoyl peroxide).The mixture was extruded and then polymerized by applying heat. Thus, the obtained rigid material was used as a support for the preparation of supported noble metal catalysts.

The composites display the combined action of a hydrophilic material (alumina) and a hydrophobic one (polymer). This is an ideal combination for the preparation of catalysts with small thicknesses of active metal surface layer. The combination of these dual properties makes the process of preparation of metal/composite catalysts easier, faster, cheaper, and more repetitive, in comparison to the preparation of catalysts of metals supported over common supports. Conventional methods using these common supports need a strict control over the viscosity, contact time, and pH of the impregnating solution, the temperature of impregnation, drying, and calcination. **Figure 22** shows a graphical description of composites catalysts synthesis procedure.

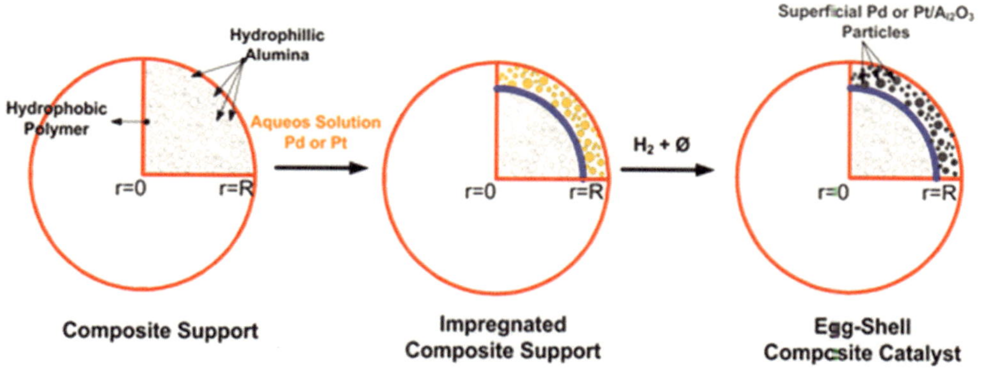

Figure 22. Graphical description of composites catalysts synthesis procedure.

Egg-shell supported metal catalysts have advantages over the catalysts with other metal distributions. They have a lower intraparticle mass diffusion resistance and enable a better control of the temperature on the surface of the catalyst, thus enabling an overall better control of the reaction and the reactor.

The results of hydrogenation of styrene using pellet catalysts and powder catalysts containing more than 1 wt% Pd showed that the composites UTAl and BTAl had higher values of the effectiveness factor (η) in comparison to the rest of the catalysts. These results can be attributed to the small thickness of the metal layer formed over the surface of the composite support, lower than the thickness in the other catalysts prepared using conventional supports as alumina or an activated carbon.

The Pd/composite catalysts of low metal content were compared to other commercial samples, LD265 (Axens) and ENGELHARD, in the reaction test of styrene hydrogenation. The Pd/

composite catalysts had slightly better catalytic properties than the commercial Pd catalysts, a fact pointing to the possibility of industrial use of these new materials.

In the tests of selective hydrogenation of terminal alkynes to terminal alkenes, the activity of the catalysts 0.3PdAl, 0.3PdCNR, 0.3PdUTAl, and Lindlar were evaluated using 1-heptyne as reactant. It was seen that at 180 min, Lindlar and 0.3PdUTAl achieved total conversion of the alkyne, with selectivity to 1-heptene of 85% in the case of the Lindlar catalyst and 97% in the case of 0.3PdUTAl. The 0.3PdAl and 0.3PdCNR catalysts suffered in this strong reaction deactivation. This was more noticeable from 1-heptyne conversion values of 10–20%.The 0.3PdUTAl catalyst had a similar activity as the commercial Lindlar catalyst. Some advantages of the new materials were evident. The Lindlar catalyst was available in powder form and was very expensive due to the high metal content (5 wt% Pd).The Pd/composites were pelletized and only had 0.25 wt% Pd content.

The good results of conversion and selectivity obtained during the reaction of hydrogenation of 3-hexyne, a nonterminal alkyne, to 3-hexene, over the 0.3PdUTAl catalyst, indicates that these composite catalysts can also be used in reactions comprising consecutive steps.

The reactions of hydrogenation presented were also favored by the use of composite supports because of their lower acid strength. This is an advantage for suppressing undesirable reactions, such as the oligomerization of alkenes and alkynes, leading to the formation of carbon deposits or gums over the catalysts. These and other reactions affect the selectivity and lifetime of the catalysts.

In the case of the reaction of synthesis of 3-hydroxybutane-2-one from 2,3-butanedione, which is of commercial interest for the perfume industry, it was seen that the Pd/composite catalysts could be used with values of conversion and selectivity close to 100%.The use of composites in the slurry reactors enabled sparing the filtering of the liquid phase at the end of the reaction. The difference in activity found between the PdBTAl and PdUTAl catalysts could be due to the differences in metal dispersion or to differences in the electronic properties of the surface Pd species.

During the enantioselectivity hydrogenation reaction of ethyl pyruvate using cinchonidineas a quiral modifier, both PtUTAl and PtBTAl composite catalysts were active and selective to (R)-Ethyl Lactate. PtUTAl catalyst had higher total conversion than PtBTAl but had less enantiomeric excess (ee). This activity and enantioselectivity behavior could be related, at least partly due to the electronic effects of the presence of chloride species that prevents the adsorption of the quiral modifier causing a decrease in the enantiomeric excess (ee), although geometrical effects of $Pt\delta^{+}O_{x}Cl_{y}$ species could not be discarded.

The advantage of the egg-shell catalysts that makes them to be preferred over other catalyst types is that they have smaller intraparticle mass transfer limitations. Having most of the metal phase on the outer surface layer, makes also the heat transfer more efficient, enabling a better control of the reaction and the reactor temperature.

A last advantage is related to the mechanical properties. The composites displayed higher values of diametrical and longitudinal resistance than alumina and silica and also suffered less attrition. This translates into a higher crush resistance when used in packed beds and a lower tendency to the formation of fines when used in slurry stirred reactors.

In summary, it can be said that the composite catalysts synthesized in this work had an egg-shell metal distribution with very small metal surface layer thickness and proved to be superior to common metal catalysts prepared with commercial supports. This was attributed to a combination of the following properties: good mechanical resistance, simplicity of preparation procedure for support and catalyst, small thickness of the active phase, uniform distribution of the metal over the support surface, ease of preparation procedure scale-up, and possibility of achieving any desired pellet shape.

Acknowledgements

The experimental assistance of F. Coloma-Pascual is greatly acknowledged. The financial assistance of CAI+D (UNL), CONICET (PIP 410), and ANPCyT are also acknowledged.

Author details

Nicolás Carrara[1], Juan Manuel Badano[1*], Carolina Betti[1], Cecilia Lederhos[1], Mariana Busto[1], Carlos Vera[1,2] and Mónica Quiroga[1,2]

*Address all correspondence to: jbadano@fiq.unl.edu.ar

1 Institute of Catalysis and Petrochemistry Research INCAPE (UNL-CONICET), Santa Fe, Argentina

2 Faculty of Chemical Engineering, National University of the Littoral, Santa Fe, Argentina

References

[1] Moulijn JA, van Leeuwen PWNM, van Santen RA, editors. Studies in Surface Science and Catalysis, Vol 79. 1st ed. The Netherlands: Elsevier; 1993, 465 p.

[2] Hagen J. Industrial Catalysis: A Practical Approach. 3nd ed. Weinheim: Wiley-VCH Verlag GmbH&Co. KGaA; 2006, 513 p. DOI: 10.1002/3527607684.ch1

[3] Patent US 3969273.Inv: Brown SM, Wallace DN; W.R. Grace & Co; 1976.

[4] Patent US 4120826.Inv: Ebel RH, Spitzer DP, Bambrick WE; American Cyanamid Company; 1978.

[5] Patent US 4810688.Inv: Ewert WM, Kubicek DH, Drake CA; Pillips Petroleum Company; 1989.

[6] Patent US 5118901.Inv: Drake CA; Pillips Petroleum Company; 1992.

[7] Patent US 5202298.Inv: Schubert PF, Bonnell RE, Freeman NL, Fentress DC, Mitchell KE, Lowrey RE, Kubicek DH, Ewert WM; Pillips Petroleum Company; 1993.

[8] Patent US 4051072.Inv: Burton RE, Berg M; General Motors Corporation; 1977.

[9] Patente US 7288686.Inv: Ryu JY; Catalytic Distillation Technologies; 2007.

[10] Sheng-Yi L, Rutherford A. The distribution of active ingredients in supported catalysts prepared by impregnation. Catal. Rev. Sci. Eng. 1985; 27: 207–340.

[11] Ismagilov IZ, Ekatpure RP, Tsykoza LT, Matus EV, Rebrov EV, de Croon MHJM, Kerzhentsev MA, Schouten JC. Optimization of anodic oxidation and Cu–Cr oxide catalyst preparation on structured aluminum plates processed by electro discharge machining. Catal. Today .2005; 105: 516–528. DOI: 10.1016/j.cattod.2005.06.053.

[12] Gulková D, Kaluža L, Vít Z, Zdražil M. Preparation of MoO3/MgO catalysts with eggshell and uniform Mo distribution by methanol assisted spreading: Effect of MoO3 dispersion on rate of spreading. Catal. Commun. 2006; 7: 276–280. DOI: 10.1016/j.catcom.2005.10.017.

[13] Zhuang YQ, Claeys M, van Steen E. Novel synthesis route for egg-shell, egg-white and egg-yolk type of cobalt on silica catalysts. Appl. Catal., A. 2006; 301: 138–142. DOI: 10.1016/j.apcata.2005.11.029.

[14] Alves JA, Bressa SP, Martínez OM, Barreto GF. Kinetic study of the liquid-phase hydrogenation of 1-butyne over a commercial palladium/alumina catalyst. Chem. Eng. J. 2007; 125: 131–138. DOI: 10.1016/j.cej.2006.08.006.

[15] Drelinkiewicza A, Pukkinen A, Kangas R, Laitinen R. Hydrogenation of 2-ethylanthraquinone over Pd/SiO2 and Pd/Al2O3 in the fixed-bed reactor.The effect of the type of support. Catal. Lett. 2004; 94: 157–170. DOI: 10.1023/B:CATL.0000020540.90536.6d.

[16] Lepage JE, Corty P, Freund E, Franck JP, Jacquin Y, Juguin B, Marcilly C, Martino G, Miquel J, Montarnal R, Sugier A, Landeghem HV. Applied Heterogeneous Catalysis Design Manufacture Use of Solid Catalysts. Institute Francais du Petrole, Editions Technip: Paris; 1987. 329 p.

[17] Iglesia E, Soled SL, Baumgartner JE, Reyes SC. Synthesis and catalytic properties of eggshell cobalt catalysts for the Fischer-Tropschsynthesis. J. Catal. 1995; 153: 108–122. DOI: 10.1006/jcat.1995.1113.

[18] Takehira K, Shishido T, Shoro D, Murakami K, Honda M, Kawabata T, Takaki K. Preparation of egg-shell type Ni-loaded catalyst by adopting "Memory Effect" of Mg–Al hydrotalcite and its application for CH4 reforming. Catal. Commun. 2004; 5: 209–213. DOI: 10.1016/j.catcom.2004.02.004.

[19] Takehira K, Kawabata T, Shishido T, Murakami K, Ohi T, Shoro D, Honda M, Takaki K.Mechanism of reconstitution of hydrotalcite leading to eggshell-type Ni loading on Mg single bond Al mixed oxide. J. Catal. 2005; 231: 92–104. DOI: 10.1016/j.jcat.2005.01.025.

[20] Qiu Y, Chen J, Zhang J. Influence of nickel distribution on properties of spherical catalysts for partial oxidation of methane to synthesis gas. Catal. Commun. 2007; 8: 508–512. DOI: 10.1016/j.catcom.2006.08.003.

[21] Galarraga C, Peluso E. Eggshell catalysts for Fischer–Tropsch synthesis: Modeling catalyst impregnation. Chem. Eng. J. 2001; 82: 13–20. DOI: 10.1016/S1385-8947(00)00352-1.

[22] Lekhal A, Glasser BJ, Khinast JG. Impact of drying on the catalyst profile in supported impregnation catalysts. Chem. Eng. Sci. 2001; 56: 4473–4487. DOI: 10.1016/S0009-2509(01)00120-8.

[23] Lin Q, Ji Y, Jiang Z-D, Xiao W-D. Effects of precursors on preparation of Pd/α-alumina catalyst for synthesis of dimethyl oxalate. Ind. Eng. Chem. Res. 2007; 46: 7950–7954. DOI: 10.1021/ie070640b.

[24] Patent WO 00/72961; Inv: Hu Y, Dent AL. 2000.

[25] Patent US 6355596. Hu Y, Dent AL. 2002.

[26] Patent US 4051072. Burton RE, Berg M. 1977.

[27] Patent US 7288686. Ryu JY. 2007.

[28] Patent US 7462751. Lowe DM, Molinier M. 2008.

[29] L. Siqin, M. Xiaotang. Patent application publication US 2001/0056036 A1. 2001.

[30] Gavriilidis A, Varma A, Morbidelli M. Optimal distribution of catalyst in pellets. Catal. Rev. Sci. Eng. 1993; 35: 399–456. DOI: 10.1080/01614949308013912.

[31] Badano JM, Betti C, Rintoul I, Vich-Berlanga J, Cagnola E, Torres G, Vera C, Yori J, Quiroga M. New composite materials as support for selective hydrogenation; egg-shell catalysts. Appl. Catal. A Gen. 2010; 390: 166–174. DOI: 10.1016/j.apcata.2010.10.008.

[32] Carrara N, Badano JM, Betti C, Lederhos C, Rintoul I, Coloma-Pascual F, Vera C, Quiroga M. Selective hydrogenation by novel composite supported Pd egg-shell catalysts. Cat. Comm. 2015; 61: 72–77. DOI: 10.1016/j.catcom.2014.12.012.

[33] Carrara N, Badano J, Bertero N, Torres G, Betti C, Martínez-Bovier L, Quiroga M, Vera C. Kinetics of the liquid phase selective hydrogenation of 2,3-butanedione over new composite supported Pd catalysts. Chem. Technol. Biotechnol. 2014; 89: 265–275. DOI: 10.1002/jctb.4113.

[34] Badano J, Quiroga M, Betti C, Vera C, Canavese S, Coloma-Pascual F. Resistance to sulfur and oxygenated compounds of supported Pd, Pt, Rh, Ru catalysts. Catal. Lett. 2010; 137: 35–44. DOI: 10.1007/s10562-010-0336-x.

[35] Badano J, Lederhos C, Quiroga M, L'Argentière P, Coloma-Pascual F. Low metal loading catalysts used for the selective hydrogenation of styrene. Quim. Nova. 2009; 33: 48–51. DOI: 10.1590/S0100-40422010000100010.

[36] Lederhos CR, L'Argentière PC, Fígoli NS. 1-heptyne selective hydrogenation over Pd supported catalysts. Ind. Eng. Chem. Res. 2005; 44: 1752–1756. DOI: 10.1021/ie040187t.

[37] Lederhos CR, L'Argentière PC, Coloma-Pascual F, Fígoli NS. A study about the effect of the temperature of hydrogen treatment on the properties of Ru/Al2O3and Ru/C and their catalytic behavior during 1-heptyne semi-hydrogenation. Catal. Lett. 2006; 110: 23–28. DOI: 10.1007/s10562-006-0085-z.

[38] Lederhos CR, Badano JM, Quiroga ME, Coloma-Pascual F, L'Argentière PC. Influence of Ni addition to a low-loaded palladium catalyst on the selective hydrogenation of 1-heptyne. Quim. Nova. 2010; 33: 816–820. DOI: 10.1590/S0100-40422010000400010.

[39] Lederhos CR, Maccarrone MJ, Badano JM, Coloma-Pascual F, Yori JC, Quiroga ME. Hept-1-yne partial hydrogenation reaction over supported Pd and W catalysts. Appl. Catal. A. 2011; 396: 170–176. DOI: 10.1016/j.apcata.2011.02.011.

[40] Maccarrone MJ, Torres G, Betti C, Lederhos CR, Coloma-Pascual F, Quiroga ME, Yori JC. Partial hydrogenation of 3-hexyne over low-loaded palladium mono and bimetallic catalysts. Appl. Catal. A. 2012; 441: 90–98. DOI: 10.1016/j.apcata.2012.07.016.

[41] Bartowsky EJ, Henschke PA. The 'buttery' attribute of wine-diacetyl-desirability, spoilage and beyond. Int. J. Food Microbiol. 2004; 96: 235–252. DOI: 10.1016/j.ijfood-micro.2004.05.013.

[42] Calbert HE, Price WV. A study of the diacetyl in cheese. I. Diacetylcontent and flavor of Cheddar cheese. J. Dairy Sci. 1949; 32: 515–520. DOI: 10.3168/jds.S0022-0302(49)92074-3.

[43] Whittaker P, Clarke JJ, San RHC, Begley TH, Dunkel VC. Evaluation of the butter flavoring chemical diacetyl and a fluorochemical paper additive for mutagenicity and toxicity using the mammalian cell gene mutation assay in L5178Y mouse lymphoma cells. Food Chem. Toxicol. 2008; 46: 2928–2933. DOI: 10.1016/j.fct.2008.06.001.

[44] Rylander PN. Catalytic Hydrogenation in Organic Syntheses. Academic Press Edition. New York. 1979. 90 p. DOI: 10.1016/B978-0-12-605355-5.50016-1

[45] Blaser HU, Jalett HP, Monti DM, JReber JF, Wehrli JT. Modified heterogeneous platinum catalysts for the enantioselectivehydroge-nation of α-ketoesters. Stud. Surf. Sci. Catal. 1988; 41: 153–163. DOI:10.1016/S0167-2991(09)60810-7.

[46] Wehrli JT, Baiker A, Monti DM, Blaser HU. Enantioselective hydrogenation of α-ketoesters: Preparation and catalytic behavior of different alumina-supported platinum

catalysts modified with cinchonidine. J. Mol. Catal. 1990;61: 207–226. DOI: 10.1016/0304-5102(90)85156-C.

[47] MeheuxPA, Ibbotson A, Wells PB. Enantioselective hydrogenation: II. Variation of activity and optical yield with experimental variables in methyl pyruvate hydrogenation catalyzed by cinchona-modified platinum/silica (EUROPT-1). J. Catal. 1991;128: 387–396. DOI:10.1016/0021-9517(91)90297-H.

Chapter 10

Effect of Absorption and Desorption of Hydrogen in Ti and Ti Alloys

Alejandra López-Suárez

Additional information is available at the end of the chapter

http://dx.doi.org/10.5772/64921

Abstract

A study of the hydrogen (H) storage capacity of pure Ti and the Ti-6Al-4V alloy is presented. The importance of an accurate quantification of oxygen and hydrogen in the materials used for hydrogen storage, the activation process and the cyclic hydrogenations, the effect of surface quality during hydrogen uptake, the improvement of hydrogen absorption using ion irradiation, and the hydrogen desorption are some topics discussed in this chapter.

Keywords: metal hydrides, hydrogen storage, absorption, desorption, Ti, Ti-6Al-4V, ERDA, RBS

1. Introduction

The history of civilization in terms of energy goes from the use of wood and coal to the use of oil and natural gas. By mid-twentieth century, oil became the world leader in the energy field. Notwithstanding the great advantages of the hydrocarbon fuel, it is a non-renewable fossil resource, which is running out and its extraction is becoming more expensive. Besides, their excessive use has led to serious problems of global pollution, which have brought severe climate change on Earth. For this reason, the next step in the energy scale could be the hydrogen (H).

Hydrogen is the most abundant element in nature and can be used as fuel in liquid or gaseous form, but its storage is risky and expensive. However, these drawbacks hydrogen has an alternative solution for storage: metal hydrides (MH), which are chemical compounds formed by a metal (M) and hydrogen (H). One of the main advantages of the use of metal hydrides as energy sources is that unlike the non-renewable ones, they do not cause pollution when burned, because its by-product is water. For this reason, hydrogen can be considered as a clean fuel.

Hydrogen can be stored in fuel cells to produce electricity and can be used in applications as diverse as transportation or electronics industry in batteries for laptops and cell phones applications. Undoubtedly, one of the most important challenges that the hydrogen-based economy has is its storing. In order to use hydrogen as a fuel, it needs to be safely stored in a medium that allows absorption, storage, and desorption, besides an easy transport. There are several ways of storing hydrogen, among them are metal hydrides, hydrides ceramics, carbon-based materials, etc.

As a consequence of the so called "Oil Crisis," in the 1970s, metal hydrides started to be considered as good candidates for using in hydrogen energy storage, due to their large capacity to accommodate an extremely high density of hydrogen in their structures. It is possible to pack more hydrogen into a metal hydride than into the same volume of liquid. The reason is that when a metal that forms a hydride is brought in contact with gaseous hydrogen, the hydrogen molecules are adsorbed onto the surface of the material. If enough energy is given to the system, the hydrogen molecules can dissociate into hydrogen atoms, which tend to enter to the crystal lattice of the metal and occupy interstitial sites. As the energy given to the system increases, hydrogen atoms are forced into the crystal until the metal becomes saturated with hydrogen. At this stage, the material goes into a new phase: the metal hydride, which allows the material to absorb hydrogen in larger amounts.

2. Metal hydrides for hydrogen storage

Metal hydrides, such as palladium, zirconium, or titanium, form an important group of reversible-sorbing materials. The main advantage of metal hydrides over other hydrogen storage media, such as activated carbon, carbon nanotubes, or zeolites, is that they can be reversible hydride, and they can also release hydrogen of the metal hydrides in an extremely pure way, which is an important factor when considering hydrogen for mobile applications [1].

The process of hydrogen absorption (or desorption) by metals is reversible and involves the surface and the bulk of the material, through several steps. In the gas phase, the reaction involves the following mechanism:

1. Surface physisorption (desorption) of molecular hydrogen (H_2). In this step, the hydrogen molecules adsorb (desorb) at (from) the surface of the metal (M).

$$M + H_2 \leftrightarrow M - H_2$$

2. Surface dissociative chemisorption (recombination). This leads to the formation (desorption) of atomic hydrogen at the surface of the metal.

$$2M - H_2 \leftrightarrow 2M - H$$

3. Surface absorption (desorption). During this step, the adsorbed atomic hydrogen is absorbed (desorbed) by (from) the metal in its sub-surface.

$$H_{ad} \leftrightarrow H_{ab}$$

4. Diffusion. The atomic hydrogen is transported from subsurface to the bulk region.

$$H_{sub} \leftrightarrow H_{bulk}$$

5. Phase transformation. Leads to the precipitation (dissolution) of the metal hydride phase.

$$H_{bulk} \leftrightarrow MH$$

Pure titanium is considered to be an element with a high affinity to hydrogen; however, high temperature has to be used in order to absorb hydrogen in this material [2]. To avoid this inconvenience, some titanium alloys, such as TiFe, Ti_2Ni, $TiMn_2$, or Ti-6Al-4V, have attracted interest for storage hydrogen because they can absorb and release hydrogen in large amounts and at lower temperature than pure titanium.

The maximum hydrogen capacity of the material depends on the number of hydrogen sites that are available for hydrogen occupation. For instance, the Ti-6Al-4V alloy is composed by a biphasic structure: the beta phase (Body-centered cubic –BCC-structure) surrounding the alpha phase [hexagonal closed packed (HCP) structure]. The fact that this alloy has a BCC structure eases the hydrogen absorption, due to the way hydrogen atoms enter the crystalline structure and fit in the interstitial sites of the crystal. In the case of the beta-phase of the Ti-6Al-4V alloy, the BCC structure has 6 octahedral sites and 12 tetrahedral sites per unit cell where hydrogen can fit. The octahedral and tetrahedral sites in a BCC structure are three times more than the ones found in the HCP and face-centered cubic (FCC) lattices [3]. It has also been observed that the rate of hydrogen diffusion in a BCC structure is several orders of magnitude higher than in a HCP or a FCC structure [4]. **Figure 1** shows the scanning electronic microscopy (SEM) micrographs of (a) pure titanium and (b) the Ti-6Al-4V alloy. Pure titanium has a hexagonal closed packed (HCP) crystalline structure; meanwhile, the Ti-6Al-4V alloy shows an alpha phase (HCP) composed by Ti and Al, which looks as a dark zone in **Figure 1b**, surrounded by the beta phase (BCC) composed by Ti and V.

2.1. The importance of an accurate quantification of oxygen and hydrogen in the materials used for hydrogen storage

The initial mechanisms of hydrogen absorption depend on the crystalline structure of the metal, but also on its surface quality. The uptake rate of hydrogen can be reduced by a combination of metal oxides and hydroxides on the metal surface, acting as a diffusion barrier [5]. In this way, accurate oxygen and hydrogen measurements must be conducted in order to correlate the storage capacity of the metal with sample oxidation. In the literature, several techniques are used to detect oxygen and hydrogen in materials; nevertheless, these methods are destructive and have limitations for the determination of concentrations and depth profiles. The ion beam analysis (IBA) techniques, such as Rutherford backscattering spectrometry (RBS) [6] and elastic recoil detection analysis (ERDA) [7], are non-destructive nuclear techniques that allow performing a very accurate measurement of the concentration and depth profile of oxygen and hydrogen, respectively.

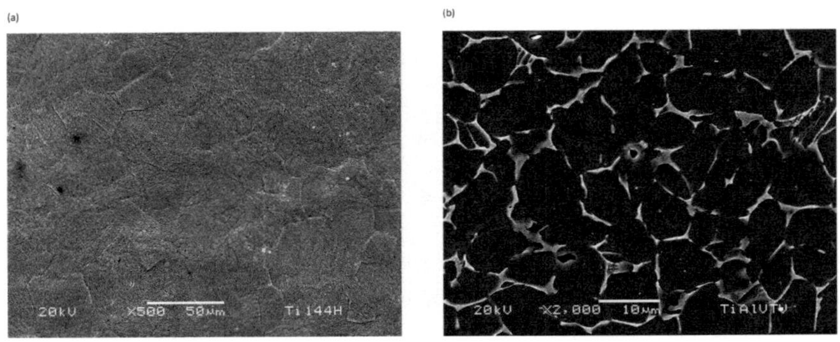

Figure 1. SEM micrographs of (a) pure titanium and (b) Ti-6Al-4V alloy [8].

The physical principle of RBS consists of impinging a beam of collimated and monoenergetic particles (usually light ions such as hydrogen or helium) on the material to be analyzed. As the result of the interaction, part of the energy of the incident particle is transferred to the nucleus of the atom in the sample; so that the backscattered particle contains information of the target, because the reduction in energy of the incident particle depends on the mass of the projectile and the nucleus.

The ERDA technique is used to quantify the concentration of light elements, such as hydrogen and carbon. During an ERDA experiment, a beam of collimated and monoenergetic ions heavier than those who are to be detected, impinges on a material. The projectiles collide elastically with the nuclei of the sample, and as they are lighter than the projectiles, they acquire enough energy to leave the material. Simultaneously with the light ions, heavy projectiles could leave the sample and can reach the detector. For this reason, it is essential to place an absorber of Mylar or aluminum before the detector in order to stop the heavier projectiles and only let to lighter recoils to reach it.

In both techniques, the element that will be analyzed is identified by the energy and the type of the emitted particles that comes from the RBS or the ERDA reactions.

In the case of oxygen measurements, an oxygen resonance energy must be used in order to quantify the element concentration. This is because the Rutherford cross section of the oxygen is quite low, causing its RBS signal to be veiled by the signal of other elements present in the material. Oxygen resonances are presented at different energies, being the most common the ones located at 3.045 and 6.585 MeV. These resonances increase the cross section of the element by several times, making possible to improve the oxygen signal and quantify its concentration.

In the next section, the 6.585 MeV oxygen resonance was used during the RBS experiment. This energy allowed obtaining the carbon signal as well, which is usually completely veiled by other elements.

2.2. The activation process and the cyclic hydrogenations

Since the metals or metal alloys used as hydrogen storage materials do not absorb hydrogen at room temperature, they must be activated. Several activation modes, including activation at high temperatures, are attempted in order to obtain a material capable to absorb hydrogen. After the thermal process, the number of paths for hydrogen to diffuse into the material increases, creating micro cracks at the grain boundaries that produces large amounts of clean surface, easing the hydrogen absorption. On the other hand, if the surface of the material is not completely clean or is covered with a surface passivation layer (combination of metal oxides and hydroxides), the material will not absorb until the contaminants or the passivation layer are removed. The activation process in this instance can also occur by increasing the temperature of the material.

The metals that were used as hydrogen storage materials in this chapter are pure titanium and the titanium alloy Ti-6Al-4V, which were manufactured by Goodfellow. The samples were polished and then were ultrasonically cleaned in acetone, followed by rinsing with deionized water. The later cleaning was used in order to remove any kind of impurity from the surface of the sample. Once the cleaning process was completed, the samples were hydrogenated in a 50% hydrogen and 50% argon atmosphere, at 1 atm pressure and a flux of 50 cc/min, during 2 h. The temperature of the hydrogenation process ranged from 150 to 650°C. When the hydride is formed during the first hydrogenation cycle, some micro-cracks are revealed in both metals, as is shown in **Figure 2** [8]. This figure shows a SEM micrograph of a Ti-6Al-4V alloy sample taken after the activation process. These micro-cracks are the result of the stress induced at the grain boundaries of both materials, since the density of the hydride is less than the metal. As a consequence of the hydrogenation reaction, the volume of the grains increases producing cracking in the grains and exposing new surfaces. This process is called the activation of the material. After activation, the particles have a higher ratio of surface area to volume than before the hydrogenation process, so the metals are able to react easily with hydrogen [9].

The activation process also reduces the metal oxides and the hydroxides deposited on the metal surface. This behavior is observed in **Figure 3**, where two RBS spectra of a Ti-6Al-4V alloy are shown before (**Figure 3a** [8]) and after (**Figure 3b**) activation. The RBS spectra show the normalized yield or number of counts of the elements present in the sample versus energy. During the RBS experiment a collimated 6.585 MeV alpha particle beam was used to measure the amount of oxygen in the samples. The 6.585 MeV oxygen resonance was used in order to improve the sensitivity of the element during the measurement. The resonances are regions where the scattering cross section is enhanced over the Rutherford cross section at the same energy and are very useful for measuring light elements, such as oxygen and carbon. After comparing the spectra of **Figure 3**, a high reduction in the oxygen concentration after activation is clearly observed. This result confirms that activation process plays an important role in cleaning the surface of the metals and preparing them for future hydrogenations.

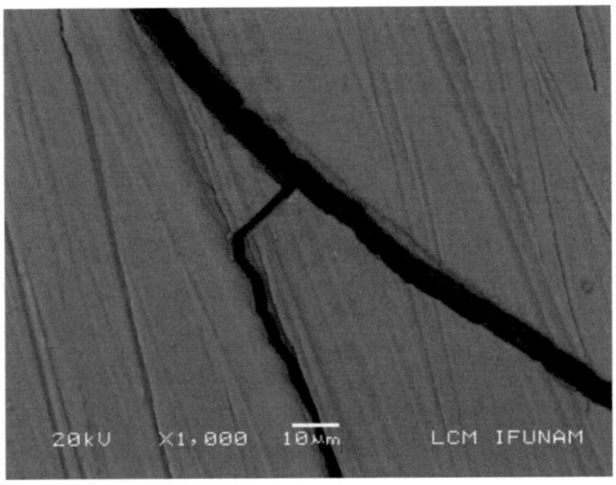

Figure 2. SEM micrograph of a Ti-6Al-4V alloy sample after activation [8].

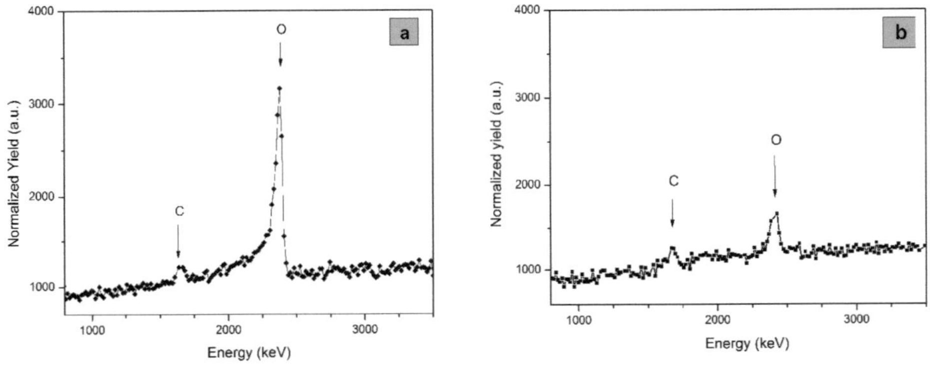

Figure 3. RBS spectra of Ti-6Al-4V alloy before (a) [8] and after (b) activation. The oxygen concentration is reduced considerably after the activation process.

Once the samples are activated, the hydrogen concentration is measured using the ERDA technique. In order to measure the hydrogen content, the samples were irradiated with a collimated 3 MeV alpha particle beam. The surface of the samples was placed at an angle of 15° with respect to the incoming beam; meanwhile a surface barrier detector was placed at an angle of 30° to the beam direction in order to detect the hydrogen recoils. A 12 μm Mylar foil was placed in front of the detector to stop elastically scattered ions heavier than the recoil hydrogen. Equation (1) was used to calculate the hydrogen concentration before and after the hydrogenation process. This equation relates the yield (Y_r) of the recoil atoms detected at channel E_d with channel width δE_d, with the atomic density of the recoil hydrogen atom (N_r) at the depth x.

$$Y_r = \frac{QN_r\sigma_r\Omega\delta E_d}{cos\theta_1\delta E_d dE_d / dx} \tag{1}$$

where Q is the incident projectile fluence, σ_r is the recoil differential scattering cross section, Ω is the detector solid angle, and dx is the increment of depth at x corresponding to an increment in energy dE_d and θ_1 is the incident angle.

The scattering cross section is given by the equation:

$$\sigma_r = \frac{\left[Z_1 Z_2 e^2 \left(M_1 + M_2\right)\right]^2}{\left[2M_2 E_0\right]^2 cos^3\phi} \tag{2}$$

where Z_i and M_i are the atomic number and the atomic mass of the i element respectively, E_0 is the incident energy of the projectile, and φ is the recoil angle.

In order to study the behavior of the activation temperature in pure titanium and the titanium alloy Ti-6Al-4V, the hydrogen doping was performed at temperatures ranging from 150 to 650°C. According to Eq. (1), the hydrogen concentrations in pure titanium and the Ti-6Al-4V alloy are shown in **Figure 4** [2]. In both cases, the ERDA results show an increment in hydrogen concentration at temperatures higher than 550°C. The figure shows that hydrogen is not absorbed until a threshold temperature is reached, which is the activation temperature. In pure titanium, this threshold temperature is closed to 550°C, but in the alloy, it is located between 550 and 600°C. It is also observed in **Figure 4** that above this temperature, the hydrogen concentration reaches a value close to 3×10^{22} H atoms/cm^3 in both materials, which is the maximum concentration that the material is able to get after activation. It will be noticed later that in both materials this hydrogen concentration can easily increase during cyclic hydrogenations.

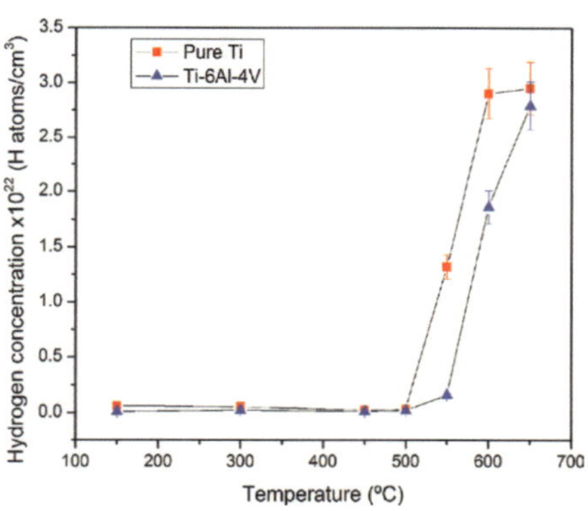

Figure 4. Hydrogen concentration versus hydrogenation temperature for pure Ti and the Ti-6Al-4V alloy [2].

Once the samples are activated, they can be cyclically hydrogenated in order to study the hydrogen absorption of the materials. **Figure 5** shows the ERDA spectra of a Ti-6Al-4V alloy sample hydrogenated three times after activation, where *"hydrogenation 1,"* in **Figure 5**, corresponds to the activation process. After the activation process, the samples were hydrogenated every 2 months and after that, their hydrogen concentration was measured, as it is shown in **Figure 5**. The curve labeled as *"reference sample"* corresponds to a Ti-6Al-4V alloy with any hydrogenation process. The tiny and superficial peak that it shows is due to the hydrogen contained in the material after preparation and before the hydrogenation process. This value was subtracted from the hydrogenated curves in order to obtain an accurate hydrogen concentration. The results in this figure also show that hydrogen content increases after each hydrogenation process, which make us assume that the Ti-6Al-4V alloy can be considered a good material for hydrogen storage purposes.

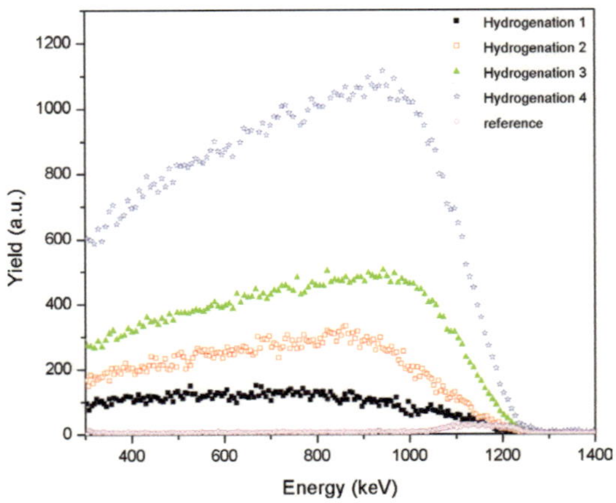

Figure 5. ERDA spectra of Ti-6Al-4V alloy after cyclic hydrogenations [8].

One way to compare the hydrogen storage of the materials is by means of its gravimetric storage capacity (wt%). This amount gives the percentage weight of hydrogen relative to the metal. The gravimetric storage capacity of Ti and Ti-6Al-4V alloy samples after each hydrogenation cycle was calculated by using Eq. (3).

$$wt = \frac{wt(\mathrm{H})}{wt(\mathrm{H+M})} \times 100 \tag{3}$$

where H and M represent the hydrogen atoms and all the metal atoms in the material, respectively.

Table 1 shows the gravimetric storage capacity of Ti and Ti-6Al-4V alloy after cyclic hydrogenations [8]. One can appreciate that the storage capacity increases after each cycle and that after activation (first hydrogenation) the storage capacity of both materials increases considerably. After the second cycle, the increment of hydrogen in the metal and the alloy is lower, but it continues increasing. According to **Figure 2** and the titanium SEM micrographs not presented in this chapter, we can assure that after activation, both materials present microcracks that ease the hydrogen diffusion through the materials, increasing the storage capacity after each hydrogenation cycle.

Hydrogenation cycle	Gravimetric storage capacity wt% (Ti)	Gravimetric storage capacity wt% (Ti-6Al-4V)
First (activation)	3.02 ± 0.21	2.93 ± 0.20
Second	3.39 ± 0.23	3.46 ± 0.24
Third	3.49 ± 0.24	3.55 ± 0.25
Fourth	3.61 ± 0.25	3.64 ± 0.25

Table 1. Gravimetric storage capacity of Ti and Ti-6Al-4V after cyclic hydrogenations [8].

On the other hand, **Table 2** presents the gravimetric storage capacity of some other materials used as hydrogen storage media [10]. According to the data of **Table 2**, the storage capacity of our materials after the fourth hydrogenation, has similar values to those materials that are commercially used as hydrogen storage materials. This fact makes us think that pure titanium and the Ti-6Al-4V alloy are good candidates as storage materials.

Storage material	Gravimetric storage capacity (wt%)
LiH	12.7
Mg_2NiH_4	3.5
$NaAlH_4$	3.7
MgH_2	7.7
V(0.85)Ti(0.1)Fe(0.05) + H	3.7
$Li_2B_{12}H_{12}$	6.5
Na_3AlH_6	1.9

Table 2. Gravimetric storage capacity of different materials used as hydrogen storage materials [10].

2.3. The improvement of hydrogen absorption by using ion implantation

Pure titanium is an element with a high affinity to hydrogen; however, it reacts with hydrogen at high temperatures, so in recent years some investigations have been carried out in order to improve the activation properties of this and other metals used as hydrogen storage materials. These methods include glow discharge in Al [11], ion mixing in TiFe [12], ion implantation in Ti [13], Pd [14], and MgH_2 [15].

The implantation process goes back to the nineteenth century and has been continually refined ever since. In the late 1940s and 1950s Robert Van de Graaff was the pioneer of accelerator construction, and the high-voltage technology, which was the base for building the first commercial ion implanters. One of the most common applications of ion implanters are the integrated circuit manufacturing, where doping or modifying silicon and other semiconductor wafers is the main goal.

Ion implantation is a process that involves generating an ion beam and impinging it into a substrate, so that the ions come to rest inside the material. During implantation, the ions interact with nucleus and electrons of the substrate producing physical, electrical, and chemical changes in the material by transferring their energy and momentum to the electrons and atomic nuclei of the target material. When an energetic ion collides with a solid, it loses energy by two processes: (1) by inelastic or electronic process in which the electrons of the material are excited and (2) by elastic or nuclear collisions with the target atoms, producing structural changes, such as interstitials and vacancies. During the nuclear collisions, an ion can transfer enough energy to a matrix atom, becoming a projectile inside the material, and producing collision cascades. After losing its energy, when the ion or the released atom does not find any vacant space in the lattice, they tend to occupy any space in the solid known as interstitial. When the ions are implanted into a metal or metal alloy, the structural change is mostly produced in the surface of the material, creating pathways by which hydrogen can migrate into the metal in an easier way.

In this section, the results of a study of hydrogenation of pure titanium after ion implantation are presented.

The material used to carry out the study of hydrogen storage was pure titanium, manufactured by Goodfellow with 99.6% purity. The samples consisted of slices cut from titanium rods, which were polished, cleaned in acetone, and rinsed with deionized water. After the cleaning process, the materials were implanted at room temperature with a Colutron ion gun. The titanium samples were implanted with 5 keV H ions at a fluence of 1×10^{14} ions/cm^2. With this process, we intentionally induced defect zones in the surface region that may accelerate the diffusion of hydrogen. After implantation, the samples were hydrogenated at 300, 450, and 600°C in a 50% hydrogen and 50% argon atmosphere, at 1 atm pressure, and a flux of 50 cc/min, during 2 h. We intentionally implanted hydrogen ions in order to avoid any element contamination in the material. The 5 keV energy was chosen in order to assure that the biggest damage and the highest number of vacancies produced during implantation, would be produced in surface.

Figure 6 shows the hydrogen depth profile of the implanted and the nonimplanted samples hydrogenated at 450°C [13]. It can be observed in the figure that the hydrogen absorption of titanium without implantation is almost null, as can be verified in **Figure 4**; however, when a previous 5 keV H ions implantation is achieved to the metal, the scene is completely different and the hydrogen storage capacity of the metal is highly improved. Both curves in **Figure 6** show a small hydrogen peak at the surfaces of both titanium samples. These peaks are related to the superficial hydrogen in the nonimplanted sample (blue circles) and to the combination of the superficial hydrogen and the implanted hydrogen in the irradiated sample (red squares), where the hydrogen signal is slightly higher as a consequence of the implanted hydrogen. It can be also appreciated in **Figure 6** that the nonimplanted material does not absorb hydrogen beyond its surface. This can be confirmed in the ERDA graph as an almost null hydrogen signal beyond the 0.15 μm. In the case of the implanted materials, they show the same superficial hydrogen signal as the nonimplanted samples, but they also present a major hydrogen signal from 0.15 to 0.83 μm. In order to verify if the hydrogen continues beyond 0.83 μm into the bulk, ERDA measurements were performed from the back of the samples. The results showed that hydrogen is present through all the material.

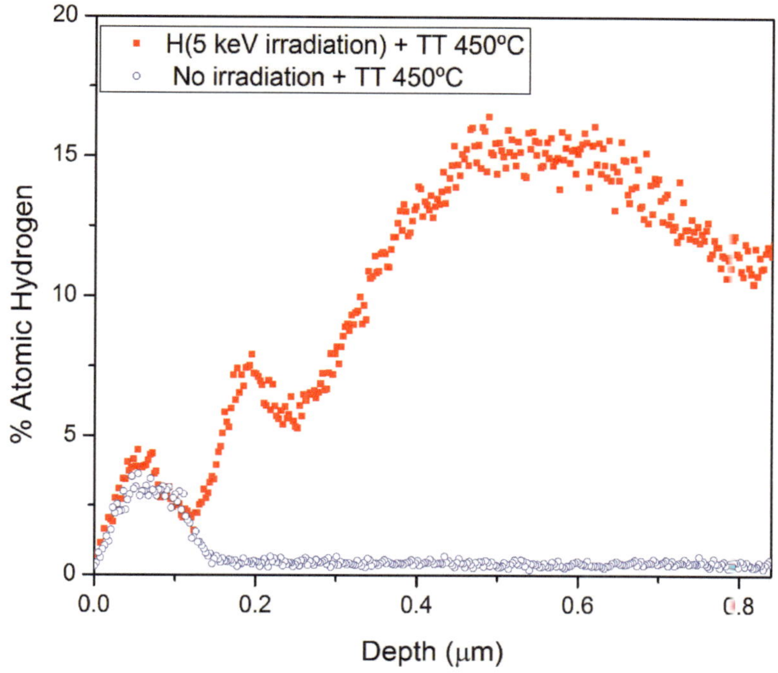

Figure 6. Hydrogen depth profile of pure titanium hydrogenated at 450°, with (red squares) and without (blue circles) implantation [13].

The high continuum hydrogen signal that can be observed in the implanted samples can be explained by means that during implantation strains in the surface region change, as well as structural changes are produced in the surface of the metal, introducing defects and dislocations and creating pathways in which hydrogen atoms can migrate into the metal. These paths may accelerate the diffusion of hydrogen through the metal, improving its storage capacities.

In **Figure 7**, a comparative between hydrogen concentration of the implanted [13] and the nonimplanted (see **Figure 4**) titanium samples after being hydrogenated at different temperatures is observed. These results were obtained directly by ERDA spectra using Eq. (1) and comparing with a reference sample of TiH_2. For 300°C, there is no difference between the hydrogen absorbed by the implanted and the nonimplanted samples; however, for temperatures above 450°C, the behavior is completely different, showing an improvement in the hydrogen content for the titanium samples previously implanted. In this way, when hydrogen ions are implanted into the pure titanium material, its activation temperature is reduced.

Table 3 shows the gravimetric storage capacity of pure titanium samples after being implanted and hydrogenated [13]. The results show that as well as the hydrogen concentration, the storage capacity was improved after the 5 keV H ion implantation. When comparing with **Table 1**, it can be noticed that the storage capacity increased from 3.02 to 3.77% in the sample hydrogenated at 600°C during the activation cycle. This confirms that implantation process does accelerate the activation process of pure titanium. We can assure that cyclic hydrogenations must increase considerably the hydrogen absorption in pure titanium, as well as in the Ti-6Al-4V alloy.

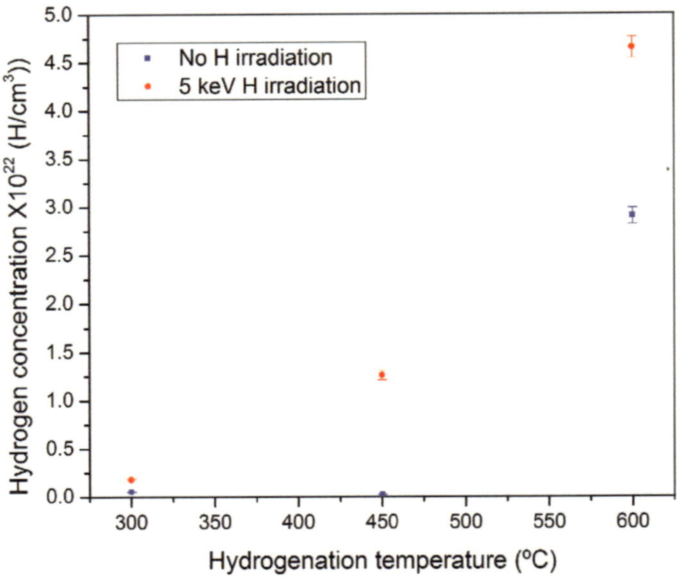

Figure 7. Hydrogen concentration of the hydrogenated samples, before and after implantation [13].

Hydrogenation temperature (°C)	Gravimetric storage capacity (wt%)
300	0.097 ± 0.007
450	0.94 ± 0.066
600	3.77 ± 0.264

Table 3. Gravimetric storage capacity of titanium samples after implantation and hydrogenation [13].

In order to check if the hydrogen absorbed by the metal is forming hydrides, XRD analysis was conducted [13]. **Figure 8** shows the results of the XRD of the implanted and hydrogenated samples, as well as a pure titanium reference. The results show the diffractions of the titanium alpha phase. The titanium dihydride phase (TiH_2) is also observed, showing the (100) and (200) lines, which correspond to the higher diffractions. The results also show a decrease in the intensity in the pure titanium (101) reflection as the temperature of hydrogenation increases. This behavior could be the consequence of a structural change or loss of crystallinity produced during implantation, hydrogenation, or both. This figure also shows that the relative intensities of the reflections (100) and (002) in the HCP titanium were modified with the presence of hydrogen. This indicates that during hydrogen absorption a change in crystal orientation was induced. An important result obtained from this figure is that the implanted hydrogen does not produce hydrides in the metal, so the hydride phase is formed during hydrogenation.

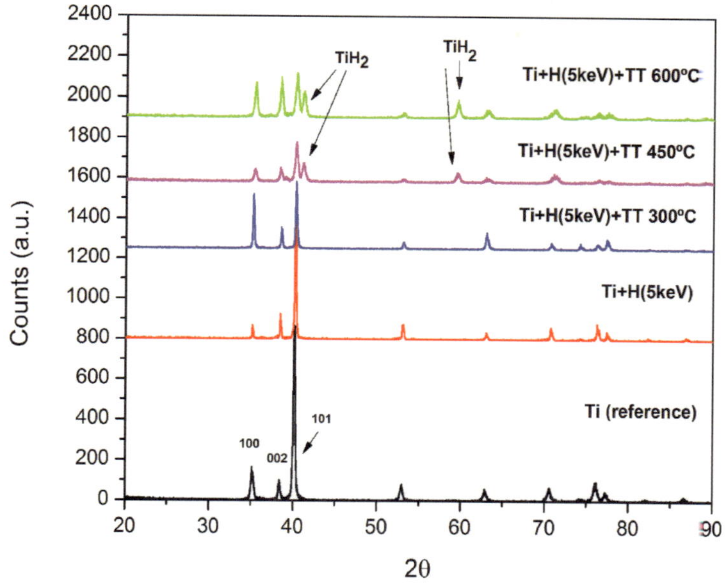

Figure 8. Comparative XRD patterns of five titanium samples: pure titanium (reference), titanium implanted with 5 keV H ions (Ti + H(5 keV)), and the three patterns in the top that correspond to samples implanted with 5 keV H ions and hydrogenated at different temperatures (Ti + H(5 keV) + TT) [13].

2.4. Hydrogen desorption of Ti and Ti-6Al-4V

If metal hydrides must considered as reliable fuel cells, it would be mandatory for them to become stable, which means that the materials must be enough stable for not releasing hydrogen when the metal is being stored. For this reason it is important to study the release of hydrogen under conditions of low temperature and atmospheric pressure. Studies were performed to determine the degree of release of hydrogen at atmospheric pressure and room temperature for samples of pure titanium and the Ti-6Al-4V alloy. Both materials were first hydrogenated using the same cleaning and hydrogenated experimental details as previously. After that, the samples were kept stored during 4 months; meanwhile their hydrogen desorption was measured using the ERDA technique.

The results showed in both materials that hydrogen was naturally released, which means that hydrogen is desorbed from materials, without providing any additional energy to the system. In the case of pure titanium, it was observed that hydrogen concentration decreases slower than the Ti-6Al-4V alloy does, showing a decrease of 30% of its total hydrogen concentration, after 130 days that remained stored. Furthermore, the alloy showed a greater decrease in hydrogen content, decreasing up to 40% after 130 days of storage. The decrements in hydrogen after both materials were stored for 1 day is almost null. These results suggest that these materials are good candidates for storing hydrogen, since they are quite stable. In terms of applications to the automotive industry, it means that the vehicle could remain stop during 4 months and it will only lose 30% of its fuel if a titanium cell were used.

2.5. The effects of hydrogen absorption and desorption in Ti and the Ti-6Al-4V alloy

As has been already mentioned, the absorption of hydrogen by metals, such as pure titanium and the Ti-6Al-4V alloy, can produce changes in their metallic structures, such as in its hardness.

Hardness measures the resistance of a material when a force is applied to it. In this way, hydrogen is an element that changes hardness when it is introduced into a metal. Hardness is defined as the ratio of the applied load P to the indentation area A. In this way, Vickers microhardness (H_v) follows the relation

$$H_v = 1854 \frac{P}{d^2} \tag{4}$$

where d is the diagonal of the indentation area.

Some works have been carried out in order to study the influence of hydrogen in hardness properties of pure titanium and the Ti-6Al-4V alloy [16]. Vickers microhardness tests were carried out on both materials at hydrogenation temperature ranging from 150 to 650°C. The results show that the mechanical properties of the metals were modified when hydrogen is absorbed into the material.

Figure 9 shows the Vickers microhardness behavior of pure titanium versus hydrogenation temperature. As was mentioned before, hydrogen concentration is increased from 550 to 650°C and after that temperature hydrogen is kept constant. The same behavior is observed when microhardness measures were performed after the metal was activated, as is shown in **Figure 9** (red squares), where pure titanium shows a tendency to become harder as hydrogen increases. This fact is a direct consequence of the hydrogen that remains in interstitial sites inside the metal and was verified when hydrogen is desorb from the material. **Figure 9** also shows two other curves that correspond to the microhardness measures of the titanium after being stored for 4 months (green circles) and 10 months (blue triangles). As was mentioned below, hydrogen is lost with time after the samples had kept stored for several months. The larger hydrogen concentration that correspond to the sample that was hydrogenated at 650°C, show average decreases in the Vickers microhardness of 8 and 27% after being kept stored for 4 and 10 months, respectively. The samples that contained smaller hydrogen concentrations, such as the metals hydrogenated between 150 and 500°C, maintained their microhardness values almost constant with time. In the case of the Ti-6Al-4V alloy the scene is repeated, showing an average decrease in hardness of 7 and 20% after 4 and 10 months, respectively. In the same way, the samples with smaller hydrogen amounts did not show a significant loss of micro-hardness with time.

The XRD patterns of the hydrogenated samples also showed that the titanium hydride peaks did not present any change in form nor size when hydrogen was released in natural way. This observation can be understood as the loss of hydrogen concentration was only due to the interstitial hydrogen that is released while the samples are kept stored and it is not due to the metal hydride that has formed inside the materials. This result shows that the interstitial hydrogen is the main contributor that increases and decreases hardness in pure titanium and the Ti-6Al-4V alloy. In this way, it is corroborated that hydrogen plays an important role in changing the mechanical properties of a material.

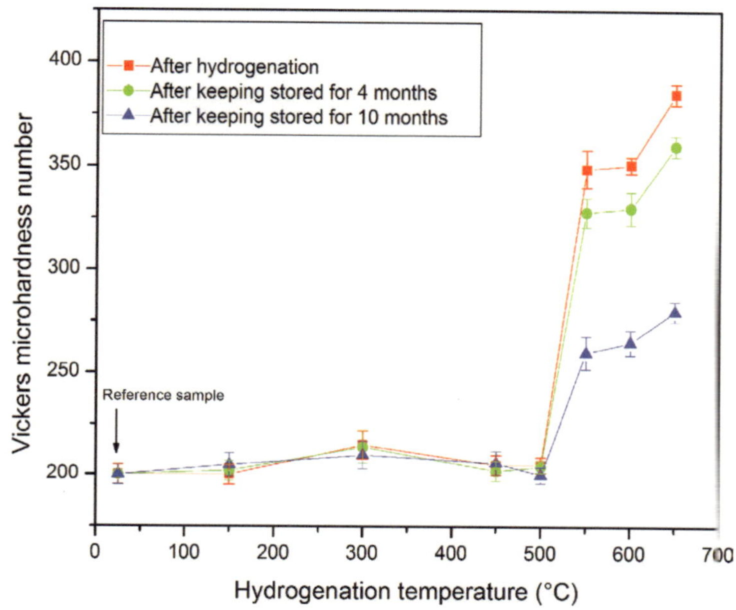

Figure 9. Vickers microhardness versus hydrogenation temperature for pure titanium samples. A reduction in hardness as the samples loose hydrogen with time can be observed [16].

3. Conclusions

The potential for storing hydrogen in metal hydrides is an area of material research that must be kept active. This chapter showed that pure titanium and the alloy Ti-6Al-4V are good candidates as hydrogen storage materials, which can increase their storage capacities after ion implantation. Ion implantation improves activation easing the diffusion of hydrogen through the whole material. It was also noted that the surface quality of the metal is important during hydrogenation in order to get better results during hydrogen absorption.

Acknowledgements

The author wishes to acknowledge the technical assistance of K. López, J.C. Morales, F. J. Jaimes, A. Morales and J.A. Lara-Velázquez. The support of DGAPA-PAPIIT project under contract IN100-216 is also acknowledged.

Author details

Alejandra López-Suárez

Address all correspondence to: chipi@fisica.unam.mx; chipi72@gmail.com

Physics Institute, National Autonomous University of Mexico, Mexico City, Mexico

References

[1] Gray MacA E. Hydrogen storage status and prospects. Adv. Appl. Ceram. 2007; 106: 25–28. doi:10.1179/174367607X152380

[2] López-Suárez A, Rickards J, Trejo-Luna R. Analysis of hydrogen absorption by Ti and Ti-6Al-4V using the ERDA technique. Int. J. Hydrogen Energy. 2003; 28: 1107–1113. doi: 10.1016/S0360-3199(02)00202-1

[3] Yu XB, Yang ZX, Liu HK, Grant DM, Walker GS. The effect of a Ti-V based BCC alloy as a catalyst on the hydrogen storage properties of MgH_2. Int. J. Hydrogen Energy. 2010; 35: 6338–6344. doi:10.1016/j.ijhydene.2010.03.089

[4] Alefeld G, Völkl J, editors. Hydrogen in Metals I. 1st ed. Germany: Springer Verlag; 1978. 428 p. doi:10.1007/3-540-08705-2

[5] Blackburn JL, et al. Measurement of the reversible hydrogen storage capacity of milligram Ti-6Al-4V alloy samples with temperature programmed desorption and volumetric techniques. J. Alloys Compd. 2008; 454: 483–90. doi:10.1016/j.jallcom. 2007.01.006

[6] Chu WK, Mayer JW, Nicolet MA. Backscattering Spectrometry. 4th ed. New York, USA: Academic Press; 1978. 384 p. doi:10.13140/RG.2.1.1948.0807

[7] Tesmer JR, Nastasi M, editors. Handbook of Modern Ion Beam Materials Analysis. 2nd ed. Pennsylvania, USA: MRS; 2010. 370 p. ISBN 9781605112176.

[8] López-Suárez A. Influence of surface roughness on consecutively hydrogen absorption cycles in Ti-6Al-4V alloy. Int. J. Hydrogen Energy. 2010; 35: 10404–10411. doi:10.1016/ j.ijhydene.2010.07.163

[9] Lee SM, Perng TP. Effect of the second phase on the initiation of hydrogenation of $TieFe_{1-x}M_x$ (M = Cr, Mn) alloys. Int. J. Hydrogen Energy. 1994; 19: 259–263. doi: 10.1016/0360-3199(94)90095-7

[10] Yang J, Sudik A, Wolverton C, Siegelw DJ. High capacity hydrogen storage materials: attributes for automotive applications and techniques for materials discovery. Chem. Soc. Rev. 2010; 39: 656–675. doi:10.1039/b802882f

[11] Saitoh H, Hjima Y, Tanaka H. Hydrogen diffusivity in aluminium measured by a glow discharge permeation method. Acta Metall. Mater. 1994; 42: 2493–2498. doi: 10.1016/0956-7151(94)90329-8

[12] Suda T, Ohkawa M, Sawada S, Watanabe S, Ohnuki S, Nagata S. Effect of surface modification by ion implantation of hydrogenation properties of TiFe alloy. Mat. Trans. 2002; 43: 2703–2705. doi:10.2320/matertrans.43.2703

[13] López-Suárez A, Valencia CE, López-Patiño J, Vargas MC, Fuentes BE. Improvement of titanium hydrogenation by low energy ion irradiation. Int. J. Hydrogen Energy. 2015; 40: 4194–4199. doi:10.1016/j.ijhydene.2015.01.166

[14] Abe H, Uchida H, Azuma Y, Uedono A, Chen ZQ, Itoh H. Improvement of hydrogen absorption rate of Pd by ion irradiation. NIMB. 2003; 206: 224–227. doi:10.1016/S0168-583X(03)00734-1

[15] Grbovic-Novakovic J, Matovic LJ, Drvendzija M, Novakovic N, Rajnovic D, Siljegovic M, Kacarevic-Popovic Z, Milovanovic S, Ivanovic N. Changes of hydrogen storage properties of MgH_2 induced by heavy ion irradiation. Int. J. Hydrogen Energy. 2008; 33: 1876–1879. doi:10.1016/j.ijhydene.2008.02.008

[16] López-Suárez A, Rickards J, Trejo-Luna R. Mechanical and microstructural changes of Ti and Ti–6Al–4V alloy induced by the absorption and desorption of hydrogen. J. Alloys Compd. 2008; 457: 216–220. doi:10.1016/j.jallcom.2007.03.031

Chapter 11

Identification of Some New Generation Additives for Polymers Obtained in the Catalytic Hydrogenation Process

Euzebiusz Jan Dziwiński, Bartłomiej Bereska,
Jolanta Iłowska, Józef Lach, Agnieszka Bereska and
Michał Szmatoła

Additional information is available at the end of the chapter

http://dx.doi.org/10.5772/65408

Abstract

The identification of new generation additives such as plasticizers and hardeners applied mainly to some polymers and resins and obtained in the catalytic hydrogenation processes is presented. The new plasticizers di(n- and isononyl)cyclohexane-1,2-dicarboxylates (DINCH) were obtained by catalytic hydrogenation of the planar aromatic rings of the mixture of di(alkyl(C_9))phthalates in the presence of the modified Ni catalyst. They may be used as a substitute for improving the flexibility of polymers, such as PVC, and also for strongly reducing the toxic effects on human health of the 1,2-di(alkyl)phthalates, which is easily released from the polymer to the environment. The identification of these types of compounds by GC/MS enables to determine the structures of the main products such as *cis* and *trans* isomers of the cyclohexane-1,2-dicarboxylates. GC/MS may be applied to the identification of some DINCH constituents extracted from polymers. Also, the analysis of these compounds by ESI/MS gives information about their mass fragmentation and enables their detection, although without differentiation between individual *cis* and *trans* isomers.

Branched poly(amines) may be used as the cross-linking agents for the most effective crosslinking of the epoxy resins. Final product from the synthesis of the poly(amines) may be easily analyzed by GC/MS and spectroscopic techniques.

Keywords: hydrogenation, identification, plasticizers, di(n- and isononyl)cyclohexane-1,2-dicarboxylates, hardeners, branched poly(amines)

1. Introduction

Di(n- and isononyl)cyclohexane-1,2-dicarboxylate (DINCH) isomers are considered to be relatively safe substitutes of the corresponding phthalates, especially when used in the manufacturing of various medical devices and toys. For this reason there is a great interest in the analysis of their isomers by different analytical methods.

In this chapter, we present the results of the application of the different modern chromatographic and spectrometric analytical techniques, such as gas chromatography (GC), electrospray-mass spectrometry (ESI/MS), Fourier transform infrared spectroscopy (FTIR), nuclear magnetic resonance (NMR), gas chromatography-mass spectrometry (GC/MS), and others similar ones to the identification of the new generation additives.

They were obtained in the catalytic hydrogenation reactions in the presence of the Ni catalyst and used as plasticizers and cross-linking agents to some polymers, such as poly(vinyl chloride) (PVC) or epoxy resins.

2. Plasticizers

2.1. Di(alkyl)cyclohexane-1,2-dicarboxylates

Di(alkyl) esters of 1,2-benzene dicarboxylic acid of higher molecular weight, such as di(2-ethylhexyl)phthalate (DEPH) or di(isononyl)phthalate (DINP), are widely used as plasticizers in the processing of various types of polymers, especially poly(vinyl chloride). These types of compounds are not chemically bound to the polymers, so they are gradually released from them by volatilization from the surface into the air, or by migration due to contact with a solid, or by extraction from the polymer into a liquid also due to direct contact [1]. These compounds are becoming challenging environmental pollutants with a strong impact on the human health. The toxicity effects of these compounds have been intensively investigated, and within the last decade some of those compounds have been classified as endocrine disruptors [2] and potential carcinogens [3–6]. The metabolism of DINP in animals [7] and in humans [8, 9] has been studied. In these studies, where deuterium-labeled DINP was used, samples of animal and human urine were found to contain monoester of mono-iso-nonylphthalate and its oxidized isomers containing hydroxy, oxo, and carboxy functional groups as metabolites of DINP.

The new di(alkyl)cyclohexane 1,2 dicarboxylates (DINCH) plasticizers may be used to improve the flexibility of some polymers, mainly PVC. They also have less toxic effect on human health when compared to di(alkyl)phthalates.

2.2. Synthesis of di(n- and isoalkyl(C_4–C_9))cyclohexane-1,2-dicarboxylates

Cyclohexanedicarboxylic esters are produced through the catalytic hydrogenation of corresponding phthalates or through the Diels-Alder reaction of maleic acid esters with ethylene followed by hydrogenation using supported nickel catalyst [10].

Di(n- and isoalkyl(C_4–C_9)) phthalates were synthesized in the esterification reaction of phthalic anhydride with appropriate aliphatic alcohol using an optimized procedure [11, 12]. Hydrogenation of these esters, after their prior purification by distillation, were carried out in a high pressure reactor in the presence of the Ni catalyst on aluminosilicate support at 150°C for 3.0 h and under 9.0 MPa hydrogen pressure, according to the following scheme:

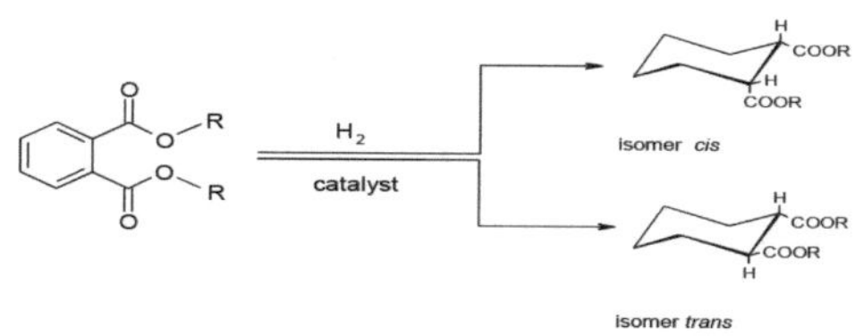

where R = alkyl (C4 – C9).

During the hydrogenation reaction of di(alkyl)phthalates, *cis* and *trans* isomers of the di(alkyl(C_4–C_9))cyclohexane-1,2-dicarboxylates are formed as the main products with a yield of 98.0%.

2.3. Identification of di(n- and isoalkyl(C_4–C_9))cyclohexane-1,2-dicarboxylates

Identification of *cis* and *trans* isomers of the synthesized di(n- and iso-nonyl)cyclohexane-1,2-dicarboxylates was done by chromatographic and spectrometric methods of GC/MS and ESI/MS.

2.3.1. GC/MS analysis

Chromatographic separation of the compounds investigated and registration of their "electron impact" mass spectra (EIMS) was done by use of a gas chromatograph HP 6890 Series GC System (Hewlett-Packard, Palo Alto, CA, USA) equipped with HP 5973 Network quadrupole mass selective spectrometric detector (Agilent Technologies, Palo Alto, CA, USA). Each sample was dissolved in $CHCl_3$ to 5,0% solution. Then 0.1 μL of that solution was injected by using Hamilton microsyringe to the split/splitless injector (split mode 100:1) kept at 250°C. The fused silica capillary column HP 50+ (30.0 m length, internal diameter 0.2 mm and 0.2 μm phase film thickness) was heated in the range of 70–290°C with programmed temperature ramp 7°C/min. As a carrier gas helium (ultra-pure, 99,999%) was used.

GC/MS was applied to analyze the synthesized *cis* and *trans* isomers of di(n- and isononyl) esters of 1,2-cyclohexanedicarboxylic acid in order to get their good chromatographic separation enabling to record their electron impact (EI) mass spectra and also to calculate their arithmetic retention indices I_A on the basis of their retention times t_r.

Figure 1 shows an example of a chromatogram of *cis* (compound a_1) and *trans* (compound a_2) of the di(3,5,5-trimethylhexyl) esters of cyclohexane-1,2-dicarboxylic acid as the reaction products of di(3,5,5-trimethylhexyl)phthalate hydrogenation. Values of the retention times t_r

of the *cis* and *trans* isomers of the esters are always lower than those of the corresponding phthalates.

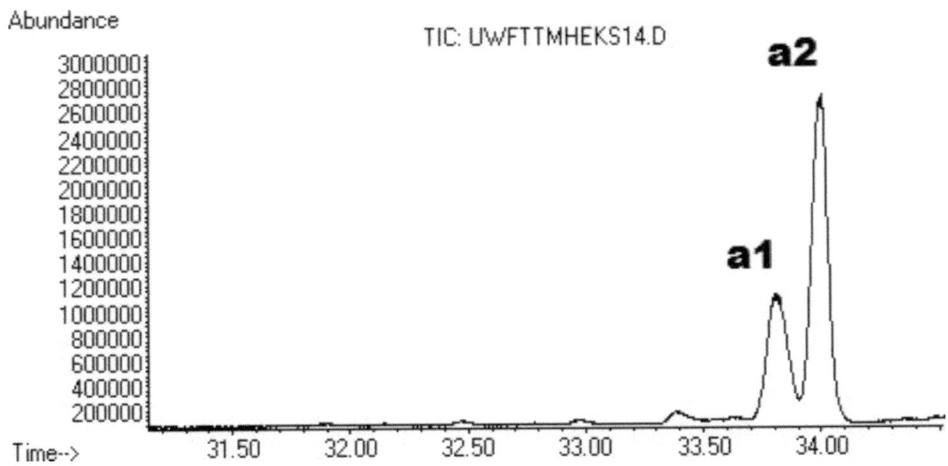

Figure 1. GC/MS chromatogram of *cis* (a_1) and *trans* (a_2) isomers of di(3,5,5-trimethylhexyl)cyclohexane-1,2-dicarboxylate.

On the basis of the retention times t_r of all analyzed *cis* and *trans* isomers and the retention times of a standard mixture of *n*-alkanes C_{20}–C_{40}, the arithmetic retention indices (I_A) were calculated using the following formulae (1) [13]:

$$I_A = 100z + 100\frac{T_i - T_z}{T_{z+1} - T_z} \quad (1)$$

where T_i, T_z, and T_{z+1} are the retention times of the analyzed component and neighboring *n*-alkanes containing *z* and *z* + 1 carbon atoms, wherein $T_z < T_i < T_{z+1}$

The obtained values of I_A for the analyzed compounds are given in **Table 1**.

The linear relationship was found between the values of arithmetic retention indices I_A of di(n-alkyl(C_4–C_9))phthalates and the number of carbon atoms present in the alkyl substituents of esters obtained during the hydrogenation of appropriate phthalates [12].

The values of I_A of di(n-alkyl(C_4–C_9)) phthalates are higher than those of their hydrogenation products. In both cases there is also a regularity according to which the esters with the longer alkyl substituents have the greater values of the arithmetic retention indices I_A. Whereas in the case of the esters with branched alkyl chains of the substituents, their retention times t_r and arithmetic retention indices I_A have lower values compared to those of the corresponding esters with straight chain substituents. The values of their retention times are arranged in the following order: $t_{ra1,a2} < t_{rb1,b2} < t_{rc1,c2}$ and, similarly, the arithmetic retention indices are arranged in the following order: $I_{A1,A2} < I_{B1,B2} < I_{C1,C2}$.

The obtained t_r and I_A reference GC parameters for the analyzed *cis* and *trans* isomers of the di(n- and isononyl)cyclohexane-1,2-dicarboxylates can be used for the unambiguous determination of the chemical structure of this type of organic compounds, particularly when both are present in the reaction product of the hydrogenation of di(n- and isononyl)phthalic acid esters. They may also be very useful for ongoing optimization of the technological parameters of the hydrogenation process using only the GC method.

Compound	Name of compound	Mol. weight [g/mol]	Retention time, t_r [min]	Arithmetic retention index, I_A
a_1	Di(3,5,5-trimethylhexyl)cyclohexane-1,2-dicarboxylic acid (*cis* isomer)	424	33.81	2780.0
a_2	Di(3,5,5-trimethylhexyl)cyclo-hexane-1,2-dicarboxylic acid (*trans* isomer)	424	34.00	2798.1
A	Di(3,5,5-trimethylhexyl)phthalate	418	35.94	2976.9
b_1	Di(2-methyloctyl)cyclohexane-1,2-dicarboxylic acid (*cis* isomer)	424	34.51	2846.7
b_2	Di(2-methyloctyl)cyclohexane-1,2-dicarboxylic acid (*trans* isomer)	424	34.68	2862.9
B	Di(2-methyloctyl)phthalate	418	37.05	3067.5
c_1	Di(n-nonyl)cyclohexane-1,2-dicarboxylic acid (*cis* isomer)	424	37.14	3074.6
c_2	Di(n-nonyl)cyclohexane-1,2-dicarboxylic acid (*trans* isomer)	424	37.33	3089.7
C	Di(n-nonyl)phthalate	418	39.85	3257.6

Table 1. Retention times t_r and arithmetic retention indices I_A of di (n- and iso-nonyl)cyclohexane-1,2-dicarboxylates.

The good chromatographic separation in the GC/MS analysis of the *cis* and *trans* isomers of di(n- and isononyl)cyclohexane-1,2-dicarboxylates enabled recording of their low-resolution EI mass spectra, which are different for each isomer, and as such could be used for unambiguous identification of their chemical structures.

Table 2 gives the relative intensities of the most characteristic peaks of the molecular and fragment ions of all analyzed *cis* and *trans* isomers of di(n- and isononyl)cyclohexane-1,2-dicarboxylates. In all mass spectra of these isomers there are low intensity peaks of their molecular ions $[M]^{+\bullet}$ present at *m/z* 424, which allow to unambiguously determine the molecular weights of these compounds. Also, their mass spectra have a number of peaks referring to the characteristic fragment ions. The most intense characteristic peak of the fragmentation ion is easily recognized in the mass spectra of all analyzed esters (except the *cis* and *trans* isomers of di(3,5,5-trimethylhexyl)cyclohexane-1,2-dicarboxylates—compounds a_1 and a_2). It occurs at *m*/*z* 155 and corresponds to the structure of the protonated anhydride of cyclohexane-1,2-dicarboxylic acid [11, 12]. However, in the mass spectra of the *cis* and *trans* isomers of compounds a_1 and a_2, in which isoalkyl substituents are more branched, as compared to other isomers—compounds b_1, b_2, c_1, and c_2 (**Tables 1** and **2**), the main peak corresponding to the ion at *m*/*z* 57 has the structure of $[C(CH_3)_3]^+$. It is formed as a result of a homolytic cleavage of the single C–C bond located at the branched carbon atom of the alkyl substituent. The cleavage of the weaker C–C bond occurs more easily than in the other molecular ions $[M]^{+\bullet}$ of the *cis* and *trans* isomers of the esters investigated, and it is the reason of the formation of the less intense ion at *m*/*z* 155 (also formed from the $[M{-}_9H_{19}]^+$ ion as a result of the elimination of alkoxy radical $^\bullet OC_9H_{19}$ from the molecular ion $[M]^{+\bullet}$ of these compounds). In general, the peak at *m*/*z* 155 may be used as a diagnostic peak for the unique identification of such type of esters, similarly as the peak at *m*/*z* 149, the main ion peak of di(alkyl)phthalates (except di(methyl)phthalate). The ion at *m*/*z* 149 has the structure of the protonated anhydride of phthalic acid [14–16].

Fragment ion	*m*/*z*	Relative intensities of fragment ions (%)					
		Isomer *cis* a_1	Isomer *trans* a_2	Isomer *cis* b_1	Isomer *trans* b_2	Isomer *cis* c_1	Isomer *trans* c_2
$[M]^{+\bullet}$	424	0.01	0.19	0.07	0.04	0.09	0.28
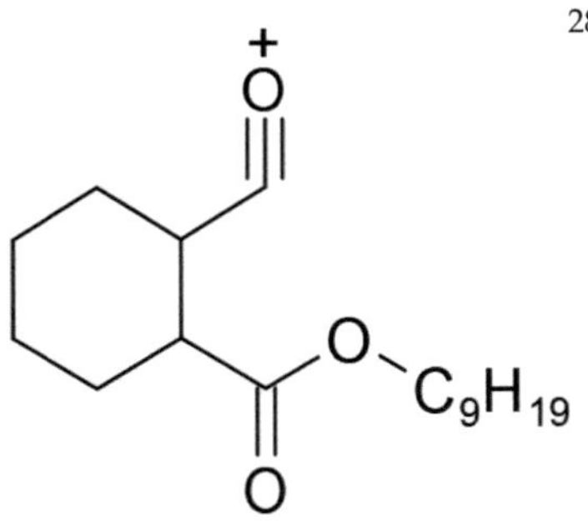	281	11.20	4.35	3.28	3.62	20.74	9.88
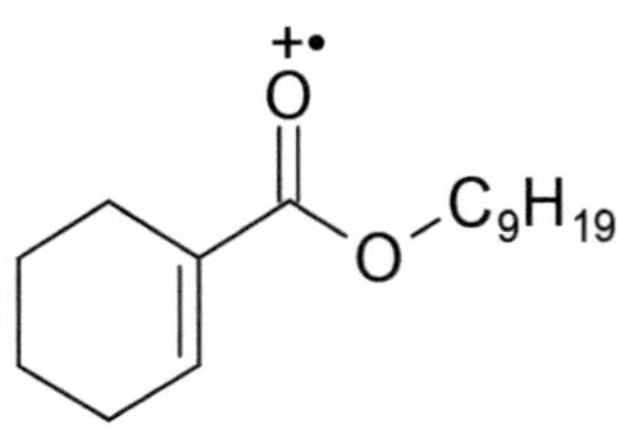	252	0.42	3.47	0.03	0.23	1.18	10.00
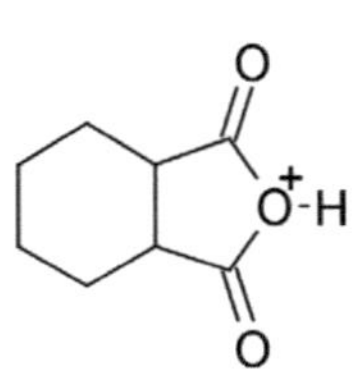	155	49.83	5.70	100.00	100.00	100.00	100.00

Fragment ion	m/z	Relative intensities of fragment ions (%)					
		Isomer *cis* a_1	Isomer *trans* a_2	Isomer *cis* b_1	Isomer *trans* b_2	Isomer *cis* c_1	Isomer *trans* c_2
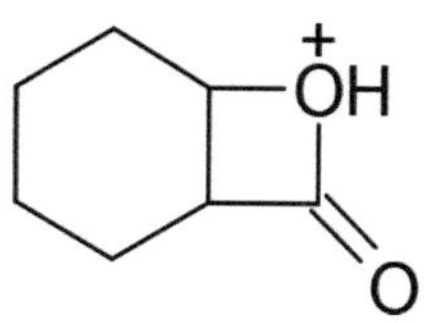	127	18.95	8.16	6.70	6.41	17.89	9.80
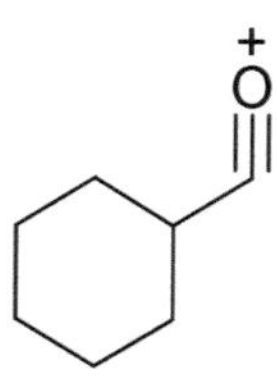	111	9.31	11.80	0.77	0.97	0.76	–
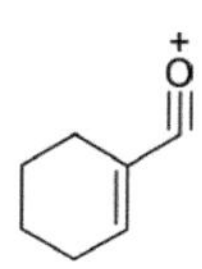	109	11.30	10.41	12.97	12.82	23.06	19.83
$[C_6H_{13}]^+$	85	2.02	1.81	7.02	7.17	6.53	6.59
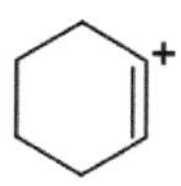	81	16.64	7.00	19.59	18.93	29.40	29.60
$[C_4H_9]^+$	57	100.00	100.00	13.60	15.39	16.68	16.80
$[C_3H_7]^+$	43	15.24	15.22	26.63	28.98	37.79	39.07

Table 2. Relative ion intensities [A1] of *cis* and *trans* isomers of di(n- and iso-nonyl)cyclohexane-1,2-dicarboxylates in their mass spectra.

Other peaks present in all the MS spectra of the analyzed *cis* and *trans* isomers—compounds a_1, a_2, b_1, b_2, c_1, and c_2—correspond to the characteristic fragment ions formed by the cleavage of single C—C and CO bonds of their molecular ions $[M]^{+\bullet}$. These fragmentation reactions are accompanied by the transfer of hydrogen atoms in the McLafferty rearrangement together with an elimination of the neutral molecules of H_2O and CO (**Table 2**).

Figures 2a and **2b** show the mass spectra of isomers *cis* and *trans* of di(3,5,5-trimethylhexyl)cyclohexane)-1,2-dicarboxylic acid.

The differences between EI mass spectra observed in all the analyzed compounds result both from different structures of the alkyl substituents of carboxyl groups of cyclohexane-1,2-dicarboxylic acids and from *cis* and *trans* isomerization being the result of the presence of cyclohexane ring in these esters.

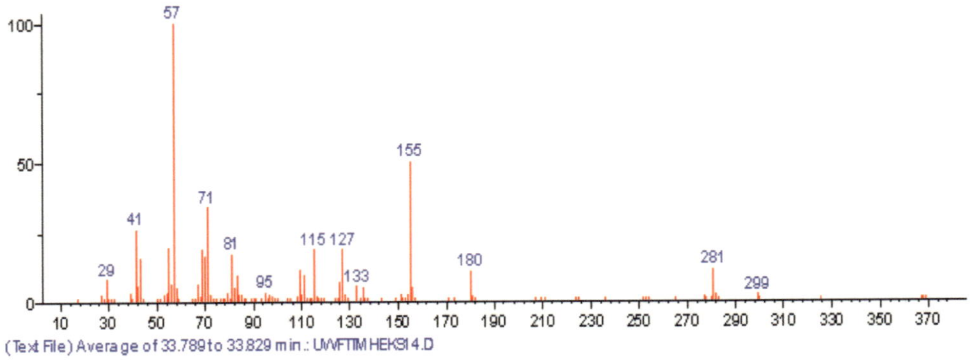

Figure 2a. Mass spectrum of compound a_1 (**Figure 1**): di(3,5,5-trimethylhexyl)cyclohexane-1,2-dicarboxylate – *cis* isomer.

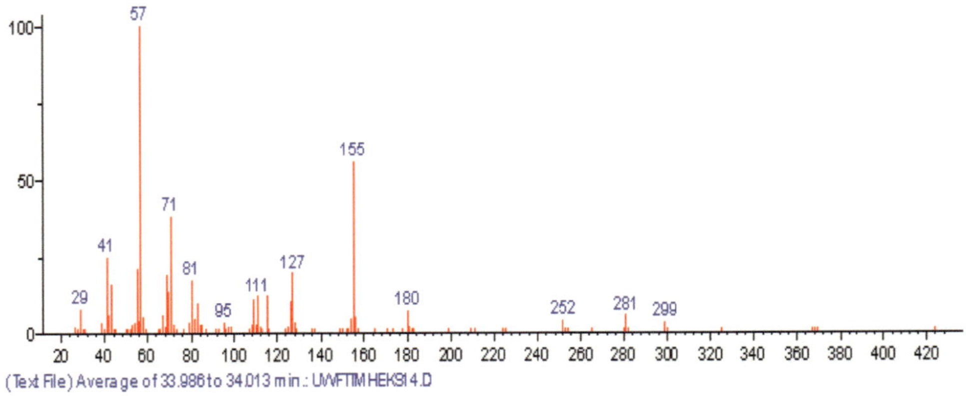

Figure 2b. Mass spectrum of compound a_2 (**Figure 1**) di(3,5,5-trimethylhexyl)cyclohexane-1,2-dicarboxylate – *trans* isomer.

2.3.2. *ESI/MS analysis*

Electrospray ionization, being a "soft" ionization technique, was used in mass spectrometry (ESI/MS) for the identification of volatile samples of di(3,5,5-trimethylhexyl)-, di(2-methyloctyl)-, and di(n-nonyl)cyclohexane-1,2-dicarboxylates. Each ESI mass spectrum of the compound investigated represents a mixture of *cis* and *trans* isomers of the same compound, and for this reason it cannot be used for their individual identification.

In ESI/MS analysis the samples of di(n- and isononyl)cyclohexane-1,2-dicarboxylates were dissolved in methanol (HPLC grade, J.T. Baker) and diluted with the same solvent to 1:40,000 (which corresponds to ca. 10 μM). All measurements were performed on an AB

Sciex Q-TRAP® 4000 series hybrid quadrupole mass spectrometer equipped with electrospray ion source. Collision gas was nitrogen at the nominal pressure 3.2×10^{-5} Torr.

The typical ESI mass spectrum of these compounds is shown in **Figure 3**. It presents a mixture of the fragment ion peaks of *cis* and *trans* isomers of di(3,5,5-trimethylhexyl)cyclohexane-1,2-dicarboxylates and only has fewer peaks of mass fragmentation ions in comparison with the individual EI mass spectra (**Figures 2a** and **2b**).

In all the ESI mass spectra of these types of compounds, the peaks of quasi-molecular ions $[M+H]^+$ are present at m/z 425 and they are formed by the addition of hydrogen cation. They arise from both *cis* and *trans* isomers. More information about the fragment ions could be obtained from the interpretation of ESI mass spectra in which they derived from mass transitions: from $[M+H]^+$ ions at m/z 281, after the cleavage of one of the unbranched or branched alkyl(C_9) substituents and following neutral loss of H_2O molecule leading to a fragment ion of m/z 155. The first transition is the most sensitive, and therefore it may be used to quantify di(n- and isononyl)cyclohexane-1,2-dicarboxylates. The second transition was used only to confirm the results of the first one.

The choice of selected characteristic ions for this type of compounds, for example, with an m/z of 155, present in other MS spectra of di[n- and isoalkyl(C_4–C_9)]cyclohexane-1,2-dicarboxylates, may be very helpful for their detection during ESI/MS analysis of one or more of the investigated compounds in complex matrices (e.g., PVC) and may also be very useful for quantitative assessment of the level of their actual impact on the human health.

Figure 4 shows a general mass fragmentation scheme of molecular ions $[M]^+$ of di(n- and isononyl)cyclohexane-1,2-dicarboxylates developed on the basis of the data obtained from their mass transitions. It describes a specific type of fragmentation reactions for this type of compounds as contrasted to the phthalic acid esters. The basic knowledge about their behavior during ionization of this type of compounds will make the interpretation of their mass spectra easier and will enable optimization of the methods of their quantification during the analysis of their release from polymers.

The good chromatographic separation during GC/MS analysis of *cis* and *trans* isomers of di(n- and isononyl)cyclohexane-1,2-dicarboxylates investigated enables the recording of their low-resolution EI mass spectra, which are different from each other and thus can be used for unambiguous identification of their individual chemical structures. Also, chromatographic data, such as the values of retention times t_r and arithmetic retention indices I_A, are very useful in their identification, even when the reference substances are not available. The ESI/MS mode was shown to be successful in the determination of *cis* and *trans* isomers of the analyzed esters present in complex matrices of a polymer.

The presented GC, GC/MS, and ESI/MS results for the representatives of the *cis* and *trans* isomers of some of the DINCH group of compounds may also be helpful in the determination of the chemical structures of their metabolites.

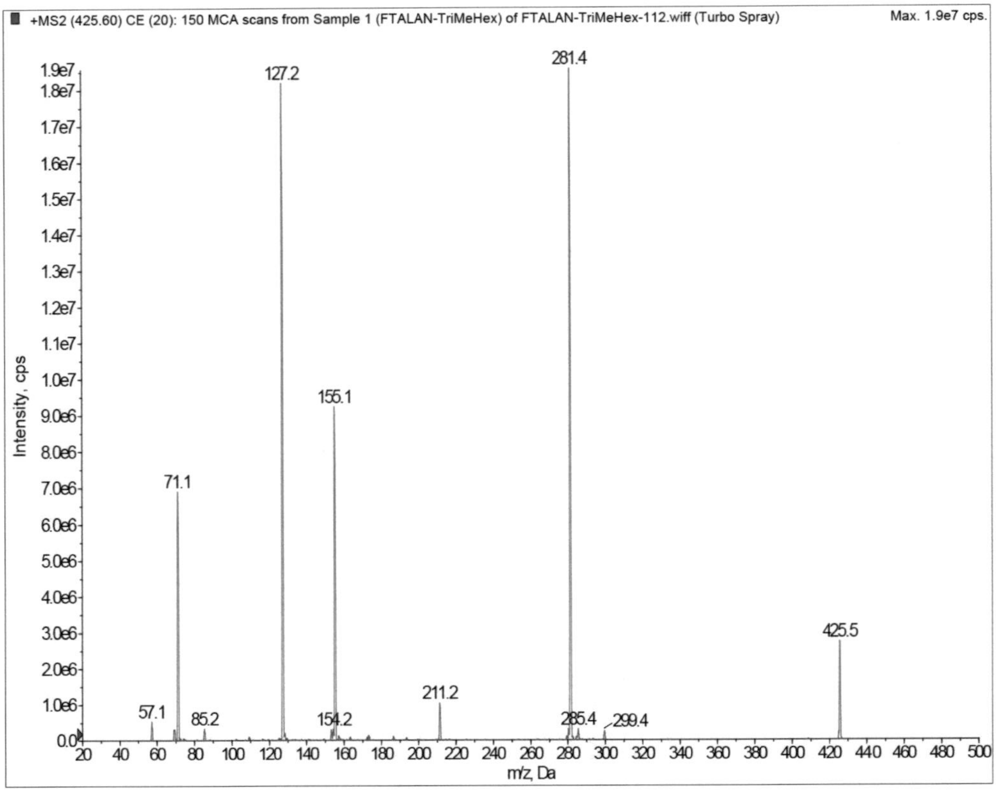

Figure 3. ESI mass spectrum of *cis* and *trans* isomers of di(3,5,5- trimethylhexyl)cyclohexane-1,2-dicarboxylate.

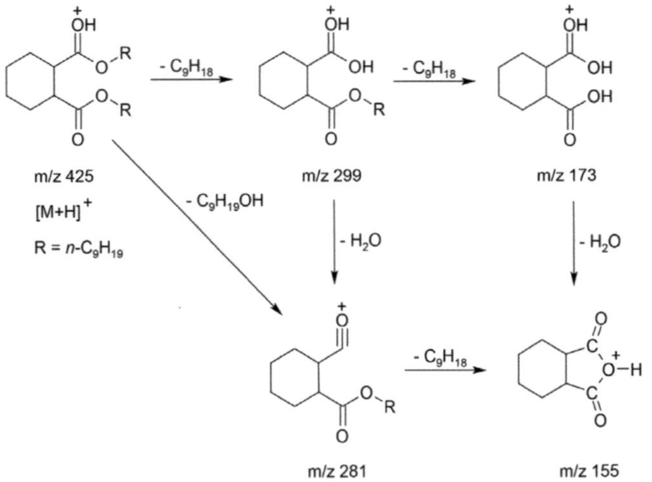

Figure 4. General mass fragmentation scheme of molecular ions $[M]^+$ of di(n- and isononyl)cyclohexane-1,2-dicarboxylates.

3. Cross-linking agents

3.1. The branched poly(amines)

Epoxy resins due to their excellent mechanical and dielectric properties along with relatively low shrinkage, high resistance to elevated temperatures, and chemical media, but also the ease of processing are extensively applied in many fields, such as adhesives, coatings, and construction materials. This flexibility of properties of epoxy- derived materials is achieved due to the resins' tendency to undergo various modifications, usually by reactive and nonreactive fillers and different types of cross-linking agents [17, 18]. The curing of epoxy resins involves the formation of rigid three-dimensional network by reaction with cross-linking agents possessing usually more than two functional groups. High cross-linking density of epoxy systems is responsible for high impact strength of hardened epoxy resins and also at the same time for their inherent brittleness. Recently, lots of valuable papers have been published on modifications of epoxy resins, where incorporation of highly branched flexible modifiers considerably improved the mechanical behavior of epoxy materials [19–21]. Currently, hyperbranched polymers (HBPs) are of the biggest interest among all polymeric modifiers for epoxy resins. Branched structures of these resins are attractive due to their low viscosity and good ratio of reactive groups to molecular mass, especially when they are compared to their linear homologs. This makes HBPs perfect potential candidates as cross-linking agents for epoxy resins. Knowing the influence of branching and functionality of hyperbranched polymers on material's characteristics can be used to tailor its final properties [22, 23].

The hyperbranched poly(amines) may be used as specific additives as modifiers of the physicochemical properties of the epoxy resins. Especially, branched poly(amines) such as (N,N,N-tri(3-aminopropyl)amine and (N,N,N′,N′-tetra(3-aminopropyl) ethylene-diamines may be used as specific additives—cross-linking agents for this type of resins.

3.2. Synthesis of branched poly(amines)

Hyperbranched polymers can be prepared by various stepwise and repetitive chemical routes within which convergent and divergent methods are of the biggest practical meaning [24–26]. One of the easiest and the most effective way to synthesize the hyperbranched polyamines is cyanoethylation of primary amines followed by hydrogenation of resulting nitriles to generate primary amines. This process consists of an initial Michael's addition of a core amine to acrylonitrile, and then in the presence of Raney nickel hydrogenation to core amine, which can be further processed in a stepwise manner. Katriztky et al. [27] synthesized various nitriles with the addition of acetic acid as a catalyst with yields reaching 90% but then failed to obtain pure primary amines in hydrogenation process over Raney nickel catalyst with the addition of ammonia in methanol, preventing from formation of secondary amino functionality. Buhleier et al. [28] by means of $NaBH_4{\cdot}CoCl_2$ reduced nitrile in 24% yield. After 15 years, this process was modified by using diisobutylaluminium hydride in a mixed solvent system of THF and hexane [29]. Worner and Mülhaupt performed hydrogenation of nitriles over Raney nickel in the presence of sodium hydroxide as a cocatalyst, which led to a decrease in reaction yield (ca. 70%) due to the retro-Michael reaction caused by the action of a strong base [30].

De Brabander-van der Berg and Meijer reported successful hydrogenation over Raney cobalt catalyst in water under H_2 pressure of 30–70 bar. Under these conditions no side products were obtained and 99.5% selectivity level per conversion was achieved [31].

The cross-linking agents (N,N,N-tri(3-aminopropyl)amine and (N,N,N′,N′-tetra(3-aminopropyl)ethylenediamines were obtained in the catalytic hydrogenation process of appropriate poly(nitriles) in the presence of modified silica support Ni catalyst.

Preparation of N,N,N-tricyanoethylamine (TCA) and N,N,N′,N′-tetracyanoethyl-1,2-ethylenediamine (TCED) was performed by means of bimolecular Michael addition of acrylonitrile to ammonia and ethylenediamine, respectively. The exotermic reaction of poly(nitriles) intermediates TCA and TCED starts with an addition of excessive acrylonitrile to the appropriate amine under ambient conditions.

The poly(nitriles) used for hydrogenation process were synthesized in the reaction of 30% water solution of ammonia or 1,2-ethylenediamine containing 1,4-dioxane and ionic liquid with an excessive amount of acrylonitrile. When the reaction was completed at 60°C, poly(nitriles) were separated by means of a separatory funnel and dried over magnesium sulfate in order to remove any traces of water.

The synthesized poly(nitriles) are the substrates in the hydrogenation reaction to obtain the primary poly(amines). This reaction was carried out in the presence of the Ni catalyst in the high-pressure autoclave giving the final product of the polyamines—(N,N,N-tri(3-aminopropyl)amine) (TAA) or (N,N,N′,N′-tetra(3-aminopropyl)-ethylenediamine)) (TAED) with 82.0% and 88.0% yields, respectively.

Poly(nitriles) used for hydrogenation reaction were obtained in a cyanoethylation process, as shown in **Figure 5** without purification procedures.

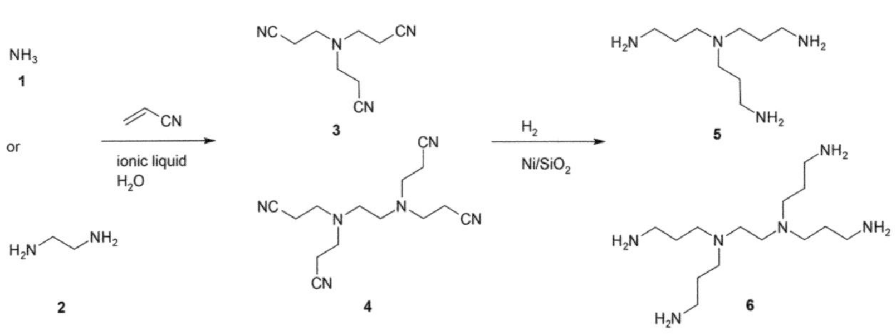

Figure 5. Cyanoethylation of ammonia and 1,2 ethylenediamine to N,N,N tricyanoethylamine (TCA) — 3 and N,N,N′,N′ tetracyanoethylo 1,2 ethylenediamine (TCED) 4 followed by their hydrogenation reaction gives (N,N,N tri(3 aminopropyl)amine) (TAA) — 5 and (N,N,N′,N′ tetra(3 aminopropyl)ethylenediamine)) (TAED) — 6, respectively.

The hydrogenation reactions were carried out at 135°C, under 10 bars for 6 h. In the case of TAA synthesis, yield of this reaction reached 85.7%, while for TAED exceeded 82%.

In case of hydrogenation of TCA for 0.14 mol of nitrile, 0.42 mol of H_2 is needed for total conversion of nitrile groups into primary amino groups. It was determined that to achieve full conversion of poly(nitriles) to the poly(amines) seven cycles of refilling hydrogen were necessary, which equaled to 0.42 mol of H_2. Exactly the same procedure was applied for hydrogenation of TCED, but full nitrile reduction was achieved after 12 cycles of hydrogen refill. In this case 0.72 mol of hydrogen was needed for the completion of the hydrogenation process (**Figure 6**).

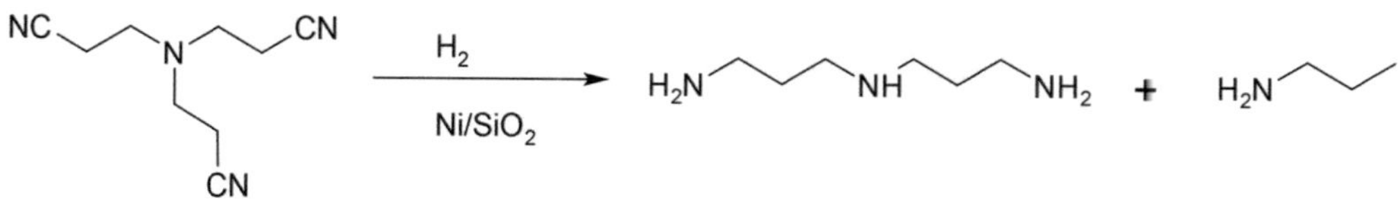

Figure 6. Hydrogenolysis of the TCA to N,N-diaminepropylamine and n-propylamine.

3.3. Identification of branched poly(amines)

In order to identify poly(amines) as the hydrogenation reaction products of the poly(nitriles), GC/MS, FTIR, and ^{1}H NMR instrumental analytical techniques were used.

3.3.1. GC/MS analysis

GC/MS analyses of the branched polyamines were done by use of a gas chromatograph HP 6890 Series GC System (Hewlett-Packard) equipped with HP 5973 Network quadrupole mass selective spectrometric detector (Agilent Technologies) in a similar way as it was described in Section 2.3.1.

Hydrogenation process of TCED leads to the formation of N,N,N′-triaminepropyl-1,2-ethylenediamine with retention time t_r =19.61 min and N,N,N′,N′-tetraaminepropyl-1,2-ethylenediamine (TAED) with t_r = 23.60 min and small amounts of undesired by-products, which peaks are shown in **Figure 7**. In this reaction mixtures, besides the main products—components B and C, n-propylamine (component A) with t_r =2.31 min is present. The mechanism of its formation probably is either by hydrogenolysis or retro-Michael reaction in which main products decomposes into molecules with lower molecular weights.

3.3.2. FTIR analysis

The standard Fourier Transform Infrared Spectrometer Nicolet 6700 type (FTIR) equipped with operating software Omnic from Thermo Company was used for the qualitative determination of the functional groups occurring in the hydrogenation reaction products. FTIR spectra were registered in the range of 4000–650 cm^{-1} with the resolution of 4 cm^{-1}.

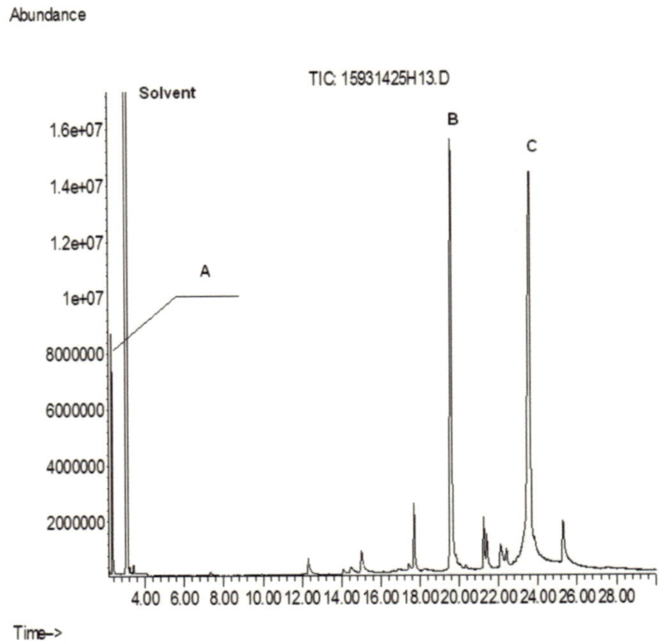

Figure 7. Chromatogram of the typical mixture of the hydrogenation products of N,N,N′,N′ tetracyanoethyl 1,2 ethylenediamine (TCED). GC peaks refer to: (A) n-propylamine; solvent — 1,4-dioxane; (B) N,N,N′ triaminepropyl-1,2-ethylenediamine; (C) N,N,N′,N′ tetraaminepropyl-1,2-ethylenediamine.

In the FTIR spectra (**Figure 8**) of the TCA and TCED there is apparent peak at 2250 cm^{-1} assigned to C≡N group. In their FTIR spectra, after hydrogenation reaction of TCA and TCED nitriles in these groups almost completely disappeared, and instead broad peaks around 3350 cm^{-1} assigned to N—H bonds in primary amino groups are observed.

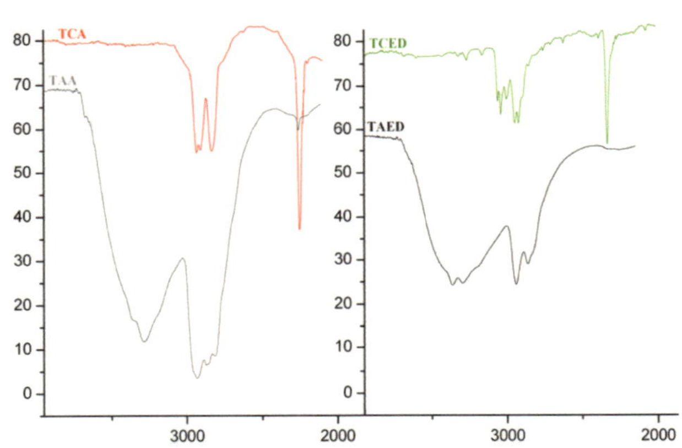

Figure 8. FTIR spectra of nitriles derived from ammonia (TCA) and ethylenediamine (TCED) and their hydrogenated products TAA and TAED.

http://dx.doi.org/10.5772/65408

3.3.3. ¹H NMR analysis

For ¹H NMR analyses the NMR Bruker Ultrashield was used. All spectra were taken at the frequency of 400 MHz, $CDCl_3$ with 0.03% TMS (v/v) was used as a solvent. Qualitative analyses were performed by using the Bruker Topspin 1 software.

The structures of final products after hydrogenation reactions were confirmed by use of ¹H NMR techniques. On the spectra recorded for TAA and TAED, there are shifts that could be assigned to protons corresponding to amine groups, meaning that the "core" molecules, ammonia and ethylenediamine, respectively, were successfully subjected to cyanoethylation reaction by acrylonitrile molecules.

The signals observed in the ¹H NMR spectra of TAA and TAED compounds (**Figure 9**) were taken to confirm their structures.

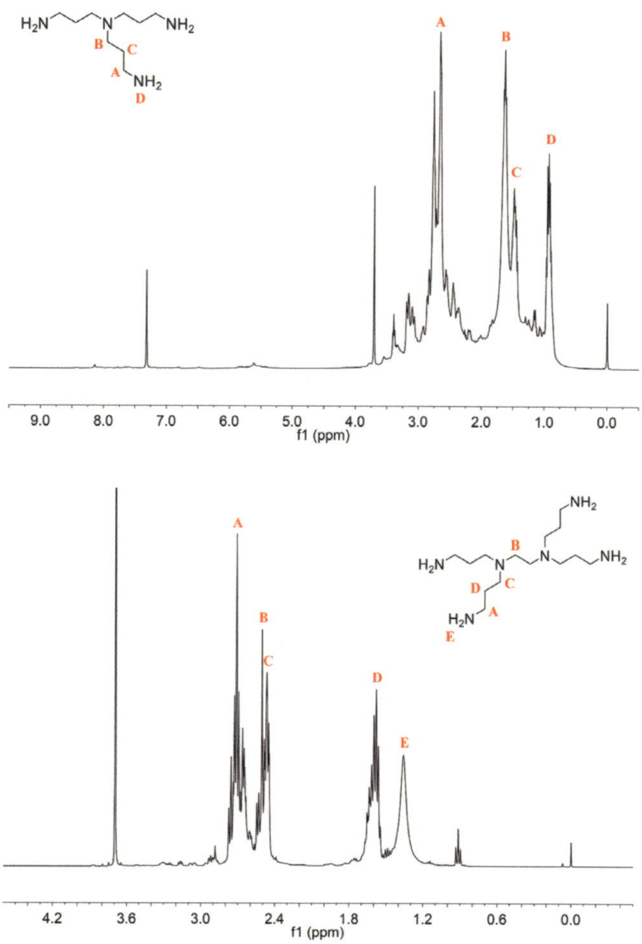

Figure 9. ¹H NMR ($CDCl_3$) spectra of TAA (above) and TAED (below).

In the case of TAA: 0.98 ppm, s, 6H, $-NH_2$; 1.50 ppm, p, $-CH_2-CH_2$; 1.70 ppm, t, 6H, $N-CH_2-CH_2$, 2.70 ppm, t, 6H, $-CH_2-NH_2$.

In the case of TAED: 1.36 ppm, s, 8H, $-NH_2$; 1.60 ppm, p, $-CH_2-CH_2$; 2.46 ppm, t, 8H, $N-CH_2-CH_2$, 2.50 ppm, s, 4H, $N-CH_2-CH_2-N$, 2.75 ppm, t, 8H, $-CH_2-NH_2$.

During hydrogenation process undesired reactions take place, leading to the formation of low-molecular weight products. After hydrogenation of poly(nitriles), besides the main products being TAA and TAED, the n-butylamine is one of the most abundant by- products present in the reaction mixture. It is formed by cleavage of bond between nitrogen heteroatom and β-carbon atom from acrylonitrile.

4. Conclusions

New generation additives for polymers such as plasticizers and cross-linking agents were synthesized in the catalytic hydrogenation process in the presence of Ni catalyst.

All hydrogenation products were analyzed using different analytical instrumental techniques such as GC/MS, ESI/MS, and spectroscopic techniques (e.g., ^{1}H NMR).

The new generation of plasticizers such as di(n- and isononyl)cyclohexane-1,2-dicarboxylates (DINCH components) are the new group of specific and safe plasticizers applied as the substitutes of di(alkyl) phthalates, especially di(2-ethylhexyl) phthalate (DEHP) and di(n- and isononyl) phthalates in polymers.

The identification of these compounds permits us to determine both the structures of the main products and their by-products as well. *Cis* and *trans* isomers of the cyclohexane-1,2-dicarboxylates may be applied to the identification of some DINCH constituents extracted from polymers and also in the elucidation of their structures in the human metabolites. ESI/MS mode analysis of these compounds and the knowledge about their mass fragmentation enable their detection, although without differentiation between individual *cis* and *trans* isomers. The obtained identification results concerning the determination of the individual chemical structures of some *cis* and *trans* di(n- and isoalkyl(C_4–C_9))cyclohexane-1,2-dicarboxylates isomers (DINCH) maybe used in their determination of exposure and risk assessments.

Also, poly(amines) possessing usually more than two functional groups are very useful as cross-linking agents and are used in the modification of physicochemical properties of the epoxy resins by the most effective chemicals.

The obtained results from the analysis of the final product of hydrogenation reaction of both *cis* and *trans* di(n- and isoalkyl(C_4–C_9))cyclohexane-1,2-dicarboxylates isomers (DINCH) and branched poly(amines) are useful in the optimization of processes of their production in an industrial scale.

http://dx.doi.org/10.5772/65408

Acknowledgements

The part of this work was realized within the INNTECH Project "New generation of cross-linking agents for epoxy resins." Research was sponsored by The National Centre for Research and Development under grant no INNOTECH-K1/IN1/49/150947/NCBR/12. The authors would like to thank Dr. B. Poźniak and I. Semeniuk (M.Sc.) for spectrometric analyses of some hydrogenation products.

Author details

Euzebiusz Jan Dziwiński*, Bartłomiej Bereska, Jolanta Iłowska, Józef Lach, Agnieszka Bereska and Michał Szmatoła

*Address all correspondence to: dziwinski.e@icso.com.pl

Institute of Heavy Organic Synthesis "Blachownia", Kędzierzyn-Koźle, Poland

References

[1] Marcilla A, Garcia S, Garcia-Quesada JC. Migrability of PVC plasticizers. Polymer Testing. 2008; 27:221–233. DOI: 10.1016/j.polymertesting.2007.10.007

[2] LaFleur AD, Schug KA. A review of separation methods for the determination of estrogens and plastics-derived estrogen mimics from aqueous system. Anal. Chim. Acta. 2011; 696:6–26. DOI: 10.1016/j.aca.2011.03.054

[3] European Union, Regulation (EC) No. 1272/2008 of the European Parliament and of the Council on Classification, Labelling and Packaging of Substances and Mixtures, Amending and Repealing Directives 67/548/EEC and 1999/45/EC, and Amending Regulation (EC) No. 1907/2006, 2008.

[4] Bernard L, Decaudin B, Lecoeur M, Richard D, Bourdeaux D, Cueff R, Sautou V. Analytical methods for the determination of DEHP plasticizer alternatives present in medical devices: a review. Talanta. 2014; 129:39–54. DOI: 10.1016/j.talanta.2014.04.069

[5] David RM. Exposure to phthalate esters. Environ. Health Perspect. 2000; 108:A440. DOI: 10.1016/j.cppeds.2007.11.001

[6] Chiellini F, Ferri M, Morelli A, Dipaola L, Latini G. Perspectives on alternatives to phthalate plasticized poly(vinyl chloride) in medical devices applications. Progr. Polym. Sci. 2013; 38:1067–1088. DOI: 10.1016/j.progpolymsci.2013.03.001

[7] Silva MJ, Kato K, Wolf C, Samandar E, Silva SS, Gray EL, Needham LL, Calafat AM. Urinary biomarkers of di-isononyl phthalate in rats. Toxicology. 2006; 223:101–112. DOI: 10.1016/j.ijheh.2006.11.008

[8] Koch HM, Angerer J. Di-iso-nonylphthalate (DINP) metabolites in human urine after a single oral dose of deuterium-labelled DINP. Int. J. Hyg. Environ. Health. 2007; 210:9–19. DOI: 10.1016/j.ijheh.2006.11.008

[9] Koch HM, Müller J, Angerer JJ. Determination of secondary, oxidized di-iso-nonylphthalate (DINP) metabolites in human urine representative for the exposure to commercial DINP plasticizers. J. Chromatogr. B. 2007; 847:114–125. DOI: 10.1016/j.jchromb.2006.09.044

[10] Zhao J, Xue M, Huang Y, Shen J. Hydrogenation of dioctyl phthalate over supported Ni catalyst. Catal. Commun. 2011; 16:30–34. DOI: 10.1016/j.catcom.2011.08.035

[11] Dziwinski EJ, Tasarska A, Lach J. GC/MS identification of some dibutyl phthalates hydrogenation products (in Polish). Przem. Chem. 2010; 89:1305–1308.

[12] Dziwinski EJ, Idczak J, Lach J. Gas chromatographic identification of some products of hydrogenation of di(n-alkyl(C_4–C_{12})) phthalates (in Polish). Przem. Chem. 2012; 91:1947–1951.

[13] Van den Dool H, Kratz PJ. A generalization of the retention index system including linear temperature programmed gas-liquid partition chromatography. J. Chromatogr. 1963; 119:463–472. DOI: 10.1016/S0021-9673(01)80947-X

[14] Dziwinski EJ, Hetper J, Kołodenny RJ. Gas chromatography-mass spectrometry of C1-C10 alkyl benzyl phthalates. J. Chromatogr. 1984; 288:221–226. DOI: 10.1016/S0021-9673(01)93700-8

[15] Smith RM. Understanding Mass Spectra. A Basic Approach. 2nd ed. Wiley-Interscience, Hoboken, NJ; 2004. 254 p.

[16] Dziwinski EJ. Mass spectrometry determination of esters of dicarboxylic acids applied as plastifiers (in Polish). Przem. Chem. 1992; 71:351–353.

[17] Ellis B, editor. Chemistry and Technology of Epoxy Resins. Chapman & Hall, London; 1993. pp. 37–71. DOI: 10.1007/978-94-011-2932-9

[18] Tanaka Y, editor. Epoxy Resins. Chemistry and Technology. Marcel Dekker, New York; 1973. DOI: 10.1007/BFb0035357

[19] Bereska B, Iłowska J, Czaja K, Bereska A. Hardeners for epoxy resins. Przem. Chem. 2014; 93:443–448. DOI: dx.medra.org/10.12916/przemchem.2014.443

[20] Sangermano M, Malucelli G, Bongiovanni R, Triola A, Harden A. Investigation on the effect of the presence of hyperbranched polymers on thermal and mechanical properties of an epoxy UV-cured system. Polym. Int. 2005; 54: 917–921. DOI: 10.1002/pi.1789

[21] Oh JH, Jang S, Lee H. Environmentally friendly high performance homopolymerized epoxy using hyperbranched epoxy as a modifier. Polymer 2001; 42: 8339–8347. DOI: 10.1039/C5RA22474H

[22] Voit BJ. New developments in hyperbranched polymers. Polym. Sci., Part A. Polym. Chem. 2000; 38:2505–2525. DOI: 10.1002/1099-0518(20000715)38:14<2505::AID-POLA10>3.0.CO;2-8

[23] Yates CR, Hayes W. Synthesis and applications of hyperbranched polymers. Eur. Polym. J. 2004; 40:1257–1281. DOI: 10.1016/j.eurpolymj.2004.02.007

[24] Tomalia DA, Naylor AM, Goddard WA III. Starburst dendrimes: molecular-level control of size, shape, surface chemistry, topology and flexibility from atoms to macroscopic matter. Angew. Chem. 1990; 102:119–157. DOI: 10.1002/anie.199001381

[25] Frauenrath H. Dendronized polymers-building a new bridge from molecules to nanoscopic objects. Prog. Polym. Sci. 2005; 30:325–384. DOI: 10.1016/j.progpolymsci.2005.01.011.

[26] Gao C, Yan D. Hyperbranched polymers: from synthesis to applications. Prog. Polym. Sci. 2004; 29:183–275. DOI: 10.1016/j.progpolymsci.2003.12.002

[27] Katriztky AR, Yang B, Tack RD. Synthesis dendramines, dendramides and their new application in the oil industry. J. Chin. Chem. Soc. 1997; 44:575–580.DOI: 10.1002/jccs.199700088

[28] Buhleier E, Wehner W, Vögtle F. "Cascade"- and "nonskid-chain-like" syntheses of molecular cavity topologies. Synthesis. 1978; 2:155–158. DOI: 10.1055/s-1978-24702

[29] Moors R, Vögtle F. Dendrimere polyamine. Chem. Ber. 1993; 126:2133–2135. DOI: 10.1002/cber.19931260925

[30] Wörner C, Mülhaupt R. Polynitrile- and polyamine- functional poly(trimethylene imine) dendrimers. Angew. Chem. Int. Ed. Engl. 1993; 32:1306–1308. DOI: 10.1002/anie.199313061

[31] de Brabander-van der Berg EMM, Meijer EW. Poly(propyleneimine) dendrimers: large-scale synthesis by heterogeneously catalyzed hydrogenations. Angew. Chem. Int. Ed. Engl. 1993; 32:1308–1311. DOI: 10.1002/anie.199313081

Chapter 12

Organometallics: Exploration Tool for Surface Phenomena in Heterogeneous Catalysis

Martin Kindl, Ondřej Matuška and Petr Kačer

Additional information is available at the end of the chapter

http://dx.doi.org/10.5772/65337

Abstract

Homogeneous and heterogeneous catalyses have sufficient common features that are based on the same molecular phenomena, thus permitting them and their benefits to be directed to the same objective. Moreover, conditions that apply to one of them can yield benefits to the other. In this chapter, the progress of combining these advantages is demonstrated on a particular topic of testing a suitable structural model from the field of organometallic compounds, whose structure resembles complexes adsorbed on the surface of heterogeneous catalysts. The adsorbed complexes represent key structures in the conversion of reactants to products, while their demanding structural characterization is one of the main persistent topics of heterogeneously catalyzed reactions. These include compounds procuring a precious mechanistic interpretation of such complex concepts as the catalytic activity and selectivity. The knowledge of the structure of an adsorbed complex leads to a better understanding of the complex surface processes and more precisely defined heterogeneous catalysts. The understanding of processes on the molecular level allows discussing the structure-activity. Analogies and differences between molecular (homogeneous catalysis) and surface (heterogeneous catalysis) chemistry are addressed in this review. The work demonstrates that these two fields are complementary, presenting a large potential for the development of novel and superior catalysts when the advantages of both sub-disciplines are combined.

Keywords: catalysis, molecular modeling, hydrogenation, molecular structure effect, platinum

1. Introduction

Traditionally, the area of catalysis has been divided into three fields: heterogeneous, homogeneous, and enzymatic catalyses. While heterogeneous catalysis has offered a great advantage in terms of the process design, the separation of products from a catalyst and its regeneration, it has usually suffered from a lack of understanding of the so-called "active site." This fact has plausibly impeded the rational development of these systems. On the contrary, homogeneous catalysis can be designed in a more rational way as its properties are easily tuned via ligand designs, providing a substantial comprehension of elementary steps. Nevertheless, these processes have often required undertaking many technical advances to be competitive to heterogeneous catalysts. From the industrial point of view, this led in the final practice to the preference of heterogeneous from homogeneous catalysis. However, the development of enhanced heterogeneous catalysts has been hindered typically owing to their content of numerous variable active sites and their low concentration. Homogeneous catalysts, on the contrary, have been well-defined systems that could be easily characterized and studied. Comparative studies of homogeneous and heterogeneous catalyses, providing a successful implementation of appropriate homogeneous models with molecular modeling, can yield a new insight into the complex of processes accompanying heterogeneous catalysis. In order to perceive the involved complex molecular events, it is also essential to construct a clear-cut active site, test its catalytic performance, and assess the relationship between its structure and activity. Finally, the acquired picture can be utilized to design a new generation of catalysts applicable "as in homogeneous catalysis."

Typically, the determining characteristic of heterogeneous catalysis lays in the structural composition of its active sites and molecular structures, in the case of a metal catalyst and the reactant, respectively [1, 2]. In other words, the chemical and physical properties of the surface of the active sites and the molecule entering the reaction generate a generous number of parameters decisive for the catalytic mechanism. In order to predict the catalytic behavior, geometric and electronic structural properties have been traditionally studied [2], particularly based on "Theory of Electronic Effects" [2, 3]. The core of this study consists in the active site (catalyst side)-reaction center (substrate side) whose behavior is determined by their interdependent electronic properties, with each component either donating or accepting electrons. Orbitals of both of the reaction players are thus directed by relatively strongly dependent interactions of their affinity and repelling forces. Last but not the least, geometric degree of freedom may have decisive impact on the surface complex (active site-reactant interaction) [2, 4]. However, the above list of parameters is not complete as other potentially significant aspects could contribute to the problem: thermodynamics, reaction conditions, and hydrogen-catalytic surface interaction [5].

Electronic effects and spatial geometry are reckoned to play the most prominent role in heterogeneously catalyzed processes. Catalytic properties, related particularly to the metal behavior, are listed below by their descending importance [2, 6–11]:

1. Form and constitution of metal particles.
2. Metallic surface organization as well as other components (as in surface alloys).
3. Surface complexes.
4. Substrate effects (strong metal-support interaction (SMSI), hydrogen spillover, and surface area).

5. Metallic presence in support pores.

The record of metals having been successfully applied to heterogeneously catalyzed reactions would be considerably extensive. Additionally, no general treatment modifying their properties (electronics, morphology, etc.) to a desired performance in individual reactions is available. As a result, the selection criteria obviously depend on rather empirical decisions.

Transition metals represent an extraordinary group of elements finding a widespread utilization in the applied heterogeneous catalysis. From the point of view of the band theory [12], all of them contain positively charged ions having shells with partially occupied *d*-bands, in which electrons freely fluctuate. Predictably, these form bonds with adjacent atoms using *dsp* orbitals with inerratic *d*-function. As a result, the structure of the metal surface is nonuniform, which is further pronounced [12] by depositing the metal particles on an internal surface of a suitable high surface area support (active carbon, zeolites, etc.). Catalytic particles are rarely uniform as evidently determined by their X-ray diffraction diagrams. Elemental metal crystallites are composed of crystallographic planes diverse in their surface coordination and architecture. There is a direct implication [6, 13] of this phenomenon in a variable degree of saturation of coordination sites on the metal surface. As a consequence, a different catalytic behavior is macroscopically observed [6, 14, 15].

Current methods of molecular modeling offer a wide array of applications even to heterogeneous catalysis. However, it is still necessary to significantly simplify the studied systems in comparison with the reality. Although modeling requires many compromises, its use allows obtaining very interesting results, which provide information on the mechanism of catalytic reactions on the surface. Owing to the development of computer technology, computer modeling and simulation have been increasingly recognized as important tools for the study and development of catalytic systems. These tools have the potential to penetrate into the reaction mechanisms, can predict properties of catalysts not yet synthesized, and provide information obtained by various computational techniques, which together with experimental results procure a comprehensive picture of the system. Nevertheless, it is always essential to maintain a proficient collaboration of computational chemists with researchers. In addition, in order to validate the new modeling techniques, theoretical results must be sufficiently confirmed experimentally. Therefore, it is advisable to have a comparison of various experiments, such as kinetic studies of reaction rates, thermodynamic information on adsorption, and spectroscopic data on the level of molecular structure. The double-feedback technique (dual-feedback mode) seems to be the "most prolific" strategy for the utilization of modeling in catalytic research where experiments are employed to confirm the results of modeling and the modeling is used to explicate the experimental results, to design new experiments, and perhaps in the future also to replace experiments with theoretical screening of different catalysts and reaction conditions.

One option to simplify the heterogeneously catalyzed reactions for the application of molecular modeling is finding a suitable model, which first needs to be designed, then mathematically and experimentally tested and if its quality is proved, the method is validated and the limits of its applicability are determined. In the case of heterogeneous platinum catalysts designated for hydrogenation reactions, well-structurally defined compounds of (olefin)Pt(PPh_3)$_2$ seems to offer a suitable model of adsorbed surface complexes as the key structures in the transformation of a reactant to products. These compounds are platinum organometallic compounds containing coordinated olefins, which in a sense resemble the morphology of the metal surface and chemisorptions occurring on it. Although these covalent compounds cannot catalyze homogeneous hydrogenation, such as the rhodium analog Rh(PPh_3)$_3$Cl, their structure very well approximates the adsorbed state on heterogeneous platinum catalysts. The model of organometallic compounds (olefin)Pt(PPh_3)$_2$ represents an attractive alternative for the application to molecular modeling in heterogeneous catalysis. By comparing values with a similar significance (adsorption *vs.* dissociation energy) or the differences in geometric data, very good qualitative (corresponding order of stabilities) and even quantitative (correlating E_{ads} and E_{diss}) correlations can be achieved.

2. Model of adsorbed complex: (olefin)Pt(PPh_3)$_2$ complex

The compounds of the type (olefin)Pt(PPh_3)$_2$ belong to a group of intensively studied substances. One of the reasons for their study are the binding ratios in these molecules. These compounds, are characterized by their structure, which is noticeably similar to the adsorbed state of olefins on the catalyst surface based on transition metals (di-σ-adsorbed complex). The adsorbed complex, as the key structure of a heterogeneously catalyzed reaction, and its problematic structural characterization in comparison to the characterization of (olefin)Pt(PPh_3)$_2$ complexes, is the reason why the described group of substances has been gaining much interest of "catalytic chemist" despite the fact that the mentioned complexes do not exhibit the catalytic activity in of a homogeneous hydrogenation catalyst. The concept of the similarity of these compounds with the adsorbed di-σ-complexes was initiated from the following experimental observations:

- The organization of alkenes has significantly altered when coordinated with platinum(0). Unsaturated carbonic atoms, diverging from the plane of the double bond near the platinum atom, cause a deviation identical to the decrease in strength and extending the carbonic double bond.

- The olefin double bond in platinum(0) complexes diverges in a minor angle from the triangular plane arranged by platinum, midpoints of the olefinic double bond and the remaining two ligands.

- The platinum(0)-olefin bond appeared firm in the complexes.

- Proton nuclear magnetic resonance (^{1}H NMR) spectrum of platinum(0)-olefin complexes revealed that during coordination of olefin, the starting sp^2 hybridization of the carbon atom was near the sp^3 hybridization, which could be interpreted as a result of the production of three-membered ring (cyclopropane type) (**Figure 1**).

http://dx.doi.org/10.5772/65337

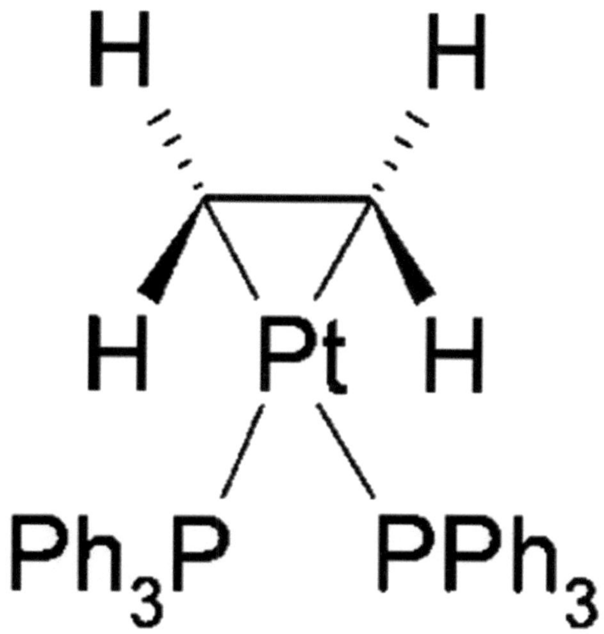

Figure 1. Three-membered ring (cyclopropane type) formed by coordination of alkene to Pt(0).

The bonding mode in Pt(0)-olefin complexes has usually been explained by means of a modified Dewar-Chatt-Duncanson approach, which assumed two simultaneous and partly mutually dependent interactions of the orbitals of the metal and alkenes, namely donation and back donation (**Figure 2**). The significance of donation can be sufficiently described using the classical donation of $\pi_{C=C}$ alkene electrons to vacant *d*-orbitals of platinum(0). The back donation has been perceived as the interaction of occupied orbitals of platinum (in the band theory of solids, these were the bands laying below the Fermi energy level) with $\pi_{C=C}^*$ olefin orbital. The reverse donation has a significant impact on the stability of the complex.

The lower is the electron density on C=C double bond (causing a stronger withdrawal of electrons), the lower is the level of $\pi_{C=C}^*$ orbital and thus a lower energy difference between $\pi_{C=C}^*$ alkene orbital and the occupied *d*-orbitals of Pt. As a result, a stronger interaction occurs between $\pi_{C=C}^*$ and occupied Pt (*d*) orbitals, which constructs a stronger bond of Pt-alkene. If a common alkene is described as ethylene-bearing electron-acceptor substituents, then the substituents have the tendency to stabilize the complex by shifting highest occupied molecular orbital/lowest unoccupied molecular orbital ((HOMO)/LUMO) energy levels in alkene toward the planes of $Pt(PPh_3)_2$. To the contrary, the groups increasing their electron density destabilize the complex to such an extent that the process of preparation could be rendered impossible. **Figure 3** shows the general order of stability in the group of ethylene derivatives bearing typical smaller substituents.

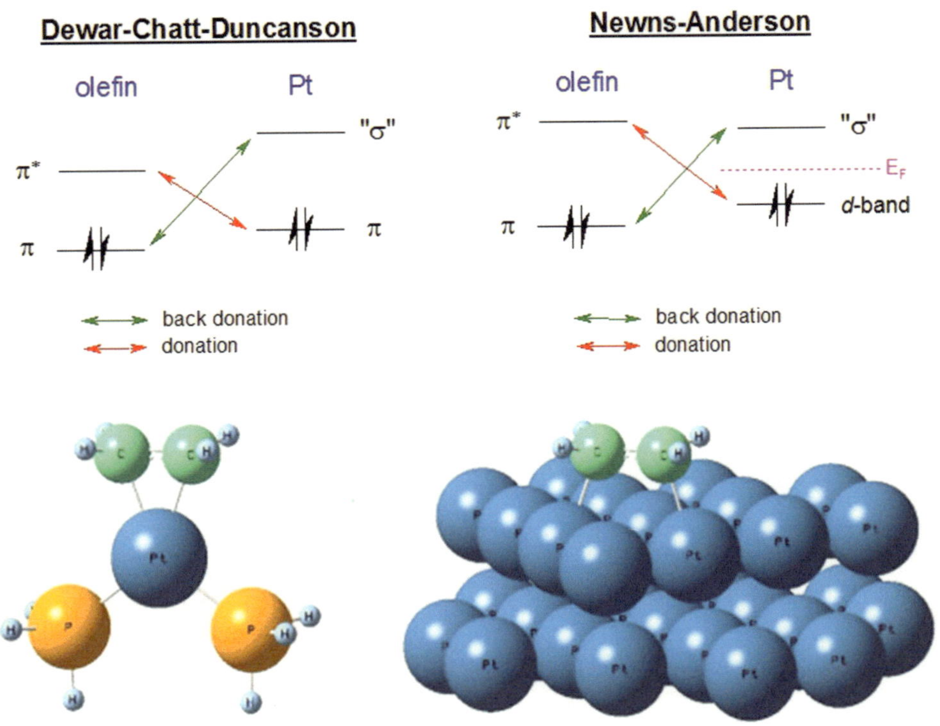

Figure 2. Orbital description of the bonds in the olefin-metal models: (1) Dewar-Chatt-Duncanson model derived for organometallic compounds; (2) Newns-Anderson model derived for surface complex substrate-catalyst.

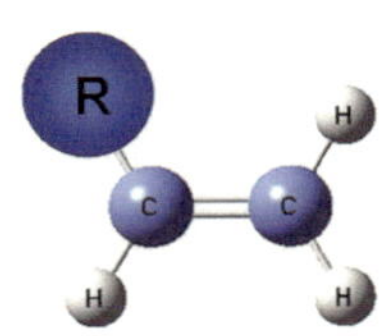

R = CF_3 > CHO > CN > CO-R > CO-OR > F > Cl > Ph > H > SiR_3 > $CH_2(OH)R$ > R

Figure 3. General order of coordinated olefins stability.

3. Synthetic procedures of (olefin)Pt(PPh_3)$_2$ complex

The literature has described a number of synthetic methods [11, 17] that led to the production of complexes of the type (olefin)Pt(PPh_3)$_2$. There are few basic procedures for the synthesis of these organometals. The usual starting substance is either chloride of bis(triphenylphos-

http://dx.doi.org/10.5772/65337

phine)platinum(II), ethylene-*bis*(triphenylphosphine)platinum(0), tris(triphenylphosphine) platinum, or tetrakis(triphenylphosphine) platinum. In particular, the following reactions are involved: the reduction reaction of Pt(II)-complexes (e.g., chloride of bis(triphenylphosphine)platinum) and the exchange of alkenic ligands in Pt(0)-coordination compounds, such as $[Pt(PPh_3)_2(C_2H_4)]$. The majority of such synthesized complexes, however, exhibited a very low stability (**Figure 4**).

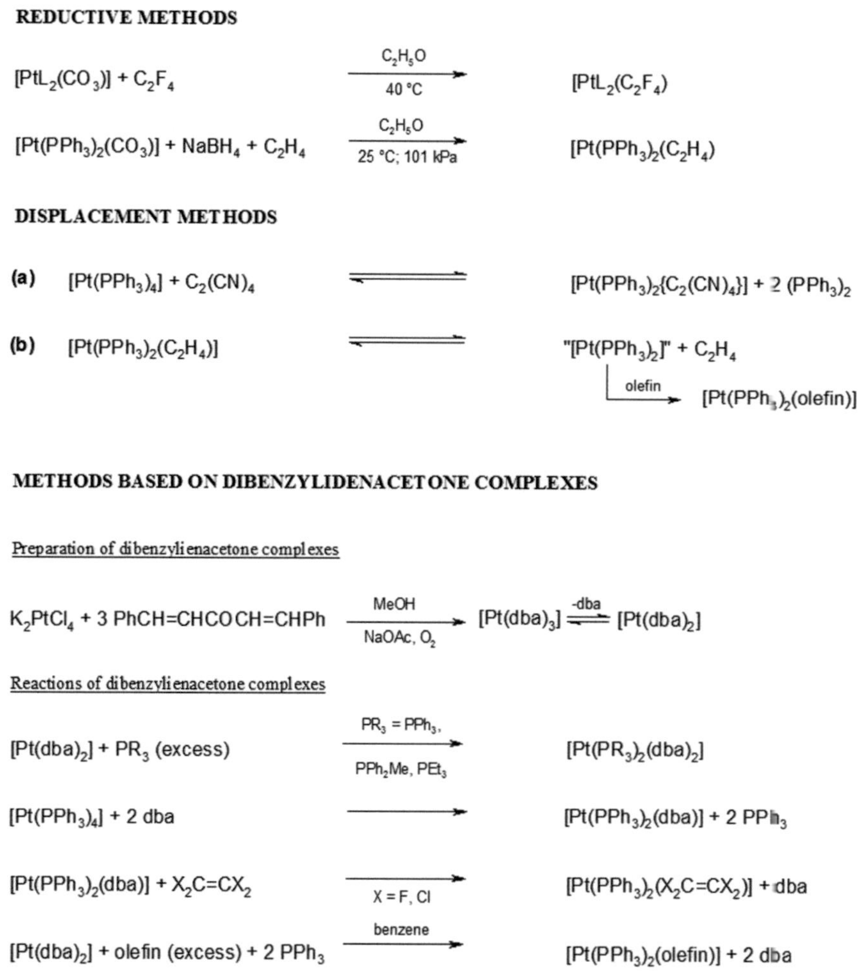

Figure 4. A scheme of preparation methods of $(olefin)Pt(PPh_3)_2$ complexes.

A series of model complexes of $(olefin)Pt(PPh_3)_2$ type were prepared. The preparation procedure consisted in reductive reactions of Pt (II)-complexes (*cis*-dichloro-bis(triphenylphosphane) platinum(II)), alternatively in the exchange of olefinic ligands in Pt-(0)-coordination compounds η-ethylene-*bis*(triphenylphosphane) platinum(0)). The syntheses were carried out in the absence of atmospheric oxygen and moisture, using the so-called Schlenk techniques

(**Figure 5**). In **Figure 6**, the ligands of the synthetically prepared complexes of the type (olefin)Pt(PPh_3) are depicted, and clearly imply that the prepared complexes could be characterized as substituted ethylene with electron-accepting substituents, which tend to stabilize the produced complex by shifting the energy levels of HOMO/LUMO in olefin toward the levels of Pt$(PPh_3)_2$. By contrast, the groups increasing the electron density in this place destabilize the complex to such a degree that its preparation is not possible. The example could be complexes (alkenic alcohol)Pt(PPh_3) that have never been spectroscopically proven in the reaction mixtures. The reason could be sought in a low stability of these complexes probably caused by a hydroxyl group present in molecules of an alkenic ligand, which by increasing the electron density on C=C bond destabilizes the potentially produced complex. **Figure 6** captured the theoretically calculated stability order of the group of ethylene derivatives bearing typical smaller substituents, showing a good correlation with the stability of the prepared complexes, which were experimentally verified by NMR *"in situ"* competitive experiments.

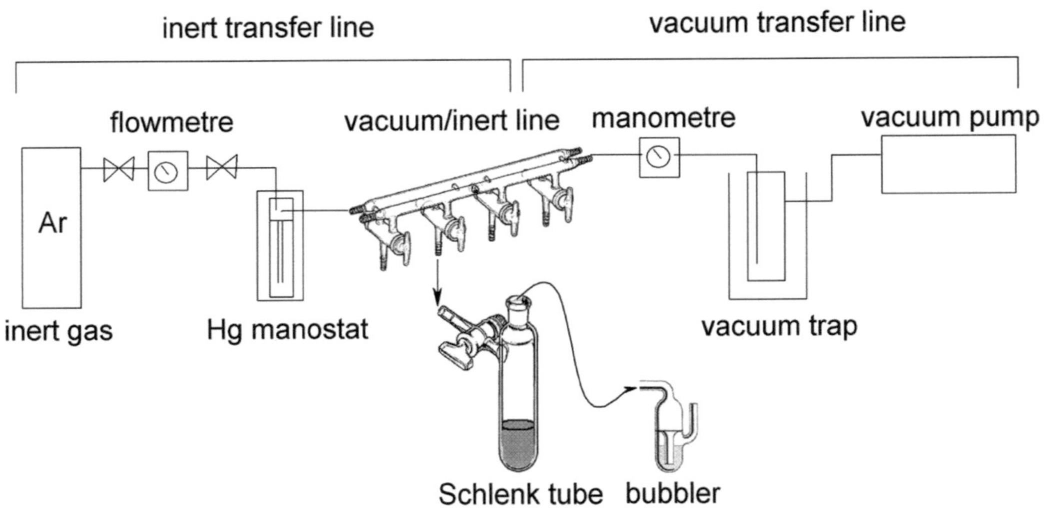

Figure 5. The apparatus used for the synthesis of (olefin)Pt-(0)-$(PPh_3)_2$ complexes.

NMR spectroscopy is one of the most important techniques for the study of substances in a solution. It may therefore be somewhat surprising to combine NMR in the liquid phase with heterogeneous catalysis, but this connection is possible and the results obtained by NMR liquid phase can be used for the interpretation of phenomena related to heterogeneous catalysis. However, it is necessary to find a suitable system by which the interaction of active site-reaction center can be approximated, and which also satisfies the condition of NMR experiments in liquid phase. If this condition is met, it is possible to monitor what is happening at the atomic level, thereby obtaining valuable information for heterogeneous catalysis.

http://dx.doi.org/10.5772/65337

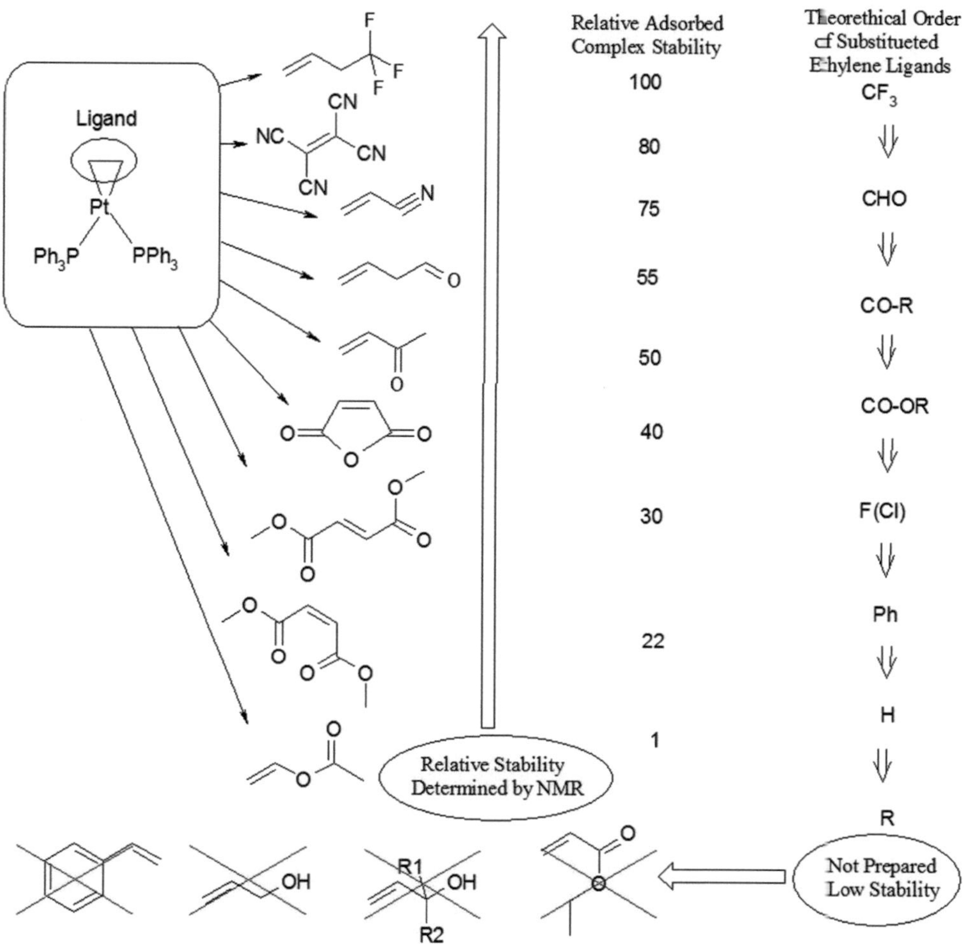

Figure 6. Synthetically prepared complexes of (olefin)Pt(PPh_3) type and their relative stability determined by NMR *"in situ"* experiments.

The developed NMR *"in situ"* method of competitive production of model (olefin)Pt(PPh_3) complexes is a significant progress toward assessing the stability of adsorbed complexes, particularly allowing to express the relative stability of a complex to be produced. The homogeneous complex of ethylene *bis*(triphenylphosphine)platinum(0) was used as the suitable model for studying the interaction of active site-reaction center, which was capable of forming coordination compounds with ligands of the CC double bond. As mentioned above, ethylene coordinated to platinum could be exchanged for other unsaturated substrates. To test this concept, experiments with esters of fumaric acid and maleic were primarily conducted, that is, with dimethyl fumarate and diethyl maleate. Using these substrates, the influence of *cis*/*trans* isomerism on the CC multiple bond was studied. Both substrates were able to displace ethylene from the complex and form a stable complex with platinum, which was manifested by a splitting of the previous singlet signal of olefinic (**Figure 7**).

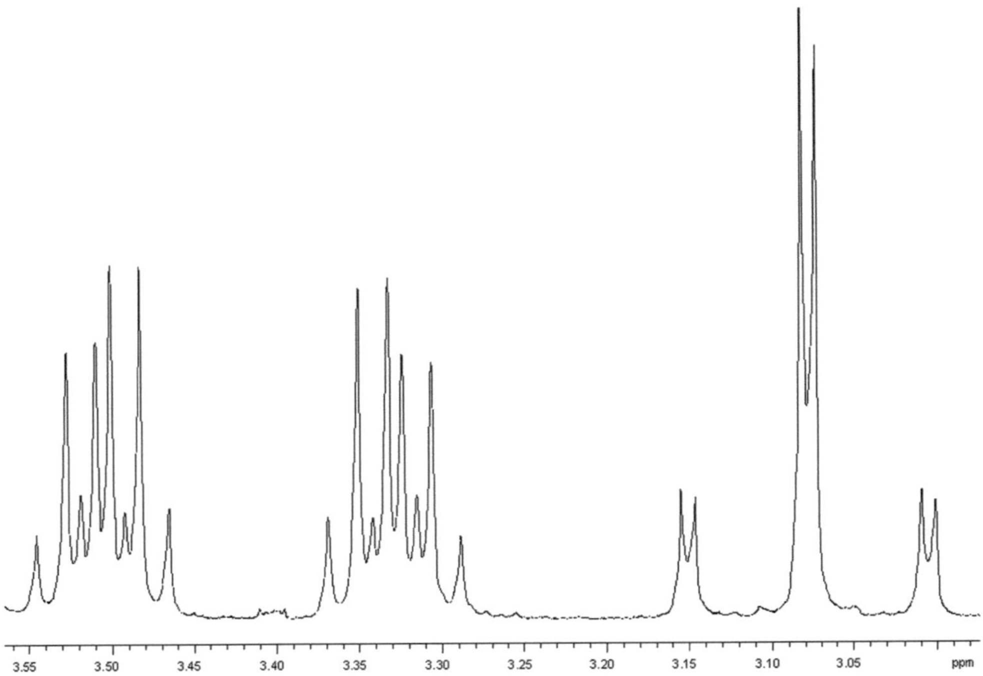

Figure 7. Fragment of ^{1}H NMR spectra of diethyl maleate bis(triphenylphosphine)platinum(0) complex.

Wherever the equimolar mixture of both ligands was added to the solution of ethylene *bis*(triphenylphosphine)platinum(0), the formation of the equimolar quantities of both complexes of esters with platinum was prevented, but a complex with dimethyl fumarate was preferentially produced. It could be assumed that the resulting ratio reflected the different adsorption properties of both ligands. In order to verify that it was the case of a behavior resulting from an inter-displacement of ligands, an experiment was conducted, in which first the complex with diethyl maleate was prepared and subsequently dimethyl fumarate added. Even under these circumstances, the ligand (diethyl maleate) displacement took place as well as the stabilizing of a ratio corresponding to the situation of a simultaneous addition of both ligands (**Figure 8**). This method has a certain limit related to a small dynamic range of NMR measurements, and thus it is necessary to compare substrates as described above with near adsorptivities. On the other hand it implies that even minor changes in the absorptivity of substrate molecules can be registered using this procedure. In a series of synthesized complexes of (olefin)Pt(PPh_3), the complexes with a near stability have always been selected, and NMR *"in situ"* competitive measurements performed. Based on them, a relative stability of the prepared complexes was determined. As shown below, this sequence very well correlated with calculated bonding energies of the listed complexes.

http://dx.doi.org/10.5772/65337

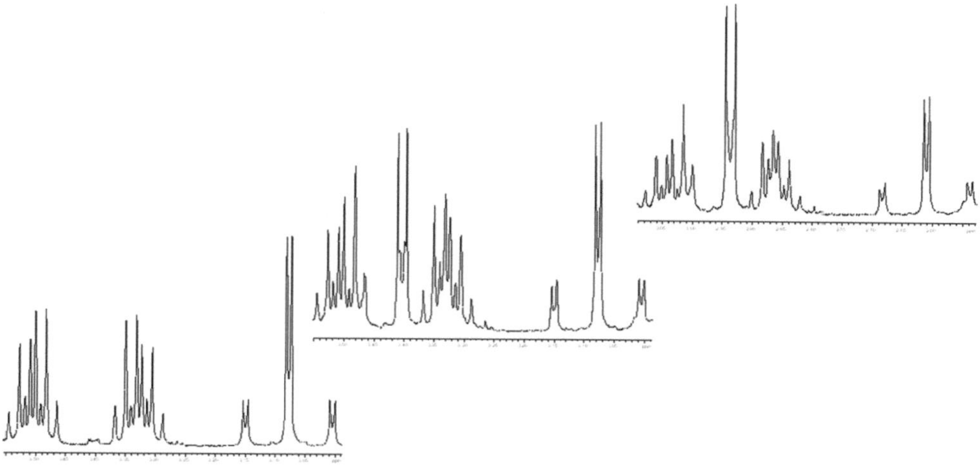

Figure 8. Selected ^{1}H NMR spectra in the course of competitive displacement of ligands (diethyl maleate and dimethyl fumarate).

4. Theoretical study of complexes of (olefin)Pt(PPh_3)$_2$

The purpose of these studies was to find a calculable correlation relationship between the structure and the stability of the complexes of the type (olefin)Pt(PPh_3)$_2$. The stability comparison was carried out by numerous available computational methods, that is, using both structural parameters (changes in the binding length and angles after coordination), then the quantification of the bond orders in alkene subunits, and finally the estimation of the binding energy between an alkene and Pt(PPh_3)$_2$. It was soon discovered and experimentally confirmed that the chemical nature of the bond metal-alkene is rather nearer the description of cyclopropane ring (see above) than the potential π-bond. These models assume that within the formalism of valence bonds, the compound Pt(PPh_3)$_2$-alkene can be considered as a derivative of platinum(II) rather than platinum(0), providing the change in hybridization of carbon atoms from sp^2 to sp^3. Many computational approaches to compounds of the type [(olefin)Pt(PPh_3)$_2$] require to tackle one important parameter, that is, a structural simplification leading to a significant reduction in the computing time, which is carried out on the assumption that phenyl groups are simply regarded as hydrogens, and in particular triphenylphosphine groups are regarded as phosphines. This simplification reduces the number of atoms in the calculation by 60 and even more significantly reduces the number of basic functions, which leads to a major acceleration of the computation. It is known that this simplification does not affect the structural and energetic results, although the steric hindrance on the platinum atom is relatively substantially altered.

The results of molecular modeling in the given group of substances were generally confirmed by the experimentally observed geometry of molecules, while any potential discrepancies have been notably dependent on the parameterization. Moreover, some aspects based on the theoretical approach allowed evaluating also such quantities as the reduction of values in the order of C-C bond, the binding energy of Pt-olefin, and other parameters. The contributions of donation and reversed donation to the bond platinum-olefin were also addressed on the theoretical basis of the orbital interaction method CDA (*charge decomposition analysis*). As a result, it was demonstrated that the π-reversed donation (Pt to LUMO olefin) was the main binding contribution and thus affected the strength of Pt-C bonds much stronger than the donation (**Figure 2**). In the presented study, a considerable attention has been devoted to quantum-mechanical calculations on a group of $(PPh_3)_2Pt$(olefin) complexes, which were sufficiently reliably approximated by complexes with the structure $(PH_3)_2Pt$(olefin), **Figure 9**. This approximation led to a marked simplification of calculations; the number of atoms in resolved molecules was thus reduced by 60 and the number of base functions by 552 to 106 bases (while it is known that the computational requirements of chemical structures increase with the number of atoms in a molecule approximately exponentially). In the search of a suitable parameterization for calculating the structure, the convergence of binding energies in molecules was preferred from their geometrical properties (bond lengths of Pt-C and Pt-P and the angles of P-Pt-P and C-Pt-C). For the geometrical optimization, the method of B3LYP (functional density functional theory (DFT) hybrid) was selected with the base 6-31G(d) for the lighter elements (Cl, P, S, O, N, F, C, and H) and the base LanL2DZ with the pseudopotential (ECP, *effective core potential*) LanL2 for Pt based on the study of *Frison* and *Grützmacher* [16], who utilized only a slightly larger "Basis Set II." To quantify the total energy of molecules, MP2 method was selected as a compromise solution for a better interpreting ability, and simultaneously for only medium requirements on the computing time. For the calculations of Hessian (estimation of infrared (IR) spectra), the same parameterization was used as for the geometrical optimization, that is, B3LYP/6-31G(d):LanL2DZ. The binding energy of olefin-platinum bonds in the molecules of (olefin)$Pt(PH_3)_2$, where the olefin was $R^1R^2C{=}CR^3R^4$ and bound by two σ-bonds with both carbons to platinum, has been interpreted as the dissociation energy of these bonds, that is, the enthalpy of decomposition reaction of an organometallic complex. Numerically, the binding energy is thus equal to the difference between the energies of a complex splitting (olefin and $(PH_3)_2$), and the complex (olefin)$Pt(PH_3)_2$, the final and the initial state, respectively:

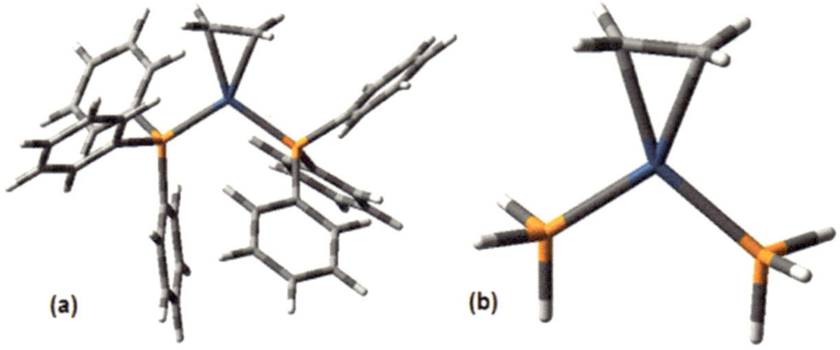

Figure 9. Molecular structure of $(PPh_3)_2Pt(CH_2{=}CH_2)$ (**a**) and $(PH_3)_2Pt(CH_2{=}CH_2)$ (**b**).

http://dx.doi.org/10.5772/65337

$$E_B = E\left[Pt\left(PH_3\right)_2\right] + E\left[olefin\right] - E\left[\left(olefin\right)Pt\left(PH_3\right)_2\right]. \quad (1)$$

During the calculations, the binding energy of Pt-olefin bonds was enumerated. Therefore, it had primarily been necessary to carry out geometrical optimization on both free planar olefins $R^1R^2C{=}CR^3R^4$ and then to evaluate their geometric properties. In olefins with a more complicated conformational behavior (e.g., unsaturated alcohols), a random inspection of the conformation space was performed related especially to those degrees of freedom where a free rotation around CC bonds occurred only with a low barrier (**Figure 10**). then the equilibrium geometry of olefins was introduced to the models of organometallic compounds $(PH_3)_2Pt(R^1R^2C{=}CR^3R^4)$, and these models were optimized without any restrictions on any degrees of freedom. The fragment (H_3P)-Pt-(PH_3) assumed the optimal geometry in the linear arrangement. The structure of the fragment (Ph_3P)-Pt-(PPh_3) was calculated as well, in which the linear structure of P-Pt-P has been analogically identified. These structures have been considered to be metastable. Although their observations have been reported, no credible spectroscopic evidence was provided. IR spectra have been estimated and optimized for different models of olefins and their complexes. The algorithm of molecular modeling of the complex (olefin)$Pt(PPh_3)_2$ is summarized in **Figure 11**.

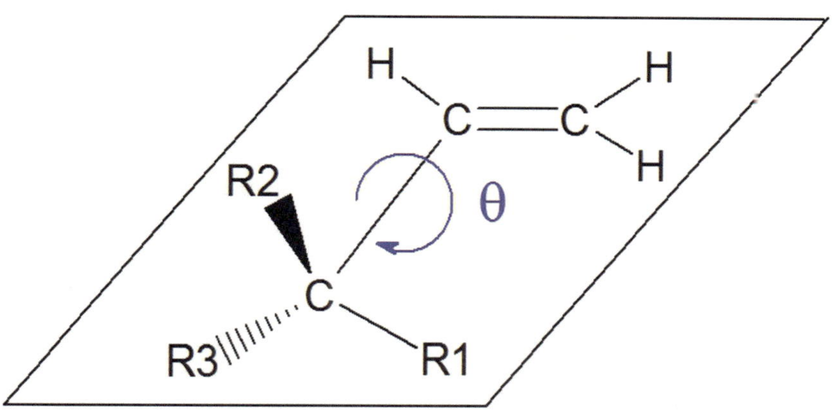

Figure 10. Alkene $CH_2{=}CH{-}CR^1R^2R^3$ with allowed rotation around C-C bond.

4.1. Structure-based effects

As a general rule, the double bonds of carbons associated with platinum(0) manifested their extension compared to the other functional groups. This has also been supported by the determined significant alterations in the constitutional and electronic properties of the olefins. The angle between P-Pt-P and C-Pt-C markedly differs among various organometallic substances due to dissimilar properties of each olefin. Platinum(0) was bound with the olefin functional groups via two σ(Pt-C) bonds. Among the listed olefins, all were arranged in ηνfashion to platinum with no free rotation, to the contrary of rotations described in the case of ethylene in platinum(II) derivatives such as Zeise's salt $K[Cl_3Pt(CH_2{=}CH_2)]$.

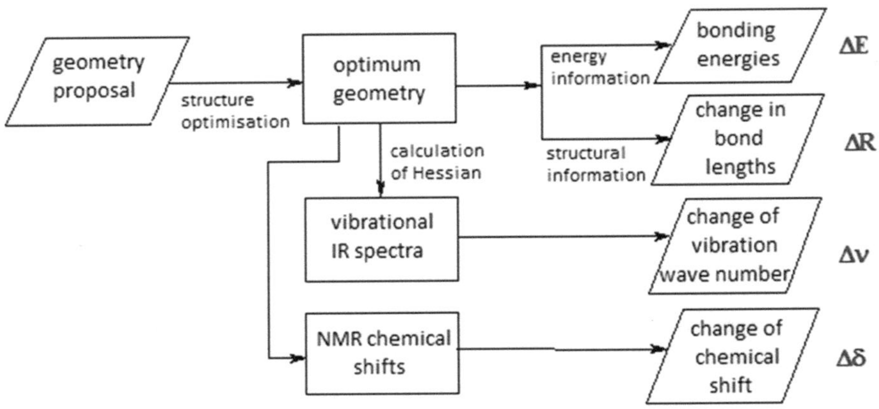

Figure 11. Flowchart of utilization of methods of molecular modeling in the group of "species" $(PH_3)_2Pt(R^1R^2C{=}CR^3R^4)$, (H_3P)-Pt-(PH_3) and $R^1R^2C{=}CR^3R^4$.

The implicit solvation method has not been applied any further, as the continuum solvation had negligible effects, the parameters as the vibrational frequencies, optimal geometries, energies, and NMR magnetic-shielding tensors, etc., as proven by a separated computation (pulse-code modulation (PCM) method, dielectric constant of dichloromethane). **Table 1** shows the global assessment of energetic and structural outcomes resulted from the final computations. The values of binding energy *EB* were applied as to assess the thermodynamic properties of Pt-olefin electron-overlap interaction: $\Delta\nu$ indicates the change in the vibrational wave number of C=C-stretching/contracting mode following the complex production, whereas ΔR is the extension primarily generated by the strength of metal-olefin complex. The pyramidalization δ was determined as an average angle of R1-C=C-R3 and R2-C=C-R4 dihedrals; thus, it has been applicable also to asymmetrical olefins, to the contrary of the previous papers [4, 14], where the pyramidalization angle was determined merely for $R_2C{=}CR_2$, that is, olefins shaping complexes in C_{2V} symmetry.

In order to make comparison and consistency with the previous studies, a reference compound was selected, that is, ethylene, whose results are further discussed in detail. Yates [19] carried out broad computations related to the consistency and attained the binding energy of 111.9 kJ/mol in $(PH_3)_2Pt(CH_2{=}CH_2)$ at MP2/6-31G(d):LANL2DZ level, however, remarkably lower quantities for the arrangements containing PPh_3 ligands. Nunzi et al. [16] determined the "bond dissociation energy," amended with BSSE (basis set superposition error), as 102 kJ/mol at local density approximation (LDA)-DFT level of theory using mixed ζ-quality Slater-type orbital (STO) basis sets. Nevertheless, the published experimental values [17] for $(PPh_3)_2Pt(CH_2{=}CH_2)$ of Mortimer (152 ± 18 kJ/mol, gas-phase) and Kirkham et al. (11.6 ± 1.6 kJ/mol) appeared to be relatively inaccurate.

http://dx.doi.org/10.5772/65337

Group/compound		E_B [kJ/mol]	Δv [cm^{-1}]	ΔR [Å]	δ [°]
Ethylene complex	(ethylene)Pt(PH_3)$_2$	**116.8**	508.41	0.0943	28.84
	(ethylene)Pt(PH_3)$_2$	–	–	0.0989	30.81
Propadiene, Butadiene and Vinylacetylene	(propadiene)Pt(PH_3)$_2$	**136.5**	245.47	0.1087	–
	(buta-1,3-diene)Pt(PH_3)$_2$	**115.5**	507.70	0.0906	28.61
	(vinylacetylene)Pt(PH_3)$_2$[a]	**121.2**	476.83	0.0973	29.87
	(vinylacetylene)Pt(PH_3)$_2$[a]	**125.8**	404.41[g]	0.0795[g]	–
Unsaturated Alcohols	(prop-2-en-1-ol)Pt(PH_3)$_2$	**107.6**	513.64	0.0944	30.52
	(2-methylbut-3-en-2-ol)Pt(PH_3)$_2$	**110.9**	503.45	0.0971	32.01
	(hex-1-en-3-ol)Pt(PH_3)$_2$	**114.8**	506.43	0.0925	29.90
	(hept-1-en-4-ol)Pt(PH_3)$_2$	**98.9**	501.68	0.0965	29.93
Cyanoethylenes	(methacrylonitrile)Pt(PH_3)$_2$	**127.2**	369.05	0.1038	32.36
	(acrylonitrile)Pt(PH_3)$_2$	**131.2**	491.22	0.1039	31.20
	(*E*-1,2-dicyanoethylene)Pt(PH_3)$_2$	**149.3**	488.76	0.1112	32.58
	(*Z*-1,2-dicyanoethylene)Pt(PH_3)$_2$	**148.0**	479.93	0.1104	31.11
	(tricyanoethylene)Pt(PH_3)$_2$	**166.8**	377.86	0.1176	32.24
	(tetracyanoethylene)Pt(PH_3)$_2$	**181.1**	317.12	0.1229	31.29
Compounds containing α–carbonyl group(s)	(but-1-en-3-on)Pt(PH_3)$_2$	**132.0**	466.08	0.0922	27.24
	(methyl acrylate)Pt(PH_3)$_2$	**122.1**	485.95	0.0949	28.46
	(maleic anhydride)Pt(PH_3)$_2$	**154.1**	279.99	0.1011	32.24
	(acrylic aldehyde)Pt(PH_3)$_2$	**136.9**	467.88	0.0906	25.93
	(cinnamic aldehyde)Pt(PH_3)$_2$	**121.0**	438.84	0.0879	27.32
	(acryl amide)Pt(PH_3)$_2$	**125.4**	482.82	0.0955	28.71
	(diethyl fumarate)Pt(PH_3)$_2$	**121.4**	494.10	0.0944	27.42
Halogen containing compounds	(tetrachloroethylene)Pt(PH_3)$_2$	**101.1**	433.82	0.1303	53.49
	(tetrafluoroethylene)Pt(PH_3)$_2$	**110.1**	444.84	0.1254	45.02
	(octafluoro-*E*-but-2-ene)Pt(PH_3)$_2$	**145.4**	359.36	0.1224	42.07
	(*E*-1,1,1,4,4,4-hexafluorobut-2-ene)Pt(PH_3)$_2$	**165.2**	555.34	0.1134	35.14
	(3-chlorobut-1-ene)Pt(PH_3)$_2$	**120.0**	521.20	0.1042	33.70
Strained olefins	(3-methylcyclopropene)Pt(PH_3)$_2$	**153.7**	414.73	0.1236	–
	(1,2-dimethylcyclopropene)Pt(PH_3)$_2$	**144.9**	377.94	0.1175	–

Group/compound		E_B [kJ/mol]	Δv [cm^{-1}]	ΔR [Å]	δ [°]
	(bicyclo[2.2.0]hex-3(5)-ene)Pt(PH_3)$_2$	**165.3**	461.09	0.1248	–
	(bicyclo[4.2.1]non-1(8)-ene)Pt(PH_3)$_2$	**145.8**	261.94	0.0992	–
Various compounds	(vinyltrimethylsilane)Pt(PH_3)$_2$^{c}	**128.3**	463.25	0.0938	29.35
	(Z-1,2-bis(phenylsulfonyl)ethylene)Pt(PH_3)$_2$^{d}	**118.5**	332.58	0.1140	39.20
	(carbon disulfide)Pt(PH_3)$_2$^{e}	**74.8**	341.57^h	0.1291^h	–
	(dioxygen)Pt(PH_3)$_2$^{f}	**194.3**	637.49^i	0.1849^i	–

[a]Coordination via C=C double bond; [b]Coordination via C≡C triple bond, vibrational shift for C≡C stretch; [c–f]Syntheses of complexes, see Refs. [9, 12, 13, 16], respectively; [g]Vibrational shift of the C≡C stretch and elongation of the C≡C bond; [h]Vibrational shift of the C=S stretch and elongation of the C=S bond; [i]Vibrational shift of the O=O stretch normal mode and elongation of the O=O bond.

Table 1. The calculated parameters of coordination compounds of type $(PH_3)_2Pt$ (olefin).

In our study, for a completely optimized complex $(PH_3)_2Pt(CH_2{=}CH_2)$ (**Figure 12**), the binding energy of Pt-ethylene bonds was calculated as 116.8 kJ/mol at MP2/6-31G(d):LANL2DZ level of theory. The relatively minor discrepancy from the outcomes of Yates [15] could have been produced by our experimental setup, which has underwent a minor optimization. The binding energy for our entirely optimized model of $(PPh_3)_2Pt(CH_2{=}CH_2)$ was determined at B3LYP/6-31G(d):LANL2DZ level, yielding a value of only 9.9 kJ/mol (in comparison to 51.9 kJ/mol for the PH_3 analog). Regarding the geometrical parameters, the C=C bond in $(PH_3)_2Pt(CH_2{=}CH_2)$ stretched from 1.331 (free) to 1.425 (coordinated) by 0.094 Å. For the PPh_3 analog, the coordinated value was 1.430 Å, which was nearer to 1.434 Å as measured in the experiment. In the complex $(PH_3)_2Pt(CH_2{=}CH_2)$, having C_{2V} symmetry, both Pt-P bonds were of the same length 2.157 Å, while in $(PPh_3)_2Pt(CH_2{=}CH_2)$ [C_S symmetry], two different Pt-P bond lengths were calculated (2.356 and 2.366 Å). The mean experimental value was 2.268 Å. The σ-Pt-C bonds in $(PH_3)_2Pt(CH_2{=}CH_2)$ were 2.157 Å long. Regarding the PPh_3 analog, the bond length was calculated 2.153 Å (mean value of 2.168 and 2.138 Å). Both of these achieved results were overestimated in comparison to 2.112 Å (average) as determined by the experiment [18]. The next aim was to determine the similarity of Pt(0)-olefin organometallics from olefins coordinated on metal surfaces. In our study, the vertical distance of platinum and C=C bond in $(PH_3)_2Pt(CH_2{=}CH_2)$ was 2.036 Å, while for ethylene adsorbed in di-σ-bridge position on the Pt(111) surface, comparable values were obtained. Moreover, regardless of the calculations performed at different theory levels, the bond dissociation energies were 116.8 kJ/mol in the case of organometallic compounds (MP2 value, this study) and 106.1 kJ/mol in the case of C_2H_4 on p(2x2)-Pt(111) (generalized gradient approximation (GGA)-DFT calculation, [15]). Thus, the values were comparable within these arrangements. Additionally, the thermodynamics of the complete reaction of molecular oxygen with platinum(0) complex has been examined. It has been recognized that ethylene in the complex of $(C_2H_4)Pt(PR_3)_2$ is substituted by dioxygen producing $(O_2)Pt(PR_3)_2$. This reaction proceeded well, providing the exposure of $(C_2H_4)Pt(PR_3)_2$ to the ambient air. Having substituted **H** for **R**, the reaction enthalpy was determined according to the following schema:

http://dx.doi.org/10.5772/65337

$$(C_2H_4)Pt(PR_3)_2(g) + O_2(g) = (O_2)Pt(PR_3)_2 + C_2H_4(g), \quad (2)$$

finally obtaining the value of $\Delta H_r^{298} = -77.4$ kJ/mol, a relatively average figure for an exothermic reaction.

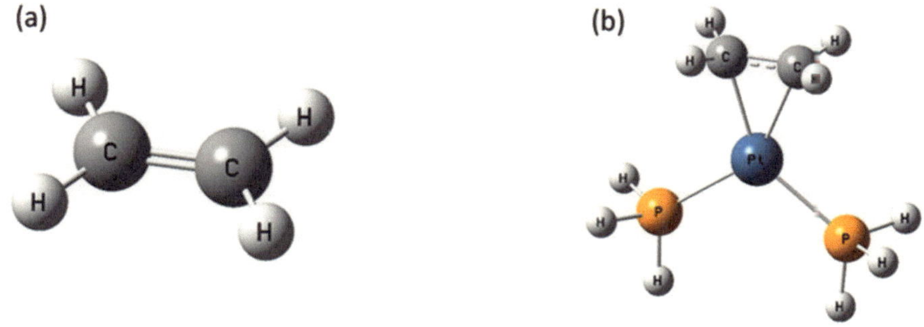

Figure 12. Molecular structure of ethylene (a, left) and (ethylene)Pt(PH_3)$_2$ (b, right).

4.1.1. Propadiene, butadiene, and vinylacetylene

Even though these hydrocarbon groups contain only alkenic or alkynic functions on the subject multiple bond, the computation disclosed that they retain enough strength upon coordination. It can be deduced that the functions CH=CH_2 (vinyl) and C≡CH (ethinyl) apparently favored accepting electrons. Moreover, interrupting the conjugation of multiple bonds is an energy-requiring procedure which has an inversely proportional effect on the dissociation energy. Different bond strengths were recorded for vinylacetylene (but-1-ene-3-yne). Compared to buta-1,3-diene or propadiene, either double or triple bond coordinated to the platinum(0) core atom with different bond strengths. As described in preceding computations and assessments, the triple bond arrangement was the ultimately favored constitution. This observation has been also substantiated during the examining of a relatively well-proceeding substitution reaction of olefinic complexes with substances bearing acetylenic group. It could have been initiated by the triple bond, a stronger electron-acceptor versus the C=C double bond acting as the substituent. Allene (propadiene) formed robust bonds to Pt(0), accompanied by a high-bond dissociation energy, and only a subtle wave number drop of C=C stretch (~275 cm^{-1}) of the bound double bond.

4.1.2. Unsaturated alcohols

Unsaturated alcohols with –OH group in α- or β-position to the double bond manifested a lower stability in the complex than ethylene. Apparently, the electron donor-acceptor characteristics associated with the hydroxy group in the complexes are rather shifted toward the electron donor. The decline in C=C stretch wave numbers was comparable to the one determined for the ethylene complex. The hydroxyl group bound in α- position to the C=C bond yielded complexes with a lower stability. Other alcoholic complexes, such as hex-1-en-3-

ol and allyl alcohol (prop-2-en-1-ol) gave slightly lower values of dissociation energy than for ethylene, suggesting their relative instability. Furthermore, allyl alcohol complexes were described to have been synthesized [19], yet verified only preliminarily by NMR. Nevertheless, this discovery indicates the potential for the production of alike derivatives. Having mentioned the minimal stability of the complex of hept-1-en-4-ol, a substantial decrease of the C=C stretch wave number was determined as well. The methyl groups in 2-methylbut-3-en-2-ol did not appear to have a noticeable impact on the complexes mainly owing to their pronounced sterical hindering (bulkiness) taking rather an aloof effect. On the other hand, they displayed a slightly higher stability compared to allyl alcohol complexes.

4.1.3. Cyanoethylenes

Table 1 depicts the tendency that can be read as the higher the amount of cyano groups on ethylene, the higher the stability of the complexes. Nitrile groups, strongly attracting electrons, supersede the double bond effect of sterical hindrance, and cause the energy level of HOMO orbitals in cyanoethylenes to be decreased in direction to the vacant orbitals of platinum. Previous studies of charge decomposition analysis predominantly indicated that π-back donation largely contributed to the final production of the platinum-olefin complex. The methyl group in methacrylonitrile, abundant with electrons, apparently caused the final complex to be less stable than the complex coordinated with acrylonitrile. Regarding 1,2-dicyanoethylenes, *E*-isomer (fumaric acid dinitrile) compared to *Z*-isomer (maleic acid dinitrile) displayed a relatively higher stability (ΔE = 1.3 kJ/mol).

4.1.4. Compounds containing α-carbonyl group(s)

The length of an alkyl group in ester products of acrylic acid had a significant impact on the electron-acceptor role of the carbonyl group. Hence, the adjoining Pt-C bond was relatively strong and unaffected. Fumaric acid esters displayed an analogous drift, that is, diethyl esters were observed to have their stability only marginally lower compared to dimethyl esters. Moreover, it is prudent to take into account any disruption of conjugated bonds in α,β-unsaturated compounds. Plain ketones, for example, but-1-en-3-one, produce complexes with a reasonably high stability. The simplest unsaturated aldehydes are a palpable example, for example, acrylic aldehyde (prop-2-en-1-al), which retains even a higher stability (i.e., higher dissociation energy). Other products of acrylic acid (amide, methylester) were demonstrated to have an average stability (125.4 and 122.1 kJ/mol, respectively). The highest dissociation energy (154.1 kJ/mol) was found in maleic acid anhydride owing to an intense π-electrons attraction by the C=C bond and only a minor sterical constrains due to the absence of any massive substituents.

4.1.5. Halogen containing compounds

Halogenated hydrocarbons present a useful model for the review of donor/acceptor electronic impacts on the C=C double bond in olefins. For example, 3-chlorobut-1-ene produced a reasonably rigid complex, which was noteworthy given its high pyramidalization (33.7°). Perfluorinated ethylenes, when bond to bis(triphenylphosphine)platinum, yielded a considerably elevated stability compared to perchlorinated products, probably owing to a geometric hindrance of chlorine atoms. Furthermore, carbon-halogen bonds are evidently more stretched after the coordination in C_2F_4 than in C_2Cl_4, feasible due to a reduced conjugation of fluorine atoms with C=C bond. Moreover, trifluoromethyl groups in *E*-CF_3CF=$CFCF_3$ incline to coordinate to a greater stability compared to the products of direct fluoro-substitution in C_2F_4, which led to obtaining a dissociation energy among the highest observed for the Pt-C_2 bonding (165.2 kJ/mol). Halogen atoms located in a position influencing the C=C double bond plausibly contribute to producing relatively stable organometallic complexes.

4.1.6. Strained olefins

The C=C bond in strained olefins coordinated to extraordinarily highly stable complexes, as the sp^2-carbon atoms apparently well-sp^3-hybridized. The modified compounds were nominated based on available information found in the literature illustrating their feasible preparation and analytical detection. The complexes of each selected strained olefins revealed high-binding energies. Pyramidalization of substituents was assessed; however, unsatisfactorily high figures were obtained. Subsequently, this extent was determined for acyclic substitution on the C=C double bond. Nevertheless, once the complex settled down, the deformation of substituents from the previous position was exceptionally significant within this group of complexes. **Figure 13** illustrates one of the instances: bicyclo[2.2.0]hex-3(5)-ene complex.

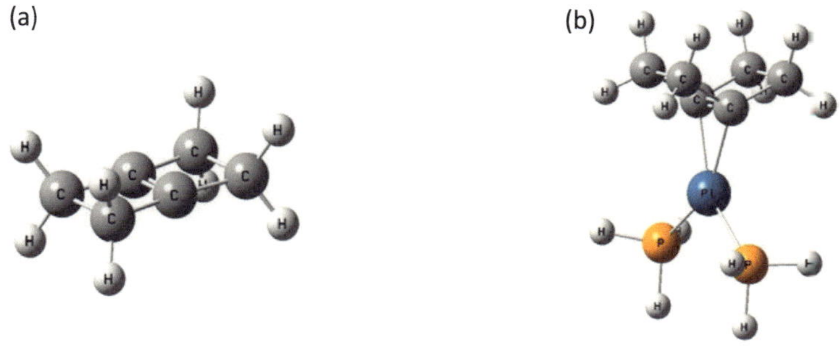

Figure 13. Bicyclo[2.2.0]hex-3(5)-ene free form (a, left) and complex (b, right).

4.1.7. Various compounds

Selected compounds (further elaborated in various publications) contained different types of double bonds in their structure and their stability in the coordinated state was calculated. For example, in the case of organosilicon compounds, particularly vinyltrimethylsilane, CH_2=CH-$Si(CH_3)_3$ delivered average quantities for E_B and further features assessed, inferring that the electronic properties of silicon atoms induced comparable effects as carbon. On the other hand, *rac*-*Z*-1,2-bis(phenylsulfonyl)ethane [15], yielded surprising data as the obtained binding energy for the coordination of the computed products of (*R*,*R*-isomer) was minor compared to ethylene. Nonetheless, the complex was prepared from $(PPh_3)_2Pt(CH_2$=$CH_2)$ by substituting the subject olefins. In order to assess the impact of the racemates, it is advisable to figure out the average of binding energies of all enantiomers. The previous tests revealed that carbon disulfide (CS_2) adduct was in a solution of the olefin complex produced with a swift reaction rate, while the calculated binding energy was unexpectedly minimal. The cause could have been the solvent-stabilization effect in a carbon disulfide solution, since adduct supported the induction of dipoles in a near vicinity of the molecule. The preparation of dimeric polynuclear complexes could be dismissed by means of X-ray analyses [20]. Following the reaction of olefin complexes with molecular oxygen, dioxygen adduct $(O_2)Pt(PH_3)_2$ was noticeably produced without a reversible option, yielding by far a complex with the maximum computed stability value as the core metallic atom and its oxidation state assumed the form of Pt(II) [21].

5. Conclusion

Molecular modeling in heterogeneous catalysis is a very effective tool for investigating phenomena on the atomic level. As a key factor, an appropriate model for the approximation of the system under study appears to be the determination. The model then opens up new possibilities for predicting the behavior of molecules in particular reaction type. In the present work, models for heterogeneously catalyzed hydrogenations on platinum catalysts were designed which were tested and found limits of their applicability. The proposed approach to the study of molecular structure effects represents another step toward the development of an appropriate procedure for theoretical design of catalysts "tailored" to investigate not only the process but mainly the interaction-reaction center (substrate)—active site (catalyst).

Acknowledgements

This work was realized within the Operational Programme Prague – Competitiveness" (CZ. 2.16/3.1.00/22197) and "National Program of Sustainability" (NPU I LO1215) MSMT - 34870/2013); The Operational Programme Prague – Competitiveness (CZ.2.16/3.1.00/21537) and "National Program of Sustainability" (NPU I LO1601) MSMT-43760/2015); The Operational Programme Prague – Competitiveness (CZ.2.16/3.1.00/24501) and "National Program of Sustainability" (NPU I LO1613) MSMT-43760/2015) and GACR (GA15-08992S).

Author details

Martin Kindl, Ondřej Matuška and Petr Kačer*

*Address all correspondence to: Petr.Kacer@vscht.cz

Department of Organic Technology, University of Chemistry and Technology, Prague, Czech Republic

References

[1] Taylor, H.S., *Fourth report of the committee on contact catalysis*. The Journal of Physical Chemistry, 1925. 30(2): p. 145–171. DOI: 10.1021/j150260a001.

[2] Ertl, G., Knozinger, H., and Weitkamp, J., *Handbook of Heterogeneous Catalysis*. 1997, VCH, Weinheim.

[3] Thomas, J.M., Thomas, W.J., and Salzberg, H., *Introduction to the principles of heterogeneous catalysis*. Journal of The Electrochemical Society, 1967. 114(11): p. 279C–279C. DOI: 10.1149/1.2426448.

[4] Boudart, M., *Catalysis by supported metals*. Advances in Catalysis, 1969. 20: p. 153–166. DOI: 10.1016/S0360-0564(08)60271-0.

[5] Zaera, F., et al., *Hydrogen Effects in Catalysis: Fundamental and Practical Applications*. 1988, Marcel Dekker, New York.

[6] Singh, U.K. and Vannice, M.A., *Kinetics of liquid-phase hydrogenation reactions over supported metal catalysts—a review*. Applied Catalysis A: General, 2001. 213(1): p. 1–24. DOI: 10.1016/S0926-860X(00)00885-1.

[7] Wells, P.B., *The influence on selectivity of the environment of catalyst sites: III. The role of hydrogen occlusion in group VIII metals*. Journal of Catalysis, 1978. 52(3): p. 498–506. DOI: 10.1016/0021-9517(78)90355-X.

[8] Boudart, M., *Heterogeneous catalysis by metals*. Journal of Molecular Catalysis, 1985. 30(1–2): p. 27–38. DOI: 10.1016/0304-5102(85)80014-6.

[9] Coq, B. and Figueras, F., *Structure–activity relationships in catalysis by metals: some aspects of particle size, bimetallic and supports effects*. Coordination Chemistry Reviews, 1998. 178: p. 1753–1783. DOI: 10.1016/S0010-8545(98)00058-7.

[10] Ponec, V., *Catalysis by alloys in hydrocarbon reactions*. Advances in Catalysis, 1983. 32: p. 149–214. DOI: 10.1016/S0360-0564(08)60440-X.

[11] Mallat, T. and Baiker, A., *Selectivity enhancement in heterogeneous catalysis induced by reaction modifiers*. Applied Catalysis A: General, 2000. 200(1): p. 3–22. DOI: 10.1016/S0926-860X(00)00645-1.

[12] Bond, G.C., *Catalysis by Metals*. 1962, Academic Press, New York. DOI: 10.1126/science.138.3537.129.

[13] Oudar, J., *Hydrogenation reactions on platinum single crystals*. Zeitschrift für Physikalische Chemie, 1996. 197: p. 125–136. DOI: 10.1524/zpch.1996.197.Part_1_2.125.

[14] Frison, G. and Grützmacher, H., *Coordinated olefins, H_2C=CHR, and phosphanes, Ph_2 R: a theoretical study of the R substituent effect*. Journal of Organometallic Chemistry, 2002. 643: p. 285–291. DOI: 10.1016/S0022-328X(01)01276-1.

[15] Yates, B., *The platinum–ethylene binding energy in $Pt(PL_3)_2(C_2H_4)$*. Journal of Molecular Structure: THEOCHEM, 2000. 506(1): p. 223–232. DOI: 10.1016/S0166-1280(00)00414-0.

[16] Lebedev, S., Kobliansky, G., and Yakubchik, A., *LXII.—The relative rates of catalytic hydrogenation of different types of unsaturated compounds. Part I. Aliphatic ethylenic derivatives*. Journal of the Chemical Society, Transactions, 1925. 127: p. 417–440.

[17] Campbell, K.N. and Campbell, B.K., *The addition of hydrogen to multiple carbon-carbon bonds*. Chemical Reviews, 1942. 31(1): p. 77–175. DOI: 10.1021/cr60098a003.

[18] Hájek, J., et al., *High-selectivity hydrogenation of cinnamaldehyde over platinum supported on aluminosilicates*. Research on Chemical Intermediates, 2006. 32(9): p. 795–816. DOI: 10.1163/156856706778938482.

[19] Rymeš, J., et al., Pt combustion catalysts prepared from w/o microemulsions, in *Studies in Surface Science and Catalysis*, Gaigneaux E., De Vos, D.E., Grange, P., Jacobs, P.A., Martens J.A., Ruiz, P., Poncelet, G., 2000, Elsevier, Louvain-la-Neuve, p. 121–129.

[20] Mortimer, C., *Platinum-ligand and palladium-ligand bond energies*. Reviews in Inorganic Chemistry, 1984. 6(3): p. 233–257.

[21] Kirkham, W.G., Lister, M.W., and Poyntz, R.B., *Relative olefin—metal bond strengths in some platinum (O) compounds*. Thermochimica Acta, 1975. 11(1): p. 89–93. DOI: 10.1016/0040-6031(75)80041-4.

Chapter 13

Recent Developments in the Use of Flow Hydrogenation in the Field of Medicinal Chemistry

Cecilia C. Russell, Jennifer R. Baker,
Peter J. Cossar and Adam McCluskey

Additional information is available at the end of the chapter

http://dx.doi.org/10.5772/65518

Abstract

This chapter focuses on recent applications of flow hydrogenation in medicinal chemistry. Flow reactors can enhance laboratory safety, reducing the risks associated with pyrophoric catalysts, due to their containment in catalyst cartridges or omnifit columns. Flow hydrogenation reduces the risks arising from hydrogen gas, with either hydrogen generated in situ from water, or precise management of the gas flow rate through tube-in-tube reactors. There is an increasing body of evidence that flow hydrogenation enhances reduction outcomes across nitro, imine, nitrile, amide, azide, and azo reductions, together with de-aromatisation and hydrodehalogenation. In addition, olefin, alkyne, carbonyl, and benzyl reductions have been widely examined. Further, protocols involving multistage flow reactions involving hydrogenation are highlighted.

Keywords: hydrogenations, flow technologies, flow synthesis, reduction, multistage, flow hydrogenation, chemoselective, catalyst

1. Introduction

In 2013, 25% of marketed drugs required at least one hydrogenation step in their production [1, 2]. Hydrogenation mediated manipulation of nitro, imine, nitrile, amide, azide, and azo moieties, as well as de-aromatisation, hydrodehalogenation, olefin, alkyne, carbonyl, and benzyl reductions are fundamental to drug discovery and development programmes [1, 3].

Flow hydrogenation offers the benefits of improved safety, yield, selectivity and reduced purification over traditional hydrogenation approaches. Flow hydrogenation through the use of contained pyrophoric catalysts, replacement of hydrogen reservoirs with in situ hydrogen generation, improved temperature control, and smaller solvent volumes all contribute to an increase in hydrogenation safety [4]. Flow technologies have improved hydrogenation outcomes by increasing substrate-gas-catalyst interactions and permitting stringent control of reaction parameters (temperature, flow rate, and pressure) with a commensurate reduction in undesirable side product and improved selectivity. Combined with optimised reaction conditions, this generally means very little or no further purification is required after the reaction [1].

This chapter details key recent development in functional group transformation, multistep synthesis utilising flow hydrogenation and technology advances [1, 3].

2. Instrumentation

Unlike batch reactions where gas-solvent contact is limited by diffusion of gas into the bulk solvent, flow hydrogenation rapidly saturates the solvent with hydrogen using two different approaches [5]. The first, used by the ThalesNano H-cube®, employs in-line mixing of hydrogen with the solvent under pressure, which prevents outgassing and rapid solvent stream saturation (**Figure 1A**) [6]. The second approach, used by the Vapourtec Gas/Liquid reactor (**Figure 1B**) and Gastropod Gas Liquid Module, employs gas permeable membranes in a tube-in-tube reactor. These systems enable solvent stream saturation by passing hydrogen gas under pressure through a gas porous polymer and into the solvent [7–10].

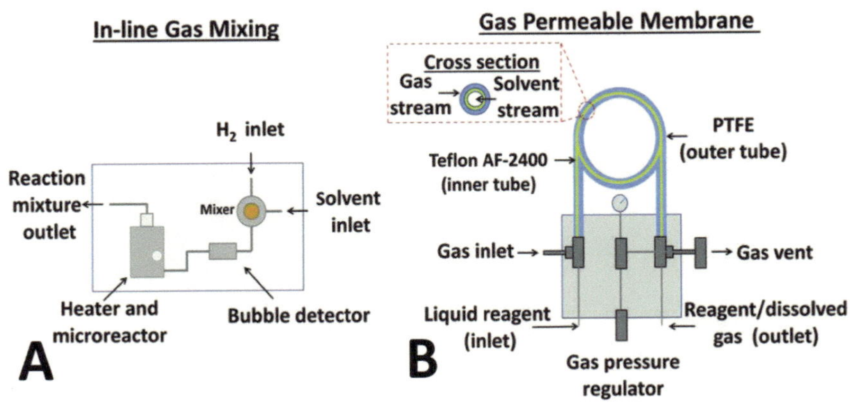

Figure 1. Schematic of the mechanical mixing setup in (**A**) the ThalesNano H-Cube® and (**B**) schematic of the gas permeable membrane (tube-in-tube) technology [1].

The Thalesnano H-cube® was the first commercial flow hydrogenator. Together with the use of in-line gas mixing, the H-cube uses exchangeable 30 or 70 mm heterogeneous catalyst cartridges and a HPLC pump. Hydrogen gas is generated in situ through water electrolysis. The system is capable of heating to 100°C and 150 bar, with a flow rate range from 0.5 to 5.0

mL min^{-1} [11]. Tube-in-tube reactors require specialised materials displaying high gas permeability while being impermeable to (nonfluorinated) liquids and corrosive chemicals, e.g., Teflon AF-2400 [12]. The Vapourtec Gas/Liquid system uses 'plug-in reactors' and the Gastropod Gas Liquid Module can be equipped with a small gas cylinder or attached to any custom flow systems [9, 10].

3. Functional group transformations

3.1. Nitro reductions

Flow nitro reductions, using palladium, platinum, and Raney Ni catalysts, under optimised conditions have been shown to provide both increased yield and simplified work up [1, 3]. Abdel-Hamid et al's recent synthesis of 1,8-naphthalimide derivatives illustrates this with an increase in yield (86–98%) and purification simplification (chromatographic to extractive) (**Figure 2**) [13].

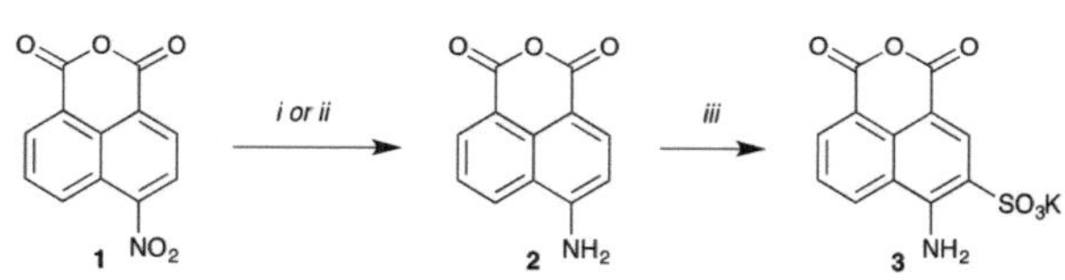

Figure 2. Synthesis of 1,8-naphthalimide derivatives. Reagents and conditions: (i) ThalesNano H-Cube®, 10% Pd/C, THF, 40°C, 10 bar, 1 mL min^{-1}, 2 cycles; (ii) $SnCl_2$, HCl, ethanol, reflux 2 h; (iii) (a) fuming sulfuric acid, 50°C, 3 h; (b) saturated aq. KCl, room temperature.

The utility of the flow nitro reduction extends across pyrrolidine, carboxylate ester, phenyl propanoate, benzothiopene, benzofurans, and indole-carboxylate scaffolds. These reactions employed either Raney Ni or 10% Pd/C catalysts from 25 to 65°C and atmospheric (atm)—20 bar, respectively, providing excellent reaction outcomes (**Table 1**).

3.2. Alkene reductions

Flow hydrogenation is particularly useful in the reduction in alkene and alkyne bonds as is evident from the examples shown in **Table 2**. Gericke et al. developed ruthenium-nitrogen-doped carbon nanotubes (NCNT) and ruthenium-hyperbranched polystyrene-supported (HPS) catalysts, providing a more sustainable process [24]. Gericke et al. suggested that HPS- and NCNT-supported catalysts are a suitable alternative to Raney Ni and have an increased production rate per mole of catalyst compared to Raney Ni. Multiple similar alkene and alkyne reductions have been reported (**Table 2**). These hydrogenation catalysts were found not to be limited to the hydrogenation of alkenes and alkynes and have been applied in the reduction in glucose **4** to sorbitol **5**, which traditionally has relied on expensive catalysts such as Raney Ni (**Figure 3**).

Starting material	Product	Yield	Catalyst	Conditions	References
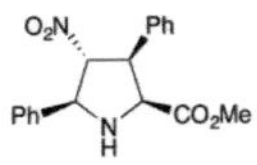	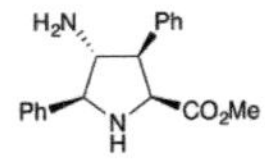	90%	Ra-Ni	65°C, 20 bar, 1.0 mL min^{-1}	[14]
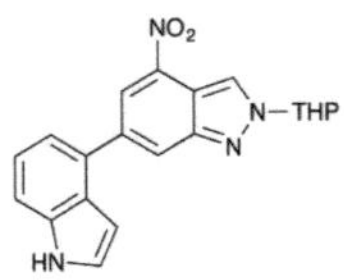	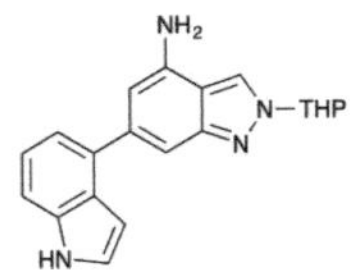	100%	10% Pd/C	25°C, 1–30 bar	[15]
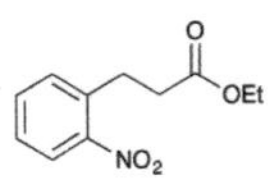	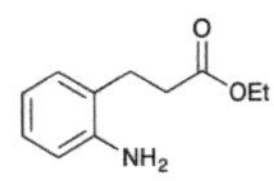	100%	10% Pd/C	50°C, 1 bar, 1.0 mL min^{-1}	[16]
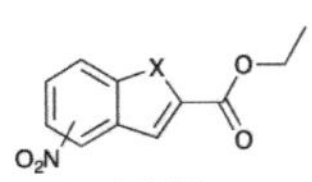 X = O, S, NH	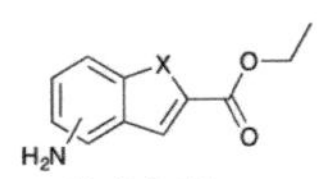 X = O, S, NH	88–98%	10% Pd/C	40–60°C, 1 bar, 0.5–1.0 mL min^{-1}	[17]

Table 1. Flow nitro reduction of selected analogues.

Starting material	Product	Yield	Catalyst	Conditions	References
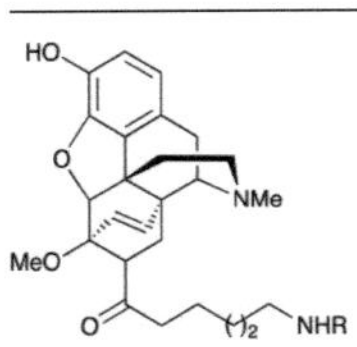	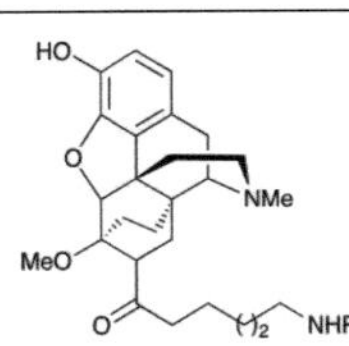	86%	10% Pd/C	25°C, 10 bar, recirculate 2 h	[18]
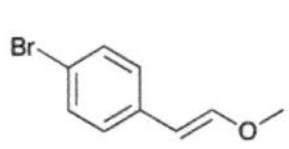	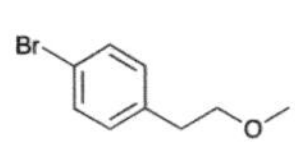	21%	5% Rh/C	70°C, 1 bar, 1.5 mL min^{-1}, 3 cycles	[19]
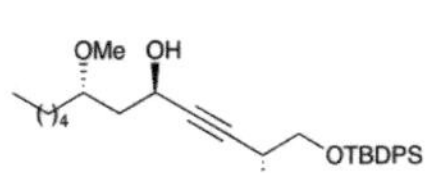	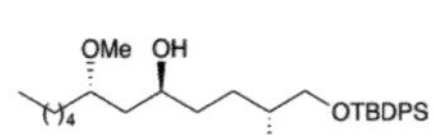	98%	10% Pd/C	40°C, 1 mL min^{-1}	[20]
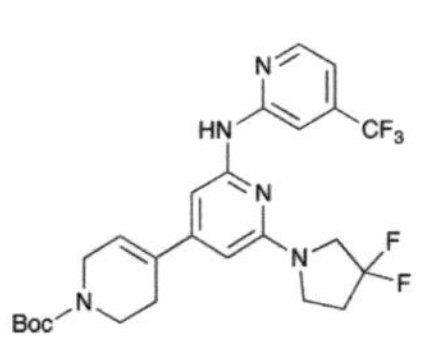	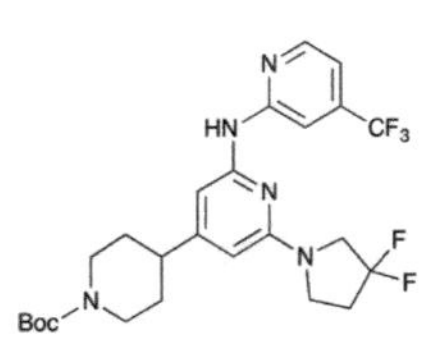	84%	10% Pd/C	30°C, 40 bar, 1 mL min^{-1}, 2 cycles	[21]
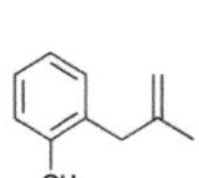	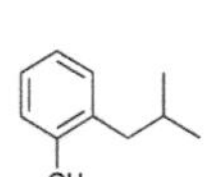	99%	RaNi	60°C, 60 bar, 1 mL min^{-1}, 24 h	[22]
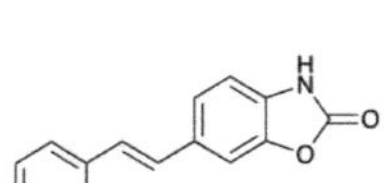	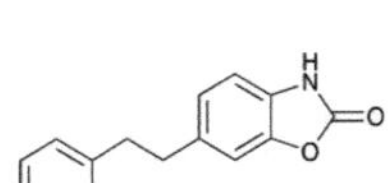	Quant.	10% Pd/C	50°C, 1 bar	[23]

Table 2. Flow reduction of selected alkenes and alkynes.

http://dx.doi.org/10.5772/65518

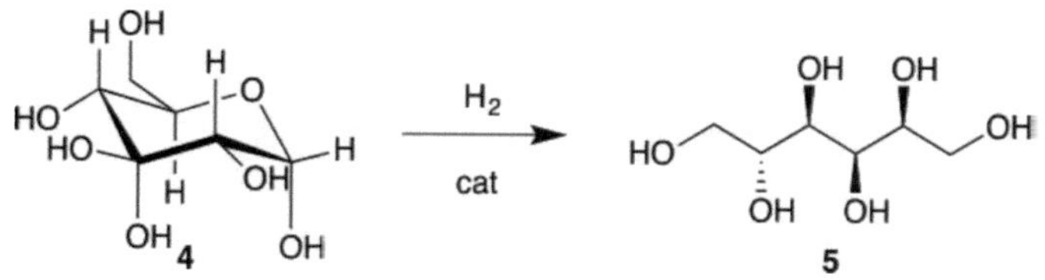

Figure 3. Reaction scheme for the hydrogenation of D-glucose (**4**).

Initial attempts by Yadav et al. under batch reaction conditions to access alcohol **6** afforded an 8:2 mixture of **6** and ketone **7** [20]. The use of the H-cube® and relatively mild reducing conditions (10% Pd/C, 40°C, and 6 bar) gave exclusively **6** in a near-quantitative yield (**Figure 4**).

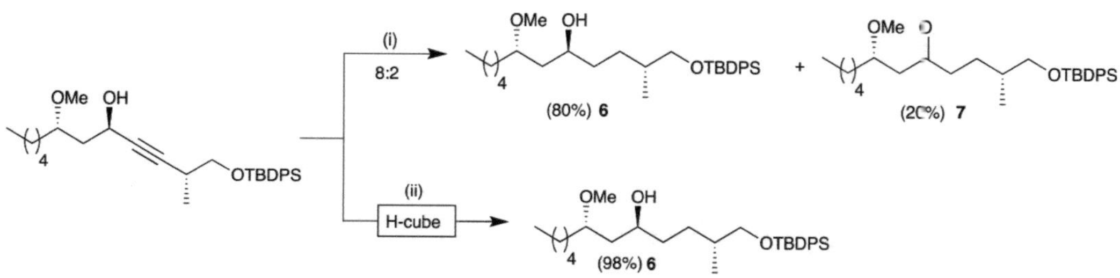

Figure 4. Synthesis of the marine macrolide sanctolide **6** via batch and flow hydrogenation. Reagents and conditions: (i) H_2, Pd/C (10%), EtOAc, rt, 8 h; (ii) H-cube®, Pd/C (10%), MeOH, 40°C, 6 bar.

Trobe and Breinbauer highlighted the use of flow methodologies to improve reaction yields (**Figure 5**) [22]. The trifluoroether **11** was accessed through a conventional traditional Witting/hydrogenation approach in a 32% yield. A modified access via a Claisen rearrangement and flow hydrogenation was developed leading to **11** in a 71% yield.

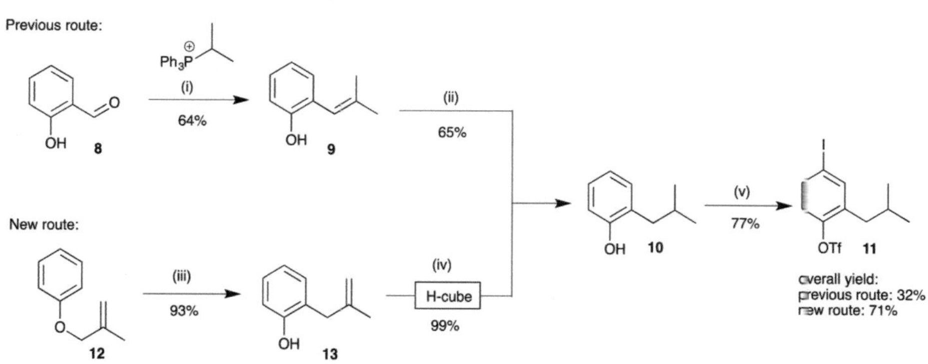

Figure 5. Improving the yield of a synthetic route from 32% to 71% with the aid of flow hydrogenation. Reagents and conditions: (i) salicylaldehyde, toluene, 80°C, 8 h; (ii) H_2, Pd/C, MeOH, 22°C, 3 h; (iii) DMAc, 190°C, 5 d; (iv) H-cube®, Ra-Ni, 60°C, 60 bar; (v) (i) ICl, AcOH, 22°C, 24 h; (ii) Tf_2O, pyridine, 0°C, 2 h.

3.3. Reductive amination

Traditionally, borohydride reagents such as $NaCNBH_3$, $NaBH(OAc)_3$, or pyridine-BH_3 have been used for reductive amination [25]. However, flow hydrogenation offers considerable advantages over transfer hydrogenation, such as improved atom economy, reduced environmental impact, simple reaction workups, and reduced exposure to toxic or reactive starting materials [26].

Flow reductive aminations are generally conducted using 10% Pd/C or 20% $Pd(OH)_2/C$, with the temperatures and pressures used substrate-dependent [1]. However, the use of an Au/Al_2O_3 catalyst has facilitated a cascade nitro reduction and direct reductive amination to afford secondary amine **16** (**Figure 6**). Unlike many conventional Pd- and Ni-based catalysts, the Au/Al_2O_3 catalyst showed selective reduction in the nitrobenzene **14** over benzaldehyde **15**, aiding imine formation and subsequent reductive amination. Under optimised conditions (1:1.5 nitrobenzene **14**: benzaldehyde **15**, 80°C and 50 bar), the desired *N*-benzylaniline **16** was generated in a 91% yield (**Figure 6**) [27].

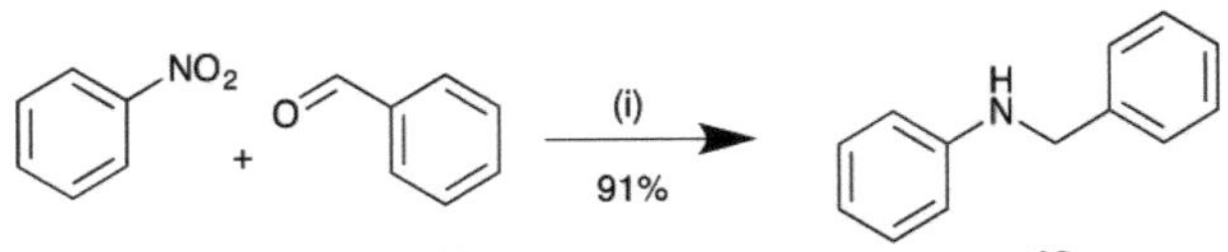

14 15 16

Figure 6. Flow reductive amination to afford *N*-benzylaniline **16**. Reagents and conditions: H-cube® Pro, 0.05 M **15** in EtOH, Au/Al_2O_3 (70 mm), 125°C, 10 bar, 0.3 mL min^{-1}.

Treatment of phenethylamine (**18**) and levulinic acid (**17**) in 2-methylfuran under the hydrogenation conditions of 85 bar H_2 pressure, 150°C and carbon-supported Fe/Ni yielded pyrrolidine **19** with a 91% conversion via a sequential reductive amination and cyclisation process (**Figure 7**) [28].

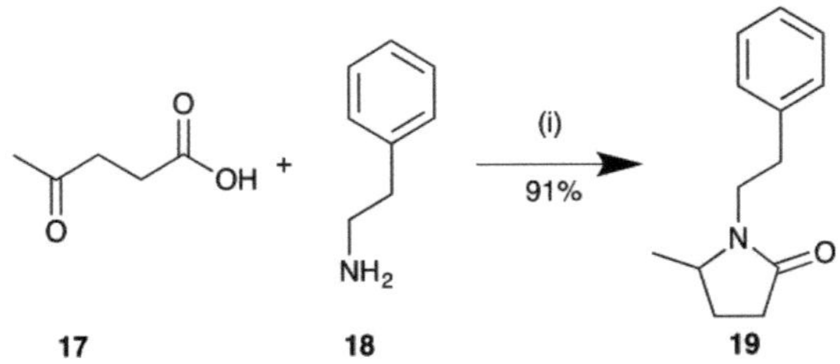

17 18 19

Figure 7. Flow reductive amination with carbon-supported Fe/Ni (C-Fe/Ni) to form pyrrolidine **19**. Reagents and conditions: H-cube® Pro, 0.025 M **18** in 2-methylfuran, C-Fe/Ni alloy (70 mm), 150°C, 85 bar, 0.3 mL min^{-1}.

3.4. Protecting group manipulation

The synthesis of carbohydrate and nucleoside mimics has led to the development of C-nucleosides and C-glycosides as antibiotic, anticancer, and antiviral agents [29]. Using flow

http://dx.doi.org/10.5772/65518

chemistry, Redpath et al. were able to access the deprotected 2-deoxy-C-galac-topyranosyl-benzoic acid **26** (**Figure 8**) [29]. The final stage of the multistep reaction, including the hydrogenation, provided **26** in a 39% yield over five steps. This route was found to provide access to galactoside and mannoside type C-nucleosides incorporating functionality analogous to the biologically important benzamide riboside through the use of an oxazoline protecting group, which had been previously inaccessible using a transmetallation/inter molecular Sakura condensation approach.

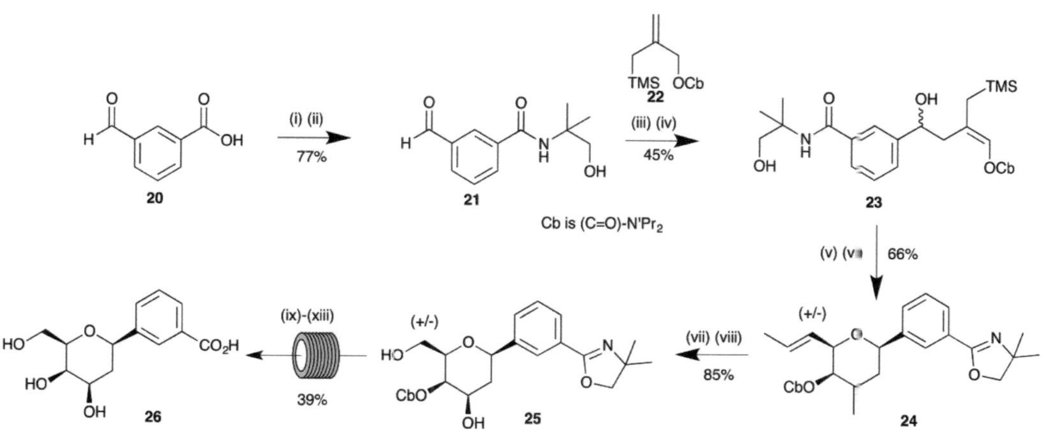

Figure 8. Synthesis of (D/L)-deoxy-β-galactopyranosyl-benzoic acid (**26**). Reagents and conditions: (i) $SOCl_2$, toluene; (ii) $H_2NC(CH_3)_2CH_2OH$, CH_2Cl_2; (iii) *sec*-BuLi, TMEDA, Et_2O; (iv) $Ti(O^iPr)_4$; (v) crotonaldehyde, $BF_3{\cdot}OEt_2$, CH_2Cl_2; (vi) MsCl, Net_3, CH_2Cl_2; (vii) O_3, CH_2Cl_2/MeOH then Me_2S; (viii) $NaBH_4$; (ix) $LiAlH_4$, THF; (x) BnBr, TBAI, NaH, 15-crown-5, THF; (xi) MeI, $MeNO_2$; (xii) 20% KOH, MeOH; (xiii) H_2, 10% Pd/C, MeOH.

Pd-catalysts and mild (RT, 1 bar) to moderate (45°C, 10 bar) conditions have been employed for the removal of benzyloxy carbamate (CBz) and benzyl (Bn) protecting groups (**Table 3**).

3.5. Multistep synthesis

A number of integrated multistep flow syntheses, with hydrogenation a key step, have been reported and are typically characterised by the reduced need for purification between synthetic steps.

Previous batch syntheses of the kinase inhibitors CTx-0152960 and CTx-029488 required the use of Boc-piperidine in the key S_NAr coupling with 1-fluoro-4-nitrobenzene to prevent formation of unwanted side products. Flow approaches removed this requirement facilitating rapid access to the Boc-free analogues in high yields (**Figure 9**) [33]. Of note, the flow hydrogenation of both the S_NAr adducts of piperidine and morpholine (**27**) required no purification. Microwave coupling of 4-morphilinoaniline and 4-(piperazine-1-yl)aniline with 2-(2,5-dichloropyrimidine-4-ylamino)-*N*-methylbenzamide afforded access to the desired **31a** and **31b**. This hybrid approach reduced the number of synthetic steps, enhanced product yield, and increased atom economy through step reduction and minimal requirement for chromatographic purification, relative to the original batch approach [33].

Starting material	Product	References
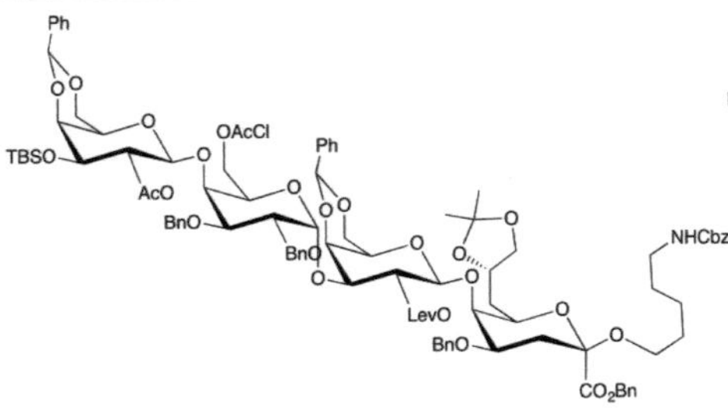	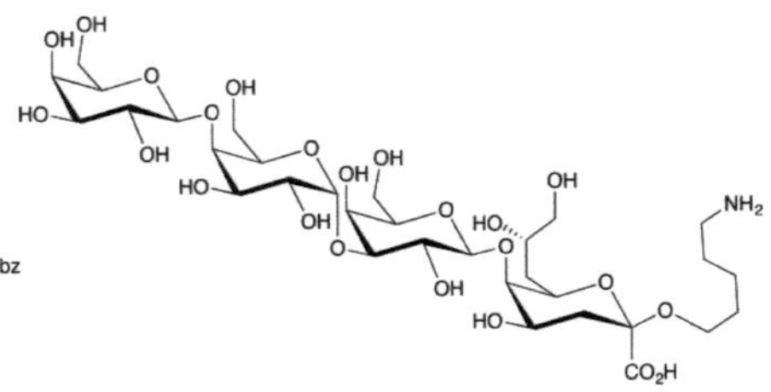	[30]
CBz protecting group	20% $Pd(OH)_2/C$; 45°C, 10 bar, 1.0 mL min^{-1}, 1 cycle; 67%	
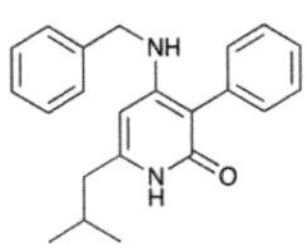	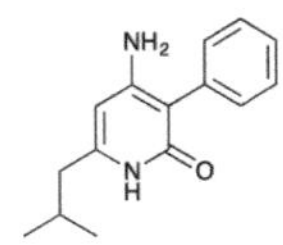	[31]
Benzyl protecting group	10% Pd/C; 65°C, full H_2 mode, 1.0 mL min^{-1}, 1 cycle; 60%	
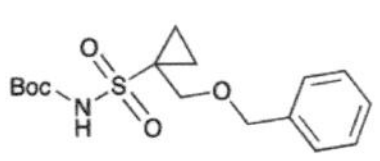	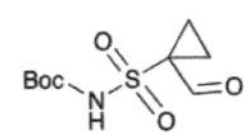	[32]
Benzoyl protecting group		
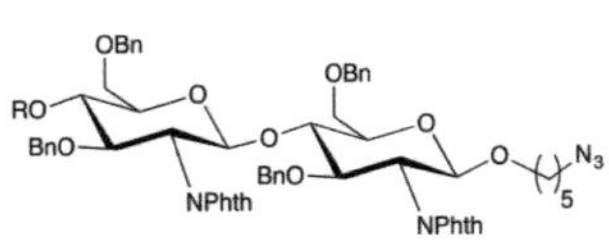	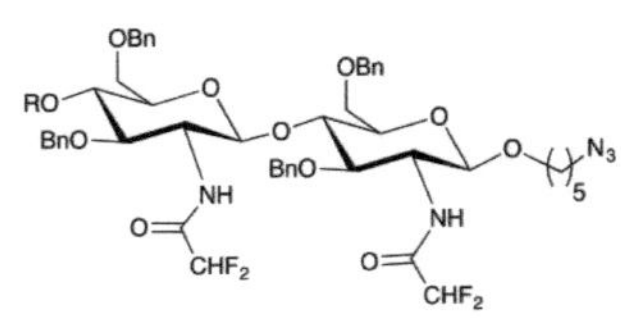	[56]
Phthalimide protecting group	10% Pd/C; 50°C, 1 bar, 50°C, 0.5 mL min^{-1}; 56–81%	
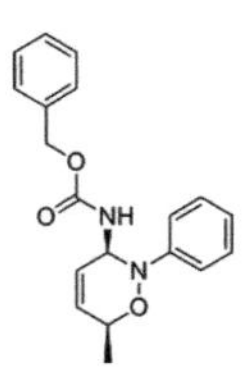	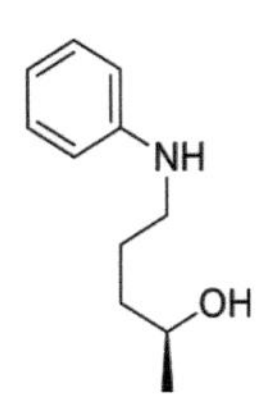	[57]
Cbz protecting group	10% Pd/C; RT, 1 bar, 50°C, 1.0 mL min^{-1}; 99%; 97% ee	

Table 3. Flow reduction and removal of protecting groups.

The modular nature of flow chemistry instrumentation has allowed Ghislieri et al. by simple manipulation of the module order and selection of staring material to produce five active pharmaceutical ingredients (APIs) across three structural classes (γ-amino acids, γ-lactams, β-amino acids). From benzyl alcohol eight compounds of interest including the drugs Lyrica and Gabapentin were synthesised in good overall yields (49–75%) (**Figure 10**) [34].

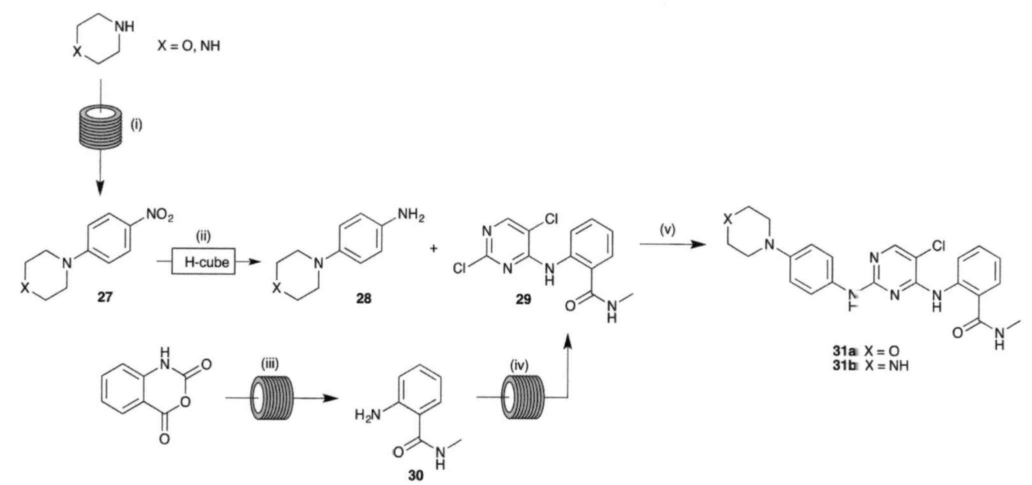

Figure 9. Synthesis of broad kinase inhibitors **31a** and **31b** by multistep flow synthesis. Reagents and conditions: (i) Vapourtec R2+, 4M piperidine or morpholine in DMF, 2 M 1-fluoro-2-nitrobenzene in DMF, 8 bar, 5 mL min^{-1}; (ii) H-CubePro, 0.05M in MeOH, 10% Pd/C CatCart® (70 mm), 50 bar, 50°C, 1.0 mL min^{-1}; (iii) Syrris FRX-100, 40% w/w aq. $MeNH_2$, 0.5 mL min^{-1}, 0–19°C, 19 h; (iv) Vapourtec R2+, 2,3,5-tri-chloropyrimidine, ^iPrNEt, ^iPrOH, 4 bar 100°C; (v) *n*-BuOH, 4 M HCl in dioxane (cat), 150°C, μW, 20 min.

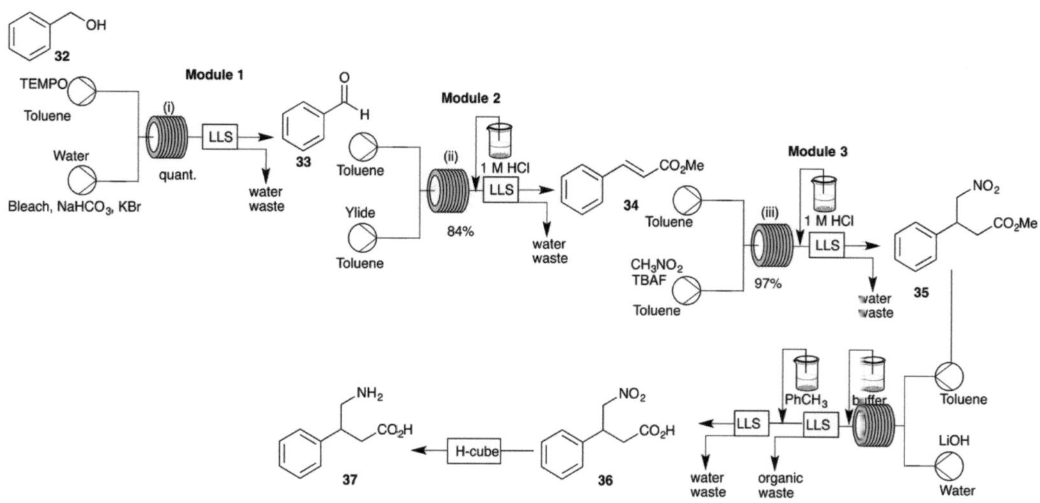

Figure 10. Divergent multistep flow synthesis of γ-amino acid derivatives. Reagents and conditions: (i) bleach (2.5 eq.), TEMPO (0.05 eq.), $NaHCO_3$ (0.3 eq.), KBr (0.2 eq.), 0°C; (ii) triethylphosphonoacetate (1.1 eq.), *t*BuOK (1.1 eq.), 50°C; (iii) CH_3NO_2 (11 eq.), TBAF (1.3 eq.), 50°C; (iv) H-cube®, Pd/C (10%), 60°C, 60 bar; (v) LiOH (3 eq.), 50°C. LLS = liquid-liquid separator. Yields for individual modules determined upon isolation.

The flow chemistry modules above have also been used for the efficient synthesis of a number of known APIs (**Figure 11**). Within this multistage process, module five employed the use of the H-cube for the preparation of β-amino acids from unsaturated α-nitrile ester (90 bar, 100°C,

Raney Ni). For nitro reductions, Pd/C and Raney Ni catalysts were favoured and afforded the desired compounds in good-to-excellent yields.

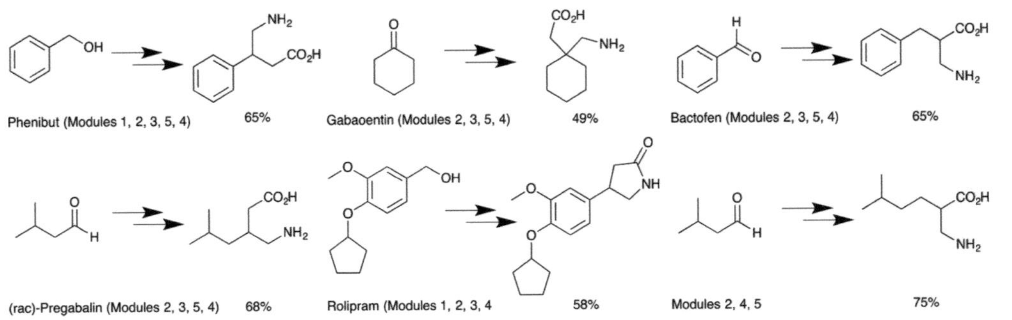

Figure 11. APIs prepared via the convergent multistep synthesis exemplified in **Figure 10**. Yields are reported for full processes without immediate purification over the 3–5 steps.

Both (*R*)- and (*S*)-rolipram were generated by using flow approaches and required no isolation of intermediates or purification, a significant step towards the automated manufacture of APIs [35]. The overall process used is outlined in **Figure 12**. While commercially available Ni and Pd catalysts failed in the case of aliphatic nitro compounds, however, a newly developed dimethylpolysilane-supported palladium/carbon (Pd/DMPSi-C) catalyst afforded the desired δ-lactam in a 74% yield (94% ee).

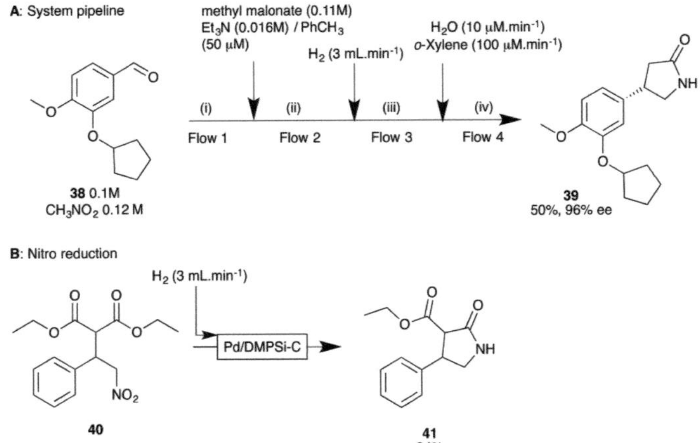

Figure 12. (**A**) Multistep flow synthesis of (*S*)-rolipram. Reagents and conditions: (i) Si-NH_2/$CaCl_2$, toluene, 75°C, 50 μL min^{-1}; (ii) PS-(*S*)-Pybox, $CaCl_2$·$2H_2O$, 0°C, 100 μL min^{-1} (total); (iii) Pd/(DMPSi-C) (1.6 mmol), 100°C, 100 μL min^{-1} (total); (iv) HOOC-silica gel, 120°C, 210 μL min^{-1} (total). The flow reaction was continued for a week and the yield and the enantioselectivity maintained. (**B**) Further details of the flow hydrogenation step.

The success of multistage flow synthesis in API production, especially the use of flow hydrogenation suggests that these approaches will continue to rapidly develop and potentially become a standard method of synthesis.

http://dx.doi.org/10.5772/65518

3.6. Scaffold formation

Flow hydrogenation has provided access to scaffolds that were inaccessible via batch hydrogenation pathways such as the 1,4-benzodiazepin-5-ones (**Figure 13**). This scaffold has been targeted in treatments for tuberculosis [36] and control of the melanocortin receptors implicated in appetite control [37] and was readily accessed under flow conditions (THF, 0.3 mL min^{-1}, 50 bar and 80°C). Isolated yields of up to 94% (**45**) requiring no purifications were noted [38].

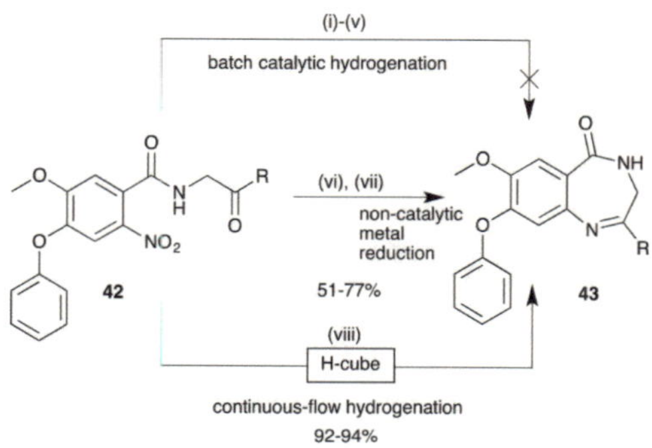

Figure 13. Synthesis of the desired 1,4-benzodiazepin-5-ones **43** via batch and flow hydrogenation. Reagents and conditions: (i) H_2, Pd/C (10%, 0.1 eq.), EtOAc:EtOH, 2:1 (0.03 M), 20°C, 1 atm; (ii) 1,4-cyclohexadiene (6 eq.), microwave mode, Pd/C (10%, 0.05 eq.), MeOH (0.1 M), 120°C; (iii) H_2, Ru/C (5%, 0.02 eq.), THF (0.03 M), 20°C, 1 atm; (iv) H_2, Ru/C (5%, 0.02 eq.), THF (0.03 M), reflux, 1 atm; (v)) H_2, Ru/C (5%, 0.04 eq.), THF (0.03 M), 20°C, 1 atm; (vi) $FeSO_4 \cdot 7H_2O$ (10 eq.), NH_4OH, EtOH, reflux; (vii) Fe (20 eq.), AcOH (0.1 M), 70°C; (viii) H-cube Pro®, Ru/C (5%), THF, 80°C, 50 bar, 0.3 mL min^{-1}.

The chiral ester (**45**) is a key intermediate in the synthesis of the angiotensin II receptor blocker sacubitril. Enantioselective flow hydrogenation using a tube-in-tube system, through two loops, provided access to the required diastereomer at 0.45 g h^{-1}. The introduction of the second loop was critical increasing the yield from 78 to 99% (**Figure 14**) [39].

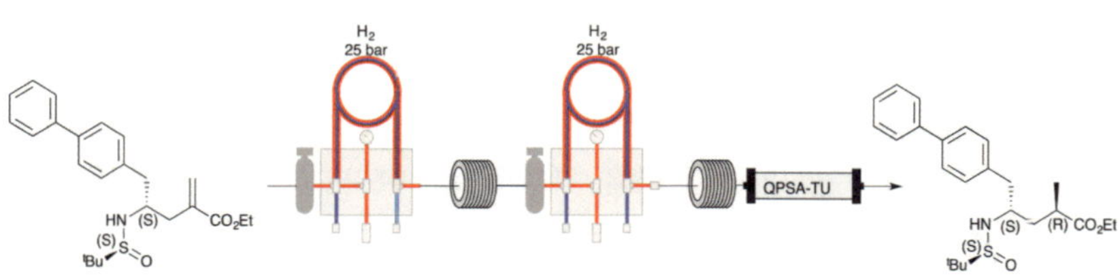

Figure 14. Enantioselective hydrogenation flow preparation of chiral ester **45**. Reagents and conditions: H_2, cat. DIPEA in EtOH (1 mol%), 20°C, 25 bar, 0.2 mL min^{-1}.

In a similar manner, H-cube mediated nitro reduction and lactam cyclisation of γ-nitro-α-amino esters with in situ cyclisation afforded, quantitatively, the corresponding γ-lactams (**47**) (**Figure 15**). Raney Ni hydrogenation (10 bar, 65°C) of the *syn-* diastereomer affords exclusively the *trans-* configuration with the *anti*-γ-nitro-α-amino esters gave the *cis* diastereomer [40].

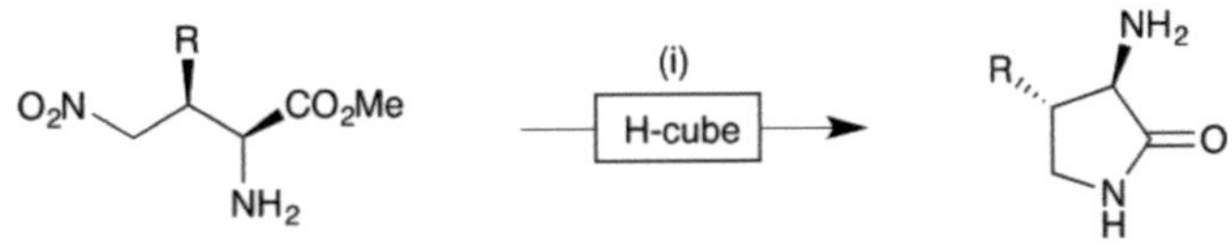

syn **46**
R = Ph, 4-F-C_6H_4, 4-MeO-C_6H_4

trans **47**
quant.

Figure 15. Formation of γ-lactams via flow hydrogenation. Reagents and conditions: (i) H-cube®, Raney Ni, MeOH (0.01 M), 65°C, 10 bar, 1.0 mL min^{-1}.

3.7. Other reactions

A variety of other reductions that are pertinent to medicinal chemistry can also be performed via flow hydrogenation as shown in **Table 4**.

Type of reaction	Starting material	Product	Conditions	References
Azide reduction	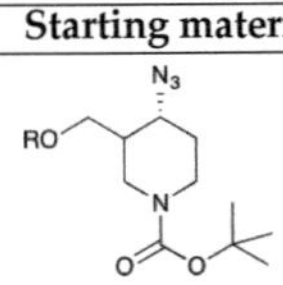	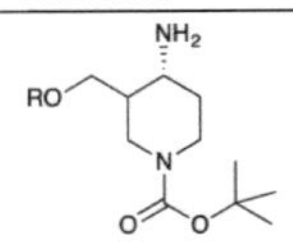	10% Pd/C, RT, 1 bar, 1.0 mL min^{-1}, 94%	[41]
Olefin reduction	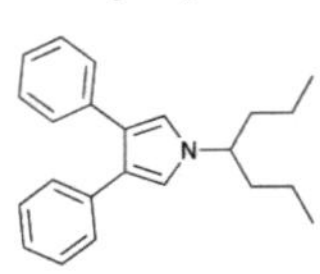	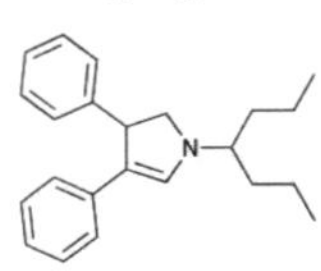	10% Pd/C, 80°C, 60 bar, 2.0 mL min^{-1}, 90%	[42]
De-aromatisation	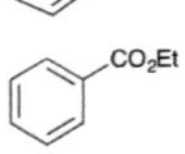	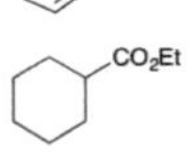	10% Ru/C or 10% Rh/C, 75–100°C, 50 bar, 1.0 mL min^{-1}, 100%	[43]
Selective reduction	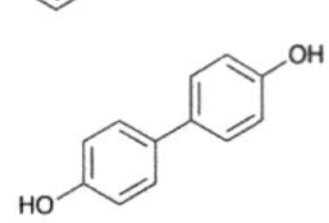	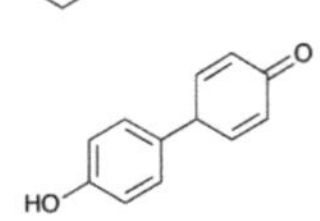	20% Pd/C, 100°C, 5 bar, 1.0 mL min^{-1}, 58%	[44]
Hydroformylation	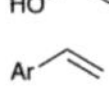	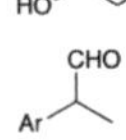 11 examples	$Rh(CO)_2(acac)$, 65°C, 25 bar, 0.6 mL min^{-1}, 69–94%	[45]

Table 4. Selected other common flow reduction reactions.

In the synthesis of the antimalarial drug, OZ439 **49**, Lau et al. optimised the hydrogenation step successfully reducing only one of the aromatic rings using 20% Pd/C (**Figure 16**) [44]. The concentration of the undesired minor by-products **50** and **51** was minimised by control of the temperature and the amount of hydrogen entering the system. This flow approach to **49** also avoided the use of genotoxic 4-(2-chloroethyl)morpholine.

3.8. Deuteration

The incorporation of a deuterium label has been used widely to probe reaction mechanisms, to probe a compound's pharmacokinetic properties, and as an internal standard in NMR and mass spectrometry [46]. The increase in bond strength (C-H versus C-D) can modify a drug's pharmacokinetic profile, and this has led to the development of deuterium containing drugs. Deutetrabenazine (SD-809) is expected to be the first deuterated drug approved by the FDA.

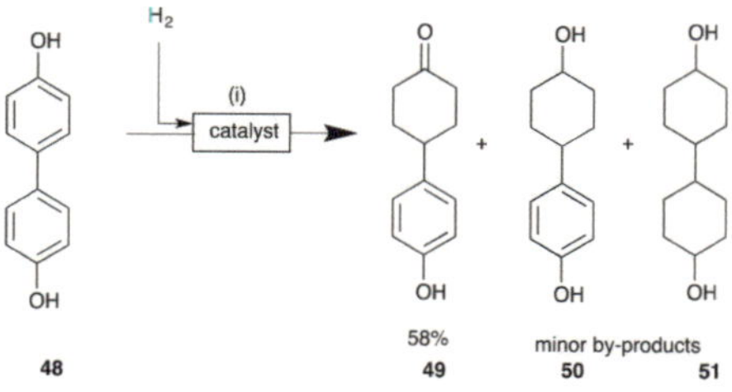

Figure 16. Selective continuous flow hydrogenation of 4-4′-biphenol **48**. Reagents and conditions: (i) H_2 (0.1 L min^{-1}), Pd/C (20%), EtOH/H_2O (1:1 v/v, 0.05 M), 100°C, 5 bar, 1.0 mL min^{-1}.

The synthesis of SD-809 is not flow mediated, but its success does suggest that the incorporation of deuterium will become a more common feature in future drugs [47]. Deuterium incorporation can be accomplished from D_2 gas and catalytic H-D exchange reactions between H_2 and D_2O. There are disadvantages to using deuterium gas on a laboratory scale, such as the handling of the gas itself, and the catalytic approaches are time consuming and do not always produce high purity D_2. However, electrolysis of D_2O by the Thales Nano H-cube® offers direct and rapid access to high purity D_2 gas and is applicable across the suite of reduction chemistries discussed above affording highly flexible incorporation of deuterium. Hsieh et al. have demonstrated this in the deuteration of a series of *trans*-chalcones (**52**) of interest for their antidiabetic activity (**Figure 17**) [48].

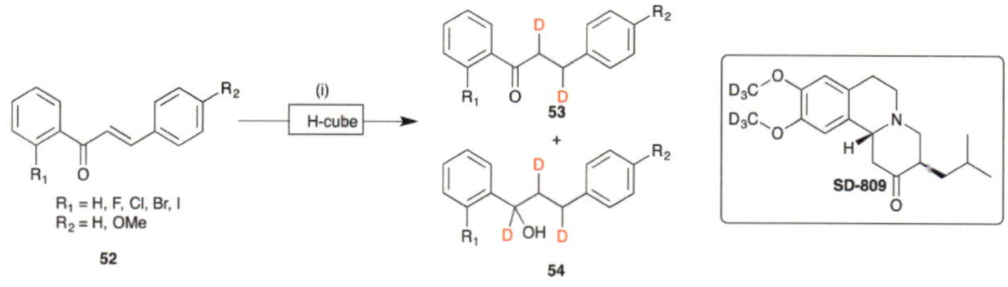

Figure 17. Deuteration of *trans*-chalcone and various derivatives. Reagents and conditions: (i) H-cube®, D_2O, 5% Pt/Al_2O_3, 100°C, 100 bar, 1 mL min^{-1}. Insert: chemical structure of SD-809.

Access to the required D_2-gas uses two separate inlet streams where the sample is introduced in an aprotic solvent and with D_2O electrolysis providing the required gas at which point the streams were combined and passaged over the deuteration catalyst (**Figure 18**).

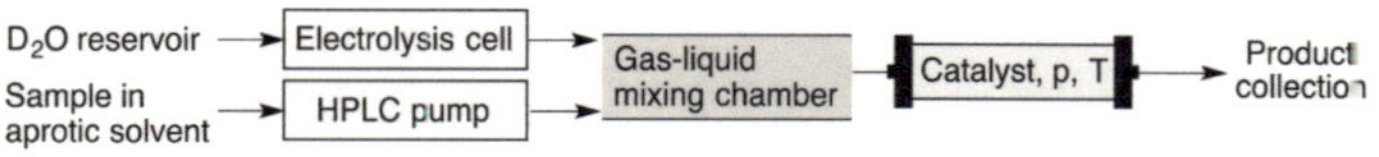

Figure 18. Schematic outline of the continuous flow reactor used to prepare deuterated compounds by Hsieh et al [48].

The C2-halogen played a significant role in determining the ratio of di- to tri-deuterated species. With all analogues except the 2-F, conversions of ≥90% and exclusive formation of the di-deuterated species (**53**) were observed with 5% Pt/Al_2O_3. However with (*E*)-1-(2-fluorophenyl)-3-(4-methoxyphenyl)prop-2-en-1-one this catalyst afforded a 5:95 ratio of **53:54** with 97% conversion. Switching to the less active $Pd/BaSO_4$ catalyst afforded 100% conversion of the 2-F analogue, with a best ratio of 89:11 (**53:54**). The presence of the fluorine had a significant effect on the deuteration of this family of chalcones.

3.9. New catalysts

As further catalysts are developed for flow hydrogenation, the specificity and robustness of the chemical transformations achievable increase. Flow hydrogenation does require the use of a rare metal catalyst, which can be both expensive and potentially environmentally unsustainable. This negates the toxicity and disposal problems related to classical reducing agents and suggests that flow hydrogenation approaches may be expensive and not environmentally benign. This has led to the development of alternative hydrogenation catalysts such as carbon-supported iron-phenanthroline complexes, nickel nanoparticles, and FeNi alloys. These new catalysts show broad spectrum reductive capabilities (**Table 5**).

Catalyst	Starting material	Product	Conditions	References
1.9% Au/Al_2O_3	Substituted nitrobenzenes (6 examples)	Substituted anilines (6 examples)	60–110°C, 10–20 bar, 0.5 mL min^{-1}, 98–100%	[49]
Pd/C (3%) Bimodal and trimodal	Alkyne, aldehyde, and halogenated species	Various	60–110°C, 1 bar, 0.5 mL min^{-1}, 24–43 Selectivity 51–100%	[50]
ARP-Pt	Olefins (24 examples) and nitrobenzenes (9 examples)		Olefin yield not reported; nitro reduction up to 99%	[51]
	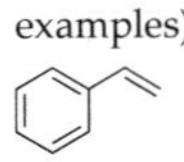	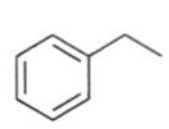	25°C, 5 bar, 2.0 mL min^{-1}, >99%	
	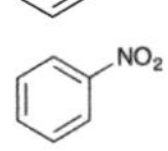	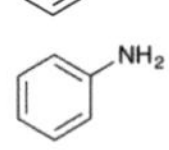	25°C, 15 bar, 1.5 mL min^{-1}, >99%	
Pd-Maghemite	Substituted nitrobenzenes (13 examples)	Substituted anilines	30°C, 1 bar, 0.5 mL min^{-1}, 86–98%	[52]
	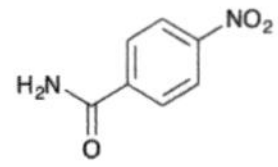	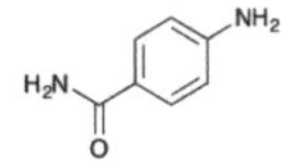	30°C, 1 bar, 94%	
	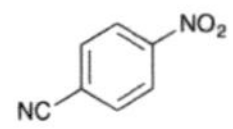	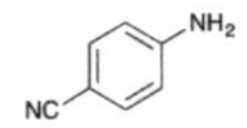	30°C, 1 bar, 0.5 mL min^{-1}, 88%	
Designed porous structure reactor (DPSR) Al_2O_3-ZnO	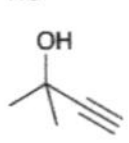	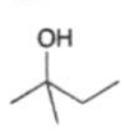	60–80°C, 6 bar, 37 kg h^{-1}, >99%	[53]
Pd^0-AmP-MCF (heterogeneous Pd nanocatalyst supported on aminofunctionalized mesocellular foam)	Michael acceptors (14 examples)	Aldehydes and alcohols		[54]
	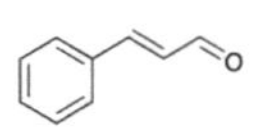	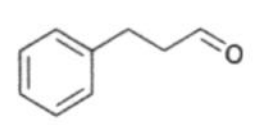	20°C, 1 bar, 1.5 mL min^{-1}, 94%	

Table 5. Use of novel catalysts in flow reduction reactions.

Dehydrohalogenation is a significant and ongoing concern in flow (and batch) reduction [55], and thus the development of new catalysts that specifically avoid this outcome is a valuable research tool. Osako et al. used platinum nanoparticles dispersed on an amphiphilic polystyrene-poly(ethylene glycol) (ART-Pt) resin as a catalyst was specifically developed to avoid reduction of –Cl, –C=O, and –CN moieties, e.g., **55a-56c** (**Figure 19**) [51].

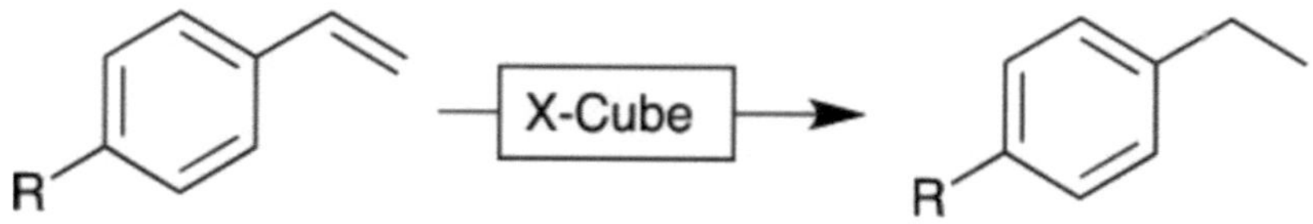

55a, R = Cl
55b, R = NHCbz
55c, R = CN

56a, R = Cl
56b, R = NHCbz
56c, R = CN

Figure 19. Alkene reductions via flow hydrogenation using the ARP-Pt catalyst. Reagents and conditions: X-cube®, H_2 (5 vol%), ARP-Pt (0.073 mmol Pt), EtOH (50 mM), 5 bar, 2mL min^{-1}.

Specialist catalyst development has often been targeted towards chemoselectivity. Fan et al. have shown that the Pd/triC catalyst was selectively reduced alkyne groups over nitro, bromo, and aldehyde groups [50]. While Rathi's palladium nanoparticles supported on maghemite were effective in reducing nitroarenes, azides, and alkenes in good to excellent yields [52].

Nagendiran et al. detail the use of aminofunctionalised mesocellular foam-supported nanopalladium in the conjugate reduction in a series of Michael acceptors [54]. Both the Vapourtec (1.0 mL min^{-1}, 0.1 M, 1 bar H_2, 20°C) and the H-cube (1.0 mL min^{-1}, 0.1 M, 40 psi H_2, 50°C) with this catalyst afforded chemoselective reduction in the olefin moiety. This approach was scalable enabling the selective reduction of cinnamaldehyde to 3-phenylpropanal on an approximately 20 g scale with no observed loss of catalysts' activity or selectivity.

4. Conclusions

The body of evidence continues to grow illustrating that flow methodologies, in particular flow hydrogenation, offer significant advantages over batch technologies in medicinal chemistry. Flow chemistry has been demonstrated to enhance yields, simplify reaction work up, improve safety, and allow in-line analysis. The coupling of modular flow systems has allowed automated and semiautomated high yield, low purification requirement synthesis of active pharmaceutical ingredients over multiple cascading steps.

Flow-based reductions continue to provide greater access to new chemical scaffolds for use in drug design and development as well as to provide efficient methods for the production of current pharmaceuticals.

Author details

Cecilia C. Russell, Jennifer R. Baker, Peter J. Cossar and Adam McCluskey*

*Address all correspondence to: Adam.McCluskey@newcastle.edu.au

Centre for Chemical Biology, Chemistry, School of Environmental Life Science, The University of Newcastle, Callaghan, New South Wales, Australia

References

[1] Cossar PJ, Hizartzidis L, Simone MI, McCluskey A, Gordon CP. The expanding utility of continuous flow hydrogenation. Organic and Biomolecular Chemistry 2015;13:7119–7130.

[2] Dormán G, Kocsis L, Jones R, Darvas F. A benchtop continuous flow reactor: a solution to the hazards posed by gas cylinder based hydrogenation. Journal of Chemical Health and Safety 2013;20:3–8.

[3] Irfan M, Glasnov TN, Kappe CO. Heterogeneous catalytic hydrogenation reactions in continuous-flow reactors. ChemSusChem 2011;4:300–316.

[4] Chandra BT, Zebrowski JP. Hazards associated with laboratory scale hydrogenations. Journal of Chemical Health and Safety 2015;23:1–10.

[5] Hizartzidis L, Tarleton M, Gordon CP, McCluskey A. Chemoselective flow hydrogenation approaches to isoindole-7-carboxylic acids and 7-oxa-bicyclio[2.2.1]heptanes. RSC Advances 2014;4:9709–9722.

[6] Jones RV, Godorhazy L, Varga N, Szalay D, Urge L, Darava F. Continuous-flow high pressure hydrogenation reactor for optimization and high-throughput synthesis. Journal of Combinatorial Chemistry 2006;8:110–116.

[7] Newton S, Ley SV, Arcé EC, Grainger DM. Asymmetric homogeneous hydrogenation in flow using a tube-in-tube reactor. Advanced Synthesis and Catalysis 2012;354:1805–1812.

[8] Madarász J, Farkas G, Balogh S, Szöllósy A, Kovács J, Darvas F, Ürge L, Bakos J. A continuous-flow system for asymmetric hydrogenation using supported chiral catalysts. Journal of Flow Chemistry 2011;1:62–67.

[9] www.cambridgereactordesign.com/Gastropod_online.html (accessed August 2016).

[10] www.vapourtec.co.uk/products/eseriessystem/reactors (accessed August 2016).

[11] www.thalesnano.com (accessed August 2016).

[12] O'Brien M, Baxendale IR, Ley SV. Flow ozonolysis using a semipermeable Teflon AF-2400 membrane to effect gas - liquid contact. Organic Letters 2010;12:1596–1598.

[13] Abdel-Hamid MK, Macgregor KA, Odell LR, Chau N, Matiana A, Whitting A, Robinson PJ, McCluskey A. 1,8-Naphthalimide derivatives: new leads against dynamin I GTPase activity. Organic and Biomolecular Chemistry 2015;13:8016–8028.

[14] Ruiz-Olalla A, Retamosa MDG, Cossío FP Densely substituted L-proline esters as catalysts for asymmetric Michael additions of ketones to nitroalkenes. Journal of Organic Chemistry 2015;80:5588–5599.

[15] Down K, Amour A, Baldwin IR, Cooper AW, Deakin AM, Felton LM, Guntrip SB, Hardy C, Harrison ZA, Jones KL, Jones P, Keeling SE, Le J, Livia S, Lucas F, Lunniss CJ, Parr NJ, Robinson E, Rowland P, Smith S, Thomas DA, Vitulli G, Washio Y, Hamblin JN. Optimization of novel indazoles as highly potent and selective inhibitors of phosphoinositide 3-kinase δ for the treatment of respiratory disease. Journal of Medicinal Chemistry 2015;58:7381–7399.

[16] Egle B, Munoz J, Alonso N, De Borggraeve W, de la Hoz A, Diaz-Ortiz A, Alcázar J. First example of alkyl–aryl Negishi cross-coupling in flow: mild, efficient and clean introduction of functionalized alkyl groups. Journal of Flow Chemistry 2014;4:22–25.

[17] Lövei K, Greiner I, Éles J, Szigetvári A, Dékány M, Lévai S, Novák Z, Túrós GI. Multistep continuous-flow synthesis of condensed benzothiazoles. Journal of Flow Chemistry 2015;5:74–81.

[18] Schembri LS, Stoddart LA, Briddon SJ, Kellam B, Canals M, Graham B, Scammells PJ. Synthesis, biological evaluation, and utility of fluorescent ligands targeting the μ-opioid receptor. Journal of Medicinal Chemistry 2015;58:9754–9767.

[19] Bartolomé-Nebreda JM, Alonso de Diego SA, Artola M, Delgado F, Delgado Ó, Martin-Martin CM, Pena MA, Tong HM, Van Gool M, Alonso JM, Fontana A, Macdonald GJ, Megens A, Langlois X, Somers M, Vanhoof G, Conde-Ceide S. Identification of a novel orally bioavailable phosphodiesterase 10A (PDE10A) inhibitor with efficacy in animal models of schizophrenia. Journal of Medicinal Chemistry 2015;58:978–993.

[20] Yadav JS, Suresh B, Srihari P. Stereoselective total synthesis of the marine macrolide sanctolide A. European Journal of Organic Chemistry 2015; 26: 5856–5863.

[21] Patel S, Cohen F, Deans BJ, de la Torre K, Deshmukh G, Estrada AA, Sengupta AS, Gibbons P, Gustafson A, Huestis MP, Le Pichon CE, Lin H, Liu W, Liu X, Liu Y, Ly CQ, Lyssikatos JP, Ma C, Scearce-Levie K, Shin YG, Solanoy H, Stark KL, Wang J, Wang B, Zhao X, Lewcock JW, Siu M. Discovery of dual leucine zipper kinase (DLK, MAP3K12) inhibitors with activity in neurodegeneration models. Journal of Medicinal Chemistry 2015;58:401–418.

[22] Trobe M, Breinbauer R. Improved and scalable synthesis of building blocks for the modular synthesis of teraryl-based alpha-helix mimetics. Chemical Monthly 2016;147:509–521.

[23] Bach A, Pizzirani D, Realini N, Vozella V, Russo D, Penna I, Melzig L, Scarpelli R, Piomelli D. Benzoxazolone carboxamides as potent acid ceramidase inhibitors: synthesis and structure-activity relationship (SAR) studies. Journal of Medicinal Chemistry 2015;58:9258–9272.

[24] Gericke D, Ott D, Matveea VG, Sulman E, Aho A, Murzin DY, Roggan S, Danilova L, Hessel V, Loeb P, Kralisch D. Green catalysis by nanoparticulate catalysts developed for flow processing? Case study of glucose hydrogenation. RSC Advances 2015;5:15898–15908.

[25] Johnstone RAW, Wilby AH, Entwistle ID. Heterogeneous catalytic transfer hydrogenation and its relation to other methods for reduction of organic compounds. Chemical Reviews 1985;85:129–170.

[26] Liu J, Fitzgerald A, Mani N. Reductive amination by continuous-flow hydrogenation: direct and scalable synthesis of a benzylpiperazine. Synthesis 2012;44:2469–2473.

[27] Artiukha EA, Nuzhdin AL, Bukhiyarova GA, Zaytsev SY, Plyusnin PE, Shubin YV, Bukhtiyarov VI. One-pot reductive amination of aldehydes with nitroarenes over an Au/Al_2O_3 catalyst in a continuous flow reactor. Catalysis Science and Technology 2015;5:4741–4745.

[28] Chieffi G, Braun M, Esposit, D. Continuous reductive amination of biomass-derived molecules over carbonized filter paper-supported FeNi alloy. ChemSusChem 2015;8:3590–3594.

[29] Redpath P, Ness KA, Rousseau J, Macdonald SJF, Migaud ME. Facile access to new C-glycosides and C-glycoside scaffolds incorporating functionalised aromatic moieties. Carbohydrate Research 2015;402:25–34.

[30] Laroussarie A, Barycza B, Andriaamboavonjy H, Kenfack MT, Blériot Y, Gauthier C. Synthesis of the tetrasaccharide repeating unit of the β-Kdo-containing exopolysaccharide from *Burkholderia pseudomallei* and *B. cepacia* complex. Journal of Organic Chemistry 2015;80:10386–10396.

[31] Ng PS, Manjunatha UH, Rao SP, Camacho LR, Ma NL, Herve M, Noble CG, Goh A, Peukert S, Diagana TT, Smith PW, Kondreddl RR. Structure activity relationships of 4-hydroxy-2-pyridones: a novel class of antituberculosis agents. European Journal of Medicinal Chemistry 2015;106:144–156.

[32] Alexandre F-R, Brandt G, Caillet C, Chaves D, Convard T, Derock M, Gloux D, Griffon Y, Lallos L, Liuzzi M, Loi AG, Moulat L, Musiu C, Parsy C, Rahali H, Roques V, Seifer M, Standing D, Surleraux D. Synthesis and antiviral evaluation of a novel series of homoserine-based inhibitors of the hepatitis C virus NS3/4A serine protease. Bioorganic and Medicinal Chemistry Letters 2015;25:3984–3991.

[33] Russell CC, Lin AYS, Hains P, Simone MI, Robinson PJ, McCluskey A. An integrated flow and microwave approach to a broad spectrum protein kinase inhibitor. RSC Advances 2015;5:93433–93437.

[34] Ghislieri D, Gilmore K, Seeberger PH. Chemical assembly systems: layered control for divergent, continuous, multistep syntheses of active pharmaceutical ingredients. Angewandte Chemie, International Edition 2015;54:678–682.

[35] Tsubogo T, Oyamada H, Kobayashi S. Multistep continuous-flow synthesis of (R)- and (S)-rolipram using heterogeneous catalysts. Nature 2015;520:329–332.

[36] Upadhyay K, Manvar A, Rawal K, Joshi S, Trivedi J, Chaniyara R, Shah A. Evaluation of structurally diverse benzoazepines clubbed with coumarins as Mycobacterium tuberculosis agents. Chemical Biology and Drug Design 2012;80:1003–1008.

[37] Joseph CG, Wilson KR, Wood MS, Sorenson NB, Phan DV, Xiang Z, Witek RM, Haskell-Leuvano C. The 1,4-benzodiazepine-2,5-dione small molecule template results in melanocortin receptor agonists with nanomolar potencies. Journal of Medicinal Chemistry 2008;51:1423–1431.

[38] Viviano M, Milite C, Rescigno D, Castellano S, Sbardella G. A continuous-flow synthesis of 1,4-benzodiazepin-5-ones, privileged scaffolds for drug discovery. RSC Advances 2015;5:1268–1273.

[39] Lau SH, Bourne SL, Martin B, Schenkel B, Penn G, Ley SV. Synthesis of a precursor to Sacubitril using enabling technologies. Organic Letters 2015;17:5436–5439.

[40] Conde E, Rivilla I, Larumbe A, Cossío FP. Enantiodivergent synthesis of bis-spiropyrrolidines via sequential interrupted and completed (3 + 2) cycloadditions. Journal of Organic Chemistry 2015;80:11755–11767.

[41] Bamborough P, Chung CW, Furze RC, Grandi P, Michon AM, Sheppard RJ, Barnett H, Diallo H, Dixon DP, Douault C, Jones EJ, Karamshi B, Mitchell DJ, Prinjha RK, Rau C, Watson RJ, Werner T, Demont EH. Structure-based optimization of naphthyridones into potent ATAD2 bromodomain inhibitors. Journal of Medicinal Chemistry 2015;58:6151–6178.

[42] Yu X, Guttenberger N, Fuchs E, Peters M, Weber H, Breinbauer R. Diversity-oriented synthesis of a library of star-shaped 2*H*-imidazolines. ACS Combinatorial Science 2015;17:682–690.

[43] Hattori T, Ida T, Tsubone A, Sawama Y, Monguchi Y, Sajiki H. Facile arene hydrogenation under flow conditions catalyzed by rhodium or ruthenium on carbon. European Journal of Organic Chemistry 2015;11:2492–2497.

[44] Lau S-H, Galván A, Merchant RR, Battiolocchio C, Souto JA, Berry MA, Ley SV. Machines vs malaria: a flow-based preparation of the drug candidate OZ439. Organic Letters 2015;17:3218–3221.

[45] Brzozowski M, O'Brien M, Ley SV, Polyzos A. Flow chemistry: intelligent processing of gas-liquid transformations using a tube-in-tube reactor. Accounts of Chemical Research 2015;48:349–362.

[46] Mullard A. Deuterated drugs draw heavier backing. Nature Reviews Drug Discovery 2016;15:219–221.

[47] Katsnelson A. Heavy drugs draw heavy interest from pharma backers. Nature Medicine 2013;19:656.

[48] Hsieh C, Ötvös S, Wu Y-C, Mándity IM, Chang F-R, Fulöp F. Highly selective continuous-flow synthesis of potentially bioactive deuterated chalcone derivatives. ChemPlusChem 2015;80:859–864.

[49] Nuzhdin AL, Moroz BL, Bukhtiyarova GA, Reshenikov SI, Pyrjaev PA, Aleksandrov PV, Bukhtiyarov VI. Selective liquid-phase hydrogenation of a nitro group in substituted nitrobenzenes over Au/Al_2O_3 catalyst in a packed-bed flow reactor. ChemPlusChem 2015;80:1741–1749.

[50] Fan X, Sans V, Sharma SK, Plucinski PK, Zaikovskii VA, Wilson K, Tennison SR, Kozynchenko A, Lapkin AA. Pd/C catalysts based on synthetic carbons with bi- and tri-modal pore-size distribution: applications in flow chemistry. Catalysis Science and Technology 2016;6:2387–2395.

[51] Osako T, Torii K, Tazawa A, Uozumi Y. Continuous-flow hydrogenation of olefins and nitrobenzenes catalyzed by platinum nanoparticles dispersed in an amphiphilic polymer. RSC Advances 2015; 5: 45760–45766.

[52] Rathi AK, Gawande MB, Ranc V, Pechousek J, Petr M, Cepe K, Varma RS, Zboril R. Continuous flow hydrogenation of nitroarenes, azides and alkenes using maghemite-Pd nanocomposites. Catalysis Science and Technology 2016;6:152–160.

[53] Elias Y, von Rohr PR, Bonrath W, Medlock J, Buss A. A porous structured reactor for hydrogenation reactions. Chemical Engineering and Process Intensification 2015;95:175–185.

[54] Nagendiran A, Sörensen H, Johansson MJ, Tai C-W, Bäckvall J-E. Nanopalladium-catalyzed conjugate reduction of Michael acceptors – application in flow. Green Chemistry 2016;18:2632–2637.

[55] Hizartzidis L, Cossar PJ, Robertson MJ, Simone MI, Young KA, McCluskey A, Gordon CP. Expanding the utility of flow hydrogenation – a robust protocol restricting hydrodehalogenation. RSC Advances 2014;4:56743–56748.

[56] Calle LP, Echeverria B, Franconetti A, Serna S, Fernández-Alonso MC, Diercks T, Cañada FJ, Ardá A, Reichardt NC, Jiménez-Barbero J. Monitoring glycan-protein interactions by NMR spectroscopic analysis: a simple chemical tag that mimics natural CH-π interactions. Chemistry – A European Journal 2015;21:11408–11416.

[57] Pous J, Courant T, Bernadat G, Iorga BI, Blanchard F, Masson G. Regio-, diastereo-, and enantioselective nitroso-Diels–Alder reaction of 1,3-diene-1-carbamates catalyzed by chiral phosphoric acids. Journal of the American Chemical Society 2015;137:11950–11953.

Chapter 14

Production of Biofuel via Hydrogenation of Lignin from Biomass

Bawadi Abdullah,
Syed Anuar Faua Ad Syed Muhammad and
Nik Azmi Nik Mahmood

Additional information is available at the end of the chapter

http://dx.doi.org/10.5772/66108

Abstract

Historically, humans have been harnessing biomass as a source of energy since the time they knew to make a fire from woods. Even today, some countries still depend on woods as a main source of energy. Biologically, biomass contains carbon-, hydrogen- and oxygen-based matters that unify in a solid material and that are potentially to be converted to fuel. Lignin is one of the components present in lignocellulosic biomass and has been actively examined to be used for biofuel production. Issues arise with the chemical characteristic and rigidity of its structure, which a setback for its viability for biofuel conversion. However, such setbacks have been counteracted with the advances of lignin-based knowledge on its conversion to chemical precursors for biofuel conversion. Recently, investigations on hydrogenation as one of the chemical processes used can be potentially utilised for efficient and viable means for biofuel production.

Keywords: biofuel, chemical precursors, lignocellulosic biomass

1. Introduction

Biomass is a potential source of fuel consists of organic materials and considered as renewable and sustainable for energy development. With fossil fuel remaining unstable in terms of cost and sustainability, the need to generate biomass-based energy is becoming more imperative and feasible in economical point of view [1, 2]. One of the main benefits of biomass-based fuel (sometimes called as biofuel) is zero carbon dioxide release as compared to fossil fuel [3]. Furthermore, the dramatic rise of carbon dioxide concentration in the atmosphere has triggered researchers to find viable alternatives to fossil-based fuels.

One type of biomass which covers mostly of plant molecules is lignocellulosic biomass. Structurally, it composes of cellulose, hemicellulose and lignin. Cellulose and hemicellulose are mainly monomeric sugar linked to form polymer [4]. However, lignin consists of phenylpropane units, which cross-linked with tight and varied chemical bonds [4]. Thus, lignin complexity makes it difficult to be degraded as compared to almost unified type of bonding of cellulose and hemicellulose. Nonetheless, a few researchers have found ways to enhance degradability of lignin through recent pretreatment technologies [4, 5].

Lignin is separated from cellulose and hemicelluloses during kraft process, and only partial is utilised in combustion application as energy source [6]. In addition, through development of bioethanol production, it is expected that more lignin is produced as by-product which is also added to the mass number of lignin. Through advances of pretreatment technology and valorisation, lignin can be used as an alternative source for fine chemicals and raw material for fuel.

In the progress of lignin utilisation for fuel production, hydrogenation of lignin becomes one of viable methods. Lignin contains functional phenolic compounds, but the difficulties of extracting the compounds remain a bottleneck to unlock this potential material for fuel production. For example, the degree of interaction between monomers (phenolic compound) varied due to the heterologous nature of each individual monomers [7]. Another reason would be the feasibility of pretreatment technologies that are needed to be strategised and to fully utilise the monomers present [7]. In addition, the monomers have to be separated from the strong linkages within lignin structure, so that conversion to fuel can be executed efficiently.

In this chapter, lignin will be discussed of its structure and its different functional phenolic compounds. In addition, lignin depolymerisation or valorisation process to obtain individual monomers will also be presented. Further discussion will also include hydrogenation of lignin and the mechanism involved in the process. At the final part of this chapter, the future perspective of lignin hydrogenation that may lead to more innovative applications is discussed.

2. Biomass conversion to fine chemicals and fuels

The present development of biomass conversion to energy and chemicals together with individual types of abundant biomass in nature has been discussed in terms of its source, availability, types of biomass and chemical characteristics.

2.1. Historical background

In the abundance of biomass in the world's nature, traditional utilisation of biomass in order to survive has long been established. Primitively, when man knows how to create fire, biomass utilisation evolved from domestic usage to even larger application such as building houses, clothes production, paper making, etc. Yet the biomass utilisation seems endless and continues to produce more relevant products.

Henry Ford was the first to design a model car that runs using ethanol, and it was also reported that Rudolph Diesel intended on using vegetable oil to power his car engine [8]. However, at the time of World War II, fossil fuels were more practical based on supply, price and efficiency which caused biofuel to be unpopular. In contrast, in the late 1990s, high intensity of research and investment were made to produce biofuel from biomass due to reverse effect on supply and price of fossil fuel which is unpredictable.

The first biomass used was in solid form—wood—which was used to produce fire and oil derived from plants and animals that has been turned to a useful medium to light up lamp [9, 10]. In the present day, biofuel is derived from food crops like sugar beets, grains and vegetable oils or, namely, first-generation biomass [11]. However, the impact on the global food supply and sustainability of land led to non-food stock for energy sources and the second generation of energy sources such as cellulosic biomass, inedible oil and algae-derived oil [11, 12]. Microalgae can reproduce in a mass rapidly per day and have been found enriched of high level of oil in total dry biomass [13]. However, regardless the high growth rates or high oil content, algae-derived oil production faces challenges that needed attention such as strain isolation, nutrient sourcing or availability, on-site management in production and handling and residual biomass or waste management [14].

Meanwhile, biobutanol, another type of fuel, has becoming popular in recent years that surpass bioethanol. Biobutanol is referred as the next generation of biofuels that offers many advantages. It can be produced directly with few combinations of physical, chemical and biological processes from food crops such as cereal crops, sugar beets and sugar cane. The first synthesis of biobutanol was achieved during 1912–1914 by acetone-butanol-ethanol (ABE) fermentation using Clostridium species as catalysts [15].

2.2. Trends in biomass conversion to fine chemicals and fuels

In developing technologies to fine tune biomass conversion to chemicals or fuels, many combinations of physical, chemical and biological approaches have been utilised. The strategy is to enhance accessibility to the main target component such as cellulose and hemicellulose or to separate main recalcitrant of the biomass which is lignin from the other component and treated individually as precursor for fuel.

For starter, all biomass should undergo pretreatment in order to achieve maximum conversion. During the past decades, the target is to produce fermentable sugar from complex lignocellulosic biomass which includes separating lignin from the complex biomass [16]. **Figure 1** shows a schematic flow of basic biomass conversion into value-added products.

During the course of chemical or biochemical conversion, an effective pretreatment is needed for maximum utilisation of the biomass. To summarise, a few criteria have been highlighted to achieve efficient pretreatment as indicated below:

i. Preservation of hemicelluloses fractions, in order to have high yield of fermentable sugars;

ii. Minimising the building up of inhibitors due to degradation;

iii. Minimising the use of high energy in the process;

iv. Cost-effective pretreatment processes.

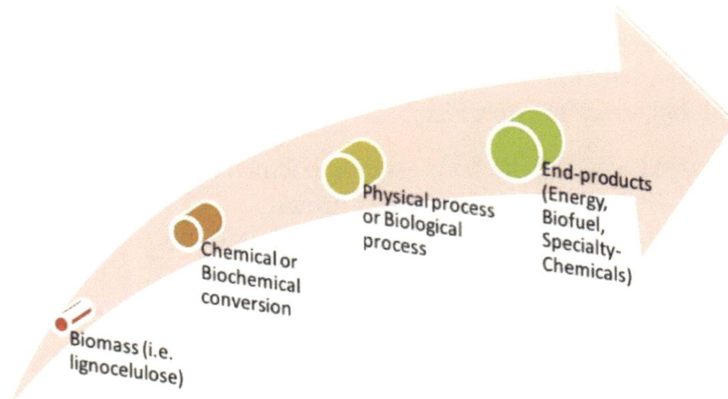

Figure 1. Schematic flow of general conversion of biomass to value-added products. Adapted from Ref. [17] with permission from Elsevier.

The above criteria become a basis for choosing the right method of pretreatment of biomass in order to maximise its efficiency. Further details on pretreatment will not be discussed here and can be found intensively discussed in other publications [18].

Meanwhile, as the complexity of the biomass dissolved through pretreatment processes, the next step is to choose the right method to directly convert the simpler form of the material to the desired product based on physical, chemical and biological processing methods.

2.2.1. *Physical-based conversion*

In this method, for some biomass, pretreatment sometimes will not be necessary. For example, woody biomass through combustion process will produce heat and electricity. The initial combustion will produce steam at high pressure and eventually the steam is used to activate turbine plants that in turn will generate electricity. Such biomass-fired steam turbine plants are located at the industrial sites that commonly where the biomass is produced. Another example is gasification process, where the biomass is directly heated and broken down into flammable gas. The gas or called 'biogas' will be drawn into filtration system to clean and refine before subjected to usage for electricity production.

The fact that major biomass components such as cellulose, hemicellulose and lignin can be fractionated based on different temperatures is exploited with the merging technology of pyrolysis [19–21]. The process involves three stages or heat-based degradation. The first stage includes water elimination, structural deformation and alkyl group formation which is also called pre-pyrolysis. The second stage involves decomposition of components and formation of pyrolysis products. Finally, the last stage produces carbon residuals and bio-oil from charred biomass. Mainly, pyrolysis dealt with cellulose and hemicellulose component conversion, but very little is known about the contribution of this process to lignin fraction. The lignin is merely converted to low concentration of phenolics and char. As it becomes more evident that large amounts of hydrolytic lignin will be produced in future bioethanol plants, lignin has gained interest as a chemical feedstock or aromatic compounds such as catechols, guaiacol, syringol, phenol, furfural, and acetic acid [22, 23].

2.2.2. *Chemical-based conversion*

In a chemical approach, cellulose in a raw biomass can be converted to biofuel. Another new process has been introduced to convert cellulose to chemical 5-hydroxymethylfurfural (HMF). Cellulose, through catalytic conversion, can be converted to HMF [24], and recently formation of HMF through catalytic conversion in ionic liquid produced more than 50% of HMF [25–28]. HMF is oxygen rich and hydrogen deficient that can be used for biofuel production through a reduction process that led to the production of 2,5-hexanedione (HD) [29]. Meanwhile, recently one report indicates that direct electrochemical method to convert HMF to HD without any means of catalysis or hydrogen. Instead HD is produced using water and metal electrode to open up furan ring in HMF and selectively in one-step mode [30].

There is another process called 'thermochemical' that involves thorough deconstruction of feedstock and develops the resulting intermediates into different types of fuels and fine chemicals. Thermochemical process consists of three processing parts which include gasification or devolatilisation to produce raw syngas, syngas conditioning and removal of other residual gases such as H_2S and CO_2 and finally conversion of syngas to fine chemicals such as olefins, aromatic mixed alcohols and fuel for transportation [31].

2.2.3. *Biological process-based conversion*

In general, biological process-based conversion involves cellulosic or hemicellulosic sugar conversion to fuel. Most popular ever known is biofuel such as bioethanol production using microorganism(s) as the catalyst in the bioconversion process. There are two-step processes which include fermentation and pretreatment to obtain the primary sugars [32]. In many publications, cellulose and hemicellulose conversion to biofuel production have been discussed thoroughly and can be read elsewhere [32, 33], and it is necessary to briefly discussed on lignin bioconversion.

Though it has inhibitory effect on any cellulose/hemicellulose conversion, if present, lignin can be converted to value-added products such as stand-alone feedstock. Recently, a *Rhodococusc* sp. can convert lignin into triacylglycerols under nitrogen-limiting conditions [34]. Furthermore, it was demonstrated that lignin valorisation can be channelled by different catabolic pathways of few aromatic catabolising bacteria to produce precursors for fuel production [3]. However, the thermal properties of each aromatic bonding in lignin need an integrated thermochemical process in order to penetrate the bonds [35].

2.3. Biomass types and characteristic

Biomass is usually derived from trees, forest residues, grasses, aquatic plants and crops. They are very important feedstock to produce valuable chemicals and fuel. These plants undergo photosynthesis process to convert carbon dioxide and water to primary and secondary metabolites. Primary metabolites include carbohydrates and lignin which presents in large concentration in biomass. The carbohydrates are simple sugar, cellulose and hemicellulose which take up more than 50% of the content of the biomass. However, some researchers divided the biomass in different moisture contents [36]. The high moisture content biomass includes aquatic plants and manures which have to undergo 'wet' processing techniques to convert it to energy. For example, conversion of these materials to biofuel commonly involves degradation of the materials using microorganisms or fermentation. Meanwhile, woody plants or dry biomass involves physical techniques such as gasification, pyrolysis or physical combustion.

3. Lignin: structural chemistry and route to conversion

Lignin is one of the major components in lignocellulosic biomass together with cellulose and hemicellulose. It intertwined with the cellulose-hemicellulose structural linkage that poses barriers for any physical or biological disturbance through the strong structure of lignin. The structural chemistry of lignin has been discussed and detailed of each of its monomers also presented. In addition, lignin conversion to fine chemicals and fuel has been added and mostly focuses on the present available data.

3.1. Lignin: structural chemistry

Through the binding of arrays of carbon-carbon and ether linkages, a single intermonomeric bonding scaffold was formed which is also known as lignin matrix [37]. This complex heterogenous structure of lignin consists essentially from three aromatic alcohols: p-coumaryl, coniferyl and sinapyl alcohols. These monolignols form phenolic substructures such as guaiacyl (G, from coniferyl alcohol), p-hydroxyphenyl (H, from coumaryl alcohol) and syringyl (S, from sinapyl alcohol) as shown in **Figure 2**. Each chemical structure confers a distinctive characteristic to lignin. Lignins composed mostly of G-units are usually softwood lignins, while lignins with different ratios of G- and S-units are hardwood lignins [37–40].

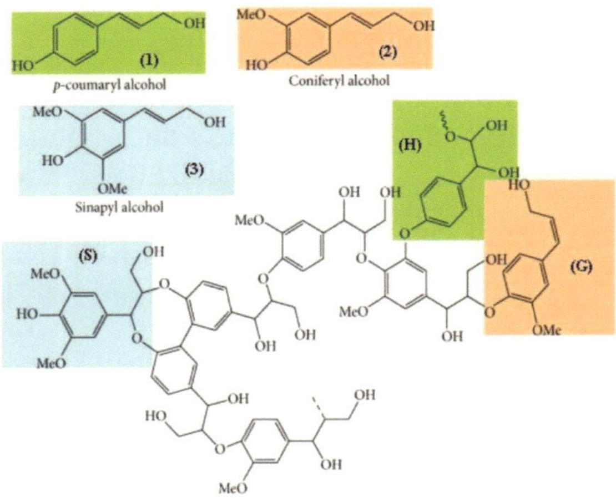

Figure 2. A segment of lignin polymer structure with monolignols involved in lignin biosynthesis: p-coumaryl alcohol (1), coniferyl alcohol (2) and sinapyl alcohol (3). Possible phenolic structures: guaiacyl (G), p-hydroxyphenyl (H) and syringyl (S). Reprinted from Ref. [38] with permission from Hindawi.

There are few types of native lignin substructures and interunit for different biomass sources as seen in **Figure 3**. The β-O-4′ (aryl ether) interunit linkage has been identified as a major lignin substructure where it provides a foundation for polymeric framework deconstruction in industrial processes. Some are known to have some resistant to chemical degradation such as β-5′, β-β′, 5-5′, 5-O-4′ and β-1′ [37, 41]. Thus, lignin usually appears in a complicated macromolecule form due to the diversity functional groups and various kinds of linkage of lignin substructures [42]. Lignin usually exists in the form of lignin-carbohydrate complexes (LCCs) resulted from dehydrogenation of G, S and H monolignols. Hence, the covalently linked to hemicellulose lignin and the occurrence of lignin cross-linked to polysaccharides provide mechanical strength to plant cell wall and additional protection to other structures from external [37].

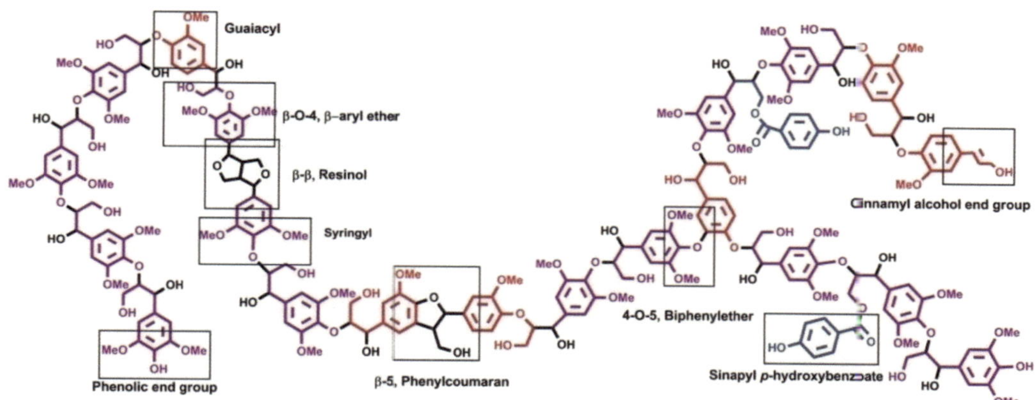

Figure 3. Common interunit linkages in lignin polymeric framework. Reprinted from Ref. [41] with permission of The Royal Society of Chemistry.

3.2. Lignin: chemical and biological conversion

High value compound can be produced through biological and chemical conversion of lignin. Processes such as gasification, hydrolysis, oxidation and pyrolysis are a well-known chemical conversion process. A wide range of polymers, chemicals and building blocks can be synthesised from chemically converted lignin [40]. Hydrodeoxygenation (HDO) is an example of chemical conversion method. It involves the removal of oxygen from oxygen-containing molecules in the presence of catalysts with high-pressure hydrogen at moderate temperature. Oxygen is removed to form water via hydrogenolysis reaction and then saturated by hydrogenation reaction [37]. Catalytic HDO of guaiacol (2-methoxyphenol), an oxygen-rich lignin model compound, has been investigated by Aqsha et al. for production of deoxygenated products. Guaiacol conversion products are mainly determined by methoxy, hydroxyl and benzene ring [43].

Alternatively, lignin converted from biological process such as enzymatic oxidation and microbial conversion involves living organisms [40]. Lignin-degrading microbes such as ligninolytic peroxidase enzymes or laccase enzymes have been exploited to oxidise aromatic units within lignin complex molecules [44].

3.3. Examples of lignin to value-added materials

Yi-Lin Chung et al. developed a catalytic and solvent-free method for synthesis of a lignin-poly (lactic acid) copolymer. The g-type poly(lactic acid) (PLA) copolymer, synthesised from graft polymerisation of lactide onto lignin, is catalysed by triazabicyclodecene (TBD) as depicted in **Figure 4**. It displays a glass transition temperature range from 45 to 85°C with multiphase melting behaviour. It also can be used to enhance UV absorption and reduce brittleness without sacrificing its elasticity [45].

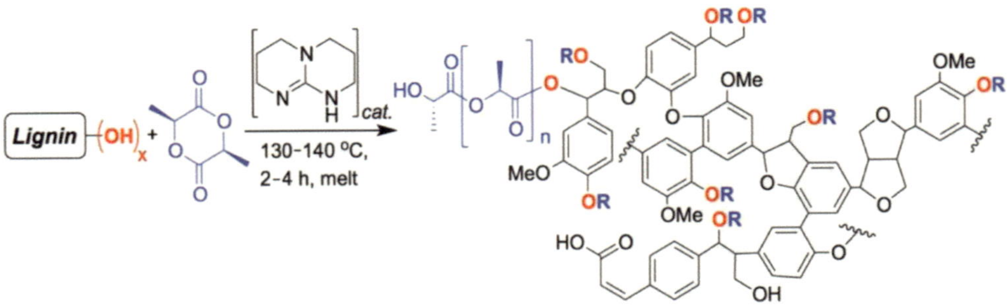

Figure 4. Ring-opening polymerisation of lactide (LA) on lignin using triazabicyclodecene (TBD) catalyst scheme. Reprinted with permission from Ref. [45]. Copyright (2013) American Chemical Society.

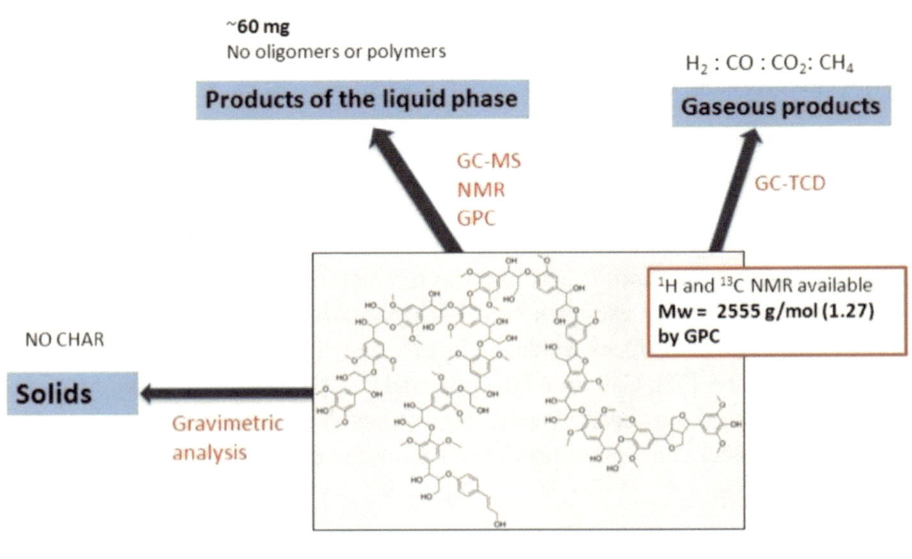

Figure 5. Quantitative analysis of product streams from organosolv lignin substrate. Reprinted with permission from Ref. [46]. Copyright (2014) American Chemical Society.

Barta and Ford devised a novel catalytic system to produce organic liquids from renewable lignocellulose feedstock. Supercritical methanol was used as reaction medium in a single-stage reactor at operating temperature of 300–320°C at 160–220 bar using copper-doped porous metal oxide. This system was tested on organosolv lignin solution (deep brown colour). As summarised in **Figure 5**, no char was formed during reaction, and the gaseous products were mainly hydrogen gas. The average molecular weight of liquid-phase product is within range of the monomer, and dimer units indicate that organosolv lignin was fully converted [46].

4. Hydrogenation of lignin

It is a continuation of previous section, but in this section, the focus is more on hydrogenation process of lignin. In this section, examples of recent literatures reported so far on the hydrogenation of lignin and details on the methods will be discussed.

4.1. Lignin valorisation

Lignin valorisation is an approach or strategy to convert lignin obtained from any sources, especially biomass, into value-added products as shown in **Figure 6** [47]. Scientists have studied this strategy for many decades through variation of thermal, catalytic and biological approaches in order to break lignin down into its constituent monomers and oligomers as well as upgrading the resulting monomers to fuels or chemicals [48].

The strategy usually involved one or more processes in order to achieve maximum lignin conversion. For example, Gianpaolo Chieffi et al. studied the efficiency of integrating smart recycle and upcycle of biomass-derived waste stream with strategies for the preparation of functional carbonaceous and composite materials using Fe and Ni (cheap and abundant metal precursors) [49].

Lignin valorisation can be roughly categorised into two sections: (1) upstream which involved separation and isolation process of lignin (e.g. pyrolysis, hydrogenolysis and hydrolysis) and (2) downstream, where valorised products are obtained through depolymerisation and chemical modification of the isolated lignin (e.g. hydrodeoxygenation, catalytic cracking, hydrogenation and dehydrogenation) [50]. In this section, we focus on the downstream process which is to convert the lignin monomers to the fuels and chemicals via hydrogenation.

4.2. Hydrogenation in general

Hydrogenation is one of the depolymerisation strategies of lignin which utilise hydrogen as a reductant to yield aromatic products with lower oxygen content and higher stability than other strategies (c.f. **Figure 6**) [47]. This process operates in mild condition affording high-value fine chemicals. However, it is extremely difficult to selectively yield products towards aromatic such as C=C, linear C=C, C=O, C=C, etc. due to the nature of catalysts and reaction conditions. In this process, the selection of the suitable catalysts is the critical parameter to ensure the desired products can be achieved.

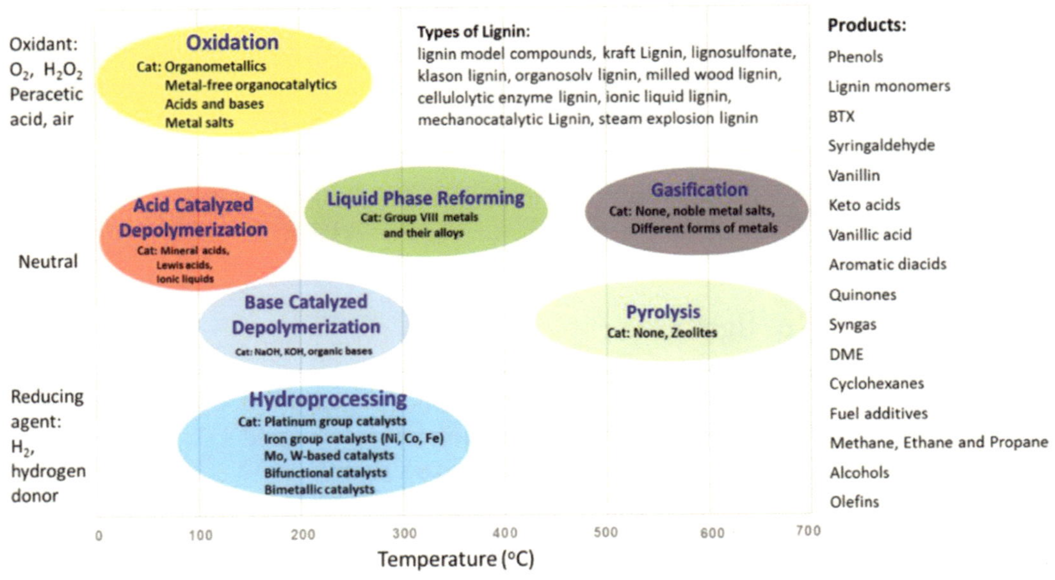

Figure 6. Summary of lignin conversion process (note: the abscissa represents the typical temperature range of the lignin conversion processes). Reprinted with permission Ref. [47]. Copyright (2015) American Chemical Society.

4.3. Examples of lignin hydrogenation

Consumption of fossil fuels would decrease significantly by implementing this approach of hydrogenating lignin to obtain the desirable products. Nevertheless, factors such as product selectivity, cost and the conversion efficiency of using commercially available catalysts are still unsatisfactory [51]. Hence, endless research has been carried out in order to overcome these difficulties.

Yu et al. [52] proposed an in situ catalytic hydrogenation system as relative to conventional method for converting lignin depolymerisation compounds to alcohols. In this work, The Raney Ni has been used for hydrogenation process. They found guaiacol conversion and cyclohexanol selectivity to be 99% and 93.74%, respectively, for 7 h time on stream (TOS). These results were obtained under the optimal conditions of 220°C, initial pressure of 3.0 MPa and mole ratio of H_2O/CH_3OH/feedstock = 20:5:0.8. Thus, this technique offers a new alternative method for hydrogenation of lignin.

The recent work of Shu et al. [53] utilised a highly efficient and selective hydrogenation process for phenolic compounds at a mild condition over step by step precipitated Ni/SiO_2 catalyst. Almost a complete conversion of guaiacol to cyclohexanol was obtained at 120°C, 2.0 MPa for 2 h TOS. The step by step precipitated Ni/SiO_2 preparation method thus significantly improved the conversion of the guaiacol (c.f. **Figure 7**). The structure of catalysts has been significantly modified by increasing the specific surface area and high Ni metal dispersion on the support that translated into high catalytic activity. Furthermore, this method also provides an appropriate acidity of catalyst and, hence, improves the catalytic performance significantly. Interestingly, this method also improves the longevity of the catalyst with an excellent recyclability.

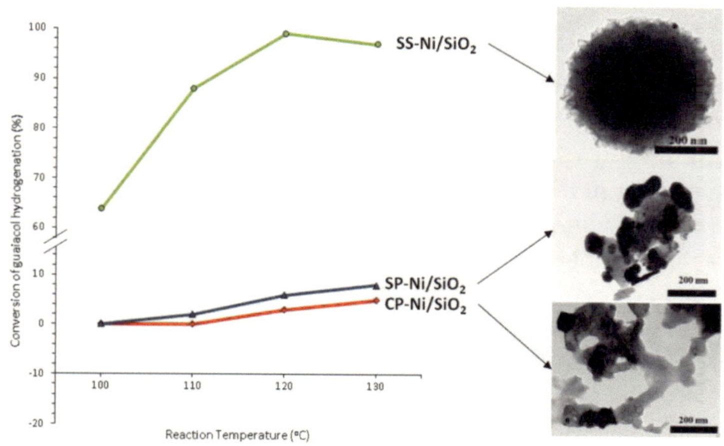

Figure 7. Conversion of guaiacol hydrogenation for different Ni/SiO_2 catalysts under 120°C, 2 MPa H_2 atmosphere for 2 h. Adapted from Ref. [52] with permission of The Royal Society of Chemistry.

Pd also exhibits a good metal to be doped with Al_2O_3 as a catalyst for hydrogenation process using wet impregnation technique. The work of Yi et al. (2016) [54] employed 3 wt.%Pd/Al_2O_3 to convert 4-ethylphenol to cyclohexanols under mild reaction conditions in an aqueous phase. The reaction exhibits 100% conversion at 60°C for 12 h TOS. The catalysts also exhibit water resistance and stability even after recycling four times. This catalyst may provide a new catalyst's formulation to selectively produce cyclohexanol at mild reaction conditions.

Overall, the choice of catalysts is very imperative steps in determining the product selectivity. For example, Raney Ni has been extensively used in hydrogenation of lignin in past decades and regarded as the most common catalysts in hydrogenation process. However, there is not much study in bimetallic catalysts that might be worth to be investigated in order to enhance the product conversion and catalyst's longevity.

5. Future trends

The development of lignin valorisation or degradation remains open for new ideas and approaches. The energy and environmental crises which modern world is experiencing are forcing to re-evaluate the efficient utilisation or finding alternative uses for natural, renewable resources and using clean technologies. In this regard, lignocellulosic biomass holds considerable potential to meet the current energy demand and to overcome the excessive dependence on fossil fuels. Further advanced biotechnologies are crucial for discovery and produce biofuels and bio-chemicals. In current scenario, future trends are being directed to lignocellulosic biotechnology and genetic engineering for improved processes and products [55]. To overcome the current energy problems, it is predicted that lignocellulosic biomass in addition of green biotechnology will be the main focus of the future research [18].

Currently, lignocellulose is processed to product through three steps that include pretreatment, saccharification and fermentation [56]. The chemical pretreatment process has shown it important in the subsequent enzymatic hydrolysis and conversion of cellulosic feedstock to valuable products in the process of fermentation. An analysis of the chemical pretreatment method result shows that the composition of biomass such as hardwood, softwood or grass is the main factor in the selection of pretreatment method [55]. However, any chemical pretreatment that requires lower-cost chemical reagents and conditions and yield more sugar is preferable. The chemical pretreatment process can be divided into six types: acidic pretreatment, alkaline pretreatment, wet oxidation, ionic liquids, oxidative delignification and organosolv [55]. However, it has been reported that chemical pretreatment process adds significantly to the cost of feedstock hydrolysates by consuming energy, expensive catalysts and chemicals while potentially hindering the downstream bioprocess [56, 57], thus requires more research effort, such as the hydrogenolysis of lignin in methanol, however, produced mostly phenols. This clearly shows that the solvent plays an essential role in directing catalytic selectivity and, thus, it must be taken into consideration in the design of catalytic systems for lignin conversion [57].

The necessity of chemical pretreatment of lignin-containing biomass represents a major barrier to downstream fermentation [56] especially involving physical and thermochemical processes that alter the physiochemical recalcitrance of biomass that enhances downstream enzyme digestibility [58–62]. In addition, it has been reported that pretreatment processes modify the polysaccharide matrix reducing overall yield of fermentable sugar or generating by-products that inhibit enzyme hydrolysis and fermentation [63, 64].

Alternatively, it has been reported that microbes with tolerance to the inhibitory compounds produced during pretreatment are of industrial interest as fermenters of sources of saccharolytic enzymes and enhances downstream fermentation capabilities, thus potentially eliminate thermochemical pretreatment steps [18]. Indeed, biological pretreatment processes are an environmental-friendly alternative to thermochemical pretreatments, improving biorefinery economics by reducing pretreatment costs, easing inhibitor formation and increasing downstream fermentation [65–67], as reported that a few fungi suitable for wheat straw biological pretreatment and increased sugar recoveries [67].

6. Conclusion

In a nutshell, the hydrogenation of lignin can potentially become a breakthrough route for production of biofuel if we can unlock the lignin functionality. With the recent findings, it is interesting to observe that the exploitation of lignin valorisation shall make this route become viable and efficient to produce biofuel via hydrogenation of lignin. Further advanced biotechnologies are crucial for discovery such as more efficient biological pretreatment system, bioprospecting and development of stable genetically engineering microorganisms, improved gene cloning and sequencing technologies, characterisation of new enzymes and production in homologous or heterologous systems and ultimately lead to low-cost conversion of lignocellulose biotechnology for improved processes and productions based on economies of scale more efficient and cost-effective conversions of lignin into value-added products. To overcome the current energy problems, it is envisaged that lignocellulosic biomass in addition of green biotechnology will be the main focus of the future research.

Author details

Bawadi Abdullah[1,*], Syed Anuar Faua Ad Syed Muhammad[2] and Nik Azmi Nik Mahmood[2]

*Address all correspondence to: bawadi_abdullah@petronas.com.my

1 Chemical Engineering Department, Universiti Teknologi PETRONAS, Seri Iskandar, Perak, Malaysia

2 Universiti Teknologi Malaysia, Skudai, Johor, Malaysia

References

[1] Climent, M.J., Corma, A., Iborra, S. Conversion of biomass platform molecules into fuel additives and liquid hydrocarbon fuels. Green Chemistry. 2014, 16, 516–47.

[2] Zhou, C.-H., Xia, X., Lin, C.-X., Tong, D.-S., Beltramini, J. Catalytic conversion of lignocellulosic biomass to fine chemicals and fuels. Chemical Society Reviews. 2011, 40, 5588–617.

[3] Linger, J.G., Vardon, D.R., Guarnieri, M.T., Karp, E.M., Hunsinger, G.B., Franden, M.A., et al. Lignin valorization through integrated biological funneling and chemical catalysis. Proceedings of the National Academy of Sciences. 2014, 111, 12013–8.

[4] Chen, H. Chemical composition and structure of natural lignocellulose. In: Biotechnology of Lignocellulose, Springer, Netherlands, 2014, pp. 25–71.

[5] Zheng, Y., Zhao, J., Xu, F., Li, Y. Pretreatment of lignocellulosic biomass for enhanced biogas production. Progress in Energy and Combustion Science. 2014, 42, 35–53.

[6] Joffres, B., Laurenti, D., Charon, N., Daudin, A., Quignard, A., Geantet, C. Thermochemical conversion of lignin for fuels and chemicals: A review. Oil & Gas Science and Technology–Revue d'IFP Energies nouvelles. 2013, 68, 753–63.

[7] Ragauskas, A.J., Beckham, G.T., Biddy, M.J., Chandra, R., Chen, F., Davis, M.F., et al. Lignin valorization: improving lignin processing in the biorefinery. Science. 2014, 344, 1246843.

[8] Schubert, C. Can biofuels finally take center stage? Nature biotechnology 2006, 24, 777–84

[9] Sussman, V. The Samaritan Oil Lamps from Apollonia-Arsuf. Tel Aviv. 1983, 10, 71–96.

[10] Russell, L.S. A Heritage of Light: Lamps and Lighting in the Early Canadian Home. Toronto, Canada, University of Toronto Press, 2003.

[11] Mohr, A., Raman, S. Lessons from first generation biofuels and implications for the sustainability appraisal of second generation biofuels. Energy policy. 2013, 63, 114–22.

[12] Rulli, M.C., Bellomi, D., Cazzoli, A., De Carolis, G., D'Odorico, P. The water-land-food nexus of first-generation biofuels. Sci Rep. 2016, 6, 22521.

[13] Rodolfi, L., Chini Zittelli, G., Bassi, N., Padovani, G., Biondi, N., Bonini, G., et al. Microalgae for oil: Strain selection, induction of lipid synthesis and outdoor mass cultivation in a low-cost photobioreactor. Biotechnology and bioengineering. 2009, 102, 100–12.

[14] Hannon, M., Gimpel, J., Tran, M., Rasala, B., Mayfield, S. Biofuels from algae: challenges and potential. Biofuels. 2010, 1, 763–84.

[15] Jones, D.T., Woods, D.R. Acetone-butanol fermentation revisited. Microbiological reviews. 1986, 50, 484.

[16] Yang, B., Wyman, C.E. Pretreatment: the key to unlocking low-cost cellulosic ethanol. Biofuels, Bioproducts and Biorefining. 2008, 2, 26–40.

[17] Menon, V., Rao, M. Trends in bioconversion of lignocellulose: biofuels, platform chemicals & biorefinery concept. Progress in Energy and Combustion Science. 2012, 38, 522–50.

[18] Anwar, Z., Gulfraz, M., Irshad, M. Agro-industrial lignocellulosic biomass a key to unlock the future bio-energy: A brief review. Journal of Radiation Research and Applied Sciences. 2014, 7, 163–73.

[19] Oasmaa, A., Kuoppala, E. Solvent fractionation method with brix for rapid characterization of wood fast pyrolysis liquids. Energy & Fuels. 2008, 22, 4245–8.

[20] Bennett, N.M., Helle, S.S., Duff, S.J. Extraction and hydrolysis of levoglucosan from pyrolysis oil. Bioresource Technology. 2009, 100, 6059–63.

[21] Moens, L., Black, S.K., Myers, M.D., Czernik, S. Study of the neutralization and stabilization of a mixed hardwood bio-oil. Energy & Fuels. 2009, 23, 2695–9.

[22] Custodis, V.B., Hemberger, P., Ma, Z., van Bokhoven, J.A. Mechanism of fast pyrolysis of lignin: studying model compounds. The Journal of Physical Chemistry B. 2014, 118, 8524–31.

[23] Brebu, M., Vasile, C. Thermal degradation of lignin—a review. Cellulose Chemistry & Technology. 2010, 44, 353.

[24] Seri, K.-i., Sakaki, T., Shibata, M., Inoue, Y., Ishida, H. Lanthanum (III)-catalyzed degradation of cellulose at 250 C. Bioresource technology. 2002, 81, 257–60.

[25] Swatloski, R.P., Spear, S.K., Holbrey, J.D., Rogers, R.D. Dissolution of cellose with ionic liquids. Journal of the American Chemical Society. 2002, 124, 4974–5.

[26] Su, Y., Brown, H.M., Huang, X., Zhou, X.-d., Amonette, J.E., Zhang, Z.C. Single-step conversion of cellulose to 5-hydroxymethylfurfural (HMF), a versatile platform chemical. Applied Catalysis A: General. 2009, 361, 117–22.

[27] Tan, M., Zhao, L., Zhang, Y. Production of 5-hydroxymethyl furfural from cellulose in $CrCl_2$/Zeolite/BMIMCl system. Biomass and bioenergy. 2011, 35, 1367–70.

[28] Zhang, Z., Zhao, Z.K. Microwave-assisted conversion of lignocellulosic biomass into furans in ionic liquid. Bioresource technology. 2010, 101, 1111–4.

[29] James, O.O., Maity, S., Usman, L.A., Ajanaku, K.O., Ajani, O.O., Siyanbola, T.O., et al. Towards the conversion of carbohydrate biomass feedstocks to biofuels via hydroxylmethylfurfural. Energy & Environmental Science. 2010, 3, 1833–50.

[30] Roylance, J.J., Choi, K.-S. Electrochemical reductive biomass conversion: direct conversion of 5-hydroxymethylfurfural (HMF) to 2, 5-hexanedione (HD) via reductive ring-opening. Green Chemistry. 2016, 18, 2956–60.

[31] Yuan, Z., Eden, M.R. Recent advances in optimal design of thermochemical conversion of biomass to chemicals and liquid fuels. Current Opinion in Chemical Engineering. 2015, 10, 70–6.

[32] Soliman, S.A., El-Zawahry, Y.A., El-Mougith, A.A. Fungal Biodegradation of Agro-Industrial Waste. In: Cellulose - Biomass Conversion, Kadla, P.J. Ed., InTech, 2013, Vol. 75.

[33] Serrano-Ruiz, J.C., Braden, D.J., West, R.M., Dumesic, J.A. Conversion of cellulose to hydrocarbon fuels by progressive removal of oxygen. Applied Catalysis B: Environmental. 2010, 100, 184–9.

[34] Kosa, M., Ragauskas, A.J. Bioconversion of lignin model compounds with oleaginous Rhodococci. Applied microbiology and biotechnology. 2012, 93, 891–900.

[35] Liu, W.-J., Jiang, H., Yu, H.-Q. Thermochemical conversion of lignin to functional materials: a review and future directions. Green Chemistry. 2015, 17, 4888–907.

[36] McKendry, P. Energy production from biomass (part 1): overview of biomass. Bioresource technology. 2002, 83, 37–46.

[37] Laskar, D.D., Yang, B., Wang, H., Lee, J. Pathways for biomass-derived lignin to hydrocarbon fuels. Biofuels, Bioproducts and Biorefining. 2013, 7, 602–26.

[38] Gordobil, O., Moriana, R., Zhang, L., Labidi, J., Sevastyanova, O. Assesment of technical lignins for uses in biofuels and biomaterials: Structure-related properties, proximate analysis and chemical modification. Industrial Crops and Products. 2016, 83, 155–65.

[39] Laurichesse, S., Avérous, L. Chemical modification of lignins: Towards biobased polymers. Progress in Polymer Science. 2014, 39, 1266–90.

[40] Lee, H., Hamid, S., Zain, S. Conversion of lignocellulosic biomass to nanocellulose: structure and chemical process. The Scientific World Journal. 2014, 2014.

[41] Lochab, B., Shukla, S., Varma, I.K. Naturally occurring phenolic sources: monomers and polymers. RSC Advances. 2014, 4, 21712–52.

[42] Crestini, C., Melone, F., Sette, M., Saladino, R. Milled wood lignin: a linear oligomer. Biomacromolecules. 2011, 12, 3928–35.

[43] Aqsha, A., Katta, L., Mahinpey, N. Catalytic Hydrodeoxygenation of Guaiacol as Lignin Model Component Using Ni-Mo/TiO2 and Ni-V/TiO2 Catalysts. Catalysis Letters. 2015, 145, 1351–63.

[44] Bugg, T.D., Rahmanpour, R. Enzymatic conversion of lignin into renewable chemicals. Current Opinion in Chemical Biology. 2015, 29, 10–7.

[45] Chung, Y.-L., Olsson, J.V., Li, R.J., Frank, C.W., Waymouth, R.M., Billington, S.L., et al. A Renewable Lignin–Lactide Copolymer and Application in Biobased Composites. ACS Sustainable Chemistry & Engineering. 2013, 1, 1231–8.

[46] Barta, K., Ford, P.C. Catalytic Conversion of Nonfood Woody Biomass Solids to Organic Liquids. Accounts of Chemical Research. 2014, 47, 1503–12.

[47] Li, C., Zhao, X., Wang, A., Huber, G.W., Zhang, T. Catalytic transformation of lignin for the production of chemicals and fuels. 2015.

[48] Beckham, G.T., Johnson, C.W., Karp, E.M., Salvachua, D., Vardon, D.R. Opportunities and challenges in biological lignin valorization. Curr Opin Biotechnol. 2016, 42, 40–53.

[49] Chieffi, G., Fechler, N., Esposito, D. Valorization of lignin waste from hydrothermal treatment of biomass: towards porous carbonaceous composites for continuous hydrogenation. RSC Advances. 2015, 5, 63691–6.

[50] Rinaldi, R., Jastrzebski, R., Clough, M.T., Ralph, J., Kennema, M., Bruijnincx, P.C., et al. Paving the way for lignin valorisation: recent advances in bioengineering, biorefining and catalysis. Angewandte Chemie International Edition. 2016, 55, 8164–215.

[51] Wang, Y.-Y., Ling, L.-L., Jiang, H. Selective hydrogenation of lignin to produce chemical commodities by using a biochar supported Ni–Mo 2 C catalyst obtained from biomass. Green Chemistry. 2016.

[52] YU, Y.-x., Ying, X., Wang, T.-j., Zhang, Q., Zhang, X.-h., Zhang, X. In-situ hydrogenation of lignin depolymerization model compounds to cyclohexanol. Journal of Fuel Chemistry and Technology. 2013, 41, 443–7.

[53] Shu, R., Zhang, Q., Xu, Y., Long, J., Ma, L., Wang, T., et al. Hydrogenation of lignin-derived phenolic compounds over step by step precipitated Ni/SiO2. RSC Advances. 2016, 6, 5214–22.

[54] Yi, J., Luo, Y., He, T., Jiang, Z., Li, J., Hu, C. High efficient hydrogenation of lignin-derived monophenols to cyclohexanols over Pd/γ-Al2O3 under mild conditions. Catalysts. 2016, 6, 12.

[55] Mtui, G.Y. Recent advances in pretreatment of lignocellulosic wastes and production of value added products. African Journal of Biotechnology. 2009, 8, 1398–415.

[56] Fisher, A.B., Fong, S.S. Lignin biodegradation and industrial implications. AIMS Bioengineering. 2014, 1, 92–112.

[57] Wang, X. Valorization of Lignin and Bio-Oil by Catalytic Hydrogenation with Ni Catalyst [PhD Thesis]. Bochum, Germany, der Ruhr Universitat; 2013.

[58] Hu, F., Ragauskas, A. Pretreatment and lignocellulosic chemistry. Bioenergy Research. 2012, 5, 1043–66.

[59] Pingali, S.V., Urban, V.S., Heller, W.T., McGaughey, J., O'Neill, H., Foston, M., et al. Breakdown of cell wall nanostructure in dilute acid pretreated biomass. Biomacromolecules. 2010, 11, 2329–35.

[60] Trajano, H.L., Engle, N.L., Foston, M., Ragauskas, A.J., Tschaplinski, T.J., Wyman, C.E. The fate of lignin during hydrothermal pretreatment. Biotechnology for Biofuels. 2013, 6, 1.

[61] Zhao, X., Cheng, K., Liu, D. Organosolv pretreatment of lignocellulosic biomass for enzymatic hydrolysis. Applied Microbiology and Biotechnology. 2009, 82, 815–27.

[62] Kim, J.-Y., Shin, E.-J., Eom, I.-Y., Won, K., Kim, Y.H., Choi, D., et al. Structural features of lignin macromolecules extracted with ionic liquid from poplar wood. Bioresource Technology. 2011, 102, 9020–5.

[63] Berlin, A., Balakshin, M., Gilkes, N., Kadla, J., Maximenko, V., Kubo, S., et al. Inhibition of cellulase, xylanase and β-glucosidase activities by softwood lignin preparations. Journal of Biotechnology. 2006, 125, 198–209.

[64] Jönsson, L.J., Alriksson, B., Nilvebrant, N.-O. Bioconversion of lignocellulose: inhibitors and detoxification. Biotechnology for Biofuels. 2013, 6, 1.

[65] Chandel, A.K., Gonçalves, B.C., Strap, J.L., da Silva, S.S. Biodelignification of lignocellulose substrates: An intrinsic and sustainable pretreatment strategy for clean energy production. Critical Reviews in Biotechnology. 2015, 35, 281–93.

[66] Wan, C., Li, Y. Fungal pretreatment of lignocellulosic biomass. Biotechnology Advances. 2012, 30, 1447–57.

[67] Salvachúa, D., Prieto, A., López-Abelairas, M., Lu-Chau, T., Martínez, Á.T., Martínez, M.J. Fungal pretreatment: an alternative in second-generation ethanol from wheat straw. Bioresource Technology. 2011, 102, 7500–6.

Chapter 15

Hydrogenation Catalysis in Biobased Ionic Liquids

Safa Hayouni, Nadège Ferlin and
Sandrine Bouquillon

Additional information is available at the end of the chapter

http://dx.doi.org/10.5772/65448

Abstract

This chapter is dedicated to hydrogenation procedures of (poly)-alkenes or unsaturated ketones in various biobased and not-biobased ionic liquids. The first part of this chapter defines the concept of biobased ionic liquids and their preparation. In the second part, hydrogenation processes performed in non-biobased ionic liquids are described. Finally, in the last part, the two themes are mixed and recent examples of hydrogenation procedures of alkenes, polyalkenes or unsaturated ketones in biobased ionic liquids are developed.

Keywords: ionic liquids, biomass, hydrogenation, transition metal-based catalysts, alkenes, (poly)-unsaturated alkenes, α,β-unsaturated ketones, mild conditions

1. Introduction

Hydrogenation reactions of unsaturated compounds have been among the most extensively studied processes in catalysis for many years [1, 2]. To process, generally metallic catalysts are required. Nevertheless, one of the principal challenges facing the field of transition metal catalysis is the efficient recycling and the possibility to reuse the catalytic species.

Ionic liquids (ILs) were rapidly considered as very promising solvents for this purpose, due to their tunable physicochemical properties and capacity to immobilise catalysts. As their name indicates, ILs are ionic species containing an organic cation and presenting usually low melting points below 100°C [3, 4]. ILs have a negligible vapour pressure, are not so viscous over a wide temperature range and many are non-flammable [5]. Chemical and physical properties of ILs can be also refined through proper choice of cation and anion. For these reasons, ILs have been intensively investigated for the last three decades as green alternatives to conventional organic solvents [6–8]. Indeed, their numerous properties confer to these compounds the opportunity to replace classical organic solvents and their fields of applications are numerous: electrochemistry, organic synthesis, catalysis, complexation, extraction, etc. [4]. However, their low biodegradability or the toxicity of their degradation products and their high (eco)toxicity led the scientific community to reduce their use or to find other greener alternatives [9–13]. Considering this aspect, biobased ionic liquids could be a good alternative to classical ionic liquids but their preparation remains relative long and costly [14].

This chapter will be dedicated to the description of hydrogenation procedures of (poly)-alkenes or unsaturated ketones in biobased ionic liquids (ILs). In order to present this specific topic, the first part of the chapter will present the preparation of various biobased ionic liquids. Next, general procedures of hydrogenation in "classical" ILs will be developed according to recently published reviews. Finally, we will show that hydrogenation processes could be performed in biobased ILs with few examples.

2. Biobased ionic liquids

Due to their biodegradability and non-toxicity, the use of renewable resources could improve the green character of ILs. Among biobased precursors, building blocks such as amino acids [15] and amino alcohols from proteins, sugars from cellulose, chitin, starch and other polysaccharides, aromatic aldehydes from lignin and other compounds like fatty acids from vegetable or algae-derived oils can be used.

Amino acids or esters have been commonly used for the preparation of cations through classical acidification reaction or esterification/anionic metathesis sequences [16, 17]. The syntheses of all these ILs are summarised in **Scheme 1**. Protic or aprotics ILs could also be prepared, but their stability under acidic conditions was not suitable. One interesting example of such IL was presented by Trivedi et al. in which the counter anion was a sodium lauryl sulphate [18].

N-Heterocyclic amino acid-derived [19] chiral imidazolium based on valine, leucine and alanine [20], 4,5-dihydrothiazolium derived from biobased amino alcohol [21] and chiral pyrrolidine-based ILs from proline [22, 23] have been also prepared by synthetic ways involving more than four reaction steps. Gathergood et al. proposed some ILs with a neutral amino acid side chain that showed improvement on biodegradability [24]; they described also very recently L-phenylalanine ethyl ester ILs and non-ionic derivatives [25], and a series of amino acid derived ionic liquids showing microbial toxicity and biodegradation [26]. In addition, one of these last ILs presented the advantage to have antimicrobial properties. Betaine or betaine derivatives (**Scheme 2**) could be also transformed into ionic liquids, which can be used in extraction processes [27] or as herbicidal compounds [28]. Recently, de Almeida Meirelles et al. reviewed the use of ILs for food and bioproducts Industries; the authors suggested that biobased ILs could be use in food processes, especially when derived from

http://dx.doi.org/10.5772/65448

amino acids or choline, and that moreover, their biocompatibility could even improve the methods [29].

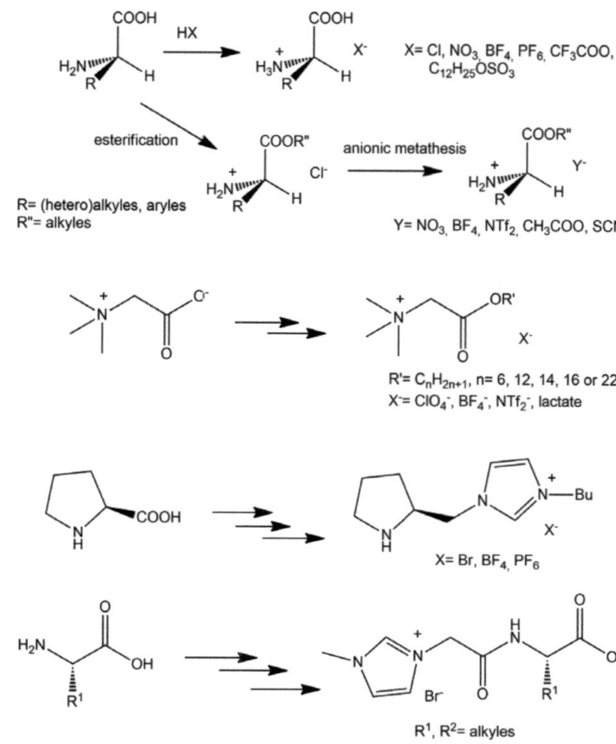

Scheme 1. Examples of aminoacids or esters-based as cations for ILs.

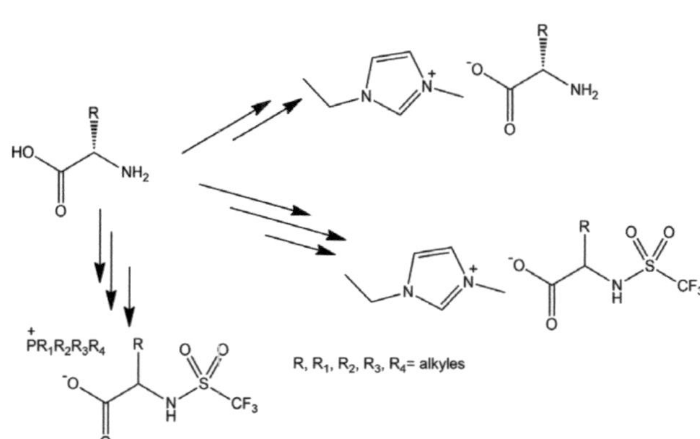

Scheme 2. Examples of aminoacids or esters-based as anions for ILs.

Aminoacids could as well be used to build the anionic part of the ILs. They are commonly associated with imidazolium, ammonium or phosphonium cations (**Scheme 2**) [30, 31].

In our group, an acido-basic method was used to form various ILs with anion from natural acids (L-lactic, L-tartaric, pyruvic, malic, malonic, succinic and osidic acids), but also L-proline and its derivatives. Even they were not readily biodegradable, these compounds showed in general lower toxicity towards various organisms than usual chlorinated and commercial ILs (**Scheme 3**) [32].

$$Bu_4N^+ HO^- + RCOOH \longrightarrow Bu_4N^+ RCOO^-$$

RCOOH= L-lactic, L-tartaric, malonic, succinic, malic, pyruvic, D-galacturonic and D-glucuronic acids

Scheme 3. Biomass-derived acid-based ILs.

Concerning the sugar family, these starting materials were essentially used to build cations. Fructose [33], glucose [34, 35], arabinose [36], isomannide [37] or isosorbide [38] have been transformed through multistep reactionnal pathways (**Scheme 4**). The resulting ILs were mainly used as chiral agents and presented in general low decomposition temperatures.

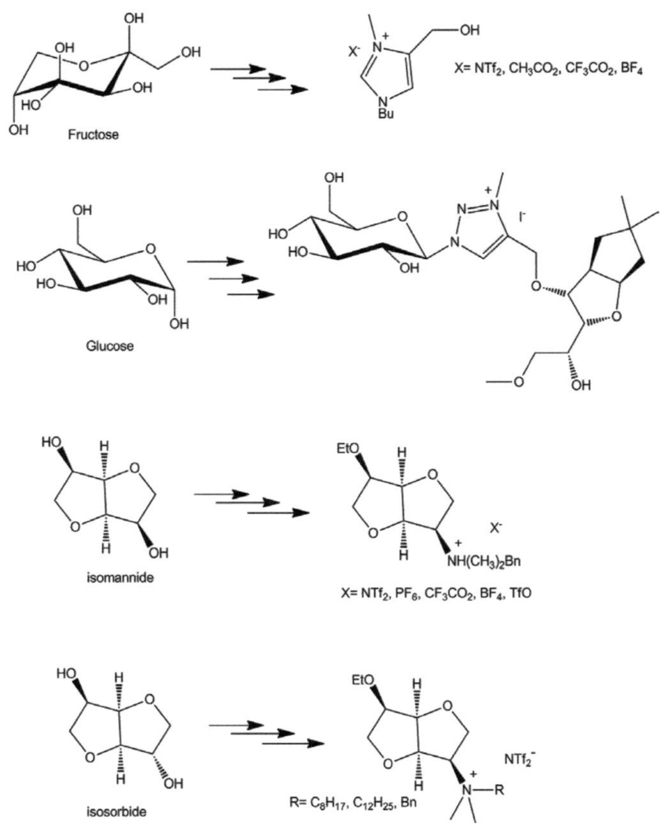

Scheme 4. Examples of sugar-based ILs.

Our group developed particularly xylose-derived ILs wearing a triazolium group. They were obtained by click chemistry between alcynated xylose and azido alkyls or benzyl, followed by methylation [39] (**Scheme 5**). Positive glass transition and low decomposition temperatures were observed, which seemed to be in relation with the presence of sugar moieties. Considering these temperatures, these ILs could only be used under mild conditions as solvents or chiral agents for chemical transformations or catalysis.

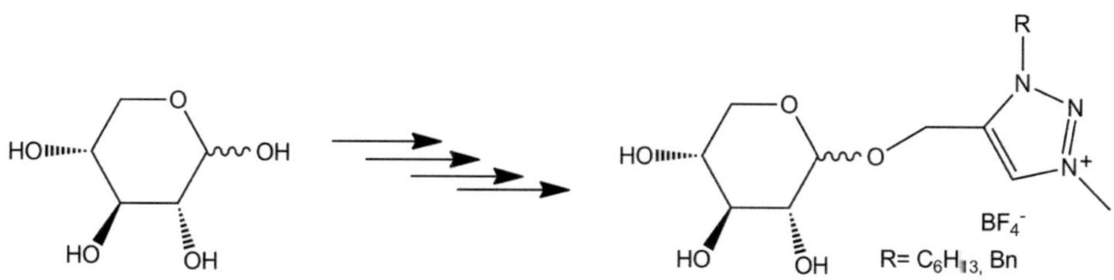

Scheme 5. Xylose-derived ILs.

Lipids represent also biosourced compounds, which could generate both cations and anions for ILs. Even if imidazolium-wearing oleic and stearic chain were easily prepared [40], the major utilisation of these lipids concerned the formation of anions, which were next associated with ammonium or phosphonium cations (**Figure 1**) [41].

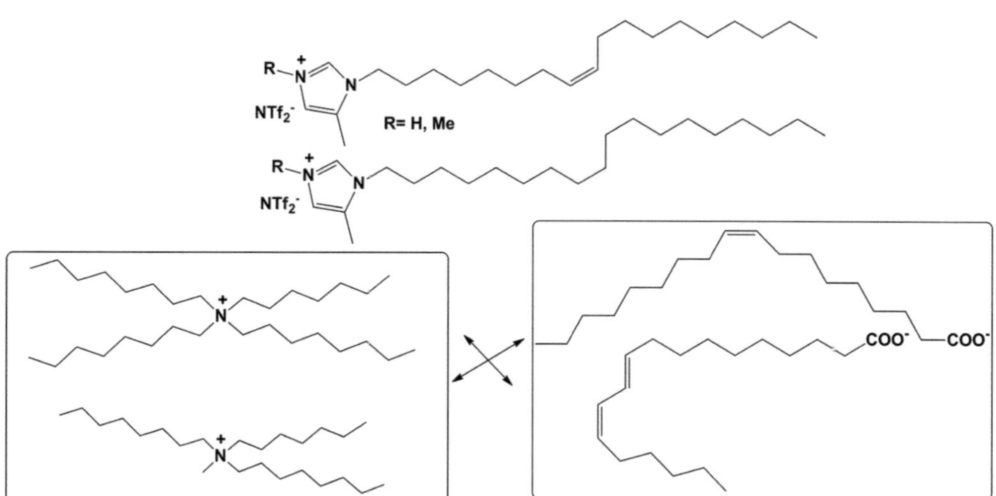

Figure 1. Lipids-based ILs.

Hulsbosch et al. described more exotic examples of bioresources for ILs as ephedrine or ampicillin in a recent and quite complete review dedicated to biobased ionic liquids for industry processing (**Scheme 6**) [14].

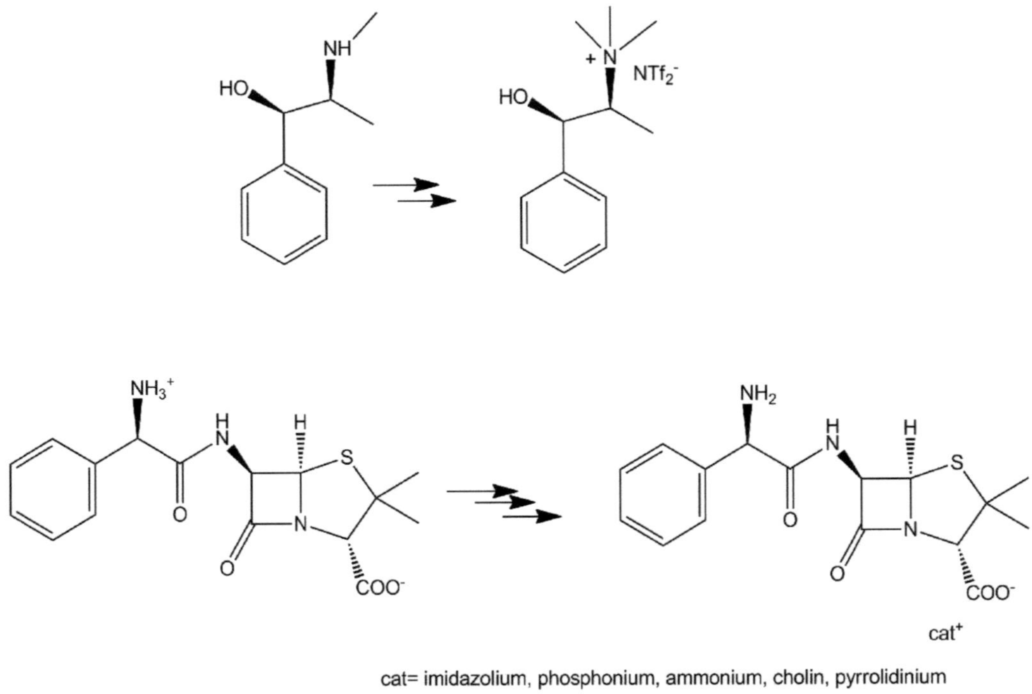

Scheme 6. Ephedrine- or ampicillin-based ILs.

3. Hydrogenation in ionic liquids

Research into catalytic hydrogenations in ILs began in 1995 with the almost simultaneous work of Chauvin [42] and Dupont [43]. Since, catalytic reactions involving metal complexes in ILs have been actively investigated. Around 300 ILs have been screened and have led to the production of useful products and intermediates [44–46]. The majority of these ILs contained heterocyclic cations, such as pyridinium, imidazolium and polyalkylammonium and recently, synthesized guanidinium, piperidinium, pyrrolium, pyrrolidinium, morpholinium, cholinium, piperazinium and thiazolium. Other ILs had bridged structures, binuclear or polynuclear, zwitterionic, hydrophobic (fluorinated) and chiral derivatives [47].

Gathergood et al. wrote, in 2011, a big chapter entirely dedicated to hydrogenation processes in ILs [48]. This chapter is quite complete. For each hydrogenation reaction reviewed, catalysts and ILs are noted, together associated with the nature of the substrate. A section describing kinetic and thermodynamic studies of hydrogenations in ILs is also presented, as well as the solubility of H_2 in many solvents and ILs under 1 atm. Some relevant examples have been identified from this chapter exclusively based on the use of H_2 and catalysts (metal complexes or nanoparticles) in ILs.

Concerning the use of transition metal-based catalyst, various Ru- or Rh-based catalysts were used for the hydrogenation of halonitrobenzenes [49], cinnamaldehyde [50], cyclohexanone [51] or hexane [52] in 1-butyl-3-methylimidazolium hexafluorophosphate (bmim PF_6) (**Scheme 7**).

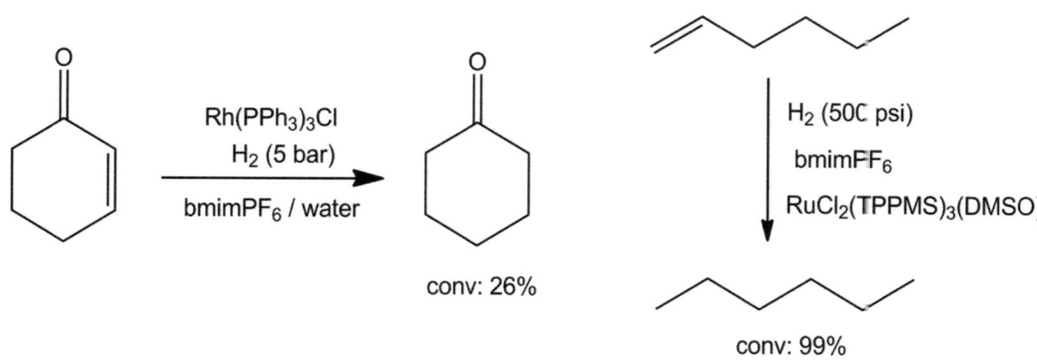

Scheme 7. Rh- or Ru-catalysed hydrogenations.

For these last compounds, Pt- or Pd-based complexes could also be used [53] as well as for the hydrogenation of benzene [54] or benzene derivatives [55] pyridinium ILs (**Scheme 8**).

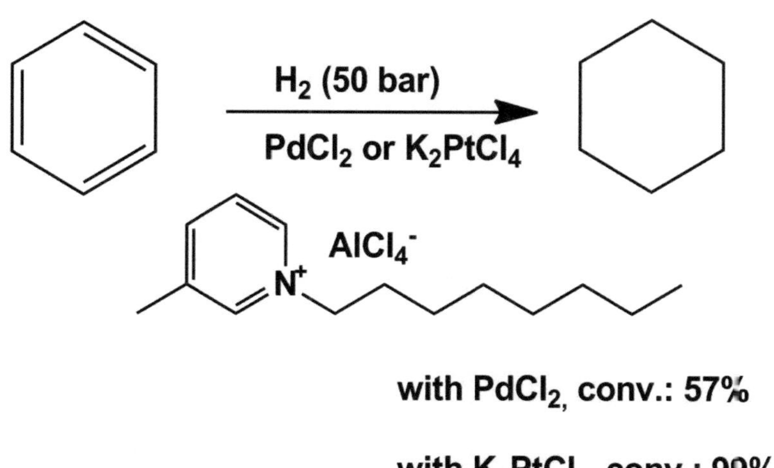

Scheme 8. Pd- or Pt-catalysed hydrogenations of benzene.

$PdCl_2$ was also used by Gathergood et al. [56] in imidazolium ILs, including a readily biodegradable IL [(3-methyl-1-pentoxycarbonylmethyl)imidazolium octylsulphate], for the selective hydrogenation of phenoxyocta-2,7-diene under mild conditions (**Scheme 9**).

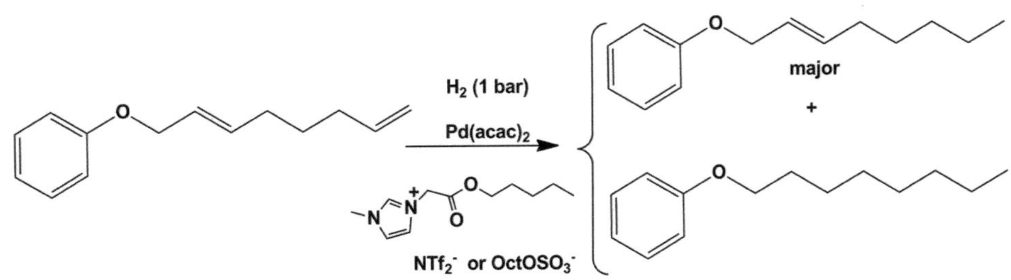

Scheme 9. Pd-catalysed hydrogenation in the presence of ester-functionalised ILs.

Concerning the asymmetric hydrogenation leading to enantiomerically pure products [57], the source of chiral induction was generally due to the presence of chiral ligands (BINAP or BINAP derivatives) coordinated to a metal catalyst [58–61], Rh- and Ru-based catalysts being generally the favourite candidates [48].

Metallic nanoparticles (NPs) could also be very useful for the hydrogenation processes. However, the knowledge about their formation and stabilisation for hydrogenation reactions in ILs is relatively new [62]. Pd [63, 64], Pt [65], Ir [66, 67], but also mixed Pd/Au NPs [68] were commonly used for (selective) hydrogenation of (poly)alkenes, while Ru-[69] and Ni-NPs [70] remained quite rare and were used for selective or complete hydrogenation of alkenes or arenes in imidazolium ILs (**Scheme 10**).

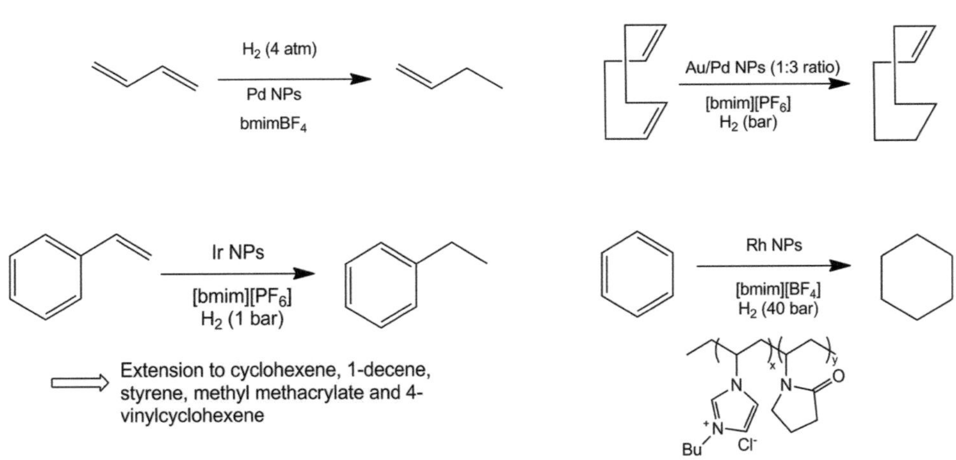

Scheme 10. NPs-assisted hydrogenations.

Scheeren et al. showed the formation of stable and isolable Pt(0)-NPs by reacting $Pt_2(dba)_3$ in 1-n-butyl-3-methylimidazolium hexafluorophosphate Bmim PF_6 with molecular hydrogen (4 atm) at 75°C [65]. These NPs were very efficient for the hydrogenation of diphenylacetylene in Si-functionalised ILs (**Scheme 11**).

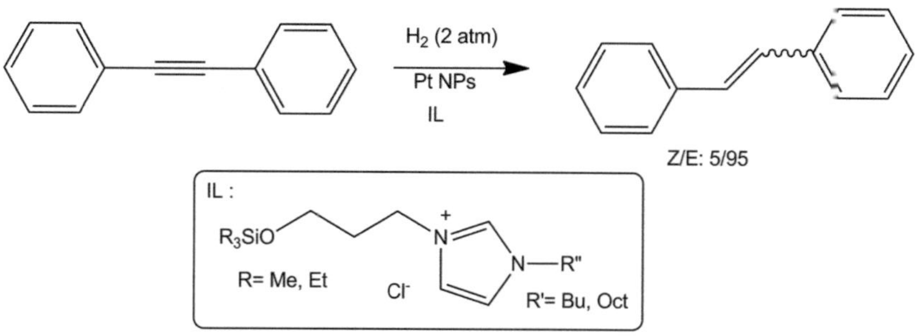

Scheme 11. Hydrogenation of diphenylacetylene in the presence of Pt NPs in Si-functionalised ILs.

4. Hydrogenation in biobased ionic liquids

In 2013, Ferlin et al. prepared easily and with good yields, biobased ionic liquids from natural organic-derived anions (L-lactate, L-tartrate, malonate, succinate, L-malate, pyruvate, D-glucuronate and D-galacturonate) by reaction between tetrabutylammonium hydroxide (TBA·OH) and an excess of the corresponding acid (**Scheme 3**) [32]. Toxicity assays towards a large panel of bacterial and fungal strains were performed. ILs containing D-glucuronate or D-galacturonate anions were the less toxic whereas TBA L-tartrate and TBA L-malate were the most toxic biomass derivatives. All ILs were less toxic to *E. coli* than TBA Br. Unfortunately, none of the salt containing biomass-derived anions passes the closed bottle test, nor tetrabutylammonium, tetraethylammonium or tetramethylammonium bromide.

These biobased ILs showed good performance and recyclability (until 10 runs without loss of activity) in catalytic selective hydrogenation of 1,5-cyclooctadiene into cyclooctene at room temperature under atmospheric H_2 pressure. In these mild conditions, they were more suitable for selective hydrogenation than commercial imidazoliums or ammonium ILs, which gave cyclooctane as major product (**Scheme 12**) [32].

Proline was also used to prepare easily chiral ionic liquids (CILs) tetrabutylammonium-(S)-prolinate, tetrabutylammonium-(R)-prolinate and tetrabutylammonium trans-4-hydroxy-(S)-prolinate from aminoacid, still by acido-basic reaction with tetrabutylammonium hydroxide (TBA·OH). While all three CILs have low antimicrobial toxicity to a wide range of bacteria and fungi, they did not pass the closed Bottle biodegradation test (**Scheme 13**) [71].

The hydrogenation of double carbon-carbon bonds of α,β-unsaturated ketones was processed under mild conditions with $PdCl_2$ as catalyst in the presence of a CIL and a co-solvent (**Scheme 14**) [71]. The best performance was achieved when isopropanol was used in a co-solvent/CIL ratio equal to 5. Total conversion of isophorone with an enantiomeric excess (ee) up to 47% was obtained, and recyclability of the system was observed for five cycles without loss of reactivity.

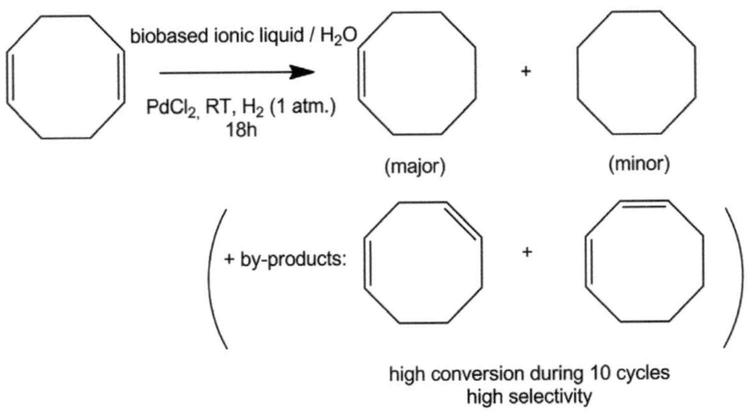

Scheme 12. Selective hydrogenation of 1,5-cyclooctadiene in the presence of ILs.

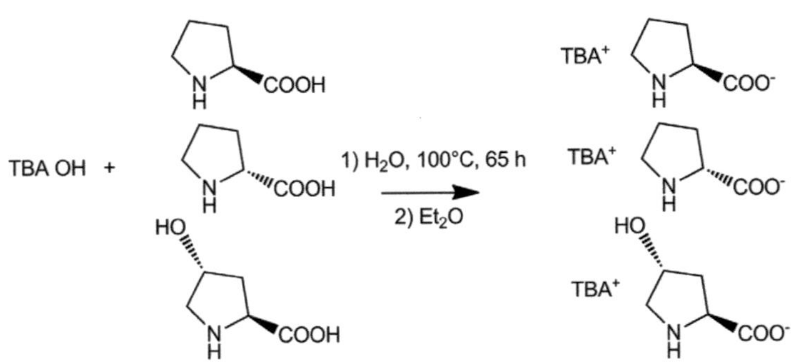

Scheme 13. Preparation of proline-based ILs.

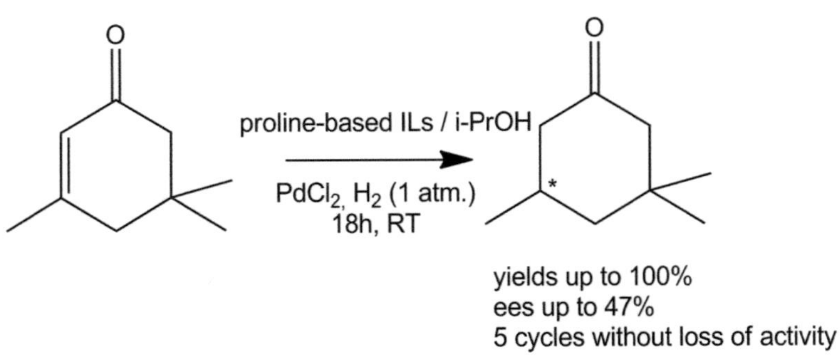

Scheme 14. Selective hydrogenation of α,β-unsaturated ketones under mild conditions in the presence of ILs.

Enantioselective hydrogenation of others α,β-unsaturated ketones was studied in the same mild conditions. Considering the particularly mild conditions, it was found that the method was very effective and competitive by comparison with previous works. Total conversion of pulegone was obtained with a good diastereoisomeric excess (de). Conversion was also complete for (R)-carvone with a very high selectivity of 90%. Hydrogenation occurred in the two carbon-carbon double bonds and an important de of 73% was observed. In the case of progesterone and 4-cholest-3-one, reaction occurred with very good yields and de, especially for 4-cholest-3-one.

Finally, concerning the field of transfer hydrogenation (so without the use of metallic species), γ-valerolactone-based ionic liquids (GVL-ILs) containing hydroxyvalerate anion were prepared according acid base reaction (**Scheme 15**) [72]. Theses ILs were successfully applied as alternative solvents for homogenous catalytic transfer hydrogenation of acetophenone and its substituted forms, but also of functionalised ketones and alkenes.

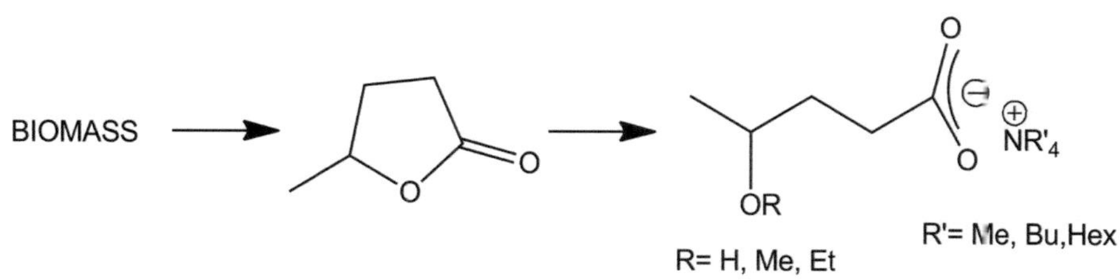

Scheme 15. Preparation of γ-valerolactone-based ILs.

A series of GVL-based ILs associated with tetraalkylammonium cations were tested. Structure of the cation had negligible influence on the catalytic activity. The potential recyclability of the catalytic system was demonstrated in four consecutive cycles, especially for the reduction of acetophenone. The highest conversions were achieved by using $[Rh(cod)_2]^+[BF_4]^-$ as catalyst precursor and formic acid as hydrogen donor. The optimal reaction conditions were 80°C and a molar ratio of HCOOH/substrate between 5:1 and 6:1 (**Scheme 16**) [73].

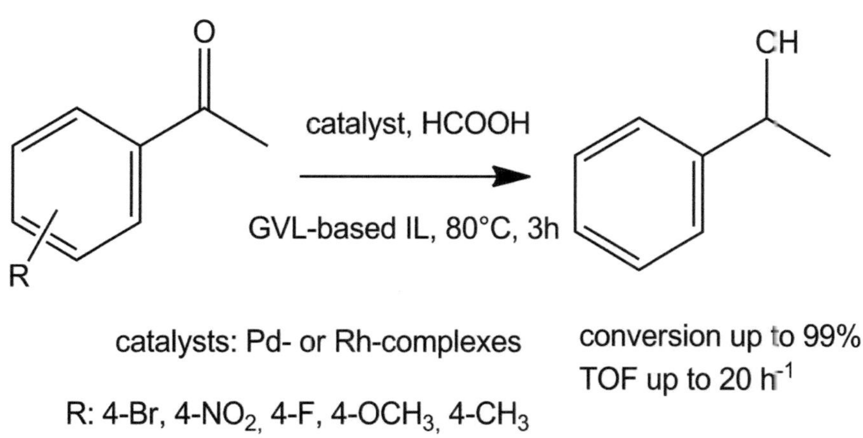

Scheme 16. Transfer hydrogenation of substituted acetophenones.

5. Conclusion

ILs are good solvents or co-solvent for metallic induced hydrogenation reactions. At first, their physical-chemical properties can be tunable depending on the anion-cation association. Consequently, the appropriate IL has to be simply chosen/found for the studied reaction. In addition, ILs are highly thermically stable, which is an advantage in the case of high-temperature hydrogenation reactions. The more important points are that they present good solubility of hydrogen and they are able to solubilise and stabilise metallic catalysts. As a consequence, they can promote hydrogenation reactions in mild conditions, sometimes the selectivity of the reaction, and/or they can allow the recyclability of the catalyst.

Hydrogenation in "usual" ILs has been performed on various unsaturated and aromatic substrates with various metallic catalysts or nanoparticles. In general, satisfying conversions and/or yields were observed. These reactions mainly used imidazoliums derivatives. They are the more common ILs, but also present (eco)toxicity and are not biodegradable.

To contour this problem, biobased and green ILs can be used. These compounds can be easily obtained from different biomass, such as amino acid, acids from bioprocess, carbohydrates or fatty oils. Despite their low biodegradability at the moment, they present in general lower (eco)toxicity than the commercials ILs. Similar to the "usual" ILs, biobased ILs are ideal solvent for hydrogenation reaction and can even bring improvements. When used as co-solvent, ammonium biobased ILs showed better selectivity and recyclability for 1,5-cyclooctadiene hydrogenation into cyclooctene in mild conditions than commercial ILs. Also biobased ILs can be chiral, brought by the chirality of their biobased building blocks, and be used for enantioselective hydrogenation reactions. With prolinate ILs, isophorone and carvone were hydrogenated in mild conditions with a good enantiomeric and diastereoisomeric excess, respectively. Transfer hydrogenation reactions were also recently investigated with biobased ILs. High conversion of acetophenone was achieved in the presence of γ-valerolactone-derived ILs, and the systems were reused four times without loss of reactivity.

Even if improvement needs to be made, especially for biodegradability of the compounds and simplification of their synthesis, biobased ILs seem to be good solvents, not only for hydrogenation reaction, but also for a wide range of chemical transformations (coupling reactions, oxidation, etc.). Moreover, with the increasing interest of valorisation of the biomass and the need to replace compounds derived from oil, new structures and applications of biobased ILs are expected for a near future.

Acknowledgements

This work was supported by the Fondation du Site Paris Reims (post doctoral fellowship for Nadège Ferlin) and the FEDER for material funds. We thank also the Tunisian Ministry of Education and Research for financial support for the cotutoring PhD of Safa Hayouni.

Author details

Safa Hayouni[1], Nadège Ferlin[1,2] and Sandrine Bouquillon[1*]

*Address all correspondence to: sandrine.bouquillon@univ-reims.fr

1 Institute of Molecular Chemistry of Reims, Sciences Faculty, University of Reims Champagne-Ardenne, Reims, France

2 School of Engineering and the Built Environment, Merchiston Campus, Edinburgh Napier University, Edinburgh, UK, Scotland

References

[1] Chaloner PA, Esteruleas MA, Joó F, Oro L. Homogeneous Hydrogenation, Catalysis by Metal Complexes, vol 15, Springer, Netherlands, 1994. 290 p. doi:10.1007/978-94-017-1791-5

[2] Oro L. Hydrogenation–Homogeneous, in Encyclopedia of Catalysis, Horváth IT editor. John Wiley & Sons, Hoboken, New Jersey, 2002. doi:10.1002/0471227617.eoc113

[3] Holbrey JD, Seddon KR. Ionic Liquids. Clean Prod Process 1999; 1: 223-236. DOI: 10.1007/s100980050036

[4] Wasserscheid P, Welton T. Ionic Liquids in Synthesis. 2nd edition, Wiley-VCH, Weinheim, 2008. 721 p. doi:10.1002/9783527621194.

[5] Stark A, Seddon KR. Ionic liquids. Kirk-Othmer Encyclopedia of Chemical Technology. Seidel A editor. John Wiley & Sons, Inc., New York. 2007;26:836–920. doi:10.1002/04712 38961.ionisedd.a01.

[6] Plechkova NV, Seddon KR. Applications of ionic liquids in the chemical industry. Chem Soc Rev. 2008;37(1):123–150. doi:10.1039/B006677J

[7] Welton T. Room-temperature ionic liquids. Solvents for synthesis and catalysis. Chem Rev. 1999;99(8):2071–2084. doi:10.1021/cr9800032t

[8] Endres F, Zein El Abedin S. Air and water stable ionic liquids in physical chemistry. Phys Chem Chem Phys. 2006;8(18):2101–2116. doi:10.1039/B600519P

[9] Deng Y, Beadham I, Ghavre M, Costa Gomes MF, Gathergood N, Husson P, Legeret B, Quilty B, Sancelme M, Besse-Hoggan P. When can ionic liquids be considered readily biodegradable? Biodegradation pathways of pyridinium, pyrrolidinium and ammonium-based ionic liquids. Green Chem. 2015;17(3):1479–1491. doi:10.1039/C4GC01904K

[10] Jordan A, Gathergood N. Biodegradation of ionic liquids—a critical review. Chem Soc Rev. 2015;44:8200–8237. doi:10.1039/C5CS00444F

[11] Docherty KM, Kulpa Jr CF. Toxicity and antimicrobial activity of imidazolium and pyridinium ionic liquids. Green Chem. 2005;7(4):185–189. doi:10.1039/B419172B

[12] Stolte S, Arning J, Bottin-Weber U, Matzke M, Stock F, Thiele K, Uerdingen M, Welz-Biermann U, Jastorff B, Ranke J. Anion effects on the cytotoxicity of ionic liquids. Green Chem. 2006;8:621–629. doi:10.1039/B602161A

[13] Pretti C, Chiappe C, Pieraccini D, Gregori M, Abramo F, Monni G, Intorre L. Acute toxicity of ionic liquids to the zebrafish (*Danio rerio*). Green Chem. 2006;8:238–240. doi: 10.1039/B511554J

[14] Hulsbosch J, De Vos DE, Binnemans K, Ameloot R. Bio-based ionic liquids: solvents for a green processing industry? ACS Sustain Chem Eng. 2016;4(6):2917–2931. doi:10.1021/acssuschemeng.6b00553

[15] Plaquevent JC, Levillain J, Guillen F, Malhiac C, Gaumont AC. Ionic liquids: new targets and media for alpha-amino acid and peptide chemistry. Chem Rev. 2008;108(12):5035–5060. doi:10.1021/cr068218c

[16] Tao G, He L, Sun N, Kou Y. New generation ionic liquids: cations derived from amino acids. Chem Commun. 2005;28:3562–3564. doi:10.1039/c0cc00028k

[17] He L, Tao GH, Parrish DA, Shreeve JM. Slightly viscous amino acid ionic liquids: synthesis, properties, and calculations. J Phys Chem B. 2009;113(46):15162–15169. doi: 10.1021/jp905079e

[18] Trivedi TJ, Rao KS, Singh T, Mandal SK, Sutradhar N, Panda AB, Kumar A. Task-specific, biodegradable amino acid ionic liquid surfactants. ChemSusChem. 2011;4(5): 604–608. doi:10.1002/adsc.201401010

[19] Wasserscheid P, Bösmann A, Bolm C. Synthesis and properties of ionic liquids derived from the chiral pool. Chem Commun. 2002;3:200–201. doi:10.1039/B109493A

[20] Clavier H, Boulanger L, Audic N, Toupet L, Mauduit M, Guillemin JC. Design and synthesis of imidazolinium salts derived from (L)-valine. Investigation of their potential in chiral molecular recognition. Chem Commun. 2004:1224–1225. doi: 10.1039/B402368D

[21] González L, Altava B, Bolte M, Burguete MI, García-Verdugo E, Luis SV. Synthesis of chiral room temperature ionic liquids from amino acids – application in chiral molecular recognition. Eur J Org Chem. 2012;26:4996–5009. doi:10.1002/ejoc.201200607

[22] Bao W, Wang Z, Li Y. Synthesis of chiral ionic liquids from natural amino acids. J Org Chem. 2003;68(2):591–593. doi:10.1021/jo020503

[23] Luo S, Mi X, Zhang L, Liu S, Xu H, Cheng JP. Functionalized chiral ionic liquids as highly efficient asymmetric organocatalysts for Michael addition to nitroolefins. Angew Chem Int Ed. 2006;45:3093–3097.doi:10.1002/anie.200600048

[24] Coleman D, Špulák M, Garcia MT, Gathergood N. Antimicrobial toxicity studies of ionic liquids leading to a 'hit' MRSA selective antibacterial imidazolium salt. Green Chem. 2012;14:1350–1356. doi:10.1039/C2GC16090K

[25] Haiß A, Jordan A, Westphal J, Evgenia Logunova E, Gathergood N, Kümmerer K. On the way to greener ionic liquids: identification of a fully mineralizable phenylalanine-based ionic liquid. Green Chem. 2016;18:4361–4373. doi:10.1039/c6gc00417b

[26] Jordan A, Haiß A, Spulak M, Karpichev Y, Kümmerer K, Gathergood N. Synthesis of a series of amino acid derived ionic liquids and tertiary amines: green chemistry metrics including microbial toxicity and preliminary biodegradation data analysis. Green Chem. 2016;18:4374–4392. doi:10.1039/c6gc00415f

[27] De Gaetano Y, Hubert J, Mohamadou A, Boudesocque S, Plantier-Royon R, Renault JH, Dupont L. Removal of pesticides from wastewater by ion pair centrifugal partition extraction using betaine-derived ionic liquids as extractants. Chem Eng J. 2016;285:596–604. doi:10.1016/j.cej.2015.10.012

[28] Pernak J, Niemczak M, Chrzanowski L, Ławniczak L, Fochtman P, Marcinkowska K, Praczyk T. Betaine and carnitine derivatives as herbicidal ionic liquids. Chem Eur J. 2016;22:12012–12021. doi:10.1002/chem.201683462

[29] Campos Toledo Hijo AA, Maximo GJ, Costa MC, Caldas Batista EA, de Almeida Meirelles AJ. Applications of ionic liquids in the food and bioproducts industries. ACS Sustain Chem Eng. 2016.doi:10.1021/acssuschemeng.6b00560

[30] Kagimoto J, Fukumoto K, Ohno H. Effect of tetrabutylphosphonium cation on the physico-chemical properties of amino-acid ionic liquids. Chem Commun. 2006;4;(21): 2254–2256. doi:10.1039/b600771f

[31] Fukumoto K, Ohno H. Design and synthesis of hydrophobic and chiral anions from amino acids as precursor for functional ionic liquids. Chem Commun. 2006:3081–3083. doi:10.1039/B606613E

[32] Ferlin N, Courty M, Gatard S, Spulak M, Quilty B, Beadham I, Ghavre M, Haiß A, Kümmerer K, Gathergood N, Bouquillon S. Biomass derived ionic liquids: synthesis from natural organic acids, characterization, toxicity, biodegradation and use as solvents for catalytic hydrogenation processes. Tetrahedron. 2013;30:6150–6161. doi:10.1016/j.tet.2013.05.054

[33] Handy ST, Okello M, Dickenson G. Solvents from biorenewable sources: ionic liquids based on fructose. Org Lett. 2003;5(14):2513–2515. doi:10.1021/ol034778b

[34] Jha AK, Jain N. Synthesis of glucose-tagged triazolium ionic liquids and their application as solvent and ligand for copper(I) catalyzed amination. Tetrahedron Lett. 2013;54 (35):4738–4741. doi:10.1016/j.tetlet.2013.06.114

[35] Poletti L, Chiappe C, Lay L, Pieraccini D, Polito L, Russo G. Glucose-derived ionic liquids: exploring low-cost sources for novel chiral solvents. Green Chem. 2007;9:337–341. doi:10.1039/B615650A

[36] Chiappe C, Marra A, Mele A. Synthesis and applications of ionic liquids derived from natural sugars. Top Curr Chem. 2010;295:177–195. doi:10.1007/128_2010_47

[37] Kumar V, Pei C, Olsen CE, Schäffer SJC, Parmar V, Malhotra SV. Novel carbohydrate-based chiral ammonium ionic liquids derived from isomannide. Tetrahedron: Asymmetry. 2008;19(6);2008:664–671. doi:10.1016/j.tetasy.2008.02.009

[38] Nguyen Van Buu O, Aupoix A, Doan Thi Hong N, Vo-Thanh G. Chiral ionic liquids derived from isosorbide: synthesis, properties and applications in asymmetric synthesis. New J Chem. 2009;33:2060–2072. doi:10.1039/B902956G

[39] Ferlin N, Gatard S, Nguyen Van Nhien A, Courty M, Bouquillon S. Click reactions as a key step for an efficient and selective synthesis of D-xylose-based ILs. Molecules. 2013;18(9):11512–11525. doi:10.3390/molecules180911512

[40] Kwan ML, Mirjafari A, McCabe JR, O'Brien RJ, Essi IV DF, Baum L, West KN, Davis Jr JH. Synthesis and thermophysical properties of ionic liquids: cyclopropyl moieties versus olefins as Tm-reducing elements in lipid-inspired ionic liquids. Tetrahedron Lett. 2013;54(1):12–14. doi:10.1016/j.tetlet.2012.09.101

[41] Parmentier D, Metz SJ, Kroon MC. Tetraalkylammonium oleate and linoleate based ionic liquids: promising extractants for metal salts. Green Chem. 2013;15:205–209. doi: 10.1039/C2GC36458A

[42] Chauvin Y, Mussmann L, Olivier H. A novel class of versatile solvents for two-phase catalysis: hydrogenation, isomerization, and hydroformylation of alkenes catalyzed by rhodium complexes in liquid 1,3-dialkylimidazolium salt. Angew Chem Int Ed. 1995;34:2698–2700. doi:10.1002/anie.199526981

[43] Suarez AZP, Dullius ELJ, Einloft S, De Souza FR, Dupont J. The use of new ionic liquids in two-phase catalytic hydrogenation reaction by rhodium complexes. Polyhedron. 1996;15:1217–1219. doi:10.1016/0277-5387(95)00365-7

[44] Sheldon RA, Arends I, Hanefeld U, editors. Green chemistry and catalysis. John Wiley & Sons, Wiley-VCH Verlag GmbH & Co, Weinheim, 2007. 448 p. DOI: 10.1002/978352 7611003

[45] Augustine RL: Heterogeneous catalysis for the synthetic chemist. CRC Press, Marcel Dekker, Inc., New York. 1995.672 p. ISBN 0-8247-9021-9.

[46] Beletskaya IP, Kustov LM. Catalysis as an important tool of green chemistry. Russ Chem Rev. 2010;79(6):441–461. doi:10.1070/RC2010v079n06ABEH004137

[47] Wasserscheid P, Keim W. Ionic liquids—new "solutions" for transition metal catalysis. Angew Chem Int Ed. 2000;39:3772–3789. doi:10.1002/1521-3773(20001103)39:21<3772::AID-ANIE3772>3.0.CO;2-5.

[48] Ghavre M, Morrissey S, Gathergood N. Hydrogenation in Ionic Liquids, Ionic Liquids: Applications and Perspectives. Kokorin A editor. CC BY-NC-SA 3.0 license. InTech; 2011. 331-392. DOI: 10.5772/14315

[49] Xu D, Hu Z, Li W, Luo S, Xu Z. Hydrogenation in ionic liquids: an alternative methodology toward highly selective catalysis of halonitrobenzenes to corresponding haloanilines. J Mol Catal A. 2005;235:137–142. doi:10.1016/j.molcata.2005.04.004

[50] Arras J, Steffan M, Shayeghi Y, Claus P. The promoting effect of a dicyanamide based ionic liquid in the selective hydrogenation of citral. Chem Commun. 2008;4058–4060. doi:10.1039/B810291K

[51] Wolfson A, Vankelecom I, Jacobs P. Beneficial effect of water as second solvent in ionic liquid biphasic catalytic hydrogenations. Tetrahedron Lett. 2005;46:2513–2516. doi: 10.1016/j.tetlet.2005.01.179

[52] Suarez T, Fontal B, Reyes M, Bellandi F, Contreras R, Ortega J, Leon G, Cancines P, Castillo B. Catalytic Hydrogenation of 1-Hexene with $RuCI_2$ $(TPPMS)_3$(DMSO). Part II: ionic liquid biphasic system. React Kinet Catal Lett. 2004;82:325–331. doi:10.1023/B%3AREAC.0000034844.21731.7f

[53] Arras J, Steffan M, Shayeghi Y, Ruppert D, Claus P. Regioselective catalytic hydrogenation of citral with ionic liquids as reaction modifiers. Green Chem. 2009;11:716–723. doi:10.1039/B822992A

[54] Geldbach T, Dyson P. Searching for molecular arene hydrogenation catalysis in ionic liquids. J Organomet Chem. 2005;690:3552–3557. doi:10.1016/j.jorganchem.2005.03.006

[55] Hardacre C, Mullan E, Rooney D, Thompson J, Yablonsky G. Comparison of mass transfer effects in the heterogeneously catalysed hydrogenation of phenyl acetylene in heptane and an ionic liquid. Chem Eng Sci. 2006;61:6995–7006. doi:10.1016/j.ces.2006.07.020

[56] Bouquillon S, Courant T, Dean D, Gathergood N, Morrissey S, Pegot B, Scammells PJ, Singer R. Biodegradable ionic liquids: selected synthetic applications. Aust J Chem. 2007;60:843–847. doi:10.1071/CH07257.

[57] Blaser HU, Malan C, Pugin B, Spindler F, Steiner H, Studer M. Selective hydrogenation for fine chemicals: recent trends and new developments. Adv Synth Catal. 2003;345: 103–151. doi:10.1002/adsc.200390000

[58] Noyori R, Kitamura M, Ohkuma T. Toward efficient asymmetric hydrogenation: architectural and functional engineering of chiral molecular catalysts. PNAS. 2004;101: 5356–5362. doi:10.1073/pnas.0307928100

[59] Bautista F, Caballero V, Campelo J, Luna D, Marinas J, Romero A, Romero I, Serrano I, Llobet A. Heterogeneization of a new Ru(II) homogeneous asymmetric hydrogenation catalyst containing BINAP and the N-tridentate bpea ligand, through covalent attachment on amorphous $AlPO_4$ support. Top Catal. 2006;40:193–205. doi:10.1038/nchem.216

[60] Chan SA, Laneman SA, Day C. Preparation and structural characterization of bis(acetylacetonato)ruthenium(II)-BINAP: an efficient route to an effective asymmetric hydrogenation catalyst precursor. Inorg Chim Acta. 1995;228:159–163. doi:10.1016/0020-1693(94)04167-T

[61] Wan DK, Davis M. Ruthenium (II)-sulfonated BINAP. A novel water-soluble asymmetric hydrogenation catalyst. Tetrahedron: Asymmetry. 1993;4:2461–2468. doi:10.1016/S0957-4166(00)82224-7

[62] Scheeren CW, Domingos JB, Machado G, Dupont J. Hydrogen reduction of Adams' catalyst in ionic liquids: formation and stabilization of Pt(0) nanoparticles. J Phys Chem C. 2008;112:16463–16469. doi:10.1021/jp804870j

[63] Umpierre A, Machado G, Fecher G, Morais J, Dupont J. Selective hydrogenation of 1,3-butadiene to 1-butene by Pd(0) nanoparticles embedded in imidazolium ionic liquids. Adv Synth Catal. 2005;347:1404–1412. doi:10.1002/adsc.200404313

[64] Huang J, Jiang T, Han B, Gao H, Chang Y, Zhao G, Wu W. Hydrogenation of olefins using ligand-stabilized palladium nanoparticles in an ionic liquid. Chem Commun. 2003:1654–1655. doi:10.1039/B302750C

[65] Scheeren C, Machado G, Dupont J, Fichtner P, Texeira S. Nanoscale Pt(0) particles prepared in imidazolium room temperature ionic liquids: synthesis from an organometallic precursor, characterization, and catalytic properties in hydrogenation reactions. Inorg Chem. 2003;42:4738–4742. doi:10.1021/ic034453r

[66] Fonseca G, Domingos J, Nome F, Dupont J. On the kinetics of iridium nanoparticles formation in ionic liquids and olefin hydrogenation. J Mol Catal A. 2006;248:10–16. doi:10.1016/j.molcata.2005.12.002

[67] Dupont J, Fonseca G, Umpierre A, Fichtner P, Teixeira S. Transition-metal nanoparticles in imidazolium ionic liquids: recycable catalysts for biphasic hydrogenation reactions. J Am Chem Soc. 2002;124:4228–4229. doi:10.1021/ja025818u

[68] Dash P, Dehm N, Scott R. Bimetallic PdAu nanoparticles as hydrogenation catalysts in imidazolium ionic liquids. J Mol Catal A: Chem. 2008;286:114–119. doi:10.1016/j.molcata.2008.02.003

[69] Prechtl M, Scariot M, Scholten J, Machado G, Teixeira S, Dupont J. Nanoscale Ru(0) particles: arene hydrogenation catalysts in imidazolium ionic liquids. Inorg Chem. 2008;47:8995–9001. doi:10.1021/ic801014f

[70] Migowski P, Machado G, Texeira S, Alves M, Morais J, Traverse A, Dupont J. Synthesis and characterization of nickel nanoparticles dispersed in imidazolium ionic liquids. Phys Chem. 2007;9:4814–4821. doi:10.1039/B703979D

[71] Ferlin N, Courty M, Nguyen Van Nhien A, Gatard S, Pour M, Quilty B, Ghavre M, Haiß A, Kümmerer K, Gathergood N, Bouquillon S. Tetrabutylammonium prolinate-based ionic liquids: a combined asymmetric catalysis, antimicrobial toxicity and biodegradation assessment. RSC Adv. 2013;3:26241–26251. doi:10.1039/C3RA43785J

[72] Strádi A, Molnár M, Óvári M, Dibó G, Richter FU, Mika L. Rhodium-catalyzed hydrogenation of olefins in γ-valerolactone-based ionic liquids. Green Chem. 2013;15 (7):1857–1862. doi:10.1039/C3GC40360B

[73] Strádi A, Molnár M, Szakál P, Dibó G, Gáspár D, Mika LT. Catalytic transfer hydrogenation in γ-valerolactone-based ionic liquids. RSC Adv. 2015;5(89):72529–72535. doi: 10.1039/C5RA12657F

Chapter 16

Dynamic Mathematical Modelling and Advanced Control Strategies for Complex Hydrogenation Process

Roxana Rusu-Both

Additional information is available at the end of the chapter

http://dx.doi.org/10.5772/65336

Abstract

Over the last past decades, the number of control system applications in the chemical and petrochemical domains has increased considerably. However, due to the diversity and particularity of chemical processes, there are still many challenges that have to be addressed like: system identification, performance enhancement, monitoring, diagnosis and more importantly closed-loop stability, robustness. Taking into account that most chemical processes are complex, nonlinear MIMO (multi-input multi-output) systems, the challenge is even greater. This book chapter is directed towards the development and the implementation of modern control algorithms for complex and high-risk petrochemical processes, the considered case study being the production of 2 ethyl-hexanol through the 2 ethyl-hexenal hydrogenation process. 2 ethyl-hexanol is mainly used in the production of plasticizers for polyvinyl chloride (PVC) manufacture. In the second part, is described the mathematical modelling of the 2 ethyl-hexenal hydrogenation process including also the simulation and validation of the developed mathematical models. The third part will focus on the design and implementation of conventional control strategy. Section four is dedicated to the design and implementation of several advanced control strategies like Internal Model Control and robust control. The conclusions section represents the last part of the chapter.

Keywords: hydrogenation process, mathematical modelling, IMC control, robust control, MIMO control strategy

1. Introduction

At present time, 2 ethyl-hexanol represents an important raw material in the production of plasticizers, solvents, oils and additives for diesel fuel, making its industrial scale production process of high importance. The industrial scale production of 2 ethyl-hexanol is made either through liquid phase 2 ethyl-hexanal hydrogenation or through gaseous phase 2 ethyl-hexanal hydrogenation. The liquid phase hydrogenation is preferred on industrial scale due to its advantages [1].

The development of complex, accurate mathematical models is an essential step in the dynamic behaviour analysis without expensive experiments and last but not least in the development and testing of control strategies. To this end, for the 2 ethyl-hexanal hydrogenation process was developed at first a distributed parameter mathematical model validated using experimental data. It consists of a system of partial differential equations based on mass (total and component) and energy conservation laws. In order to analyse the dynamic behaviour and to emphasize the interactions between the parts of a hydrogenation process a dynamic study was performed. Several scenarios have been carried out in order to evaluate the dynamic behaviour. The one presented in this chapter is the study of the catalyst deactivation effect. The effect of a variation of the input flow temperature of the streams as well as the effect of a change in the reactor load: the volumetric ratio between the 2 ethyl-hexanal flow rate and the recirculated 2 ethyl-hexanol flow rate are worth to study. The dynamic behaviour study shows the complexity of the hydrogenation process due to heat, mass and kinetic interactions, which are dependent on the operating conditions, reactor loading as well as on the trajectory from one state to another.

Unfortunately, despite their accuracy, detailed, nonlinear mathematical models are too complex for efficient use in controller design so the considered approach is the use of a simple model of the process, which describes its most important properties in combination with an advanced control algorithm which takes into account the model uncertainties, the disturbances and command signal limitations.

To this end, another mathematical model–operational model–is determined based on the main connections between input and output parameters of the process and was obtained based on the result analysis from both simulations step responses related to a distributed parameter model and the experimental data.

For the hydrogenation process presented in this case study, various methods of control where designed: conventional PID control, internal model control (IMC) and robust control in order to find the optimal solution.

The two main control objectives of all the applied control strategies are: (a) to maintain the inlet reactor temperature below an imposed critical value; (b) to ensure a high 2 ethyl-hexanol (product) concentration. From a technological point of view the reactor inlet temperature can be controlled by modifying the 2 ethyl-hexanol recirculated flow rate. The product concentration is influenced by the reactant flow rates and also by the catalyst degree of activity which acts as a variable disturbance. The catalyst degree of activity will continuously decrease as the hydrogenation reaction takes place up to the point it needs to be replaced. However, during this period, this effect can be compensated by continuously increasing the input temperature of the reactants.

For analysis purposes, all proposed control structures were implemented in MATLAB/SIMULINK environment. The simulation scenarios are presented in comparison for all designed control strategies. The two main objectives of the scenarios are the set-point tracking analysis and disturbance rejection analysis. A robustness analysis is also performed.

Finally, by analysing all the advantages and disadvantages of the designed control structures the final solution recommended for the control of the 2 ethyl-hexanal hydrogenation process is the robust control. The concluding remarks are formulated in the last section of the chapter.

2. Mathematical models of the 2 ethyl-hexanal hydrogenation process: simulation and validation

The mathematical modelling of an industrial scale process is a complex problem requiring the following steps: (a) Choosing the model structure based on physical knowledge; (b) Determining and estimating the model parameters from the available data; and (c) Accuracy evaluation of the developed model (model validation) based on experimental data.

The model validation step is closely related to the parameter estimation step. As a result of the estimation procedure it must be examined to what extent the model really explains the real plant behaviour. This aspect can be highlighted by exciting the system model with the same input signal and studying the nature and value of the difference between the model output signals and the real plant variables.

2.1. Nonlinear mathematical model

Based on laws of mass conservation and energy conservation, the mathematical model determined in this section for the hydrogenation process 2 ethyl-hexanal is nonlinear. A dynamic mathematical model can be used to simulate complex mass transfer phenomena and to understand processes occurring inside the reactor. So far there are various models proposed in the literature [2, 3] for the hydrogenation reaction kinetics, but there is no mathematical model to describe the processes taking place inside the hydrogenation reactor. There is a need to develop equations, to determine the parameters and boundary conditions that form the mathematical model consisting of differential equations with partial derivatives. The spatially distributed nature of the process is generally unnoticed or ignored and the control techniques are applied using conventional approximate models with concentrated parameters, identified by experiments input/output. Because these simple models ignore the spatial nature of the process, they often suffer from the close interaction and apparent delays due to diffusion and convection phenomena inherent in such processes. Hence, the need for a modelling procedure that generates a general model, a model that takes into account the spatial structure of the process variable and is deducted from the input and output measured data.

The production of 2 ethyl-hexanol through the liquid phase 2 ethyl-hexenal hydrogenation is depicted in **Figure 1**.

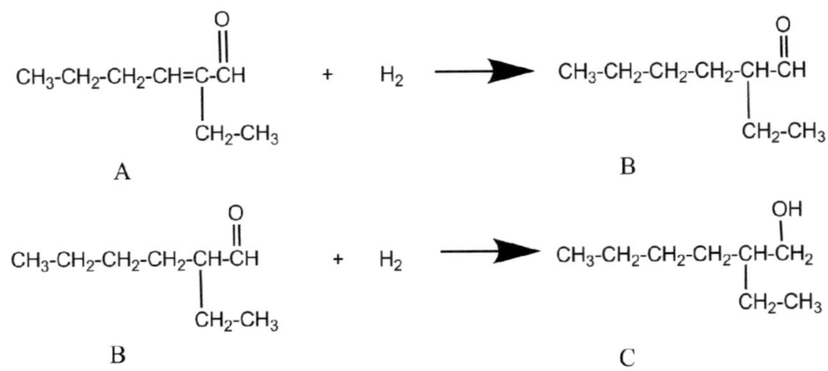

Figure 1. Hydrogenation reaction: A—2 ethyl-hexanal, B—2 ethyl-hexanal, C—2 ethyl-hexanol.

As it can be observed, the production of 2-ethyl hexanol is in fact two successive hydrogenation reactions. Thus, the reaction product of the first hydrogenation reaction—2-ethyl-hexanal—is a reactant in the next one. In the hydrogenation process n-butanol and iso-butanol can be obtained as side products of some side reactions. Also, symmetric or asymmetric C8 ethers can be obtained through the etherification reaction of butanols. These side reactions are favoured by operating parameters like: input flows through the reactor, inlet temperature. However, for mathematical modelling purposes only the main reaction will be considered and the following assumptions are made: the model parameters are considered to be constant on the radial section of the reactor; both liquid and gas velocity are considered constant; adiabatic reactor; perfect mixing is considered and the chemical reactions occur only at the catalyst surface. Also, in the reaction zone the following phenomena occur: mass transfer through the volume element dz (theoretical plate); 2 ethyl-hexanal hydrogenation on the catalyst surface and heat transfer through volume element dz.

The component mass transfer is essential in a heterogeneous reactor with several phases (gas, liquid) because the reactants have to pass from one phase to another making the modelling of gas-liquid-solid mass transfer process a crucial step. Substances from a gas phase (hydrogen) and a liquid (2 ethyl-hexanal) are transformed on the surface of a solid catalyst (nickel on silicon). One of the most important factors in the chemical reaction is the reaction rate (reaction kinetics). In the present case there are two rates of reaction: r1, which relates to the hydrogenation of 2 ethyl-hexanal and r2, which relates to the hydrogenation of 2 ethyl-hexanal (intermediate product). For this particular case the chosen catalyst is nickel (Ni) based catalyst on a silica (Si) support. Considering the literature studies [2] for the Ni catalyst supported on Si, the best model to express the reaction rate of the proposed model is:

$$r_1 = \frac{k_1 \cdot K_{enal} \cdot K_H \cdot C_{enal} \cdot C_{H_2}}{\left(1 + K_{enal} \cdot C_{enal} + K_{anal} \cdot C_{anal} + \sqrt{K_H \cdot C_{H_2}}\right)^3} \tag{1}$$

$$r_2 = \frac{k_1 \cdot K_{anal} \cdot K_H \cdot C_{anal} \cdot C_{H_2}}{\left(1 + K_{enal} \cdot C_{enal} + K_{anal} \cdot C_{anal} + \sqrt{K_H \cdot C_{H_2}}\right)^3} \tag{2}$$

where enal is 2 ethyl-hexanal, anal is 2 ethyl-hexanal, oct is 2 ethyl-hexanol, H is hydrogen, K_i (m^3/mol) is the adsorption equilibrium constant for the component *i* (*i:* enal, anal, H) , k_i (m^3/s kg) is rate constant of the surface reaction *i*,(*i*:1,2), C_i (kmol/m^3) concentration of component *i*.

In industrial scale heterogeneous reactors, the catalyst ages gradually and gets deactivated to the point it becomes inefficient and needs to be replaced. The dependence between the catalyst activity degree and the reaction rate can be expressed as follows:

$$r_i = r_i \cdot e^{-k_D t} \tag{3}$$

where k_D is the catalyst degree of deactivation.

The reaction rates expression should also include the temperature dependence of the reaction rate constants and absorption constants, which must comply with the Arrhenius law [4] given by:

$$k_i = k_0^i \cdot e^{-\frac{E}{RT}} \tag{4}$$

where *E* is the activation energy and *T* is the temperature.

The developed total and component mass balance equations are as follows:

$$\begin{cases} \frac{\partial Q_L}{\partial t} = -v_L \frac{\partial Q_L}{\partial z} + v_G \cdot S \cdot M_{H_2} \cdot a_v \cdot N_{H_G} \\ \frac{\partial Q_G}{\partial t} = -v_G \frac{\partial Q_G}{\partial z} - v_G \cdot S \cdot M_{H_2} \cdot a_v \cdot N_{H_G} \end{cases} \tag{5}$$

$$\begin{cases} \frac{\partial C_H^L}{\partial t} = -v_L \frac{\partial C_H^L}{\partial z} + a_v \cdot N_{H_G} - vph \\ \frac{\partial C_H^G}{\partial t} = -v_L \frac{\partial C_H^G}{\partial z} - a_v \cdot N_{H_G} \\ \frac{\partial C_{enal}}{\partial t} = -v_L \frac{\partial C_{enal}}{\partial z} - r_1 \cdot \rho_{sol} \\ \frac{\partial C_{anal}}{\partial t} = -v_L \frac{\partial C_{anal}}{\partial z} + (r_1 - r_2) \cdot \rho_{sol} \\ \frac{\partial C_{oct}}{\partial t} = -v_L \frac{\partial C_{oct}}{\partial z} + r_2 \cdot \rho_{sol} \end{cases} \tag{6}$$

and the heat balance equations for the liquid and gas phases are

$$\begin{cases} \frac{\partial T_L}{\partial t} = -v_L \frac{\partial T_L}{\partial z} - \sum \frac{\Delta_R H_i \cdot r_i}{c_{pL}} \\ \frac{\partial T_G}{\partial t} = -v_G \frac{\partial T_G}{\partial z} - \sum \frac{\Delta_R H_i \cdot r_i}{c_{pG}} \end{cases} \tag{7}$$

All the parameters are detailed in the nomenclature section.

An important feature of the three-phase reactors is the hydrodynamic characteristic. One of the hydrodynamics parameters with high influence on the performance of the three-phase reactor is the pressure loss of the two-phase mixture. The pressure loss in the functional layer is an important parameter, which depends on the amount of energy required to operate and which also correlates the interphase mass transfer coefficients. The friction pressure loss can be calculated using equation ERGUN [5]:

$$\delta = \frac{1-\varepsilon}{\varepsilon^3}\left(1.75 + 150\frac{1-\varepsilon}{Re_p}\right)\frac{\rho \cdot v^2}{d_p} \tag{8}$$

where ε is layer void fraction, and v, υ, ρ are considered for the phase for which the pressure loss is computed.

The dynamic process simulator was implemented in MATLAB programming environment along with graphical extension to SIMULINK. The process simulator, being represented by a mathematical model comprising a system of nonlinear partial differential equations, was implemented as a function-s: S-function.

To solve the partial differential equations the finite difference method [6] was used. According to this method, the derivatives where written as finite differences. The solution domain must be covered with a network of nodes in order to apply the proposed method. Theoretically, the approximation of the exact solution will be better if the number of nodes included in the network is greater. To approximate the concentrations, temperature and pressure over time and space (height of the reactor) 100 discretization points (the number of theoretical plates) were chosen, being considered the best choice between the complexity of the model and the accuracy of the results. This method is reasonably simple, robust and is a good general candidate for the numerical solution of differential equations.

2.2. Validation and dynamic behaviour analysis of the nonlinear mathematical model

The developed nonlinear, distributed parameter mathematical model was calibrated and tested based on experimental data acquired from a functional 2 ethyl-hexanal hydrogenation reactor at Oltchim S.A. company, Romania.

The calibration process was performed taking into account the constructive characteristics of the hydrogenation reactor. The height of the reactor is approximately 20 m, with a diameter of 1 m, while the height of the catalyst is around 18 m. The reactants (hydrogen – gaseous phase – and 2 ethyl-hexanal—liquid phase) are fed at the top of the reactor. Perfect mixing can be considered because the reactor is equipped with a fine sifter at the top. The product (2 ethyl-hexanol) is extracted at the bottom of the reactor. Part of the product is cooled to 90–100 °C in a heat exchanger and recirculated at the top to maintain the inlet temperature below 160 °C. The operating temperature of the reactants is around 100 °C. The plant is also equipped with a heater, to be able to increase the input temperature of the reactants as the catalyst gets deactivated. The flow ratio between the 2 ethyl-hexanal flow and the recirculated 2 ethyl-hexanol flow must be maintained at a specific value in order to maintain the output temperature below the critical value. The main reactor operating point is characterized by the following parameter values: hydrogen flow of 1250 – 1300 (m^3/h), 2 ethyl-hexanal flow of 4 (m^3/h), recirculated 2 ethyl-hexanol flow of 24 –32 (m^3/h).

The accuracy of the developed mathematical model was tested in different operating points and the comparison between the simulated results and the experimental data in the main operating point is presented in **Table 1**.

Parameter	Simulated value	Plant data
Hydrogen concentration (Kmol/m^3)	0.078	0.0549
2 ethyl-hexanal concentration (Kmol/m^3)	0.0021	0.0020
2 ethyl-hexanol concentration (Kmol/m^3)	5.346	5.32
Output temperature (K)	441.6	441.3

Table 1. Mathematical model validation.

The standard deviation is between 5 and 10 % in all cases. This indicates the presence of an acceptable systematic error. The evolution of the main parameters can also be observed in **Figures 2–4**.

By choosing 100 discretization points the height of the theoretical plates of the reactor is 0.2m. Taking into account the diameter of the reactor (~1m) it can be considered that the concentration of 2 ethyl-hexanol and the temperature of the product are constant along the theoretical plate. However, to prove this assumption, a study was performed on the influence of the number of discretization points on the accuracy of the results. The simulation results are presented in the **Figures 5** and **6**, considering 80, 100 and 125 discretization points. As can be observed from the figures, there are no significant changes in the accuracy of the model, the differences being only at the second decimal.

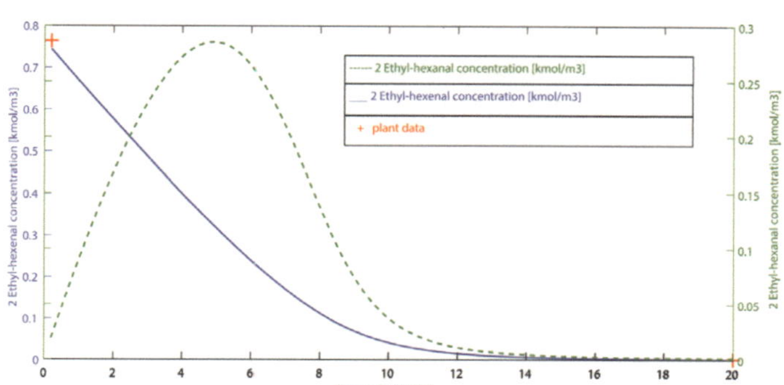

Figure 2. 2 ethyl-hexanal and 2 ethyl-hexanal concentration evolution: simulated vs. plant data.

Another important study that needs to be performed to prove that the developed mathematical model captured the hydrogenation mechanism is a dynamic behaviour study. To this end, it is necessary to evaluate that the model captures the effect of catalyst deactivation [7]. By analysing the experimental data it was concluded that the catalyst degree of activity decreases up to 50% after 4 months of continuous functioning. **Figure 7** presents the effect of catalyst deactivation on the product concentration by maintaining the reactants input flow and temperature constant.

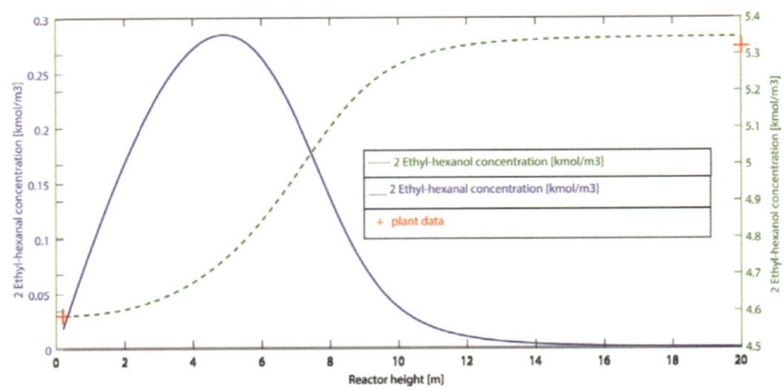

Figure 3. 2 ethyl-hexanal and 2 ethyl-hexanol concentration evolutions: simulated vs. plant data.

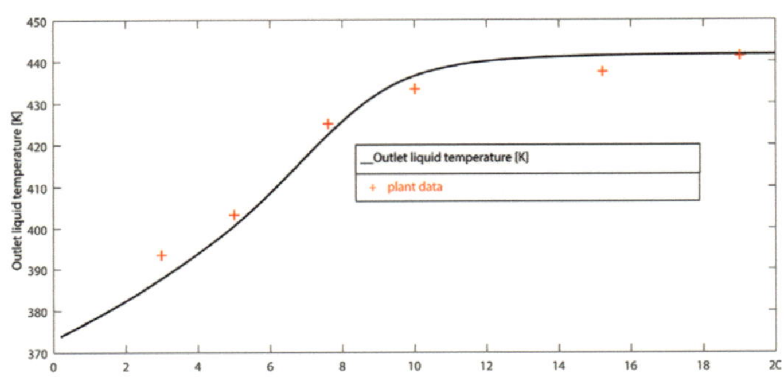

Figure 4. Product temperature evolution: simulated vs. plant data.

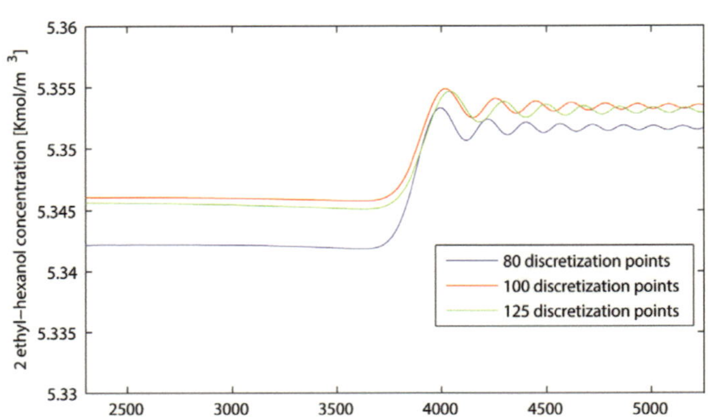

Figure 5. 2 ethyl-hexanol concentration evolution: 80, 100 and 125 discretization points.

http://dx.doi.org/10.5772/65336

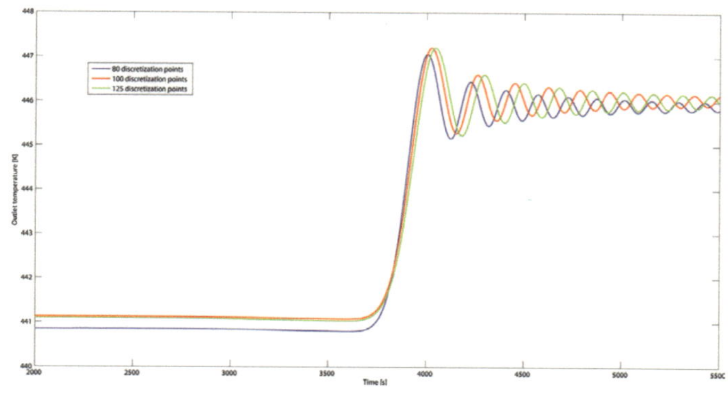

Figure 6. Outlet temperature evolution: 80, 100 and 125 discretization points.

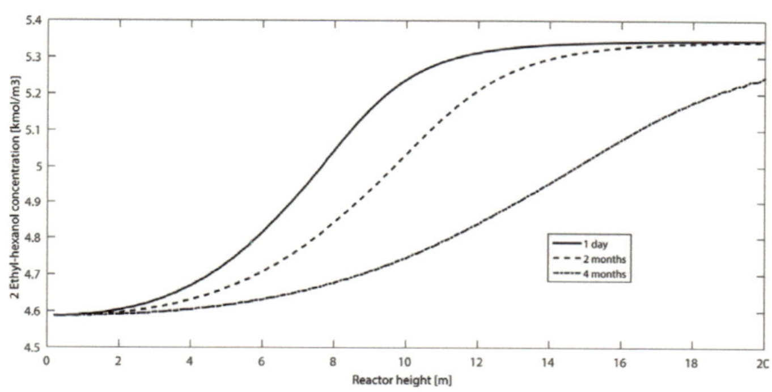

Figure 7. Catalyst deactivation effect on the product concentration.

It is obvious that the catalyst deactivation influences considerably the quality of the product increasing the production costs. This effect can be diminished if the input temperature of the reactants is gradually increased as the catalyst gets older.

2.3. Operational mathematical model: development and validation

Based on the previously presented studies, it can be concluded that the hydrogenation process dynamics ethyl 2-hexanal is very complex; thus, it is only normal that the resulting model is nonlinear, higher order with distributed parameters. The only problem is that highly complex models are difficult to use in the development of most control strategies, being more appropriate for control strategy testing. For this reason it is necessary to design a simpler, linear operation model to be used in control design. There are two possible approaches. The first one infers model reduction methods and linearization which can be troublesome. In this section is presented a more unconventional approach, developing a simpler operational model based on the main connections between input and output parameters, using experimental identification methods and the developed nonlinear mathematical model. The results will be validated by simulation.

To this end, by analysing the hydrogenation process and based on the process engineers experimental knowledge the main input variables are considered to be: input flow of 2 ethyl-hexanal, recirculated input flow of 2 ethyl-hexanol, input temperature of the reactants and hydrogen pressure. The main output variables are considered to be the 2 ethyl-hexanol concentration and the output temperature which is critical. Nevertheless, the dependence between the input and the output variables is considered to be of second order as follows:

$$\begin{vmatrix} T_{out} \\ C_{out} \end{vmatrix} = \begin{vmatrix} \frac{K1 \cdot e^{-\sigma s}}{T1 \cdot s^2 + T2 \cdot s + 1} & \frac{K2 \cdot e^{-\sigma s}}{T3 \cdot s^2 + T4 \cdot s + 1} \\ \frac{K3 \cdot e^{-\sigma s}}{T5 \cdot s^2 + T6 \cdot s + 1} & \frac{K4 \cdot e^{-\sigma s}}{T7 \cdot s^2 + T8 \cdot s + 1} \end{vmatrix} \begin{vmatrix} Q_{oct} \\ T_{in} \end{vmatrix} + \begin{vmatrix} \frac{K5 \cdot e^{-\sigma s}}{T9 \cdot s^2 + T10 \cdot s + 1} & \frac{K6 \cdot e^{-\sigma s}}{T11 \cdot s^2 + T12 \cdot s + 1} \\ \frac{K7 \cdot e^{-\sigma s}}{T13 \cdot s^2 + T14 \cdot s + 1} & \frac{K8 \cdot e^{-\sigma s}}{T15 \cdot s^2 + T16 \cdot s + 1} \end{vmatrix} \begin{vmatrix} Q_{enal} \\ P \end{vmatrix} \tag{9}$$

The parameter values are determined using experimental identification methods and considering step variations of the recirculated 2 ethyl-hexanol flow (Q_{oct}), input temperature of reactants (T_{in}). The same step variations are considered for the 2 ethyl-hexanal input flow (Q_{enal}) and the hydrogen input pressure (P) even if these parameters are considered to be constant in normal mode of operation. For validation **Figure 8** presents the simulated values obtained for the output temperature and considering a step variation in the recirculated 2 ethyl-hexanol flow.

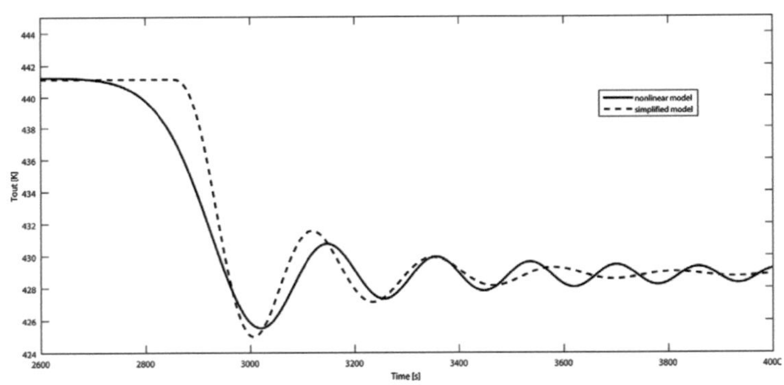

Figure 8. Output temperature evolution: operational model vs. nonlinear model.

3. Design and implementation of conventional control strategy

Conventional PID controllers are the most common control solution in the industry. Most of the research deals with mono-variable (single input single output—SISO) processes. However, most industrial processes are by their nature multi-variable (multi-input multi-output—-MIMO). Using mono-variable controllers for each output variable, even if it is a solution easy to apply, it will lead to inferior performances. It is possible that despite the fact that each individual PID loop control works, the overall PID control structure to fail. For this reason there is a demand for the development of multi-variable PID control strategies to compensate

http://dx.doi.org/10.5772/65336

the effect of functional interactions between variables from many companies that consider the interactions between variables in multi-variable systems as the main common problem in the industry.

The hydrogenation process is characterized by the presence of time delays of approximately 30 minutes. The difference between the dead time of each input-output pair is about 1–2 minutes. For this reason, the dead time is considered to be identical for all input-output pair. A typical approach to deal with time delay is the non-delayed output prediction [8, 9]. The non-delayed output may be estimated and the controller can be computed as for a process without delay. The most popular output predictor is the Smith predictor.

Currently, the 2 ethyl-hexanal hydrogenation plant is operated using the feed forward control (indirect, open-loop control). Thus, using methods sometimes simple, sometimes complicated (even in closed loop), the following parameters are controlled and adjusted to the desired level: the 2 ethyl-hexanal input flow, the recirculated 2 ethyl-hexanol flow, hydrogen input pressure and the reactant input temperature. At present time, an operator decides whether or not to manually modify the control loops set points to maintain the same process parameters and product specifications. To achieve a more effective operation of the hydrogenation process, both conventional and advanced control methods require a closed-loop control structure by including the reaction from the output.

In order to develop a multi-variable PID control, the operational model described in the previous section determined by the Eq. (9) will be used. The desired multi-variable controller matrix has the following form:

$$H_R(s) = \begin{vmatrix} H_{R11}(s) & H_{R12}(s) \\ H_{R21}(s) & H_{R22}(s) \end{vmatrix} \tag{10}$$

where H_{R11}, H_{R22} intended for direct adjustment of output variables and H_{R12}, H_{R21} are intended to counter act the interactions between input and output channels. The four controllers are computed by imposing a phase margin $\gamma_k = 60°$ [9].

The obtained controllers are described by:

$$\begin{gathered} H_{R11} = 1.3259\left(1 + \frac{1}{102.3018 \cdot s}\right); H_{R12} = 2.733\left(1 + \frac{1}{106.667 \cdot s}\right); \\ H_{R21} = 595.6621\left(1 + \frac{1}{90.4977 \cdot s}\right);\ H_{R22} = 223.8721(1 + \frac{1}{119.0476 \cdot s}) \end{gathered} \tag{11}$$

Figure 9 presents the closed loop Smith predictor control structure using a PID control.

The next step is to test and evaluate the performances of the developed control strategy by analysing its ability to reject disturbance effects, respectively, the set point tracking capability. In the first scenario, a reference variation of 10 K for the first output value (outlet temperature of the product, Tout) is considered. The simulation results are presented in **Figure 10**.

The second scenario is designed to test the control system capability to counteract the disturbance effects. Hence, a 6 % step variation in the 2 ethyl-hexanal input flow is considered for the results presented in **Figure 11**.

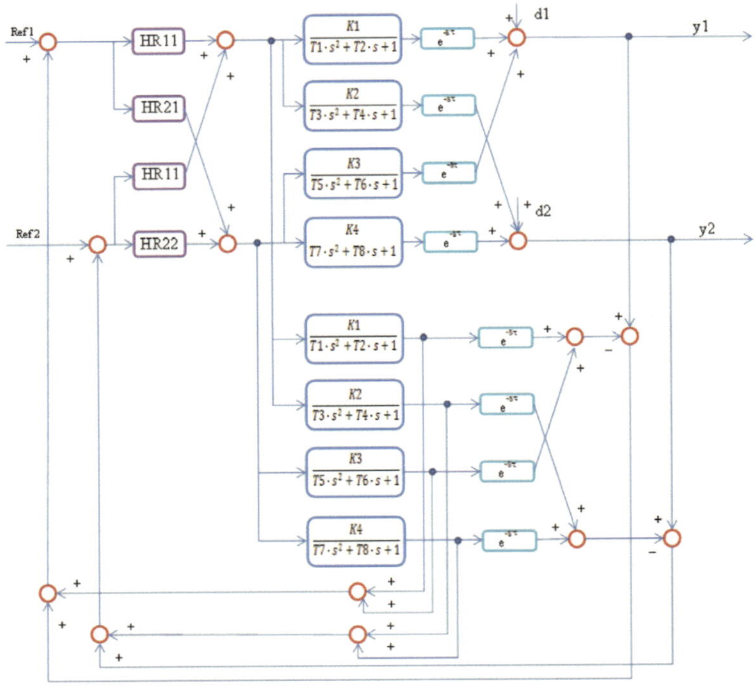

Figure 9. Closed loop control structure: conventional PID control in MIMO-SP structure.

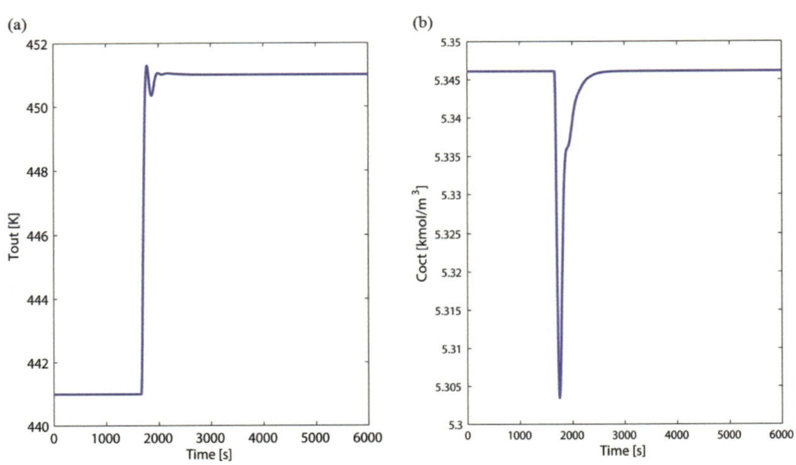

Figure 10. Output temperature reference step variation of 10 K: (a) output temperature evolution and (b) 2 ethyl-hexanol concentration evolution.

http://dx.doi.org/10.5772/65336

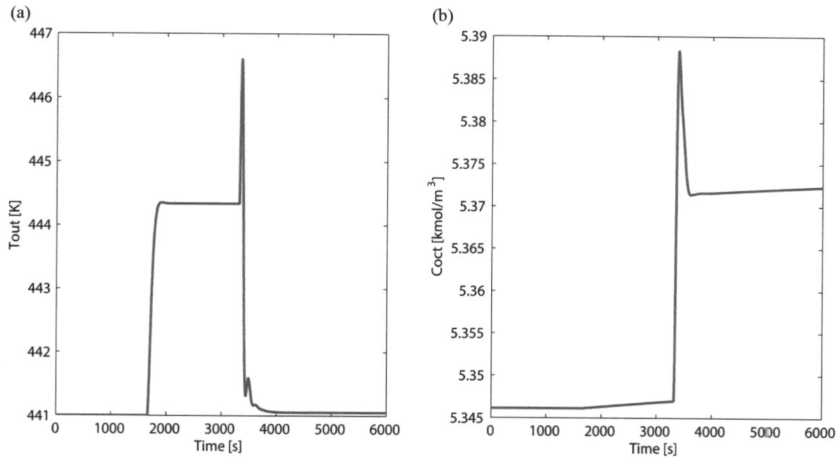

Figure 11. (a) Output temperature evolution and (b) 2 ethyl-hexanol concentration evolution considering a step variation in the 2 ethyl-hexanal input flow.

Based on the above results, the effectiveness of the proposed control is emphasized, presenting acceptable overshoot, response time and deviation values, but with opportunities for improvement. Thus, in the third scenario is considered an evolution for a period of 4 months and the catalyst activity degree is decreased up to 50%. **Figure 12** shows the temperature and 2 ethyl-hexanol concentration evolutions in this situation. It can be observed a steady error of 0.06% for the temperature and an error of 1% for the 2 ethyl-hexanol concentration.

The last simulation scenario is conceived in order to test the robustness of the designed control system for process parameter variations: gain variation and time constants variations (**Figure 13**).

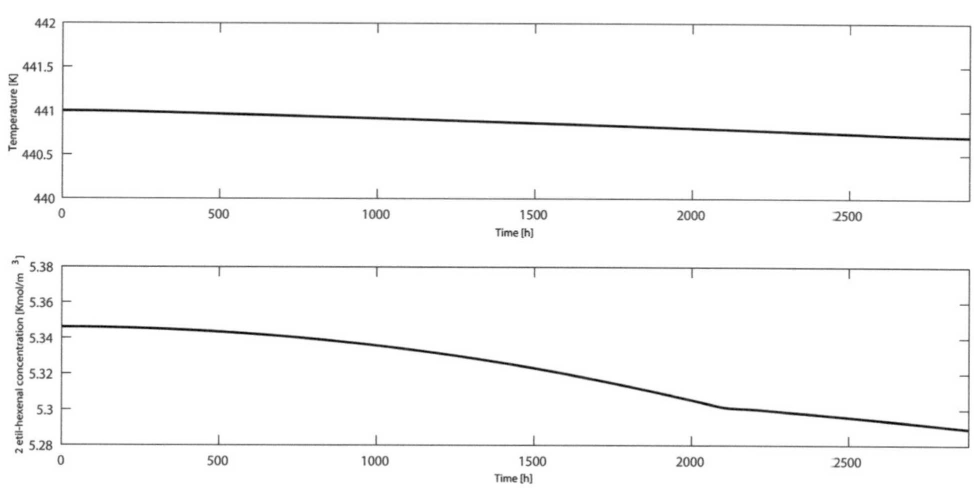

Figure 12. Catalyst deactivation: MIMO-SP PID controls.

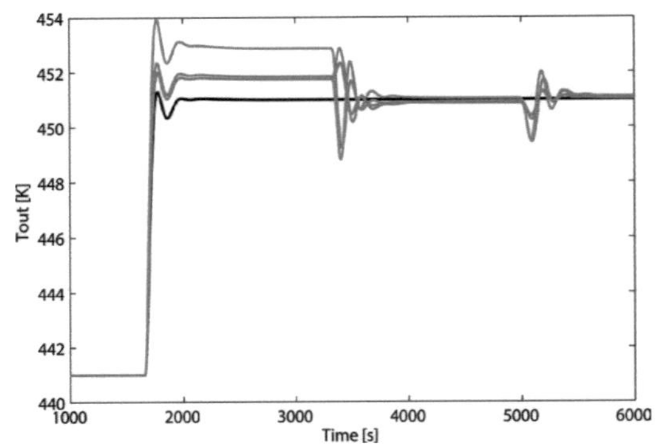

Figure 13. Output temperature evolution for reference step variation—nominal case vs. uncertain case (multivariable PID control).

A decrease in the control system performance can be observed, inferring reduced robustness, an aspect that may be improved by using advanced control strategies.

4. Design and implementation of advanced control strategies

4.1. Internal model control

The internal model control has emerged as an alternative to traditional feedback control algorithm feedback output as the simulation methods, mathematical modelling and model validation techniques developed [10]. This method provides a direct link between the process model and the controller structure. The IMC control structure is presented in **Figure 14** where p represents the transfer function describing the interconnections between process inputs u and outputs y, p_d represents the transfer function that describes the effects of disturbance on the output, p_m is the mathematical model of the process and q is the transfer function of the IMC controller.

The model of the process is assumed to be equal to the process transfer function matrix presented before in Eq. (9) inferring the need of a Smith predictor structure:

$$\begin{vmatrix} T_{out} \\ C_{out} \end{vmatrix} = \begin{vmatrix} H_{f11m}(s) \cdot e^{-\sigma s} & H_{f12m}(s) \cdot e^{-\sigma s} \\ H_{f21m}(s) \cdot e^{-\sigma s} & H_{f22m}(s) \cdot e^{-\sigma s} \end{vmatrix} \begin{vmatrix} Q_{oct} \\ T_{in} \end{vmatrix}; \qquad (12)$$

In order to ensure the process decoupling it is necessary to determine the pseudo-inverse matrix [11] of the steady state gain matrix:

$$H_m(0) = \begin{vmatrix} H_{f11m0} & H_{f12m0} \\ H_{f21m0} & H_{f22m0} \end{vmatrix} = \begin{vmatrix} -2.1 & 1 \\ -0.0049 & 0.011 \end{vmatrix} \qquad (13)$$

$$H_m^{\#} = H_m(0)^H \cdot (H_m(0) \cdot H_m(0)^H)^{-1} = \begin{vmatrix} -0.6044 & 54.945 \\ -0.2692 & 115.384 \end{vmatrix} \quad (14)$$

The decoupled process is obtained as: $H_D(s) = H_m(s)CH_m^{\#} = \begin{vmatrix} H_{f11d} & H_{f12d} \\ H_{f21d} & H_{f22d} \end{vmatrix}$. Due to the static decoupling, steady-state matrix $H_D(s=0)$ will be equal to the unit matrix, all the elements which are not on the first diagonal being equal to zero. Having the decoupled process the next step is to approximate the elements on the first diagonal of the matrix HD(s) with simple transfer functions using identification methods:

$$H_{f11d}(s) = \frac{1 \cdot e^{-1650s}}{1055 \cdot s^2 + 55.85 \cdot s + 1}; H_{f22d}(s) = \frac{1 \cdot e^{-1650s}}{2204.1 \cdot s^2 + 69.64 \cdot s + 1} \quad (15)$$

The last step consists of the IMC controller design using: $H_{RIMC}(s) = H_{f1\_d - inv}{}^{*}(s) \cdot f(s)$ where f(s) is the IMC filter:

$$f(s) = \frac{1}{(\lambda_i s + 1)^n} \quad (16)$$

where λ_i is the time constant of the filter associated to each output and n is chosen such as the final IMC controller is proper.

The obtained IMC controllers are:

$$H_{RIMC1}(s) = \frac{1055 \cdot s^2 + 55.85 \cdot s + 1}{400 \cdot s^2 + 40 \cdot s + 1}; H_{RIMC2}(s) = \frac{2204.1 \cdot s^2 + 69.64 \cdot s + 1}{100 \cdot s^2 + 20 \cdot s + 1} \quad (17)$$

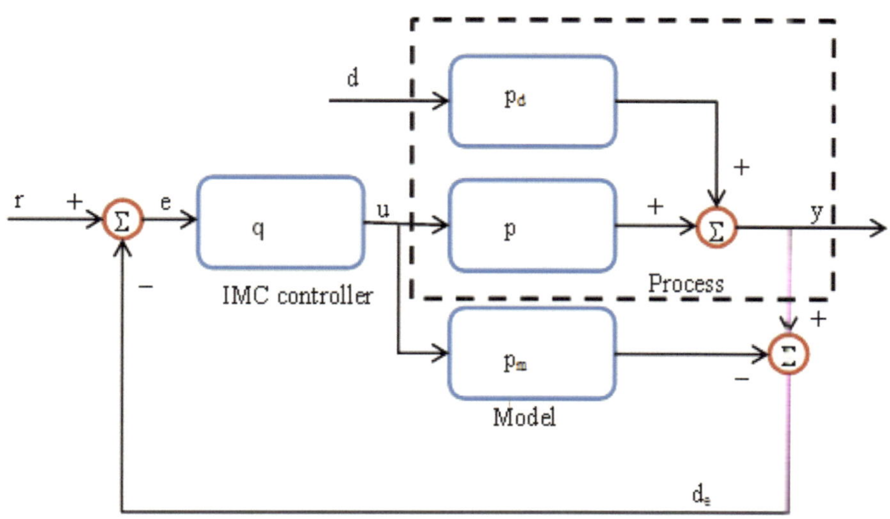

Figure 14. IMC control structure.

The IMC control system evaluation and testing are presented in comparison to the conventional multi-variable PID control strategy in order to conclude the results. To this end, the first test scenario consists in the set point tacking analysis. The simulation results are presented in **Figure 15**.

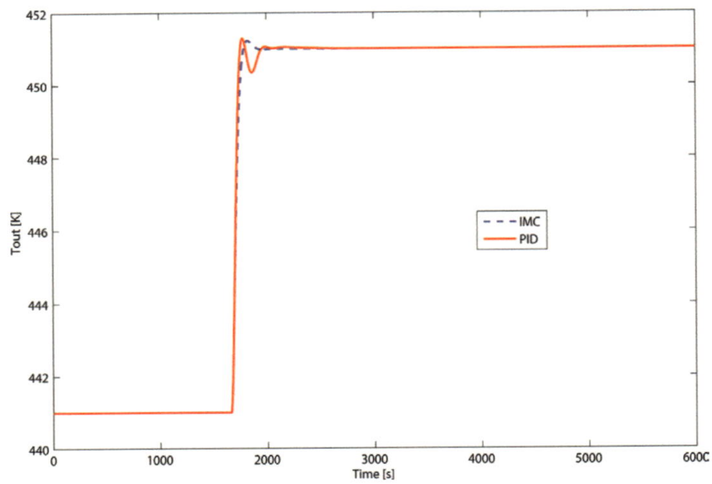

Figure 15. Output temperature evolution for reference step variation: PID vs. IMC.

The second simulation scenario is focused on the disturbance rejection analysis considering a 0.25 [Kmol/m^3] disturbance in the 2 ethyl-hexanal flow input flow (**Figure 16**).

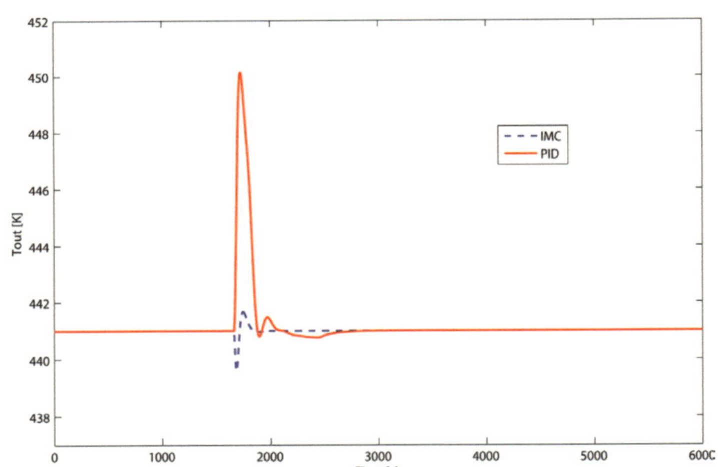

Figure 16. Output temperature evolution for a disturbance in the 2 ethyl-hexanal input flow: PID vs. IMC.

The third simulation scenario evaluates the IMC control system capability to counteract the effect of the catalyst deactivation (**Figure 17**).

http://dx.doi.org/10.5772/65336

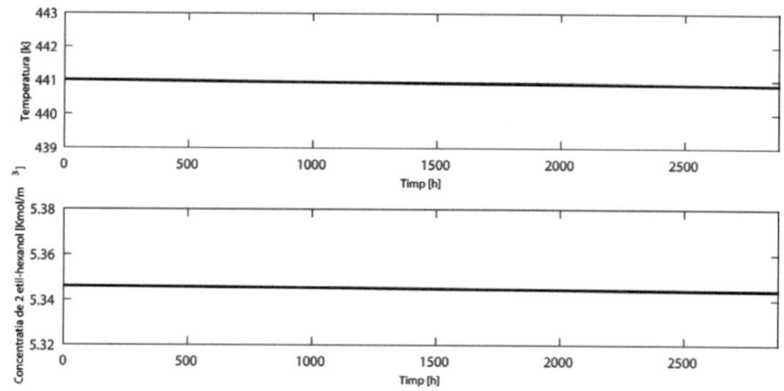

Figure 17. Catalyst deactivation: IMC control.

As in the previous case the last test scenario consists in the robustness evaluation of the IMC control system by considering the same variation of the process parameters (**Figure 18**).

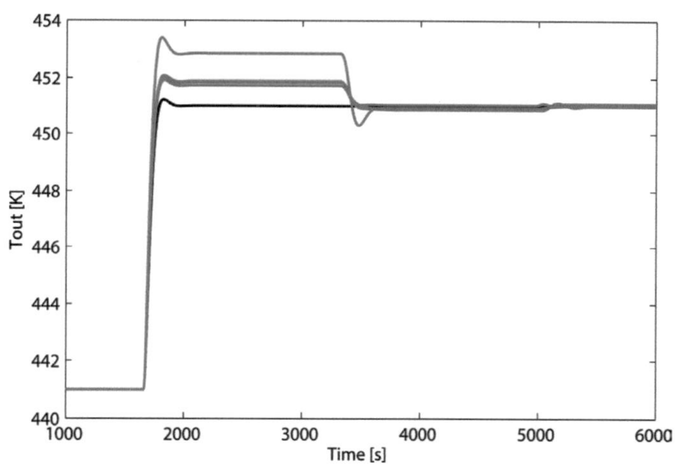

Figure 18. Output temperature evolution for reference step variation—nominal case vs. uncertain case (IMC control).

By analysing the comparative results between the MIMO SP-PID and SP-IMC control strategies it can be concluded that the SP-IMC control strategy outperforms the classical control. Another advantage is that it is easy to design and implement. However, even if the performances are clearly better the robustness to process parameter variations can be improved.

4.2. Robust control

Robust control can be defined as an attempt to control the uncertain systems (uncertainties). This approach accepts the idea of incomplete knowledge of the process, which has an uncertain dynamic and is influenced by disturbances insufficiently known. If, however, for these uncertainties can be established a mathematical norm, by using the robust control theory a robust, unique, able to meet certain performance specifications (hard or relaxed), controller can designed, respecting the uncertainty domain. Regardless of the method used to determine the mathematical model it is necessary to impose simplifying assumptions so that the obtained model is suitable for controller design. The differences between the real plant and the mathematical model represent modelling uncertainties or errors. Precisely from this view point the choice of robust control algorithms for the hydrogenation process 2-ethyl hexanal is justified. The robust controller design is a laborious task itself, but as the computational tools evolved the only difficult part left is the process parameter variation range determination. The same process operational model presented in Eq. (9) is used for multi-variable robust controller design based on H_{infinity} approach to ensure robust stability for both the nominal model and the entire class of systems that exist in a particular area of uncertainty around the nominal model. A Smith predictor MIMO structure is also necessary due to the large time delays presented by the hydrogenation process like in the previous sections.

$$\begin{vmatrix} T_{out} \\ C_{out} \end{vmatrix} = G_s \begin{vmatrix} Q_{oct} \\ T_{in} \end{vmatrix} + G_z \begin{vmatrix} Q_{enal} \\ P \end{vmatrix} \tag{18}$$

The first step was to determine the process state space representation:

$$\begin{matrix} \dot{X} = A \cdot X + B \cdot U \\ Y = C \cdot X + D \cdot U \end{matrix}; X = \begin{bmatrix} T_{out} \\ C_{out} \end{bmatrix}; U = \begin{bmatrix} Q_{oct} \\ T_{in} \end{bmatrix} \tag{19}$$

$$A=\begin{bmatrix} a11 & a12 & 0 & 0 & 0 & 0 & 0 & 0 \\ a21 & 0 & 0 & 0 & 0 & 0 & 0 & 0 \\ 0 & 0 & a33 & a34 & 0 & 0 & 0 & 0 \\ 0 & 0 & a43 & 0 & 0 & 0 & 0 & 0 \\ 0 & 0 & 0 & 0 & a55 & a56 & 0 & 0 \\ 0 & 0 & 0 & 0 & a65 & 0 & 0 & 0 \\ 0 & 0 & 0 & 0 & 0 & 0 & a77 & a78 \\ 0 & 0 & 0 & 0 & 0 & 0 & a87 & 0 \end{bmatrix}; B=\begin{bmatrix} b11 & 0 \\ 0 & 0 \\ b31 & 0 \\ 0 & 0 \\ 0 & b52 \\ 0 & 0 \\ 0 & b72 \\ 0 & 0 \end{bmatrix}$$
$$C=\begin{bmatrix} 0 & c12 & 0 & 0 & 0 & c16 & 0 & 0 \\ 0 & 0 & 0 & c24 & 0 & 0 & 0 & c28 \end{bmatrix}; D=\begin{bmatrix} 0 & 0 \\ 0 & 0 \end{bmatrix} \tag{20}$$

The nominal values of the process parameters are:

$$\begin{gathered} \overline{a11} = -0.046, \overline{a12} = -0.0225, \overline{a33} = -0.0522, \overline{a34} = -0.0279, \overline{a55} = -0.0434, \\ \overline{a56} = -0.0207, \overline{a77} = -0.0361, \overline{a78} = -0.0176, \overline{a87} = 0.0313, \overline{b31} = 0.0156, \\ \overline{b72} = 0.0156, \overline{c12} = -0.1890, \overline{c16} = 0.1657, \overline{c24} = -0.0087, \overline{c28} = 0.0124 \end{gathered} \tag{21}$$

It is a well-known fact that, in real control systems, uncertainties are unavoidable and can negatively affect the stability and the performance of the whole control system. Usually, the uncertainties can be classified in two main categories: disturbance signals (input/output disturbance, sensor/actuator noise) and dynamic perturbations (difference between the actual dynamics of the process and the mathematical model) [12]. The dynamic perturbations are usually caused by inaccurate characteristic description, torn and worn effects and shifting operating points. They are also called 'parametric uncertainties' and are represented by certain process parameter variation over a certain value range. In a control system the dynamic uncertainties can be represented in multiple ways. For this particular case the output multiplicative representation is considered showing the relative errors (between the actual system $G_p(s)$ and the nominal model $G_o(s)$) not only the absolute errors: $G_p(s) = [I + \Delta(s)]\, G_o(s)$. No matter what type of uncertainty representation is chosen, the actual, perturbed system can be represented like a standard upper linear fractional transform, where the uncertainties are lumped in a single block Δ, a diagonal matrix corresponding to parameter variations (**Figure 19**).

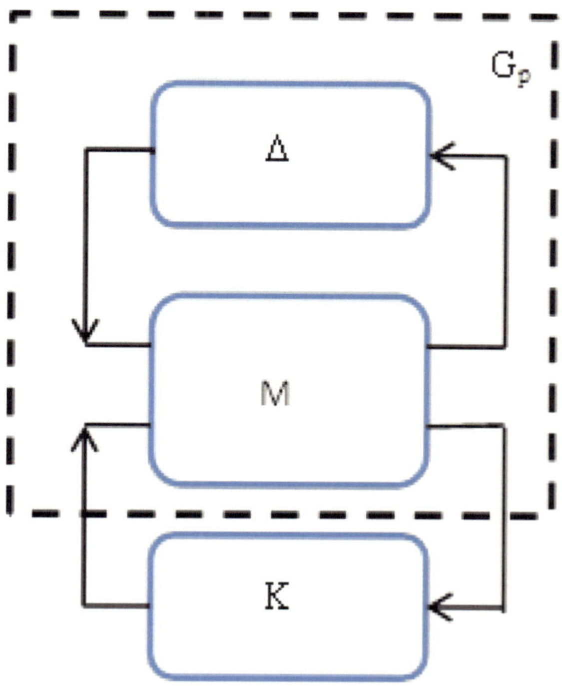

Figure 19. Generalized structure of the closed loop system.

The interconnection matrix for the considered multiplicative perturbation is:

$$M = \begin{bmatrix} o & G_0 \\ I & G_0 \end{bmatrix} \tag{22}$$

The uncertainty description is determined in an unconventional manner [13]. By using the experimental identification methods several second order models were determined using experimental data from different points of operation. In this way one can determine the interval for nominal model parameters variations.p11, p12, p33, p34, p55, p56, p77, p78, p87, pb31, pb72, pc12, pc16, pc24 and pc28 represent the computed possible, relative perturbation of the nominal process parameters. Each parameter *aij*, *bij* and *cij* ($i, j = 1.8$) may be represented as a linear fractional transformation (LFT) considering multiplicative uncertainties.

The next step is to determine the process mathematical model that takes into account also the model parameter uncertainties, G_{mds} having the following form [12]:

$$G_{mds} = \begin{bmatrix} A & | & B1 & B2 \\ - & - & - & - \\ C1 & | & D11 & D12 \\ C2 & | & D21 & D22 \end{bmatrix} \tag{23}$$

$$A=\begin{bmatrix} a11 & a12 & 0 & 0 & 0 & 0 & 0 & 0 \\ a21 & 0 & 0 & 0 & 0 & 0 & 0 & 0 \\ 0 & 0 & a33 & a34 & 0 & 0 & 0 & 0 \\ 0 & 0 & a43 & 0 & 0 & 0 & 0 & 0 \\ 0 & 0 & 0 & 0 & a55 & a56 & 0 & 0 \\ 0 & 0 & 0 & 0 & a65 & 0 & 0 & 0 \\ 0 & 0 & 0 & 0 & 0 & 0 & a77 & a78 \\ 0 & 0 & 0 & 0 & 0 & 0 & a87 & 0 \end{bmatrix};B2=\begin{bmatrix} b11 & 0 \\ 0 & 0 \\ b31 & 0 \\ 0 & 0 \\ 0 & b52 \\ 0 & 0 \\ 0 & b72 \\ 0 & 0 \end{bmatrix};C1=\begin{bmatrix} a11 & a12 & 0 & 0 & 0 & 0 & 0 & 0 \\ a21 & 0 & 0 & 0 & 0 & 0 & 0 & 0 \\ 0 & 0 & a33 & a34 & 0 & 0 & 0 & 0 \\ 0 & 0 & a43 & 0 & 0 & 0 & 0 & 0 \\ 0 & 0 & 0 & 0 & a55 & a56 & 0 & 0 \\ 0 & 0 & 0 & 0 & a65 & 0 & 0 & 0 \\ 0 & 0 & 0 & 0 & 0 & 0 & a77 & a78 \\ 0 & 0 & 0 & 0 & 0 & 0 & a87 & 0 \\ 0 & 0 & 0 & 0 & 0 & 0 & 0 & 0 \\ 0 & 0 & 0 & 0 & 0 & 0 & 0 & 0 \\ 0 & 0 & c12 & 0 & 0 & 0 & 0 & 0 \\ 0 & 0 & 0 & 0 & 0 & c16 & 0 & 0 \\ 0 & 0 & 0 & c24 & 0 & 0 & 0 & 0 \\ 0 & 0 & 0 & 0 & 0 & 0 & 0 & c28 \end{bmatrix};D12=\begin{bmatrix} 0 & 0 \\ 0 & 0 \\ 0 & 0 \\ 0 & 0 \\ 0 & 0 \\ 0 & 0 \\ 0 & 0 \\ 0 & 0 \\ 0 & 0 \\ b31 & 0 \\ 0 & b72 \\ 0 & 0 \\ 0 & 0 \\ 0 & 0 \\ 0 & 0 \end{bmatrix}$$

$$B1=\begin{bmatrix} p11 & p12 & 0 & 0 & 0 & 0 & 0 & 0 & 0 & 0 & 0 & 0 & 0 & 0 & 0 \\ 0 & 0 & 0 & 0 & 0 & 0 & 0 & 0 & 0 & 0 & 0 & 0 & 0 & 0 & 0 \\ 0 & 0 & p33 & p34 & 0 & 0 & 0 & 0 & 0 & pb31 & 0 & 0 & 0 & 0 & 0 \\ 0 & 0 & 0 & 0 & 0 & 0 & 0 & 0 & 0 & 0 & 0 & 0 & 0 & 0 & 0 \\ 0 & 0 & 0 & 0 & p55 & p56 & 0 & 0 & 0 & 0 & 0 & 0 & 0 & 0 & 0 \\ 0 & 0 & 0 & 0 & 0 & 0 & 0 & 0 & 0 & 0 & 0 & 0 & 0 & 0 & 0 \\ 0 & 0 & 0 & 0 & 0 & 0 & p77 & p78 & 0 & 0 & pb72 & 0 & 0 & 0 & 0 \\ 0 & 0 & 0 & 0 & 0 & 0 & 0 & 0 & p87 & 0 & 0 & 0 & 0 & 0 & 0 \end{bmatrix};C2=\begin{bmatrix} 0 & c12 & 0 & 0 & 0 & c16 & 0 & 0 \\ 0 & 0 & 0 & c24 & 0 & 0 & 0 & c28 \end{bmatrix};D22=\begin{bmatrix} 0 & 0 \\ 0 & 0 \end{bmatrix}$$

$$D21=\begin{bmatrix} 0 & 0 & 0 & 0 & 0 & 0 & 0 & 0 & 0 & 0 & 0 & pc12 & pc16 & 0 & 0 \\ 0 & 0 & 0 & 0 & 0 & 0 & 0 & 0 & 0 & 0 & 0 & 0 & 0 & pc24 & pc28 \end{bmatrix};D11=zeros(15,15) \tag{24}$$

The block diagram of the closed loop system in Smith predictor structure including the robust multi-variable controller and the uncertainties bloc is presented in **Figure 20**.

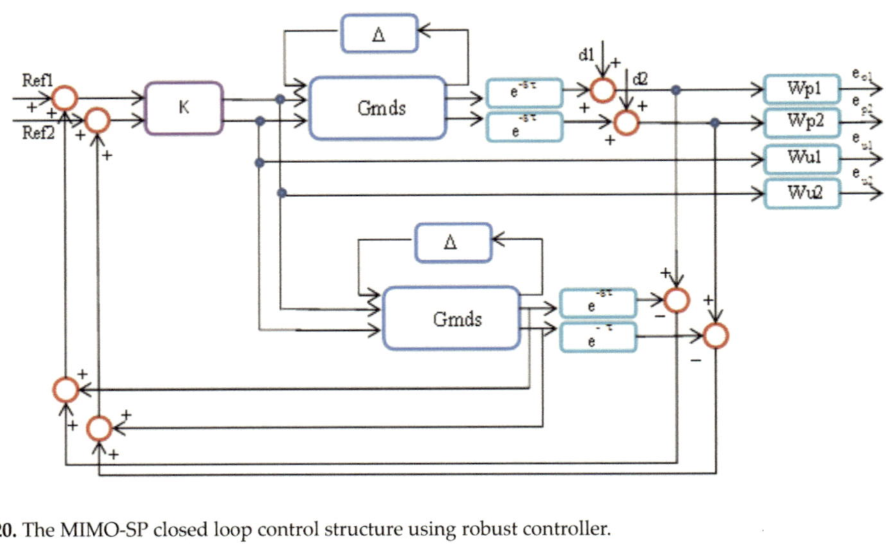

Figure 20. The MIMO-SP closed loop control structure using robust controller.

As it can be observed G_{mds} is nominal model of the process, *d*1, *d*2 represent the disturbances, Δ is the matrix uncertainties, Wu1, Wu2 and WP1, WP2 are weighting functions which reflect the relative significance of performance requirements. For good mitigation of disturbances and

http://dx.doi.org/10.5772/65336

to ensure a certain response time and overshoot Wu1, Wu2 and WP1, WP2 have the following form:

$$Wp1(s) = 0.9\frac{s^2+2s+5}{s^2+1.1s+0.02}; Wp2(s) = 0.1\frac{s+3}{s+0.1}; Wu1 = 10^{-5}; Wu2 = 10^{-5} \tag{25}$$

It should be noted that choosing the suitable weighting functions is a crucial step in the synthesis of robust controller and usually requires several attempts. By applying the presented method, using MATLAB/SIMULINK environment – *hinfsyn* function, the following robust H_8 controller was determined:

$$A_K=\begin{bmatrix} -29.50 & -1.48e4 & 8.65 & 159.78 & 25.83 & 1.29e4 & -12.5 & -227.8 & -1.62e5 & -1e535 & -8.39e3 \\ 0.313 & 0 & 0 & 0 & 0 & 0 & 0 & 0 & 0 & 0 & 0 \\ -1.83 & -923.8 & 0.48 & 9.94 & 1.61 & 809.99 & -0.78 & -14.21 & -1.016e4 & -8.43e3 & -523.69 \\ 0 & 0 & 0.0313 & 0 & 0 & 0 & 0 & 0 & 0 & 0 & 0 \\ 5.45 & 3.22e3 & 9.73 & 178.2 & -4.82 & -2.8271e3 & -14.06 & -254.1 & 3.56e4 & 2.95e4 & -9.24e3 \\ 0 & 0 & 0 & 0 & 0.0313 & 0 & 0 & 0 & 0 & 0 & 0 \\ 0.68 & 402.42 & 1.21 & 22.24 & -0.59 & -352.81 & -1.79 & -31.73 & 4.45e3 & 3.69e3 & -1.15e3 \\ 0 & 0 & 0 & 0 & 0 & 0 & 0.0313 & 0 & 0 & 0 & 0 \\ 0 & 0 & 0 & 0 & 0 & 5.55e-17 & 0 & 0 & -0.0175 & -0.0317 & 0 \\ 0 & -5.55e-17 & 0 & 0 & 0 & 5.55e-17 & 0 & 0 & 0.0317 & -1.08 & 0 \\ 0 & 0 & 0 & -8.67e-19 & 0 & 0 & 0 & 0 & 0 & 0 & -0.1 \end{bmatrix}; B_K=\begin{bmatrix} 0 & 0 \\ 0 & 0 \\ 0 & 0 \\ 0 & 0 \\ 0 & 0 \\ 0 & 0 \\ 0 & 0 \\ 0 & 0 \\ 0.0255 & 0 \\ -0.0288 & 0 \\ 0 & -0.0069 \end{bmatrix} \tag{26}$$

$$C_K=\begin{bmatrix} -9.2e3 & -4.6e6 & 2.7e3 & 4.9e4 & 8.07e3 & 4.05e6 & -3.9e3 & -7.11e4 & -5.08e7 & -4.2e7 & -2.6e6 \\ 3.4e3 & 2.01e6 & 6.8e3 & 1.11e5 & -2.9e3 & -1.76e6 & -8.79e3 & -1.58e5 & 2.22e7 & 1.84e7 & -5.7e6 \end{bmatrix}; D_K=\begin{bmatrix} 0 & 0 \\ 0 & 0 \end{bmatrix}$$

Like for the previous control strategies the first simulation test scenario consists of closed loop simulation evaluation under nominal parameter values. The same simulation scenario was performed for the whole class of systems in the uncertainty domain for a reference step variation for the output temperature. The simulation results show good performances of the developed control structure even considering the process parameter uncertainty domain (**Figure 21**).

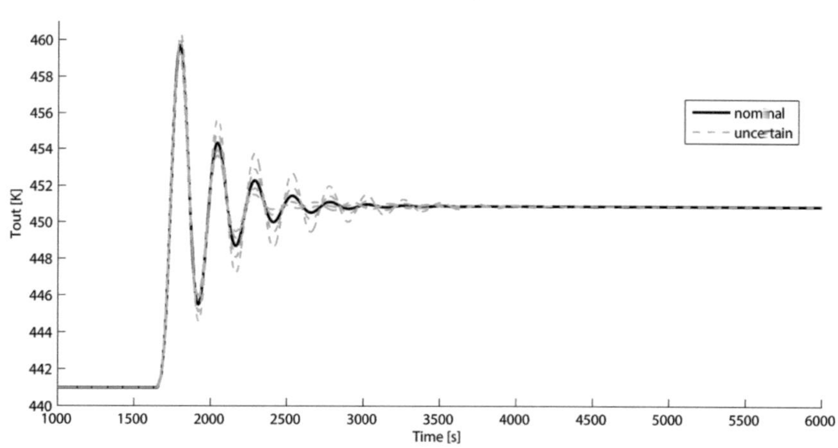

Figure 21. Output temperature evolution—robust controller: nominal vs. uncertain considering a reference step variation.

The second simulation scenario will evaluate the capability to reject the disturbance effect and also to test at the same time the robustness of the system. To this end, a step variation of the 2 ethyl-hexanal input flow is considered along with the process parameter variations. Good performances are reached even in the case of the uncertain plants (**Figure 22**).

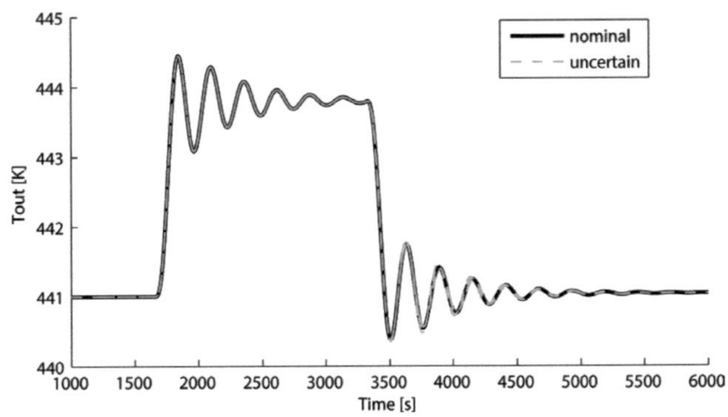

Figure 22. Output temperature evolution—robust controller: nominal vs. uncertain considering a reference step variation of the 2 ethyl-hexanal input flow.

Another performance that needs to be evaluated is the ability to counter act the catalyst deactivation effect (**Figure 23**).

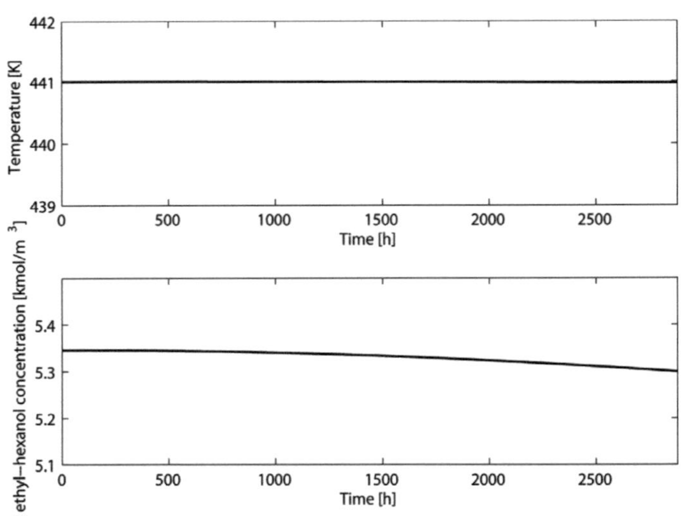

Figure 23. Catalyst deactivation: robust controller.

By analysing all the results obtained in the previous figures it can be concluded that even considering the catalyst deactivation steady-state errors of 0.006% and 0.18% are achieved for the output temperature and 2 ethyl-hexanol concentration, which are clearly within acceptable limits making the robust control strategy the most suitable for 2 ethyl-hexanal hydrogenation process control.

5. Conclusions

The developed model of the hydrogenation process, presented in this chapter, is able to represent the dynamic behaviour of the reactor during operation. Real plant data was used for mathematical model validation. From the dynamic point of view, the system behaves as an element with a large time constant and a large time delay. Hydrogenation multiphase catalytic reactors have complex behaviour and from this point of view, the use of advanced control strategies together with online optimization techniques appears to be a suitable procedure to deal with the problem of operating at high level of performance and safety.

A possibility to describe the processes which occur inside the reactor by a linear nominal transfer matrix and uncertainty description is detailed. A practical method for obtaining the uncertainty description is also presented. Three control strategies are proposed, developed, implemented and tested.

Finally, a comparison between the advantages and disadvantages of the proposed control solutions is performed.

Acknowledgements

This work was supported by a grant of the Romanian National Authority for Scientific Research and Innovation, CNCS– UEFISCDI, project number 155/2012 PN-II-PT-PCCA.

Nomenclature

Symbol	Description
C_i	molar concentration (kmol/m^3) of component *i*:enal (2 ethyl-hexanal), anal (2 ethyl-hexanal), oct (2 ethyl-hexanol)
$C_H{}^L$	concentration of hydrogen in liquid phase
$C_H{}^G$	concentration of hydrogen in gaseous phase
Q_i	volumetric flow rate (m^3/s) of component in phase *i*: L(liquid), G(gaseous)
T	temperature (K)
c_p	specific heat capacity (kJ/kg K)
a_v	specific gas-liquid contact area (m^{-1})
z	axial reactor coordinate
S	cross-sectional area
r_i	rate of surface reaction *i*: 1, 2
K_i	adsorption equilibrium constant, component i (m^3/mol)

Symbol	Description
k_j	rate constant of surface reaction j (mol/s kg)
ϱ_{sol}	solution density (kg/m^3)
N_{HG}	the flux of hydrogen transferred from gaseous phase to liquid phase [5, 7] $N_{H_G} = K_{L,H}(C_H^* - C_{H,L})$ $\frac{1}{K_{L,H}} = \frac{1}{H_H C k_G} + \frac{1}{k_{L,H}}$ $C_H^* = p_H/H_H$
vph	the transferred hydrogen flow through the liquid film, adjacent to the catalyst pellet, on to its' surface [5, 7] $vph = \frac{M_{L,H}}{2K_H}\left[\left(\frac{w\cdot k_1\cdot\eta_i}{M_{L,H}} + \frac{K_H\cdot p_H}{H_H} + 1\right) - \sqrt{\left(\frac{w\cdot k_1\cdot\eta_i}{M_{L,H}} + \frac{K_H\cdot p_H}{H_H} + 1\right)^2 - 4\frac{K_H}{M_{L,H}}\cdot\frac{w\cdot k_1\cdot\eta_i\cdot p_H}{H_H}}\right]$ $\frac{1}{M_{L,H}} = \frac{1}{a_v\cdot K_{L,H}} + \frac{1}{a_S\cdot k_{S,H}}$
$\Delta_R H_i$	the reaction heat of reaction *i:1, 2* (kJ/kmol)
p_H	partial pressure of hydrogen in the convective zone of the gas phase [atm]
H_H	Henry's law constant (atm m^3/Kmol)

Author details

Roxana Rusu-Both

Address all correspondence to: roxana.both@aut.utcluj.ro

Technical University of Cluj-Napoca, Cluj-Napoca, Romania

References

[1] ICIS. [Internet]. Available from: www.icis.com [Accessed: 2016-03-15].

[2] Smelder G. Kinetic analysis of the liquid phase hydrogenation of 2-ethyl-hexenal in the presence of supported Ni,Pd and Ni-S catalyst. The Canadian Journal of Chemical Engineering. 1989;67:51–61. DOI: 10.1002/cjce.5450670109

[3] Olah GA, Molnar A. Hydrocarbon Chemistry. 2nd ed. New York: John Wiley & Sons; 2003. 871 p. DOI: 10.1002/0471433489

[4] Bâldea I . Kinetics and Reaction Mechanisms (in Romania). Cluj-Napoca Universitary Publishing House Cluj; 2002. ISBN 973-610-130-4

[5] Bozga G, Muntean O. Chemical Reactors (in Romanian). Bucuresti: Technical Publishing House; 2001. ISBN 9733115789

[6] Popovici C.I. Numerical Analysis with Matlab (in Romanian). Iasi: Venus Publishing House; 2007. 408 p. ISBN 978-973-756-043-8

[7] Both R, Cormos AM, Agachi PS, Festilă C. Dynamic modeling and validation of the 2 ethyl-hexenal hydrogenation process. Computer & Chemical Engineering Journal. 2013;52:100–111. DOI: 10.1016/j.compchemeng.2012.11.012

[8] Melo DNC, Costa CBB, Toledo ECV, Santos MM, Maciel MRW, Filho RM. Evaluation of control algorithms for three phase hydrogenation catalytic reactors. Chemical Engineering Journal. 2008;141:250–263. DOI: 10.1016/j.cej.2007.12.026

[9] Wang QG, He Z, Cai W-J, Hang CC. PID Control for Multivariable Processes. Berlin, Heidelberg: Springer-Verlag; 2008. 266 p. DOI: 10.1007/978-3-540-78482-1

[10] Agachi PS, Nagy ZK, Cristea VM, Lucaci A. Model Based Control. Weinheim, Germany: Wiley-VCH Verlag GmbH & Co. KGaA; 2006. DOI: 10.1002/9783527609475

[11] Skogestad S, Postlethwaite I. Multivariable Feedback Control: Analysis and Design. New York: John Wiley & Sons; 2005. 590 p. ISBN 978-0-470-01167-6

[12] Gu DW, Petkov PH, Konstantinov MM. Robust Control Design with MATLAB. 1st ed. London, Springer-Verlag; 2005. 389 p. ISBN-10: 1-85233-983-7

[13] Both R, Dulf EH, Festilă C, Agachi PS. Robust control of a catalytic 2 ethyl-hexenal hydrogenation reactor. Journal of Chemical Engineering Science. 2012;74:300–309. DOI: 10.1016/j.ces.2012.02.033